CREATING A CLIMATE FOR CHANGE

Communicating Climate Change and Facilitating Social Change

The need for effective communication, public outreach, and education to increase support for policy, collective action, and behavior change is ever-present, and is perhaps most pressing in the context of anthropogenic climate change. This book is the first to take a comprehensive look at communication and social change specifically targeted to climate change.

Creating a Climate for Change is a unique collection of ideas examining the challenges associated with communicating climate change in order to facilitate societal response. It offers well-founded, practical suggestions on how to communicate climate change and how to approach related social change more effectively. The contributors of this book come from a range of backgrounds, from government and academia to non-governmental and civic sectors of society. Each chapter goes beyond posing problems or discussing the difficulties, and offers constructive suggestions for improving communication and social change efforts. The book concludes that re-envisioning communication strategies and exploring new approaches are necessary if we are to effectively facilitate action on climate change. The book is accessibly written, and any specialized terminology is explained.

Creating a Climate for Change will be of great interest to academic researchers and professionals in climate change, environmental policy, science communication, psychology, sociology, and geography.

Susanne Moser is a research scientist at the National Center for Atmospheric Research's (NCAR) Institute for the Study of Society and Environment, Boulder, Colorado. She is an Aldo Leopold Leadership Program fellow and an associate of the International Human Dimensions Program (IHDP) Core Project on Global Environmental Change and Human Security (GECHS).

Lisa Dilling is a visiting fellow at the Center for Science and Technology Policy Research of the Cooperative Institute for Research in Environmental Sciences (CIRES), University of Colorado at Boulder. She has been awarded a visiting fellowship by CIRES, a John A. Knauss National Sea Grant fellowship, and a National Science Foundation graduate fellowship.

CREATING A CLIMATE FOR CHANGE

Communicating Climate Change and Facilitating Social Change

Edited by

SUSANNE C. MOSER

and

LISA DILLING

CAMBRIDGE
UNIVERSITY PRESS

CAMBRIDGE UNIVERSITY PRESS
Cambridge, New York, Melbourne, Madrid, Cape Town, Singapore, São Paulo

Cambridge University Press
The Edinburgh Building, Cambridge CB2 8RU, UK

Published in the United States of America by Cambridge University Press, New York

www.cambridge.org
Information on this title: www.cambridge.org/9780521869232

First published 2007
This digitally printed version 2008

A catalogue record for this publication is available from the British Library

Library of Congress Cataloguing in Publication data

Creating a climate for change : communicating climate change & facilitating social change/
edited by Susanne C. Moser and Lisa Dilling.
p. cm.
Includes bibliographical references and index.
ISBN-13: 978-0-521-86923-2
ISBN-10: 0-521-86923-4
1. Climatic changes. 2. Communication in the environmental sciences. 3. Communication in
social action. I. Moser, Susanne C. II. Dilling, Lisa. III. Title.

QC981.8.C5.C767 2006
363.7'05--dc22

2006037261

ISBN 978-0-521-86923-2 hardback
ISBN 978-0-521-04992-4 paperback

Contents

Part III   Creating a climate for change

Preface

If you focus on results, you will never change.
If you focus on change, you will get results.
Jack Dixon

In 1999, the National Center for Atmospheric Research received a grant from The MacArthur Foundation to help improve the communication between scientists and non-governmental groups about climate change. We started our project in 2003 using a portion of those funds, and expanded the scope to focus more broadly on how climate change communication might more effectively facilitate social change in society more generally.

We saw real opportunity in broadly surveying existing academic knowledge and facilitating conversation across disciplines and with practitioners. Our own experience and a review of the literature suggested that the practice of climate change communication had resulted in disappointing and even counterproductive results. Thus, our central guiding objective was to understand how communicators might advance societal response to climate change by better conveying its urgency and creating a more inclusive and productive conversation.

We convened a multidisciplinary workshop with both academic and practitioner experts in communication and social change. Over 40 individuals from academia, government, non-profit advocacy groups, the business community, and other areas of the private sector met in June 2004 for a three-day workshop at NCAR. Quite deliberately, we invited not only professionals concerned with climate change but also others from unrelated fields and professional backgrounds to contribute fresh thinking on the communication – social change challenge. To facilitate communication among us, we had only two rules: let's speak in plain English and let's barn-raise! Inspired by Michael Kahn's 1974 essay ("The Seminar") on different conversation styles, we hoped that everyone would add their insights to a fuller understanding of the challenges and possible solutions. We argued that if any one discipline or field of work had all the answers to the question of effective communication, this focus of inquiry would not be needed.

And barn-raise we did. We enjoyed each other's company and learned a lot, if simply because we listened and talked to people we don't usually meet in our regular networks and gatherings. In a deeply engaged, respectful atmosphere, new insights emerged. These insights, developed further since and augmented by viewpoints not represented at the workshop (but identified as critical there), are collected in this volume.

This book does not amount to a radical departure from existing understanding and practice, but presents a snapshot and milestone on a path of change for how we talk about and respond to climate change. It reflects state-of-the-art thinking in numerous relevant disciplines, presents rich examples of current practice, and also poses new questions to the research community. At the same time, it suggests that we don't just need to get better at what we already do; we also need to do new things. We need to open up the communication process to a wider community, in which participants own the process and content of communication. Mutual empowerment and support for change then become central. Such communication will shift us away from mere persuasion and notions of information transfer to dialogue, debate, negotiation, and visioning. These more interactive forms of communication have a far greater chance of supporting individual behavior change, change in organizations and different sectors of society, but they can also help shift social norms, policies, culture, and social relations that underpin deeper societal transformations needed to address global warming.

Despite our attempts to seek balance and integration, the following chapters do not constitute a grand theory of communication for social change. It was not our intent to arrive at such an all-encompassing theory that could integrate all the pieces seamlessly. We never asked our contributors to agree with a particular perspective or position. But we did question bold claims; we did ask that other viewpoints be considered, and if possible reconciled. We did push everyone to think harder. The advances we can claim could not have surfaced without this interaction between bodies of thought.

More important to us was praxis: What has worked? What has not? What have we learned from these experiences and how do they inform those emerging from other chapters? Our collaborators continued in the spirit of barn-raising and offered suggestions on how to improve communication in support of societal response to climate change. As a result, the book's contributors have advanced our understanding — based in some cases on new research, in others on syntheses of the existing literature, in yet others on real-world experiences. All are original contributions. Their insights and suggestions moved us toward a more democratic communication process involving a broad spectrum of societal actors involved in climate change.

Our own hope in making progress on global warming was sustained in no small measure by this project, by the fact that people from different walks of life, who do not usually talk to each other, gather around an idea and start talking — and not *at* but *with* each other. It is to those who came to the workshop and those who contributed to this volume that we owe our greatest thanks. They demonstrate what is possible: a new, if challenging, but most rewarding and fruitful conversation.

Many more than those represented in this volume have helped make this dialogue possible. The project would never even have happened without Adele Simmons, then at The MacArthur Foundation, who granted the initial gift to NCAR that allowed us to meet in the June 2004 workshop. We cannot thank Bob Harriss enough for entrusting it to us and trusting our ideas. He supported us all along through his optimism and constructive critique. We received additional funding from NCAR's Environmental and Societal Impacts Group (now the Institute for the Study of Society and Environment), NCAR's Walter Orr Roberts Institute through Cindy Schmidt, and a grant from the National Science Foundation through our steady supporter, Cliff Jacobs.

Along the way, this project was supported by a fabulous Steering Committee. We thank Vicki Arroyo, Caron Chess, Sharon Dunwoody, David Gershon, Mickey Glantz, Sonia Hamel, Dale Jamieson, Doug McKenzie-Mohr, Bob O'Connor, Cindy Schmidt, Paul Slovic, Shelly Ungar, and Elaine Vaughan for all their guidance, ideas, reality checks, and encouragement. Several individuals in addition to the ones included in this volume contributed to our workshop and project, and we would like to acknowledge their valuable insights — Tim Barnes, Dennis Bray, Sarah Conn, Mickey Glantz, Deborah Lynn Guber, Sonia Hamel, Cliff Jacobs, Willett Kempton, Robert Lempert, Franz Litz, Susan Munves, Bob O'Connor, Roger Pielke, Jr., Opalanga Pugh, Carol Rogers, Blake Smith, Clive Spash, and Will Toor.

We could not have pulled off the workshop or completed this book without the exceptional administrative and other support — from travel planning and recording of conference conversations, to website design, maintenance and help with graphics, to editing and proofreading, to the invisible, but indispensable, emotional support that helps people keep going until something is finally done. Thank you to Nataly Ascarrunz, Yarrow Axford, Marilyn Averill, Rebecca Haacker-Santos, Vicki Holzhauer, Jan Hopper, Mark McCaffrey, Rebecca Morss, Jean Renz, Hillary Rosner, Sharon Shearer, D. Jan Stewart, Jason Vogel, and in particular to John Tribbia and Susan Watrous for keeping us sane during the book's completion phase.

In the spirit of the MacArthur grant, it was our goal to make this project not just an academic exercise, but to include those who in their daily work experiment with communication and social change strategies. Precisely because practitioners are busy doing just that, we did not expect that writing a book chapter would be high on their list. Several surprised us, saying they want to write their own, and we want to especially acknowledge them — we know this wasn't part of their day job. Others gratefully accepted the chance to work with either an academic colleague or with one of our three wonderful writers and editors, Natasha Fraley, Sarah Rabkin, and Susan Watrous. They enabled the voices and stories of practitioners to be included in this volume — a part of the conversation so often left out.

A big thank you also to Matt Lloyd at Cambridge University Press for recognizing the value of this eclectic compendium. He, Helen Morris, Emma Pearce, Dan Dunlavey and Imran Mirza at Keyword Group were enthusiastic supporters from the start. Their experienced and patient hand smoothed out the rough edges we didn't see and shepherded the book to completion.

Finally, we would like to thank each other. The first edited book for each of us, and our first project together, we dove into it with the enthusiasm of those who don't know what's ahead. Two years and a lot of learning later, we know that neither of us could have done a better job alone. Our different training, professional experiences, perspectives, and personalities complemented each other in essential ways. It is our hope that the result will stimulate rich new thinking on communication for social change, and maybe even encourage some to go out and be part of the change. As Eleanor Roosevelt famously said: "You must do the things you think you cannot do."

Susanne C. Moser
Lisa Dilling

Foreword

There is a remarkable and recurring shape in both art and science. Hogarth, the seventeenth-century artist, would have seen it as an "S"-shaped "line of beauty," and Verhulst, the mathematician, in 1838, as yet another example of rapid but self-limiting growth in the form of the logistic equation. And for me it is a powerful model of how social, behavioral, and technological change takes place. So whether we are charting the proportion of the public expressing concern for global warming over time or the number of people, institutions, and countries taking action to limit climate change, we hope the eventual path will be "S"-shaped. Such a curve would show a slow increase of climate change risk perceptions, mitigation or adaptation policies, and individual behaviors followed by a period of rapid growth, until finally the rate of growth slowed once a very large proportion (but not all) of people, institutions, or countries have changed.

The chapters in this volume suggest that if we were to plot public awareness of global warming or climate change we are probably high on the curve, although much of public knowledge of causes and solutions may be inaccurate by scientific standards. Yet public concern and political will have not yet turned the corner leading to an adequate response to this threat. And indeed if we use as a criterion specific actions, rather than vague ones such as "saving energy" or "helping the environment" − then it is still very early days. Overall, these exciting, stimulating, and sometimes conflicting chapters address how we can bend the flat line upwards to put us on the path to accelerated action.

Fortunately, there are many examples of such periods of rapid change following years of painful plodding. Recent history suggests that long-term trends in individual behavior can undergo dramatic change. For example, growing scientific evidence and public pressures led to the adoption of stricter laws, penalties, and enforcement measures related to smoking, seat belts, drunk driving, and littering. In turn, the implementation of these structural changes led to rapid increases in seat-belt use and actual reductions in smoking, drunk driving, and littering. In the United States, social change in the areas of civil rights, feminism, and the environment has been extraordinarily rapid following decades or even centuries of slow, incremental change.

My colleagues Anthony Leiserowitz, Tom Parris, and I have recently argued that at least four conditions are required for these accelerations in collective action. These include: changes in public values and attitudes, vivid focusing events, an existing structure of institutions and organizations capable of encouraging and fostering action, and practical available solutions to the problems requiring change. For example, the struggle for civil rights in the United States was galvanized by dramatic televised images of overt racism, which offended widely held values of justice, fairness, and equality. A variety of organizations, especially African-American churches and their leaders, skillfully forced long-ignored issues of race relations onto the national agenda. Legal solutions were readily available and quickly implemented, including the repeal of Jim Crow laws, the Civil Rights Act (1964), and the Voting Rights Act (1965). Another example is the relatively quick international adoption and implementation of the Montreal Protocol on ozone protection. Strong global values and attitudes favoring the protection of human and environmental health already existed and ozone depletion was directly linked to skin cancer. Response to the scientific discovery of the role played by chlorofluorocarbons (CFCs) in ozone depletion developed slowly until the identification of the ozone "hole" provided a vivid image and metaphor. A broad set of health, environmental, and industry non-governmental organizations was ready to respond to the emerging sense of crisis. Finally, the companies that produced CFCs were able quickly to develop substitutes and to take advantage of a new regulatory environment that phased CFCs out of production.

Most of the chapters in this book address one or more of these four conditions, such as the use of communication to change or amplify public values and attitudes, the importance of vivid imagery and focusing events, the need to understand and use the structure of existing institutions to foster action, and to share examples of practical solutions already implemented or readily at hand. For each of these, there is a strong emphasis on what each of the authors learned about how to communicate better their concerns, institutions, or solutions.

Thus this volume provides an important synthesis of ideas and approaches to accelerate the global response to climate change. But it is only a starting point. Some conclusions emerge very clearly. Make the global local. Communicate hope, not just fear. Strengthen existing values and use existing organizations. Make action attractive and efficacious. Many chapters will challenge readers to rethink their current efforts to communicate and respond to climate change. Other chapters raise more questions than they answer, but they are important. Are events such as Hurricane Katrina useful focusing events or do they encourage a wrong mental model of climate change and its impacts? Are scientists part of the problem or the solution? And ultimately, can the modest practical actions of individuals, cities, and even states change a global dynamic that requires nothing less than an end to almost all fossil fuel use?

So even if this volume doesn't answer all your questions, do read on. Let me assure you that there never has been as diverse a group of participants, as fresh a set of new voices, brought together by two splendid editors, in a single volume, at a special time, to address the urgency of what might be the grand challenge of the twenty-first century.

Robert W. Kates

List of Contributors

Julian Agyeman
Urban and Environmental Policy
  and Planning
Tufts University
Medford, MA

Susan Anderson
City of Portland Office of
  Sustainable Development
Portland, OR

Vicki Arroyo
Pew Center on Global Climate
  Change
Arlington, VA

John Atcheson
Office of Energy Efficiency and
  Renewable Energy
US Department of Energy
Washington, DC

Marilyn Averill
Center for Science and Technology
  Policy Research
University of Colorado—Boulder
Boulder, CO

Mary Catherine Bateson
Institute for Intercultural
  Studies
New York, NY

Sally Bingham
The Regeneration Project
San Francisco, CA

Ann Bostrom
School of Public Policy
Georgia Institute of Technology
Atlanta, GA

Caron Chess
Department of Human Ecology
Rutgers University
New Brunswick, NJ

Nancy Cole
Global Environment Program
Union of Concerned Scientists
Cambridge, MA

Lisa Dilling
Center for Science and Technology
  Policy Research
Cooperative Institute for Research in
  Environmental Sciences
University of Colorado–Boulder
Boulder, CO

Bob Doppelt
Resource Innovation
University of Oregon
Eugene, OR

Sharon Dunwoody
School of Journalism and Mass
  Communication
University of Wisconsin–Madison
Madison, WI

Pierre duVair
California Energy Commission
Sacramento, CA

Barbara Farhar
National Renewable Energy
  Laboratory
Golden, CO

Natasha Fraley
Pacific Grove, CA

David Gershon
Empowerment Institute
Woodstock, NY

Tina Grotzer
Harvard Graduate School of
  Education
Cambridge, MA

Robert Harriss
Houston Advanced Research Center
The Woodlands, TX

Halida Hatic
Air Quality Planning Unit
US Environmental Protection
  Agency
New England Office
Boston, MA

Orville Huntington
Koyukuk/Nowitna National
  Wildlife Refuge Complex
Huslia, Alaska

Keith James
Department of Psychology
Portland State University
Portland, OR

Dale Jamieson
Humanities and the Social Sciences
New York University
New York, NY

Branden Johnson
Division of Science, Research and
  Technology
New Jersey Department of
  Environmental Protection
Trenton, NJ

Bob Kates
Trenton, ME

Dan Lashof
Climate Center
Natural Resources Defense Council
Washington, DC

Anthony Leiserowitz
Decision Research
Eugene, OR

Rebecca Lincoln
Project Zero
Harvard Graduate School of
    Education
Cambridge, MA

Kathy Lynn
Resource Innovations
University of Oregon
Eugene, OR

Aaron McCright
Department of Sociology
    & Lyman Briggs School
    of Science
Michigan State University
East Lansing, MI

Shannon McNeeley
Department of Anthropology &
    Regional Resilience and
    Adaptation Program
University of Alaska—Fairbanks
Fairbanks, AK

David Meyer
Departments of Sociology and
    Political Science
University of California—Irvine
Irvine, CA

Laurie Michaelis
Living Witness Project
Headington, Oxford
United Kingdom

Susanne Moser
Institute for the Study of Society
    and Environment
National Center for Atmospheric
    Research
Boulder, CO

Linda Giannelli Pratt
Office of Environmental Protection
    and Sustainability
Environmental Services Department
City of San Diego
San Diego, CA

Benjamin Preston
Division of Marine and Atmospheric
    Research (CMAR)
CSIRO
Aspendale, Victoria
Australia

Sarah Rabkin
Environmental Studies Department
University of California—Santa Cruz
Santa Cruz, CA

Kathleen Regan
Department of Civil &
    Environmental Engineering
Tufts University
Medford, MA

Samuel Sadler
Renewable Energy Division
Oregon Department of Energy
Salem, OR

April Smith
Department of Psychology
Colorado State University
Fort Collins, CO

Abbey Tennis
Climate Change Program
Office for Commonwealth
  Development
Boston, MA

John Tribbia
Department of Sociology
University of Colorado-Boulder
Boulder, CO

Sheldon Ungar
Division of Social Sciences
University of Toronto—Scarborough
Whitby, Ontario
Canada

Anthony Usibelli
Energy Policy Division
Washington Department of
  Community, Trade, and Economic
  Development
Olympia, WA

Lucy Warner
Communications Department
UCAR
Boulder, CO

Susan Watrous
Santa Cruz, CA

Abby Young
ICLEI—Local Governments for
  Sustainability
Cities for Climate Protection
  Campaign
Berkeley, CA

Tables

Figures

Textboxes

Introduction

Lisa Dilling

University of Colorado—Boulder

Susanne C. Moser

National Center for Atmospheric Research

It is June 23, 1988, a sweltering day in Washington, DC, and members of the US Senate Committee on Energy and Natural Resources are settling into their seats. What they are about to hear will change the direction of American politics forever. Up to the podium steps a six-foot middle-aged scientist, a little hoarse, a little nervous, and quietly vies for the attention of the eminent body.

The timing is perfect. Over 100 degrees outside and a deadly drought gripping much of the country, James E. Hansen, chief scientist of NASA's Goddard Institute for Space Studies, is here to nail the case for global warming. His message is simple and clear. "The greenhouse effect has been detected, and it is changing our climate now." He states "with 99 percent confidence" that the evidence was in — the world was indeed getting warmer, and model projections pointed to worse heat waves and droughts in the future. As observers later recalled, "Besieged by the media afterward, [Hansen] said, 'It's time to stop waffling so much and say that the greenhouse effect is here and affecting our climate now.' Suddenly global warming — and Hansen — became world news."[1]

And world news it was. Not because of the news value of climate change — global warming had been buzzing around for a while — but because rarely if ever before did a scientist's warning set off such determined response. The June hearing was just the beginning. Seven hearings in the Senate and five in the House followed, each adding to the persuasiveness and urgency of the scientists' warning. Skeptical voices faded away in the storm of those convinced that America should take the lead in moving the world toward binding global greenhouse gas emission reductions. By 1992 world leaders signed on to the UN Framework Convention on Climate Change which the US Senate ratified shortly thereafter. The administration and Congress committed funding to the tune of hundreds of millions of dollars in incentives for renewable energy and clean technology development. Efficiency standards and emission caps

1

were instituted as a matter of course. Industry – inspired to highest performance by competition and corporate responsibility – chose not to complain or resist, but ramped up its own R&D and by 1997 outperformed not only the emission targets but its own highest hopes.

Later that year, the Houston Protocol – the document implementing the Framework Convention – codified the US example as the global goal. It was signed and shortly thereafter ratified by Congress, becoming the standard of other international agreements. Under the strong leadership of the United States, China, India, and other major developing countries immediately signed on and joined the race for the cleanest economy in the world. The ever-strengthening science did not, however, only encourage real emission reductions. It also spurred developed nations into unprecedented support for developing nations, helping them leap-frog the fossil-fuel heavy development stages and offering compassionate assistance in dealing with the first impacts, the challenges of adaptation, and with building a resilient society. In 2000, more than two thirds of the US population pledged to partake in the Millennium Challenge – a program to reduce personal emissions by half in 15 years.

In June 2005, 17 years almost to the day after his first urgent wake-up call, Hansen returned to the Senate for another hearing. Greeted with the respect of a statesman, the now-nearly-gray man appeared before the legislators with another clear and simple message: "The world has responded. I am here today to report to you of the observable progress we are making. The challenge is not over and we must continue our work. But I am here today to thank you."

This is not the story that historians will write – at least not with these dates and details. But we may yet write the history of a society heeding the ever-louder warnings about what many scientists agree is the biggest challenge humans have ever faced. The good news is that, in just the past few years more and more voices have joined those of scientists in calling for action to address climate change. And beyond just talk, signs of concrete action abound. Advocacy groups have launched new and smarter campaigns, many are coming together in novel coalitions, more and more in the business community are dropping their opposition to greenhouse gas (GHG) regulations, cities and states are taking action, and the US Congress is finally considering some modest policy proposals.

However, as the fundamental scientific consensus on human-induced climate change[2] has become stronger (Houghton *et al.*, 2001; Oreskes, 2004) and the impacts from global warming are now being regularly documented at far-flung locations around the globe (McCarthy *et al.*, 2001), carbon dioxide and other heat-trapping GHGs continue to rise inexorably in the atmosphere,[3] and people continue to lack adequate coping strategies for climate variability or change. This speaks to the magnitude of the challenge, the reality of the problem, and the lack of real progress as yet on effective solutions.

A persistent conundrum

Society at large does not appear to be deeply concerned with global warming, and as a result, is not yet acting on the ever-more urgent warnings emanating from the science and advocacy communities. Despite encouraging signs, ignorance, disinterest, apathy, and opposition are still prevalent. The resulting frustration among climate scientists and advocates runs high. They see the problem of global warming as urgent, difficult but not impossible to address, and needing immediate and substantial societal action. Yet their strategies to raise the sense of urgency in the public and among policy-makers don't seem to be working — at least not fast enough.

The familiar refrain goes something like this: "If only they understood how severe the problem is ... If only we could explain the science more clearly, train to be better communicators, become more media-savvy, get better press coverage ... The science of global warming is clear — why are we not acting as a society to combat the problem? Why are they not listening? Why is no one doing anything?"

Well, some things are being done, but not nearly enough to be commensurate with the magnitude of the problem. Thus, a persistent conundrum and challenging opportunity emerges: While the balance of available scientific evidence conveys an increasing sense of urgency, society as a whole — particularly in the United States — does not appear to view the problem as immediate, and certainly not as urgent. The often suggested remedy — by scientists and others — is the generic prescription: "better communication." Better communication is seen as essential in leading us out of this conundrum, out of political gridlock, pointing a path forward, and energizing leaders and the broader public to mobilize for effective action.

But what do we mean by "better communication"? For many, it simply means "explaining the issue more clearly" or "reaching more people." But the evidence shows that lack of a widespread sense of urgency is not the result of people not knowing about the issue. It is also not just due to not understanding it or lack of information. In fact, research has shown that the public is overwhelmingly aware of the problem of global warming. Over 90 percent of the US population has heard of it, some know the problem is related to energy use, and quite a high percentage can correctly identify impacts associated with global warming.[4] Far fewer understand the physics of the greenhouse effect, but one could argue that this level of understanding is not particularly necessary for action – even those who do not understand the basics of electricity generation still use appliances. What such survey studies also find is that while many judge the problem to be serious or very serious (Seacrest *et al.*, 2000; Brewer, 2003), only about a third of Americans find the issue personally concerning or worrisome (Stamm *et al.*, 2000)[5] – a percentage that has gone down in recent polls, rather than up (e.g., Kull *et al.*, 2004; Brechin, 2003). The disparity in these two findings – high awareness but low personal concern – shows that if creating urgency were just a matter of understanding the "facts," we would not be in the current conundrum.

So, clearly, there is something in *how* we communicate climate change that is failing to mobilize a wider audience. Simply talking about climate change in the way that has been done for the past few decades is not creating a sense of urgency or effective action. Certainly, there is an important role still for making the science of global warming accessible to the public. This function has served well in raising the issue to the high level of awareness that it already enjoys. But simply providing more information or speaking more loudly about climate change is not enough.

New research, interdisciplinary connections, and the experience of pioneers moving forward to act on the climate change problem point to a new approach. A quick glance around the United States reveals pockets of activity and success in motivating action in many different types of institutions – municipal and state governments, businesses, faith-based organizations, educational institutions, and the like. What can we learn from these examples about what works and why? How do we best draw together these lessons to inform others who do feel the problem is urgent and wish to promote appropriate action? We believe that the characteristics of the problem itself, the way people perceive and process information, and the motivators and barriers to action need to be examined through a new lens – one that integrates multidisciplinary knowledge on communication and social change. We look at what works – and what doesn't – on the ground, in different sectors, at different levels of governance, and let these practical experiences inform our communication and social change strategies and theories. Together scholarship and practice provide hope for a way out of the conundrum, a way forward towards effective communication and empowered action.

Why is climate change not perceived as urgent?

This book highlights stories of success in communicating and action on climate change, while taking a realistic look at the challenges before us. The champions we celebrate certainly have faced tough hurdles in their efforts. Without a doubt, global climate change is a difficult topic to talk about, a tough issue to spark interest among non-experts. First detected and defined by scientists, human-induced climate change has been called by many names: a carbon dioxide problem, an energy problem, global warming, an "enhanced greenhouse effect" – all abstract, benign-sounding, and utterly ... uninteresting, at least to most non-climate scientists (Clark *et al.*, 2001; Scheurs *et al.*, 2001).

In 1895, Svante Arrhenius, a Nobel laureate in chemistry laid the theoretical groundwork describing how fossil-fuel energy use could result in a warming atmosphere. As early as the 1950s, scientists in the United States, Europe, and elsewhere began to sound the alarm on climate change and potential impacts as they realized how human activities were altering the atmosphere, and therefore potentially the climate, of the entire Earth, but it would be decades before this scientifically defined problem would be more widely recognized and make it onto the public and policy agendas (Weart, 2003; Scheurs *et al.*, 2001). Why was it then, and why does it now continue to be, so difficult to make climate change relevant and important in light of the climate's central role as a life support system? The climate change problem has several characteristics that make it difficult to understand and communicate, much less to be perceived as urgent.

Lack of immediacy

Carbon dioxide and other GHGs are invisible and at atmospheric concentrations (even rising ones) have no direct negative health impacts on humans as do other air pollutants. Moreover, it has taken a while (in most places) for impacts on the environment to be detected. Most people do not connect driving their cars or flipping on a light switch with emitting CO_2 into the atmosphere. As a social problem, then, it is just not visible or experienced directly (yet) in the same way that job losses, obesity, or traffic congestion are.

Remoteness of impacts

The impacts of global warming are typically perceived as remote. Images of
ice receding in the Arctic and sea-level rise affecting distant tropical islands in
the Pacific, while dramatic, do not personally affect most of the world's
population (McCarthy *et al.*, 2001; Rayner and Malone, 1998; O'Brien and
Leichenko, 2000). And in most economically-advantaged societies, a
perception prevails, supported by much science and even more political
rhetoric, that society will be able to adapt to any adverse changes once they
arrive (e.g., Voice of America, 2004). In many less-advantaged societies that
are facing immediate, grave risks from disease, poverty, unsanitary
conditions, warfare, and so on, global warming simply cannot compete
against these direct personal threats and concerns.[6]

Time lags

The reason that scientists feel it is urgent to act on global warming involves
the enormous lags in the climate system. Over time the accumulation of
GHGs in the atmosphere will cause large-scale changes such as warming
of the ocean and changes in the climatic system that are not easily reversible
(Houghton *et al.*, 2001). The human systems that create these emissions —
such as the energy and transportation systems — also change only over
periods of decades, making it difficult to reduce GHG emissions instan-
taneously should society decide to make it a priority (Field *et al.*, 2004). But
these lags in the system that so alarm the scientific community also work
against making the problem urgent in the eyes of the general public.

Solution skepticism

The proposed solutions to solving the climate change problem also do not
engender a sense of urgency. Solutions are rarely discussed in scientific
presentations of the problem, leaving the audience to fill in their own (often
incorrect) concepts of what those solutions might be.[7] When they are
discussed, suggestions such as reducing home energy use or using public
transportation can provoke skepticism and resistance as it is hard for
individuals to see how alternatives could be made to work or how those
small actions could make any discernible difference to this global problem
(AGU, 1999; Bostrom, 2001). Similar skepticism — fed by political rhetoric,
ignorance, and some truth — prevails over international policy instruments
such as those codified in the Kyoto Protocol.

Threats to values and self-interests

At the national and international levels, solutions to global warming are seen as intensely political. In the United States, climate change remains a highly contested political issue as proposed solutions and policy mechanisms are viewed by some as conflicting with closely held values, priorities, and interests such as national sovereignty, economic growth, job security, and the "American way of life."[8] As a highly contested issue with an elusive, distant payoff, tackling climate change solutions is a challenge that most politicians would rather avoid unless political gain can be had from taking a position.[9]

Imperfect markets

The economic system of market-dominated capitalism relies on the straightforward notion of supply meeting demand, but it is well known that markets exhibit failures in accounting for externalities such as pollution.[10] These failures currently prevent the market from adequately accounting for externalized damages to the environment (and society). In addition, economic taboos such as assumptions about the role of consumption and economic growth are rarely discussed as they are central to the current conception of the economic engine.

Tragedy of the commons

The problem of global warming is maybe the ultimate "commons" problem (Hardin, 1968; NRC, 2002; Dietz, Ostrom, and Stern, 2003). The nations of the world all share one atmosphere. When GHGs are emitted from anywhere, they affect the climate of the Earth as a whole. Rules about using the atmosphere for the discharge of GHGs are only slowly being defined, while monitoring, accountability, and consequences for "overusing" the global atmospheric commons are extremely difficult to ensure and implement.

Political economy and injustice

The ethical implications of sharing one atmospheric commons go further. Some regions are disproportionately affected by climate change, and societal vulnerability to these negative impacts is also highly uneven due to differential levels of exposure and sensitivity to the risks, and differential ability to cope and adapt (Agyeman, Bullard, and Evans, 2003; Kasperson,

Kasperson, and Dow, 2001; Kasperson and Dow, 1991). Whether the decision is taken to maintain the status quo or undertake aggressive action to mitigate global warming, the burden and benefits of outcomes are unequally shared across nations and generations. Unfortunately, those who currently benefit from the status quo and who perceive themselves to be less severely impacted have little incentive to push for action (Agyeman, Bullard, and Evans, 2003; Kasperson, Kasperson, and Dow, 2001; Kasperson and Dow, 1991; Kasperson and Kasperson, 1991). Those, on the other hand, who are likely to be impacted more severely – the poor within developing and developed countries – have much incentive but little power and even fewer means to influence policy-making.

In summary, the inherent natural characteristics and deep societal roots of climate change stack the deck against the issue being recognized as an urgent and actionable problem. Communicators who have succeeded in motivating action to address this problem have been able to negotiate these challenges and still find a way to excite and engage different audiences constructively. Throughout this volume we find examples and strategies that have worked in preventing audiences from getting bogged down in these characteristics of the problem in different settings.

Communication and its impacts on the public's perception of urgency

Experience shows that the conundrum of the growing urgency of the problem vis-à-vis the lack of action is compounded by common communication practices of scientists, communicators, and advocates in the arena of climate change. Many of these are not unique to the problem of global warming – issues such as uncertainty, complexity, media practices, organized opposition, and people's mental models often play a role in controversial social issues. Those who are skilled in communicating and moving toward action have found modes of operating that recognize these pitfalls and remain focused on strategies that appeal to the constituencies they are working with. We discuss some of the most common communication pitfalls next.

Uncertain science as a political battlefield

For many years – especially in the United States, but to a lesser extent also in Europe and Australia – the rhetorical battle over the reality, causes, and solutions of global warming has been carried out within the arena of science. Scientists and others claiming authority on the issue took sides over whether or not the science itself was true or certain enough to act upon, whether the problem warranted precautionary or only adaptive action, and who should carry the financial burden. While legitimate scientific debate was and is useful and warranted, many of these "scientific" battles mask the true nature of the debate: namely one over values such as the responsibility of the present generation to future generations, the responsibility of economically advantaged nations towards less advantaged ones, the role of governments in regulating human choices over anything from energy use to development in hazardous areas, the rights of humans versus those of the non-human world, and so on (Briscoe, 2004; Sarewitz, 2004; Jamieson, 1996; Shackley and Wynne, 1996). Opponents of action on climate change have successfully organized and hired "their" experts (often called skeptics or contrarians) whose modus operandi has been to raise doubts about the overwhelming consensus on the state of the science while disproportionately highlighting the remaining unknowns (e.g., McCright and Dunlap, 2001, 2003). Even mainstream, credible scientists convinced of the seriousness of climate change have contributed to this emphasis on the unknown, often focusing more on "what we don't know yet" than on "what we do know." Scientists' professional culture, standards of conduct, and self-interest tend to emphasize uncertainty in standard communications (Briscoe, 2004; Shackley and Wynne, 1996). The result of these long-standing debates carried out on the back of science is a sad legacy: the trust in science is further eroding; those listening to the debates as media consumers are confused about the science, economics and politics; scientific uncertainty has hardened as a justification for inaction (Jamieson, 1996; Shackley and Wynne, 1996); and surveys show that the frequently partisan nature of the debate more often than not makes listeners turn away from the issue in disgust (ibid.).

Media practices and trends

Most Americans receive their information on climate change from mass media outlets such as television and newspapers. As researchers have pointed out, the tendency of the media to report two opposing viewpoints means that the mainstream consensus view is typically "balanced" by an opposing contrarian viewpoint. In practice, this amounts to a "bias" since the viewpoints of a handful of contrarians are given equal weight to the thousands of scientists who hold a general agreement with the consensus view of the IPCC (Boykoff and Boykoff, 2004; Mooney, 2004; Dearing, 1995). In addition, the number of independent outlets presenting news is dwindling, there is a sizable distrust of news sources among readers, and reporters deplore the challenges of good reporting under increasing economic pressures and the editorial policies that they give rise to (Eastland, 2005; France, 2004). Science reporting is increasing if measured by the number of stories alone (Pellechia, 1997). Yet the number of US newspapers with dedicated science sections has shrunk down to one, the number of reporters with science or environmental beats is declining, and reporters' understanding of climate science is very limited (Major and Atwood, 2004; Wilson, 2000; Bell, 1994; Wilkins, 1993).[11]

Inappropriate frames and mental models

People absorb new information through pre-existing frames of reference, or cognitive structures (so-called mental models), to order information (Kempton, 1991). They intimately affect people's understanding, perceptions, and reactions to information. For example, if climate change is reported on TV accompanied by images of weather disasters, the "weather" frame may be triggered. This frame suggests that climate change can neither be caused nor solved by humans, but is an "act of God." By focusing on large scale "weather"-like impacts, there is thus a danger that the communication may invoke a sense of helplessness or resignation — after all, who can control the weather (Morgan *et al.*, 2002; Bord, O'Connor, and Fisher, 2000; Bostrom *et al.*, 1994; Read *et al.*, 1994)?

Cultural barriers

Unlike many other socially defined problems of the twentieth century, global warming does not clearly resonate with any current cultural icons or values. There is no clear "brand" or "cultural whirlwind" defining the problem in a way that allows the public to easily relate (Ungar, 1992, 2000). It's not the subject of dinner-table conversations, and appears rarely in non-expert blogs or TV reality shows. At those recent times when it has entered popular culture, the problem is mischaracterized (either overblown or minimized) and the audience is left with additional confusion.[12]

Alarmism and other ineffective ways to create urgency

To make any issue a personal concern or even worry, it would have to affect one's own or one's family's well-being, or rise to moral significance (e.g., Hannon, 2005; Schultz, 2001). As British statesman Sir Crispin Tickell noted, it is difficult for climate change to appear urgent except in cases of catastrophe or disaster (Tickell, 2002). However, trying to create urgency by appealing to fear — of disasters, health risks, or the like — is unreliable at best in prompting behavior change. Frequently, this technique leads to the exact opposite from the desired response: denial, paralysis, apathy, or actions that can in fact create greater risks than the one being mitigated (Moser and Dilling, 2004).

Another persuasive technique commonly used is trying to shame individuals into changing their behavior. This taps into the second possible criterion needed to activate a personal worry or concern — the moral dimension. Pointing the finger at SUV owners or those who use energy in seemingly wasteful ways is ubiquitous among champions and advocates trying to promote behavior change. Yet guilt appeals, even more so than fear appeals, tend to be ineffective in generating the desired behavior. Most of us react with wild rationalizations for our behavior, with rejection, resentment, and annoyance at such manipulation attempts rather than with better behavior (Moser and Dilling, 2004; O'Keefe, 2002a,b; Nabi, 2002).

Given these many pitfalls in common communication practice that work against mobilizing action on climate change, it is no surprise that most people do not feel a personal urgency on the issue. The successful innovators in this book have found ways to communicate that recognize these pitfalls and manage to circumvent or avoid them in practice.

Barriers to action

The fundamental claim of this book is that better information dissemination, more knowledge, or more effective communication alone will not necessarily lead to desirable social changes. While we strongly believe that better understanding has an important role to play, communication that does not keep barriers to behavior and social change in mind is unlikely to be effective or sufficient. Research has demonstrated that even if participants have high levels of knowledge about the problem and the community has invested in changing their attitudes through advertising or educational campaigns, behavior is often unaltered (McKenzie-Mohr, 2000). Barriers to action can be internal to an individual (lack of knowledge on how to implement a specific act, such as replacing a thermostat) or external to an individual (e.g., lack of public transportation infrastructure). Organizations and institutions experience these obstacles to change in response to global warming as well. Successful communication that mobilizes action on climate change therefore must take into account the options that people have for action and their social and cognitive characteristics — in other words, what can they effectively do with the information they are given? The stories of this volume illustrate how effective communication for social change has taken into account these barriers and can therefore make a positive difference on climate. Some of the barriers to be aware of and overcome include the following.

Cognitive barriers

The way people think about issues and how they process information can either help or hinder making appropriate choices and taking conscious action. This begins with the metaphorical "getting the foot in the (mental) door." In the context of information overload, constant and ever-faster stimulation via TV and other news media, advertisement, email, the web, and so on, the primary barrier is to get on someone's radar screen, i.e., to cognitively register with a person. That in itself is not a given, even if exposure through the media occurs (Crane *et al.*, 1994). Getting through the information filters, triggering appropriate mental models and hence response options, and engaging people via encouraging frames of reference all play important roles (Morgan *et al.*, 2002; Bostrom *et al.*, 1994; Read *et al.*, 1994). An ability to weigh and sort out real or perceived conflicts between action choices is critical, as is the development of sufficient will to take an action. In short, individuals acting in their personal lives or making decisions for organizations face similar cognitive challenges.

What adds to these cognitive challenges is the fact that most adults' educational experiences do not prepare them well to deal with integrated, interdisciplinary problems that require agile responses and systems-thinking capabilities. In an educational system that over time evolved to emphasize the recall of details of separate subject areas rather than the connections between them, most children even today do not receive truly integrated education.

Psychological barriers

Psychological reactions to information are critical components of our processing and willingness to act; they can be as and sometimes even more powerful than the way we think about an issue (Dillard and Pfau, 2002). Certain strong emotional responses can end all further thinking — such as massive fear, despair, or a sense of being completely overwhelmed and powerless (Nicholsen, 2002; Macy and Brown, 1998).[13] Other emotions — such as guilt or other ways of feeling manipulated — can provoke staunch resistance. While emotional reactions are difficult to foresee with certainty, they are fatal to ignore when crafting an effective communications strategy.

Lack of peer support

Change is hard simply because it is a break in the routine, habit, or tradition. It triggers fear of the unknown, or aversion to risk, or simply resistance to the hassle of having to do something differently. New information, however credible, thus does not easily persuade individuals to act in new ways unless it comes from a trusted source (Mutz, 1998; Rogers, 1995). Generally, personally familiar sources are more trusted than more distant and less familiar sources; those coming from similar circumstances are believed to understand one's situation better than those coming from very different backgrounds. Often, it takes observing the actions by a neighbor, a friend, or a competing firm to spur action (Rogers, 1995). Many (behavior) change initiatives such as social marketing, weight loss, and rehabilitation programs (to name a few) employ peer support and pressure, mutual accountability, and maybe a greater sense of responsibility to great success (Cialdini, 2001).

Organizational inertia and resource constraints

Parallel to the resistance in individual behavior, organizational behavior change also often encounters active or passive resistance (Doppelt, 2003; Senge, 1990). Organizations have inertia of their own, and practices or procedures ingrained over time are often difficult to change or overcome, even with strong leadership. If the change requires extra funding, attention, or time, which may be hard to in times of limited budgets or under pressure, it is often easier for the organization to let innovations pass by. Even if a champion for action on global warming exists in one department of a city government or corporate division, for example, she may have to convince several other separate departments and key individuals in order for her organization to take action.

Lack of political will and leadership

The US political system — like many others — with its checks and balances is set up for stability. While examples can be found where long-term, intergenerational, and delayed-payoff policies and budget decisions have been and continue to be made, election cycles and accountability to constituents favor incrementalism (e.g., Hayes, 2001; Lindblom, 1959).[14]

Politicians are not rewarded — and sometimes even punished — for making tough, unpopular choices that have no immediate payoff and may even involve short-term sacrifice. In addition, interest group politics means that interests with the loudest voice are heard, while other interests are not fully represented. What politicians across the political spectrum have been able to agree on is the need for further research — hardly a sign of urgency given that the United States has been researching climate change for more than 25 years (Weart, 2003; NRC Geophysics Study Committee, 1977).

Invariably, some observers call for greater political will and see the solution to the policy impasse in aggressive leadership that could mobilize the rest of the country, while others place their hopes in the old saying "when the people lead, the leaders will follow." In fact, significant policy development is occurring in the United States at the local and state level, with recent action by governors, hundreds of mayors and cities signing on to climate protection programs, and corporations making climate change part of their strategic thinking. In the past, policies at the local and state levels have "trickled up" to the national level, and it remains to be seen how existing climate actions play out on the national scene.[15] At present, however, a sense of urgency is still lacking at the national level in the United States except in pockets of activity by a few committed champions. Former Vice President Al Gore's *An Inconvenient Truth*, movie, book and media whirlwind may help put greater pressure on political leaders.

Technological barriers

Finally, reducing emissions of carbon dioxide and other GHGs into the atmosphere is also a considerable technological challenge. All of the proposed solutions to stabilizing the amount of heat-trapping gases emitted by humans, including improving energy efficiency, decarbonization, sequestration, alternative energy sources, and various geoengineering schemes, represent major technological challenges (Hoffert *et al.*, 2002). While many of these technologies are currently available or under development, challenges still remain as to their availability at economies of scales and at reasonable costs while minimizing the negative impacts on, and tradeoffs with, the affected public (Pacala and Socolow, 2004).

A fresh approach

The relationship between the climate problem and stimulating societal responses via communication is clearly more complex than is commonly accounted for. A fuller understanding of the role of communication and how it intersects with social change is necessary. While a high level of basic awareness has been achieved, understanding can (and some would argue *should*) still be deepened significantly. But a high level of awareness and a better understanding of the science underlying climate change do not directly or necessarily translate into concern or action (e.g., voting, behavior changes, policy support, or other forms of engagement). Differently put, "better communication" goes beyond simply designing more effective ways of conveying information from an expert to a lay audience.

Yet observation of current efforts suggests just that. For better or worse, a large share of the responsibility for communicating climate change still falls to scientists and others who lay claim to scientific or technical expertise. Among many of these communicators, the tripartite conviction that (1) climate change is essentially a scientific issue, (2) experts understand it and others don't, and (3) the purpose of communication thus is to educate the ignorant is, in short, still alive and well. Communication on global warming based on these assumptions thus creates an abiding rift between listener and speaker, preventing the listener from truly gaining ownership of the problem because of its alleged purely technical nature and the implicit hierarchy of expert/lay person in which it is approached.

The discussion above has demonstrated how this traditional approach to communication has failed to motivate – the public is aware of the term "global warming," but not energized by it to act. Climate change simply does not resonate deeply with the general public; it remains disconnected from people's daily lives, from their more immediate concerns. This suggests, then, that climate change has not been communicated effectively until communicators understand how to bridge this "gap of meaning." To do so, it seems to us, is impossible without understanding the "audience" more fully.

"Know thy audience," of course, is an old adage in communication practice. But the existing literature and the chapters in this book point to something more fundamental than simply going down a checklist of audience characteristics or surveying potential recipients of climate change information for what will resonate with them. We have come to see the importance of dialogue, of the genuine exchange among other-than-scientific viewpoints and needs, and the integration of climate change with other-than-climate-change concerns. This has led us to a broader definition of communication in support of social change as a *continuous and dynamic process unfolding among people that facilitates an exchange of ideas, feelings, and information as well as the forming of mutual understanding and common visions of a desirable future.* Communication – etymologically rooted in the same Latin word as communion — points to meanings of participation and sharing, of imparting meaning, and making common (Harper, 2001).

This volume moves us toward this broader conceptualization, to a fresh approach. It takes stock of the communication and social change challenges, practices, and debates, as well as of pertinent research and practical experience to draw lessons and propose more effective strategies. Contributors offer deeper insight into why the problem of global warming is not seen as urgent, and, in turn, how to redesign communication efforts so that they can support action from the personal to the political, from the household to the national and international level, rather than, at best, be irrelevant, or, at worst, a hindrance. As such, this book offers ways to *improve on current practice* — designing communication strategies that empower rather than alienate — and ways to *envision new practices* — how to move beyond message delivery and toward dialogue and engagement.

Book organization and chapter preview

There is no dearth of research on various aspects of climate change communication (e.g., messaging, framing, the role of the media, and especially the resulting perceptions, attitudes, opinions, and levels of understanding of the issue among various audiences). Similarly, there is a vast scholarship on social change. In the realm of practice, there is also no lack of people communicating climate change, and — by trial and error — adapting their approaches or the content of their messages. And there are many who work on mobilizing various sections of the public to change their climate-relevant behaviors or to support or adopt policies that would address the growing risk of climate change. Over recent years, in fact, a movement of sorts has been building toward climate change action involving individuals, organizations, corporations and churches, cities and towns, a widening spectrum of advocacy groups, as well as states and some members of the US federal government (Isham and Waage, forthcoming).[16] This growing and more diverse involvement of different players has broadened the conversation on climate change. People have tried on different framings, forged linkages between their traditional concerns and the global, systemic ones. As a result new coalitions are being forged, which involve finding new common denominators, mutually agreeable meanings, and action strategies.

To our knowledge, this book is the first comprehensive effort to assemble the insights from all this research and experience in one place and let these insights inform each other to extract lessons and improve strategies.

The chapters that follow are organized into two major parts: one on communication, the other on social change.

In the first half of the book, we gather perspectives on communication from a variety of fields and experience such as risk perception, risk communication, framing, mental models, and message content and delivery, with reflections on the role, importance, and limitations of different messengers and communication channels. This section explores the question of which audiences have been and are left to be engaged. It also examines the emotional and cognitive reception of climate change information and people's responses to sometimes scary or overwhelming content. To understand audiences better and adjust messages accordingly, we begin by examining how people perceive the problem of global warming, what they understand, and how the problem fits into their existing beliefs (chapters by Bostrom and Lashof, and Leiserowitz). Moser examines the complementary perspective of how global warming evokes strong emotions that may inhibit effective action. Ungar examines the notion of issue culture – how it is created and fostered – and how to make climate change culturally more resonant by harnessing the powers and insights of advertising. He offers a different perspective on whether global warming has some of the critical elements needed to sustain mass action. Dunwoody then explore the challenges of trying to communicate the issue of climate change via the media and examines the advantages and limitations of communicating climate change through various mass media channels.

While the economy of scale of mass communication allows one to reach wide audiences, it may not allow one to reach – and persuade – specific audiences. The chapter by Pratt and Rabkin reports on the efforts of one city, San Diego, which sought first to elicit information on its citizens' concerns before embarking on a public education campaign in order to develop a more effective outreach plan. Agyeman *et al.* and McNeely and Huntington explore communication challenges with non-white, non-middle-class audiences, tapping into the environmental justice issues that intersect with global warming, and examining differential impacts and responsibilities. Bingham discusses the connection between climate change (science) and values in the context of communicating to religious communities. Her personal experience illuminates the tricky balance of talking about global stewardship without alienating congregations in faith-based settings.

Several chapters examine the role of scientists as messengers of climate change, including how scientific messages have reached the powerful in the past (Warner) and what lessons we can learn from that experience, and the delicate balance between scientific credibility and public engagement

(Cole with Watrous). McCright provides a glimpse of the operation and communication tactics of organized interests opposing action on global warming and offers strategies to deal with contrarians. Regan — building on interviews with scientists and non-experts — speaks to the need and challenge of broadening the conversation on global warming beyond the contentious and scientific to a new mode of dialogue. Finally, Chess and Johnson explore the need for trust in messengers and remind us of the limits of information to affect behavior, leading to some strategies for improving the chances of motivating change. Throughout, the contributors ask how communication can be directed effectively at actors and actions that could make a difference in bringing about social change required for meaningful climate change action.

In the second half of the volume, we then focus on insights on change in individual behavior, organizations, local and state governments, businesses and the market, cities and neighborhoods, and political and legal systems. Many barriers and hopes for the future with respect to social change are discussed from the cultural, psychological, legal, and economic perspectives. The first two chapters focus on the individual — what is known about the factors and barriers that affect the behavior and change of behavior of individuals. Tribbia explores the cognitive, motivational, and other person-specific factors in the context of an individual's social milieu, while Michaelis takes a cultural-theoretical perspective to explore the larger context in which individuals act. Grotzer and Lincoln as well as Bateson explore the contributions that education can make by reaching young people early and throughout their lives. They discuss necessary changes in teaching methods and foci as well as the educational system and culture more broadly that might help students grow up to be more adept global citizens in a world of rapid global and climate change.

The next two chapters move from the individual to the organizational level. Rabkin and Gershon discuss the role of peer motivation, social marketing tools, and empowerment to achieve unexpected results in a Portland, Oregon neighborhood. James *et al.* discuss the importance of actively managing organizational change to support action on global warming. Such organizational changes, of course, are typically stimulated by and embedded in larger contextual changes — in the marketplace or in the policy environment. As Arroyo and Preston, and Atcheson point out, businesses and markets have an important role to play in solutions to global warming, and they explore in depth the motivations and strategies the business community has in responding to climate change. These optimistic perspectives are somewhat tempered by Dilling and Farhar, who critically examine strategies and policies that help promote energy efficiency and renewables as a way to "make it easy" for the consumer to behave in climate-friendly ways.

The next four chapters focus on actions and strategies to combat global warming currently under way at the city, state, and regional level in the United States. Young discusses the International Council for Local Environmental Initiatives' Cities for Climate Protection campaign, a network of communities pioneering emission reduction strategies at the local level. She explores the conditions that got them involved, and the communication and mutual support among them. One example of such a forward-thinking community is Santa Monica, CA – a community that chose to address climate change without talking about it (Watrous and Fraley). We then hear from actors in the US Northeast (Tennis) and on the US West Coast (duVair *et al.*) of their leadership and institution-building efforts, and how these states and regional collaboratives have dealt with obstacles and resistance.

The last set of chapters in this part looks at tools and processes of social change at an even broader level yet. Averill discusses the role of litigation and related legal tools for promoting action and communicating global warming. Meyer reflects on how social movements arise, and how successful movements organize and communicate for their cause. The final chapter by Jamieson addresses the relationship of politics, ethics, and responsibility for global warming, and examines the political changes required to effectively address climate change in the United States.

Throughout Part II, we asked contributors to go beyond merely stating problems, challenges, or shortcomings in past efforts to foster social change, and lament the limitations in our understanding from research. We urged them to offer a clear assessment of what has worked, in what context, toward what end, and then suggest additional or promising ways to improve the communication of climate change. We asked them to address what role communication played; what was said, by whom, to whom, in what way to increase the chances that a particular behavior, institutional, policy, or other social change would take place. Those who achieved a particular change were asked to be transparent about the "how," and how particularly difficult obstacles were overcome.

The book concludes in Part III with thoughts on the way forward. Harriss extends on a well-received summary he first offered at the 2004 workshop which initiated this project. We then conclude with our own synthesis of practical next steps and research needs from listening to this diverse group of contributors.

What's left unheard and maybe unsaid

Even as diverse an anthology as this cannot cover everything. We did not intend to compile an all-encompassing collection of perspectives on climate change. We sought voices from many disciplines and experiences, from different regions, different generations, a range of philosophical convictions, and gathered diverse offerings of tools for communicating urgency and promoting social change. The common thread that weaves these voices together is collegial respect for these different perspectives and a clear recognition of the immensity of the challenge that climate change poses for society.

Yet some readers will find particular voices and insights missing such as that of federal government policy-makers and communicators, discussions of specific policy approaches and technological solutions, or — given the global scope of the problem — more on non-US activities. Others will look for a greater focus on adaptation — clearly an important and necessary complement of societal response to mitigating climate change.

Executive and legislative branches of the federal government of the United States play a key role in both researching and communicating climate change. We focus in this book on organizations and individuals who are deeply engaged and on the cutting edge of communication and social change on climate change. While the US federal government remains committed to research, it has played a less visible role in communicating about climate change in the past several years. We recognize, however, that the federal actors can play an extremely important role in both communication and social change.

With some important exceptions, the conversation about adaptation in the United States is even further behind than that on mitigation (Luers and Moser, 2006). Far less is known empirically, for example, about how people view our ability to cope with and adapt to climate change impacts, or how adaptation is communicated and heard. While many efforts are under way where climate impacts scientists are working with resource managers to raise their awareness and use of climate change information in their long-term decisions, the broader public remains largely untouched by these communicative interactions. One big exception is Alaska and other far-north regions, which are already experiencing the impacts of climate change. People in this region require less convincing of the reality and urgency of the issue, but are primarily concerned with what to do about it (this finding is clearly articulated by McNeeley and Huntington, Chapter 8). We consciously chose to focus on mitigation at this time, but suspect that many of the lessons produced here will be directly applicable to the adaptation context as well as to other highly complex global environmental change problems.

Finally, we chose to focus largely on the United States, not because it is the only country where communication of this topic is needed — although it is needed here especially — but because our expertise is greatest in this cultural context. Many lessons contained in this volume are transferable to other cultural contexts only with careful consideration, adaptation, and testing for effectiveness. Our reading of climate change communication strategies emerging in other countries, however, gives us confidence that many of the principles if not the content carry over rather readily (see, e.g., Futerra, 2005a,b).[17] For example, the need for credible messengers is as high in the United States as it is elsewhere, but what such a messenger would say to resonate with local audiences is highly context-specific.

In sum, this anthology distills the scholarship of researchers in a variety of disciplines and brings it together with the wisdom of practitioners "on the ground" in offices, communities, and states across the country. The chorus of these diverse voices helps us understand the complementary foundational elements of communication and social change that contribute to the status quo *and* point the way forward to a more productive understanding of the challenges and opportunities. Existing policies at the international and national level are insufficient to significantly slow down anthropogenic climate change and enhance people's adaptive capacity to cope with its impacts. Clearly, we need a new way forward — a new way that actively engages and empowers the public, ignites a deeper debate, creates a vision of a future worth fighting for, and develops and implements the solutions that will allow us to get there. This volume, we hope, will help light the way.

Notes

1. The beginning of this imaginary story — up to this point — draws on actual facts (Hansen, 1988). The quote is from Boyle (1999), recalling the event in a feature story for *Audobon Magazine*.
2. We use the terms global warming and climate change interchangeably, but communicators disagree — less on the different meanings and implications — but more on which terms may be more effective in reaching various audiences. Most scientists prefer climate change, or anthropogenic climate change, to encompass the many related changes in the atmosphere and global climate. The term allows, for example, for the possibility that while global average temperatures are increasing, local or regional climates may cool. It also makes room for changes in precipitation, extreme climatic events, seasonal patterns, and so on. Many in the media, most advocates, and other public communicators, on the other hand, tend to use the term global warming, which is now widely recognized by, and resonates more than climate change with, the public. Several other terms have recently come into play, such as climate disruption and climate crisis. These latter terms have not been tested for audience response, leaving us hesitant to endorse them. Our primary

concern in this book is not with finding the best term, but how to make the concept meaningful; we thus use the common and recognized phrases instead.

3. Data can be found from the US National Oceanic and Atmospheric Administration at: http://www.cmdl.noaa.gov/ccgg/insitu.html; accessed January 3, 2006.

4. Note that the exact figures differ from study to study due to differences in questions, depth of study, and temporal variance. Reviews of studies on similar questions, however, reveal similar trends and orders of magnitude in their findings. We thus cite only rough approximations to indicate levels of concern or understanding. See also Leiserowitz (2003).

5. Similar figures are found for Canadians; see Environics International (1998).

6. These primary concerns were recognized by the United Nations in its Millennium Development Goals, including eradicating extreme poverty and hunger, reducing by half the proportion of people living on less than a dollar a day, ensuring that all boys and girls complete a full course of primary schooling, reducing child mortality, and so on. While environmental sustainability is part of these goals, climate change is not mentioned as an overriding concern.

7. Research from FrameWorks Institute's "Climate Message Project" demonstrates that the messages told by environmental advocates consist predominantly of proof that the problem is real and of warnings of negative consequences. See http://www.frameworks-institute.org/clients/climatemessage.shtml; accessed June 13, 2005. Solutions are frequently not part of the communication. As a consequence, individuals fill in with their own ideas. For example, Bostrom *et al.* (1994) found that individuals who thought that the ozone hole was related to creating global warming also thought that it did so by letting more heat from space in through the "hole." Some respondents thought that perhaps NASA spacecraft were punching holes through the ozone layer, and that NASA sends its spacecrafts up through the same hole, thereby not creating new "holes" and reducing the threat of global warming. This highlights how mental models and pre-existing beliefs color one's perception of possible solutions, in the absence of alternate ways of thinking about a problem.

8. Historical data suggest a close correlation between gross domestic product (GDP) and energy use, over 85 percent of which is currently provided by fossil fuels in the United States (see DOE Energy Information Administration, 2005). Advocates for climate policy point out, of course, that there is no intrinsic necessity that this correlation between economic growth and energy be linked to the use of fossil fuels *per se* – alternative energy sources could also support economic productivity, but are currently less available (see, e.g., Union of Concerned Scientists website, "Clean Energy" (2005). Available at: http://www.ucsusa.org/clean_energy/renewable_energy/index.cfm, accessed June 9, 2005).

9. Note that not only politicians with a desire to be re-elected have a tendency to postpone tough choices. Studies repeatedly find that people in general would prefer to "discount [their] concern," as one commentator recently called it. Things in the future, in far-away places, things that can't be known for sure, that can't be experienced with the senses, or that do not affect a person directly, are generally taken less seriously than their opposites (Hannon, 2005; Hendrick and Nicolaij, 2004).

10. The notion of externalities was first introduced by Arthur Pigou (1932).

11. *The New York Times* is one of the few among major US newspapers with a dedicated weekly science section. On media ownership trends, see the *Columbia Journalism Review* at: http://www.cjr.org/tools/owners/; accessed January 4, 2006.

12. Recent examples include the blockbuster movie, *The Day After Tomorrow*, and the bestselling novel by Michael Crichton, *State of Fear*. For analyses of the effect of the movie on public perceptions of the problem in different countries, see, e.g., Leiserowitz (2004), Leaman and Norton (2004).

13. In Macy and Brown (1998) especially pp. 26–32; in Nicholsen (2002) especially Chapter 5.

14. One example of long-term policy is the establishment of social security – an intergenerational program set up in the midst of crisis. Another example is investment in basic research, much of which has no immediate, and sometimes no discernible, payoff at

all. Such decisions speak to the possibility of taking a constructive political stance on long-term matters.
15. For elaborations on this argument, see, e.g., Walker (1969), Kosloff *et al.* (2004), McKinstry (2003).
16. See also: http://www.whatworks-climate.org; accessed January 4, 2006.
17. A good example is the UK's climate change communication strategy, which highlights similar needs and principles regarding effective communication. See also http://www.climatechallenge.gov.uk/; accessed January 4, 2006.

References

Agyeman, J., Bullard, R., and Evans, B. (2003). *Just Sustainabilities: Development in An Unequal World.* London: Earthscan/The MIT Press.

American Geophysical Union (AGU) (1999). Waiting for a signal: Public attitudes toward global warming, the environment and geophysical research. Report available at: http://www.ago.org/sci_soc/attitude_study.pdf

Bell, A. (1994). Media (mis)communication on the science of climate change. *Public Understanding of Science*, 3, 259–75.

Bord, R. J., O'Connor, R. E., and Fisher, A. (2000). In what sense does the public need to understand global climate change? *Public Understanding of Science*, 9, 205–18.

Bostrom, A., Morgan, M. G., Fischhoff, B., *et al.* (1994). What do people know about global climate change? 1. Mental models. *Risk Analysis*, 14, 6, 959–70.

Bostrom, M. (2001). American attitudes to the environment and global warming: An overview of public opinion. Report prepared for the FrameWorks Institute, Washington, DC.

Boykoff, M. T. and Boykoff, J. M. (2004). Balance as bias: Global warming and the US prestige press. *Global Environmental Change*, 14, 2, 125–36.

Boyle, R. H. (1999). "You're getting warmer ..." *Audobon Magazine* (November/December). Available at: http://magazine.audobon.org/global.html; accessed June 15, 2005.

Brechin, S. R. (2003). Comparative public opinion and knowledge on global climatic change and the Kyoto Protocol: The US versus the world? *International Journal of Sociology and Social Policy*, 23, 10, 106–34.

Brewer, T. L. (2003). US public opinion on climate change issues: Evidence for 1989–2002. Available at: http://www.ceps.be; accessed January 5, 2006.

Briscoe, M. (2004). *Communicating Uncertainty in the Science of Climate Change: An Overview of Efforts to Reduce Miscommunication Between the Research Community and Policymakers and the Public.* Washington, DC: International Center for Technology Assessment.

Cialdini, R. B. (2001). *Influence: Science and Practice*, 4th edn. Boston, MA: Allyn and Bacon.

Clark, W. C., *et al.* (2001). Acid rain, ozone depletion, and climate change: An historical overview. In *Learning to Manage Global Environmental Risks*, Vol. 1, Social Learning Group, ed. Cambridge, MA: The MIT Press, pp. 21–55.

Crane, V., Nicholson, H., Chen, M., *et al.* (1994). *Informal Science Learning: What the Research Says About Television, Science Museums, and Community-Based Projects.* Ephrata, PA: Science Press.

Dearing, J. W. (1995). Newspaper coverage of maverick science: Creating controversy through balancing. *Public Understanding of Science*, **4**, 341–61.

Department of Energy (DOE), Information Administration (2005). *Monthly Energy Review*, January, Table 1.3, p. 7.

Dietz, T., Ostrom, E., and Stern, P. C. (2003). The struggle to govern the commons. *Science*, **302**, 1907–12.

Dillard, J. P. and Pfau, M. (eds.) (2002). *The Persuasion Handbook: Developments in Theory and Practice*. Thousand Oaks, CA: Sage Publications.

Doppelt, B. (2003). *Leading Change Toward Sustainability*. Sheffield, UK: Greenleaf Publishing.

Eastland, T. (2005). The collapse of big media: Starting over. *Wilson Quarterly*, (Spring).

Environics International (1998). *The International Environmental Monitor*, Vol. 1: Citizens Worldwide Want Tough Environmental Action Now. Available at: http://www.mori.com/polls/1998/iem_02.shtml; accessed January 10, 2006.

Field, C. B., Raupach, M. R., and Victoria, R. (2004). The global carbon cycle: Integrating humans, climate and the natural world. In *The Global Carbon Cycle: Integrating Humans, Climate and the Natural World*, eds. Field, C. B. and Raupach, M. R. SCOPE Report #62, Washington, DC: Island Press, pp. 1–13.

France, M. (2004). Is there a market for nonpartisan news? Media Commentary. *Business Week*, November 29.

Futerra (2005a). *UK Communications Strategy on Climate Change*. London: Futerra.

Futerra (2005b). *The Rules of the Game: Principles of Climate Change Communication*. London: Futerra.

Hannon, B. (2005). Pathways to environmental change. *Ecological Economics*, **52**, 417–20.

Hansen, J. E. (1988). Testimony before US Senate Committee on Energy and Natural Resources. Greenhouse effect and climate change, 100th Congress, 1st Session, June 23.

Hardin, G. (1968). The tragedy of the commons. *Science*, **162**, 1243–8.

Harper, D. (2001). *Online Etymology Dictionary*. Available at: http://www.etymonline.com/, accessed March 28, 2005.

Hayes, M. T. (2001). *The Limits of Policy Change: Incrementalism, Worldview, and the Rule of Law*. Washington, DC: Georgetown University Press.

Hendrick, L. and Nicolaij, S. (2004). Temporal discounting and environmental risks: The role of ethical and loss-related concerns. *Journal of Environmental Psychology*, **24**, 4, 409–22.

Hoffert, M., *et al.* (2002). Advanced technology paths to global climate stability: Energy for a greenhouse planet. *Science*, **298**, 981–7.

Houghton, J. T., Ding, Y., Griggs, D.J., *et al.* (eds.) (2001). *Climate Change 2001: The Scientific Basis*. Contribution of Working Group I to the Third Assessment Report of the Intergovernmental Panel on Climate Change. Cambridge, UK: Cambridge University Press.

Isham, J. and Waage, S. (forthcoming). *Ignition: The Birth of the New Climate Movement* Washington, DC: Island Press.

Jamieson, D. (1996). Scientific uncertainty and the political process. *Annals AAPSS*, **545**, May, 35–43.

Kasperson, R. E., Kasperson, J. X., and Dow, K. (2001). Vulnerability, equity, and global environmental change. In *Global Environmental Risk*, eds. Kasperson, J. X. and Kasperson, R. E. London: Earthscan, pp. 247–72.

Kasperson, R. E. and Dow, K. (1991). Developmental and geographical equity in global environmental change. *Evaluation Review*, **15**, 149–71.

Kasperson, R. E. and Kasperson, J. X. (1991). Hidden hazards. In *Acceptable Evidence: Science and Values in Risk Management*, eds. Mayo, D. G. and Hollander, R. D. Oxford, UK: Oxford University Press, pp. 9–28.

Kempton, W. (1991). Public understanding of global warming. *Society and Natural Resources*, **4**, 4, 331–45.

Kosloff, L. H., *et al.* (2004). Outcome-oriented leadership: How state and local climate change strategies can most effectively contribute to global warming mitigation. *Widener Law Journal*, **14**, 173–204.

Kull, S., *et al.* (2004). Americans on climate change. Menlo Park, CA: Program on International Policy Attitudes (PIPA), University of Maryland and Knowledge Networks.

Leaman, J. and Norton, A. (2004). The Day After Tomorrow – are the British too cool on climate change? London: MORI Social Research Unit, UK, Prime Minister's Strategy.

Leiserowitz, A. (2004). Before and after *The Day After Tomorrow*: A US study of climate change risk perception. *Environment*, **46**, 9, 22–37.

Leiserowitz, A. (2003). American opinions on global warming. Project report. University of Oregon.

Lindblom, C. E. (1959). The science of "muddling through." *Public Administration Review*, **19**, 79–88.

Luers, A. L. and Moser, S. C. (2006). *Preparing for the Impacts of Climate Change in California: Advancing the Debate on Adaptation*. CEC-500-2005-198-SD, Report prepared for the California Energy Commission, Public Interest Energy Research Program and the California Environmental Protection Agency, Sacramento, CA.

Macy, J. and Brown, M. Y. (1998). *Coming Back to Life: Practices to Reconnect Our Lives, Our World*. Gabriola Island, BC: New Society Publishers.

Major, A. M. and Atwood, L. E. (2004). Environmental risks in the news: Issues, sources, problems, and values. *Public Understanding of Science*, **13**, 295–308.

McCarthy, J. J., Cauziani, O. F., Leary, N. A., *et al.* (eds.) (2001). *Climate Change 2001: Impacts, Adaptation, and Vulnerability*, Contribution of Working Group II to the Third Assessment Report of the Intergovernmental Panel on Climate Change. Cambridge, UK: Cambridge University Press.

McCright, A. M. and Dunlap, R. E. (2001). Challenging global warming as a social problem: An analysis of the conservative movement's counter-claims. *Social Problems*, **47**, 4, 499–522.

McCright, A. M. and Dunlap, R. E. (2003). Defeating Kyoto: The conservative movement's impact on US climate change policy. *Social Problems*, **50**, 348–73.

McKenzie-Mohr, D. (2000). Fostering sustainable behavior through community-based social marketing. *American Psychologist*, **55**, 531–7.

McKinstry, R. B. (2003). Laboratories for local solutions for global problems: State, local and private leadership in developing strategies to mitigate the causes and effects of climate change. *Penn State Environmental Law Review*, **12**, 15–82.

Mooney, C. (2004). Blinded by science: How "balanced" coverage lets the scientific fringe hijack reality (State of the Beat). *Columbia Journalism Review*, **6**, November/December; available at: http://www.cjr.org/issues/2004/6/ mooney-science.asp, accessed June 15, 2005.

Morgan, M. G., *et al.* (2002). *Risk Communication: A Mental Models Approach.* New York: Cambridge University Press.

Moser, S. and Dilling, L. (2004). Making climate hot: Communicating the urgency and challenge of global climate change. *Environment*, **46**, 10, 32–46.

Mutz, D. C. (1998). *Impersonal Influence: How Perceptions of Mass Collectives Affect Political Attitudes.* Cambridge, UK: Cambridge University Press.

Nabi, R. L. (2002). Discrete emotions and persuasion. In *The Persuasion Handbook: Developments in Theory and Practice*, eds. Dillard, J. P. and Pfau, M. Thousand Oaks, CA: Sage Publications, pp. 289–308.

National Research Council (NRC) (2002). *The Drama of the Commons.* Commission on Behavioral and Social Sciences and Education. Washington, DC: National Academy Press.

National Research Council (NRC) Geophysics Study Committee (1977). *Energy and Climate.* Washington, DC: National Academy Press.

Nicholsen, S. W. (2002). *The Love of Nature and the End of the World.* Cambridge, MA: The MIT Press.

O'Brien, K. L. and Leichenko, R. M. (2000). Double exposure: Assessing the impacts of climate change within the context of economic globalization. *Global Environmental Change*, **10**, 221–32.

O'Keefe, D. J. (2002a). Guilt as a mechanism of persuasion. In *The Persuasion Handbook: Developments in Theory and Practice*, eds. Dillard, J. P. and Pfau, M. Thousand Oaks, CA: Sage Publications, pp. 329–44.

O'Keefe, D. J. (2002b). Guilt and social influence. In *Communication Yearbook* **23**, ed. Roloff, M. E. Thousand Oaks, CA: Sage Publications, pp. 67–101.

Oreskes, N. (2004). The scientific consensus on climate change. *Science*, **306**, 1686.

Pacala, S. and Socolow, R. (2004). Stabilization wedges: Solving the climate problem for the next 50 years with current technologies. *Science*, **305**, 968–72.

Pellechia, M. G. (1997). Trends in science coverage: A content analysis of three US newspapers. *Public Understanding of Science*, **6**, 1, 49–68.

Pigou, A. C. (1932). *The Economics of Welfare*, 4th edn. New York: Macmillan and Co.

Rayner, S. and Malone, E. L. (eds.) (1998). *Human Choice & Climate Change. What Have We Learned?* Vol. 4. Columbus, OH: Batelle Press.

Read, D., *et al.* (1994). What do people know about global climate change? 2. Survey studies of educated laypeople. *Risk Analysis*, **14**, 6, 971–82.

Rogers, E. M. (1995). *Diffusion of Innovations*, 4th edn. New York: Free Press.

Sarewitz, S. (2004). How science makes environmental controversies worse. *Environmental Science & Policy*, **7**, 385–403.

Scheurs, M., *et al.* (2001). Issue attention, framing, and actors: An analysis of patterns across arenas. In *Learning to Manage Global Environmental Risks*, Vol. 1, Social Learning Group, ed. Cambridge, MA: The MIT Press, pp. 349–62.

Schultz, P. W. (2001). Assessing the structure of environmental concern: Concern for the self, other people, and the biosphere. *Journal of Environmental Psychology*, **21**, 327–39.

Seacrest, S., Kuzelka, R., and Leonard, R. (2000). Global climate change and public perception: The challenge of translation. *Journal of the American Water Resources Association*, **36**, 2, 253–63.

Senge, P. (1990). *The Fifth Discipline*. New York: Doubleday.

Shackley, S. and Wynne, B. (1996). Representing uncertainty in global climate change science and policy: Boundary-ordering devices and authority. *Science, Technology and Human Values*, **21**, 275–302.

Stamm, K. R., Clark, F., and Eblacas, P. R. (2000). Mass communication and public understanding of environmental problems: The case of global warming. *Public Understanding of Science*, **9**, 219–37.

Tickell, C. (2002). Communicating climate change. *Science*, **297**, 737.

Ungar, S. (2000). Knowledge, ignorance and the popular culture: Climate change versus ozone hole. *Public Understanding of Science*, **9**, 297–312.

Ungar, S. (1992). The rise and (relative) decline of global warming as a social problem. *The Sociological Quarterly*, **33**, 483–501.

Union of Concerned Scientists (2005). Clean Energy. Available at: http://www.ucsusa.org/clean_energy/renewable_energy/index.cfm, accessed June 9, 2005.

Voice of America (2004). US Plans to Emphasize Climate Change Adaptation at Upcoming Conference. December 3; available at: http://www.voanews.com/english/2004-12-03-voa83, accessed June 9, 2005.

Walker, J. L. (1969). The diffusion of innovations among the American states. *The American Political Science Review*, **63**, 3, 880–99.

Weart, S. (2003). *The Discovery of Global Warming*. Cambridge, MA: Harvard University Press.

Wilkins, L. (1993). Between facts and values: Print media coverage of the greenhouse effect, 1987–1990. *Public Understanding of Science*, **2**, 1, 71–84.

Wilson, K. M. (2000). Drought, debate, and uncertainty: Measuring reporters' knowledge and ignorance about climate change. *Public Understanding of Science*, **9**, 1, 1–13.

PART ONE

Communicating climate change

1

Weather it's climate change?

Ann Bostrom
Georgia Institute of Technology

Daniel Lashof
Natural Resources Defense Council

Introduction

No sooner do we start experiencing the world than the world starts shaping our causal beliefs about it, by providing feedback on our actions, predisposing us to expect certain outcomes from particular actions, and thus to link causes to effects. It is only human to generalize and abstract stories from these. While specific actions and their specific consequences may be misremembered or forgotten (Brown, 1990; Koriat *et al.*, 2000; Loftus *et al.*, 1987), their cumulative legacy includes a set of general causal beliefs, or mental models, of how things work. Mental models are our inference engines, how we simulate sequences of events in our minds and predict their outcomes. Add to this vicarious experiences and adopted beliefs – the effects of science education, advertising, and other communications – and people can explain just about anything, including global warming, through the pre-existing lenses of their mental models.

When it comes to communicating climate change, awareness of our own mental models and those of the people we want to communicate with is key. Why? Because our mental models predispose us toward particular ways of thinking about a problem, its causes, effects, and its solutions. In other words, if we hold in our minds a mental model that wrongly captures what causes a problem, our response to the problem will be equally inappropriate. For example, a "heartburn" mental model of chest pains leads some people to take a digestive aid rather than seek timely medical care for heart attacks. The same holds true for the ways in which we might think about global warming. In fact, the opinion surveys that we review in this chapter suggest

Support from the National Science Foundation (NSF 9022738 and 9209553) and the US Environmental Protection Agency for some of the earlier studies reported here is gratefully acknowledged. We also thank Susi Moser and Lisa Dilling for their many helpful comments on earlier versions of this chapter. Any errors are our sole responsibility.

that past global warming messages have triggered misunderstandings and inappropriate responses. For climate change communication to be effective, it is critical that communicators understand what mental models their audiences hold, and to correct or replace those that are misleading.

Study after study of global warming perceptions and beliefs demonstrate that people hold a variety of mental models about the issue, some of which mislead them regarding causes and solutions (Bostrom and Fischhoff, 2001; Bostrom *et al.*, 1994; Böhm and Pfister, 2001; Kempton, 1997; Löfstedt, 1991; Read *et al.*, 1994). Certainly, Americans do not all explain global warming the same way. From the early 1990s to the present, a handful of beliefs dominate the causes people volunteer when asked to describe or explain global warming or climate change. Many attribute global warming to human activity (Bostrom, 2001b). While people may mention automobile emissions, references to carbon dioxide (CO_2) are rarely volunteered, and people are just as likely to volunteer other beliefs (Bostrom *et al.*, 1994; Böhm, 2001). Among these, two conflations are notable. Many continue to attribute global warming to the ozone hole or ozone depletion. Many also conflate global warming with natural weather cycles.[1] We will discuss below why thinking about global warming along these lines is problematic.

Mental models of hazardous processes generally include four elements: identification of the problem, causes, consequences, and controls or solutions (these may also be called recognition of symptoms, sources and pathways, effects, and mitigation or mitigating factors). An image or phrase associated with any element of the communication can trigger use of a mental model. For example, a picture of an exhaust pipe may trigger inferences about air pollution and respiratory illness. Even sounds or who communicates (the messenger) can contribute to how we think about an issue. As mentioned above, some people do not identify global warming as a problem distinct from ozone depletion. Perceived causes range from natural climate and weather variability to a variety of human actions (see below). Right or wrong, these tend to correspond to perceived controls, such as reducing air pollution. In open-ended interviews people volunteered consequences including everything from hotter summers to human health effects (Bostrom and Fischhoff, 2001). Judgments or valuations of the process (e.g., how bad it is) are often volunteered as well; some researchers believe that the images or mental representations that people hold of such processes are inherently affective – that is, positive or negative (Finucane *et al.*, 2000; see also Leiserowitz, Chapter 2, this volume). By comparison, in a line of research that is analogous to mental models studies of environmental hazards, health researchers depict "common sense models of illness" as cognitive responses to illness, parallel to and distinct from emotional responses to illness, both of which influence illness outcomes (Leventhal *et al.*, 1992; Hagger and Orbell, 2003). Strategies for coping with illness derive from beliefs in these basic concept categories, and also from beliefs about the timeline or temporal unfolding and duration of the process and consequences (see also Moser, Chapter 3, this volume).

Relatively little research has been done on perceptions of the temporal unfolding of global warming (but see, Sherman and Booth Sweeney, 2002). What we do know is that scientists use temporal unfolding (or scale) to distinguish between weather and climate, a distinction that is not obvious or salient to many.

In the following we review the evidence on conflation of weather and climate, look more closely at the role of extreme weather consequences in this conflation, then at the role of natural versus human causes. Finally we examine the policy implications of these conflations, and strategies to address them through more effective climate change communications.

Conflating weather and climate

Weather — The state of the atmosphere at a definite time and place with respect to heat or cold, wetness or dryness, calm or storm, clearness or cloudiness; meteorological conditions.

Climate — The average course or condition of the weather at a particular place over a period of many years, as exhibited in absolute extremes, means, and frequencies of given departures from these means, of temperature, wind velocity, precipitation, and other weather elements.

Bostrom et al. *(1994)*

The evidence that people conflate weather and global climate change comes from a variety of sources, including public opinion polls, focus groups, and cognitive studies. For almost two decades, both national polls and in-depth studies of global warming perceptions have shown that people commonly conflate weather and global climate change. There are likely several sources for this conflation, including confusion about or lack of familiarity with the temporal unfolding of each, and the scientific distinction between them. Moreover, the distinction is complicated by seasonal or interannual variability in the climate.

While atmospheric scientists distinguish between climate and weather (see definitions at the top of this section), this distinction is difficult for many lay people. In one study in the 1990s where lay people provided definitions of climate, climate change, and weather, 77 percent of lay climate definitions were coded as weather by a half dozen independent judges. Some of the definitions conflated the two explicitly: "Climate is the, is the weather" or "Climate is weather conditions that are on the Earth" (Bostrom *et al.*, 1994). Climate change was also defined as: "Anything out of the norm. In other words, having warm winters such as we've had. Extended summers. Hot, dry summers." In a follow-up survey, 32 percent disagreed with "Climate means average weather" and 23 percent agreed erroneously with "Weather means average climate," and 42 percent said erroneously that "Climate often changes from year to year" was true or maybe true (Read *et al.*, 1994).

Weather extremes as evidence for (or against) global warming

Weather changes such as warmer weather (28 percent, from Table 1.1), and, to a lesser extent, increased storm intensity and frequency (21 percent, summing lines three and five in Table 1.1), are the consequences most likely to come to mind in interviews about global climate change (see also Leiserowitz, Chapter 2, this volume). In a survey of 800 registered voters by the Mellman Group, Inc., sponsored by World Wildlife Fund, August 11–14, 1997, 54 percent of respondents thought more extreme weather conditions, such as droughts, blizzards, and hurricanes, were either certain or very likely to happen as a consequence of global warming. However, only 16 percent were "most concerned" about these extreme weather consequences. Similarly, nearly half (49 percent) thought longer and hotter heatwaves leading to more heat-related deaths were either certain or very likely

Table 1.1. *Responses to open-ended questions about what people remember hearing about global warming*

Phenomena associated with global warming	Percent responding in affirmative
Temperatures rising, more heat/getting hotter, climate changes	28
Melting of polar ice cap, glaciers melting	17
Change in weather/weather patterns, extreme weather conditions	14
Ozone depletion, hole in ozone layer	8
More hurricanes/tropical storms, floods, natural disasters	7

Source: Bostrom (2001a); this portion of the study was sponsored by NET, conducted by Hart Research and American Viewpoint, N = 1,502 registered voters in eight states, September 14–23, 1999.

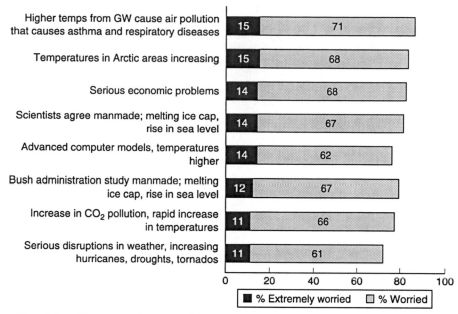

Fig. 1.1. Concerns about problems resulting from global warming (Replies in response to: "Now I am going to list some specific problems that some people say could happen as a result of global warming. After each one, please tell me how worried you are that this could happen as a result of global warming – extremely worried, very worried, worried, not that worried, or not worried at all, or do you think this is not a result of global warming?")
Source: Greenberg, Quinlan, Rosner survey of 1,008 likely voters conducted from December 14 to 20, 2004 for the Natural Resources Defense Council. The margin of error is +/−3.1 percentage points.

(Bostrom, 2001a), but only 15 percent indicated that they were most concerned about these consequences. As Figure 1.1 shows, newer data also suggest that people are less concerned about extreme weather than they are about other consequences.

Current weather is treated as evidence for or against global climate change, with anecdotes more common than not. Newspapers received questions such as, "Doesn't this winter show there's no global warming?" in the cold winter of 2003 (Williams, 2005). In what might be seen by non-scientific audiences as only subtly different from use of weather anecdotes, changes in the frequency or patterns of extreme events have long been cited by climate scientists as evidence of global warming (e.g., J. Hansen's 1988 testimony, cited in the Introduction by Dilling and Moser, this volume; Emanuel, 1997; Webster *et al.*, 2005).

Scientists have routinely said that any one extreme event or extreme weather season could not be directly attributed to climate change. Whether intentional or not, the media also contributed to this dissociation effect. In a study of the US network news, Ungar (2000) found no correlation between coverage of extreme weather events (heatwaves, droughts, hurricanes, and floods) and stories on climate change.

The heatwave in Europe in the summer of 2003, and the 2004 and 2005 hurricane seasons appear to have triggered a shift away from this public dissociation. Stott, Stone, and Allen (2004) estimated the contribution of human-induced increases in atmospheric concentrations of greenhouse gases and other pollutants to the risk of a heatwave surpassing a mean temperature threshold: the mean summer temperature in 2003 exceeded their threshold, but no other year on record did (records started in 1851). They estimated that it is very likely (with greater than 90 percent confidence) that human influence has at least doubled the risk of experiencing such an extreme heatwave.

While the scientific community and media appear to have perpetuated a dissociation between extreme weather events and climate change in the 1990s, survey data suggest that public associations between the two may have remained strong. In a pair of national surveys before and after the Kyoto debate in the fall of 1997, 69 percent and 71 percent respectively thought global warming would cause more storms (Krosnick *et al.*, 2000), which were higher percentages than for other consequences. In a 1998 poll, the likelihood that "Global warming will cause major changes in climate and weather [in the next 30 years]" was judged 0.63 on average, with 37 percent of respondents estimating the likelihood between 0.76 and 1.0.[2] This was a higher likelihood than that attached to the discovery of life on another planet (mean probability = 0.46), or a woman being elected president of the United States (mean probability = 0.59) in the next 30 years. In one recent poll, 39 percent of respondents thought the severity of recent hurricanes was the result of global climate change (by age, 47 percent of the 18−34 group thought so, compared to 35 percent of respondents age 35 or higher).[3] In another recent national poll with somewhat different wording, 58 percent of respondents thought global warming was very or somewhat responsible for making these storms worse.[4]

Causes of climate change: natural or human?

While a majority of Americans believe that people have contributed to global warming, many also still view it as natural. Weather is understood as

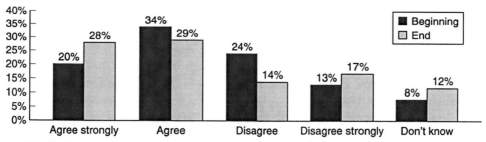

The Earth experience cycles in temperatures. What we are seeing now in increased temperature is just the beginning of another natural weather cycle.

Fig. 1.2.   Agreement that global warming is natural at the beginning and end of a survey.
Source: Based on data from Bostrom (2001b).

natural, on an immense scale that can make controlling it difficult to conceive, and conflating global warming with weather may be helping to perpetuate the perception that all global warming is natural. This has two implications, commonly mentioned by respondents: first, since it's natural, nothing can be done about it, and second, natural processes are cyclical, so we will be back to cooler times soon without having to do anything about it.

In a national survey of 688 American adults in 1995, respondents were asked to choose between three statements regarding the relative contributions of humans and nature to global warming: 42 percent said they have equal roles, 40 percent said that global warming is brought about mostly by things people do, and 18 percent thought global warming is brought about mostly by what nature does.[5] In another study in 2002, while a majority of respondents agreed with the view that "in the past 100 years human industrialization has changed concentrations of gases in the atmosphere that affect the Earth's climate. If we created it, we can fix it," fully a third of respondents had the view that fluctuations in weather and the Earth's temperature are natural and cannot be changed, with only 7 percent agreeing with both views.[6] Responses to a related question are shown in Figure 1.2, which in its wording demonstrates the proclivity for even survey designers to blur distinctions between climate and weather. Survey designers asked about respondents' agreement with the statement that global warming is a natural phenomenon. This question was provided at the beginning and end of a survey on global warming attitudes and beliefs, and illustrates polarization — that thinking about the topic increased the strength of both agreement and disagreement. Polarization was also found in a study of the impacts of the 1997 Kyoto debate on public opinion about global warming in the United States (Krosnick *et al.*, 2000).

Implications: how a weather 'framing' can inhibit behavioral and policy change

While the association between climate and extreme weather may seem like a positive in terms of drawing attention to climate change and motivating concern, it can also lead to counterproductive responses. Aubrun and Grady (2001: 2) captured the problem succinctly: "When scary weather is the problem, SUVs seem like the solution." As illustrated amply in studies of health communication, fear appeals are likely to backfire unless they include concrete help regarding what to do (Witte, 1994; Hagger and Orbell, 2003; Rimal and Real, 2003; see also chapters by Moser; and Tribbia, Chapters 3 and 12, this volume).

In psychometric studies, weather/natural disasters are perceived as less controllable and less dreaded than other risks. These attributes contribute to perceptions that global warming, like weather, is uncontrollable. Climate-related risks have been judged less controllable, less avoidable, and more acceptable than risks not related to climate (McDaniels *et al.*, 1995; Lazo *et al.*, 2000). This is consistent with risk communication theories that suggest that it is the interaction of perceived relevant threat and perceived efficacy – that is, the ability to do something about the risk, to take effective action – that precipitates action (Witte, 1994; Maibach and Parrott, 1995; Rimal and Real, 2003; and Tribbia, Chapter 12, this volume).

Strategies to improve public understanding of climate change

As suggested in the introduction of this chapter, if we want to improve the public's understanding of climate change and predispose them toward taking action (e.g., toward changing their own behaviors or supporting policy change), we need to use or develop mental models and ways of framing the climate issue that suggest the right cause(s), trigger – maybe through the description of certain consequences – an affective response that makes remedial action desirable, and also suggest appropriate actions or solutions.

Metaphor and analogy are fundamental to our ways of interpreting and reacting to the world (Gentner *et al.*, 2001; Lakoff and Johnson, 1980; Lakoff, 1987). Improving our understanding of how different people understand and use metaphors and analogies for climate change – as one in this chapter – will enable policy-makers and communicators to use them much more effectively as tools.

One alternative to the "greenhouse effect" and other common frames that trigger the ozone or weather associations has been extensively tested in recent years: a "thickening blanket of carbon dioxide" that "traps heat" in the atmosphere. When people were presented (or primed) with this image, their response and understanding improved markedly, suggesting these metaphors may solve some of these communication problems observed to date. In a study comparing the effectiveness of different explanations for how global warming works, people who saw a "heat-trapping" or "CO_2-induced heat-trapping" explanation, as compared to no explanation, a more detailed scientific explanation, or an explanation that focused on impacts, were more likely to mention relevant concepts when asked to describe global warming (Auburn and Grady, 2001).[7]

Understanding the key causal mechanism − that is, having a simple, correct mental model of the process − is critical in promoting effective action. "The key determinant of behavioral intentions to address global warming is a correct understanding of the causes of global warming" (Bord *et al.*, 2000: 205; see also O'Connor *et al.*, 1999).

As discussed above, although the evidence is substantial that a majority of the US public already supports CO_2 emission reduction policies, people still confuse weather with climate, and attribute or link global warming to stratospheric ozone depletion, which leads to relatively inappropriate or ineffective actions, including avoiding aerosol sprays, blaming chemical industries, and ignoring energy efficiency. Only 26 percent of those surveyed in Northwestern Ohio, Northwestern North Carolina and Southwestern Kansas believed that energy use causes global warming, and about the same proportion were willing to pay more for utilities to avert climate change (Cutter *et al.*, 2003). People do not refer spontaneously to the greenhouse effect; the fact that CO_2 traps heat is still missing from most people's mental models of global warming (Aubrun and Grady, 2001). In sum, the "greenhouse effect" is *not* working as an explanatory metaphor.

Aubrun and Grady (2001) compared the effects of brief exposures to the "thickening blanket of carbon dioxide" prime with those of 23 other primes, including other carbon dioxide heat-trapping primes (e.g., a "carbon dioxide envelope"), carbon dioxide dysfunction primes (e.g., "carbon dioxide build-up"), impact primes (e.g., global catastrophes), and an abbreviated scientific description of global warming. They found that the thickening blanket prime improves understanding of correct mechanisms for global warming, as well as reducing references to incorrect mechanisms. As health communications research on fear appeals predicts, focusing on the potentially catastrophic consequences of global warming without providing information about causal − and potential mitigation − mechanisms is ineffective and possibly even counterproductive (see chapters by

Chess and Johnson; and Moser, Chapters 14 and 3, this volume). In contrast, improved understanding of how global warming works is positively associated with greater engagement and seeing global warming as an urgent problem.

Thus, the research to date suggests that using a heat-trapping metaphor is likely to strengthen existing support for climate change reduction policies and suggestions for individual behavior change. Future studies will have to show how the use of a heat-trapping metaphor affects the development of public understanding and the support of "no regrets" strategies for adaptation, which many now view as necessary complements to greenhouse gas mitigation (Pielke, 2004).

Summary

For almost two decades both national polls and in-depth studies of global warming perceptions have shown that people commonly conflate weather and global climate change. Not only are current weather events such as heatwaves, droughts, or cold spells treated as anecdotal evidence for or against global warming, but weather changes such as warmer weather and increased storm intensity and frequency are the consequences most likely to come to mind to most people when thinking about climate change. Distinguishing weather from climate remains a challenge for many. The problem with this weather "framing" of global warming is that it may inhibit behavioral and policy change in several ways. Weather is understood as natural, on an immense scale, not subject to human influence. These attributes contribute to perceptions that global warming, like weather, is uncontrollable. In this chapter we presented a synopsis of the evidence for these perceptions from public opinion polls, focus groups, and cognitive studies regarding people's mental models of, and "frames" for, global warming and climate change, and the role weather plays in these. The available research suggests that priming people with a model of global warming as being caused by a "thickening blanket of carbon dioxide" that "traps heat" in the atmosphere solves some of these communications problems and makes it more likely that people will support policies to address global warming.

Notes

1. The exact numbers vary by study. For example, in Read *et al.* (1994), 22 percent of those surveyed agreed with the statement that "climate means pretty much the same thing as weather." Attributions to nature and to human actions are discussed below.
2. The Shell Poll, Millennium survey, N = 1,264 adults nationally, November 5–8, 1998.

3. ABC/*Washington Post* telephone poll conducted nationally September 23–27, 2005, after Hurricane Katrina and during Hurricane Rita, N = 1,019, margin of error +/−3 percent. Question wording: "Do you think the severity of recent hurricanes is most likely (the result of global climate change), or is it (just the kind of severe weather events that happen from time to time)?"

4. Results from Democracy Corps survey, N = 1,012 likely voters, conducted September 5–7, 2005, margin of error +/−3.1 percent. Question wording: "As you know, Katrina was a category 4 hurricane, and it has been proceeded in the last couple of years by a series of storms hitting Florida and the Gulf coast. How responsible do you believe global warming is for making these storms worse – very responsible, somewhat responsible, not very responsible or not at all responsible?"

5. Resources for the Future, funded by Ohio State University, NSF, US EPA, and NOAA; conducted by OSU Survey Research. Results based on a national sample of 688 American adults interviewed September 1–October 5, 1995 (see Bostrom, 2001a).

6. Bostrom (2001b). National Opinion Surveys, one third split of 1,000 likely voters nationwide, conducted June 19–27, 2002, sampling error +/−5 percent.

7. TalkBack study, N = 400 respondents, ethnically diverse, roughly half male, half female, ages 20 and above in Chicago and Washington, DC.

References

Aubrun, A. and Grady, J. (2001). *The Missing Conceptual Link: Talkback Testing of Simplifying Models for Global Warming.* Providence, RI: Cultural Logic.

Böhm, G. (1997). Risk perception in the area of global environmental change events. Presentation at the Annual Meeting of the Society for Risk Analysis-Europe, June 15–18, Stockholm, Sweden.

Böhm, G. and Pfister, H.-R. (2001). Mental representation of global environmental risks. *Research in Social Problems and Public Policy,* **9**, 1–30.

Bord R. J., O'Connor, R. E., and Fisher, A. (2000). In what sense does the public need to understand global climate change? *Public Understanding of Science,* **9**, 3, 205–18.

Bostrom, A. and Fischhoff, B. (2001). Communicating health risks of global climate change. In *Research in Social Problems and Public Policy,* eds. Böhm, G., Nerb, J., McDaniels, T., and Spada, H., special issue of *Environmental Risks: Perception, Evaluation, and Management,* **9**, Amsterdam: JAI, pp. 31–56.

Bostrom, A., Morgan, M. G., Fischhoff, B., *et al.* (1994). What do people know about global climate change? 1. Mental models. *Risk Analysis,* **14**, 6, 959–70.

Bostrom, M. (2001a). *American Attitudes to the Environment and Global Warming: An Overview of Public Opinion.* Study conducted for the FrameWorks Institute, Washington, DC.

Bostrom, M. (2001b). *Communicating Global Warming: An Analysis of Priming Effects.* Study conducted for the FrameWorks Institute, Washington, DC.

Brown, N. R. (1990). Organization of public events in long-term memory. *Journal of Experimental Psychology: General,* **119**, 297–314.

Cutter, S., Mitchell, J. T., Hill, A., *et al.* (2003). Attitudes toward reducing greenhouse gas emissions from local places. In *Global Change and Local Places: Estimating, Understanding, and Reducing Greenhouse Gases,* ed. The Association of American Geographers Global Change in Local Places (GCLP) Research Group, Cambridge, UK: Cambridge University Press, pp. 171–91.

Emanuel, K. A. (1987). The dependence of hurricane intensity on climate. *Nature,* **326**, 483.

Finucane, M. L., Alhakami, A., Slovic, P., *et al.* (2000). The affect heuristic in judgments of risks and benefits. *Journal of Behavioral Decision Making*, **13**, 1, 1–17.

Gentner, D., Holyoak, K. J., and Kokinov, B. K. (eds.) (2001). *The Analogical Mind*. Cambridge, MA: The MIT Press.

Hagger, M. S. and Orbell, S. (2003). A meta-analytic review of the common-sense model of illness representations. *Psychology and Health*, **18**, 2, 141–84.

Hansen, J. E. (1988). Testimony before U.S. Senate Committee on Energy and Natural Resources. Greenhouse effect and climate change, 100th Congress, 1st Session, June 23.

Kempton, W. (1991). Lay perspectives on global climate change. *Global Environmental Change*, **1**, 3, 183–208.

Kempton, W. (1997). How the public views climate change. *Environment*, **39**, 9, 12–21.

Koriat, A. and Goldsmith, M. (2000). Toward a psychology of memory accuracy. *Annual Review of Psychology*, **51**, 1, 481– 537.

Krosnick, J. A., Holbrook, A. L., and Visser, P. S. (2000). The impact of the fall 1997 debate about global warming on American public opinion. *Public Understanding of Science*, **9**, 239–60.

Lakoff, G. (1987). *Women, Fire, and Dangerous Things*. Chicago, IL: University of Chicago Press.

Lakoff, G. and Johnson, M. (1980). *Metaphors We Live By*. Chicago, IL: University of Chicago Press.

Lazo, J. K., Kinnell, J. C., and Fisher, A. (2000). Expert and layperson perceptions of ecosystem risk. *Risk Analysis*, **20**, 2, 179–93.

Leventhal, H., Diefenbach, M., and Leventhal, E. A. (1992). Illness cognition: Using common sense to understand treatment adherence and affect cognition interactions. *Cognitive Therapy and Research*, **16**, 143–63.

Löfstedt, R. E. (1991). Climate change perceptions and energy-use decisions in northern Sweden. *Global Environmental Change*, **1**, 4, 321–4.

Loftus, E. F., Banaji, M. R., Schooler, J. W., *et al.* (1987). Who remembers what? Gender differences in memory. *Michigan Quarterly Review*, **26**, 64–85.

Maibach, E. and Parrott, R. L. (1995). Introduction. In *Designing Health Messages: Approaches from Communication Theory and Public Health Practice*. eds. Maibach, E. and Parrott, R. L. Thousand Oaks, CA: Sage Publications, p. vii.

McDaniels, T., Axelrod, L., and Slovic, P. (1995). Characterizing perception of ecological risk. *Risk Analysis*, **15**, 5, 575–88.

O'Connor, R. E., Bord, R. J., and Fisher, A. (1999). Risk perceptions, general environmental beliefs, and willingness to address climate change. *Risk Analysis*, **19**, 3, 461–71.

Pielke Jr., R. A. (2005). What is climate change? *Issues in Science and Technology*, **20**, 4, 2004. Available at http://www.issues.org/issues/20.4/p_pielke.html; accessed February 2, 2006.

Read, D., Bostrom, A., Morgan, M. G., *et al.* (1994). What do people know about global climate change? 2. Survey studies of educated laypeople. *Risk Analysis*, **14**, 6, 971–82.

Rimal, R. N. and Real, K. (2003). Perceived risk and efficacy beliefs as motivators of change: Use of the risk perception attitude (RPA) framework to understand health behaviors. *Human Communication Research*, **29**, 3, 370–99.

Sherman, J. D. and Booth Sweeney, L. (2002). Cloudy skies: Assessing public understanding of global warming. *System Dynamics Review*, **18**, 2, 207–40. Available at http://web.mit.edu/jsterman/www/cloudy_skies1.pdf; accessed January 14, 2006.

Stott, P. A., Stone, D. A., and Allen, M. R. (2004). Human contribution to the European heatwave of 2003. *Nature*, **432**, 610–14.

Ungar, S. (2000). Knowledge, ignorance and the popular culture: Climate change versus the ozone hole. *Public Understanding of Science*, **9**, 297–312.

Webster P. J., Holland, G. J., Curry, J. A., *et al.* (2005). Changes in tropical cyclone number, duration, and intensity in a warming environment. *Science*, **309**, 5742, 1844–46.

Williams, J. (2005). Understanding global climate change science. *USA Today*, August 11. Available at http://www.usatoday.com/weather/climate/wclisci0.htm; accessed December 19, 2005.

Witte, K. (1994). Fear control and danger control: A test of the Extended Parallel Process Model (EPPM). *Communication Monographs*, **61**, 2, 113–34.

2

Communicating the risks of global warming: American risk perceptions, affective images, and interpretive communities

Anthony Leiserowitz
Decision Research

Introduction

Large majorities of Americans believe that global warming is real and consider it a serious problem, yet global warming remains a low priority relative to other national and environmental issues and lacks a sense of urgency.[1] To understand this lack of urgency, the study on which this chapter is based examined the risk perceptions and connotative meanings of global warming in the American mind and found that Americans perceive climate change as a moderate risk that will predominantly impact geographically and temporally distant people and places. This research also identified several distinct interpretive communities of climate change: segments of the public that conceptualize and respond to the issue in very different ways. The chapter concludes with five strategies to communicate about global warming in ways that either resonate with the values and predispositions of particular audiences or that directly challenge fundamental misconceptions.

Public risk perceptions are critical components of the socio-political context within which policy-makers operate. Public risk perceptions can fundamentally compel or constrain political, economic, and social action to address particular risks. For example, public support or opposition to climate policies (e.g., treaties, regulations, taxes, subsidies) will be greatly influenced by public perceptions of the risks and dangers of climate change.

In this context, American public risk perceptions of climate change are critical for at least two reasons. First, the United States, with only 5 percent of the world's population (US Census Bureau, 2005), is currently the world's largest emitter of carbon dioxide, the primary heat-trapping gas, alone accounting for nearly 25 percent of global emissions (Marland *et al.*, 2006). Per capita, Americans emit 5.43 metric tons of carbon each year. By comparison, the average Japanese emits 2.64 metric tons per year, while the average Chinese emits only 0.86 and the average Indian only 0.33 metric tons per year (Marland *et al.*, 2006). Second, successive US presidents and congressional leaders have been at odds with much of the world community regarding the reality, seriousness, and need for vigorous action on climate change. For example, in 1997, just prior to the Kyoto climate change conference, the US Senate passed a nonbinding resolution (95–0) co-sponsored by Robert Byrd (D) of West Virginia and Chuck Hagel (R) of Nebraska, which urged the Clinton administration to reject any agreement that did not include emission limits for developing as well as industrialized countries, arguing that to do so would put the United States at a competitive economic disadvantage (US Senate 1997). Further, in 2001 President George W. Bush renounced a campaign pledge to regulate carbon dioxide as a pollutant, withdrew the United States from the Kyoto Protocol negotiations, and proposed national energy legislation to increase drilling for oil and natural gas, and mining for coal, and build over a thousand new fossil-fuel-burning power plants (Pianin and Goldstein 2001; Revkin, 2001; United States, 2001).[2] Clearly, the American public will play a critical role, both in terms of their direct consumption of fossil fuels and resulting greenhouse gas emissions, and through their support for political leaders or government policies, to mitigate or adapt to global climate change.

Since the year 2000, numerous public opinion polls demonstrate that large majorities of Americans are aware of global warming (92 percent), believe that global warming is real and already under way (74 percent), believe that there is a scientific consensus on the reality of climate change (61 percent), and already view climate change as a somewhat to very serious problem (76 percent) (Leiserowitz, 2003; PIPA, 2005). Further, of the 92 percent of Americans who have heard of global warming, large majorities, including Republicans and Democrats, conservatives and liberals, support a variety of national and international policies to mitigate climate change, including regulation of carbon dioxide as a pollutant (77 percent), improving auto fuel efficiency (79 percent), subsidizing the development of renewable energy (71 percent), and ratifying the Kyoto Protocol (88 percent) (Leiserowitz, 2003). At the same time, however, Americans continue to regard both the environment and climate change as relatively low national priorities. For example, in a 2000 Gallup poll, the environment ranked 16th on Americans' list of most important problems facing the country. Further, global warming ranked 12th out of 13 environmental issues, just below urban sprawl (Dunlap and Saad, 2001). Thus Americans paradoxically seem highly concerned about global warming, yet view it as less important than nearly all other national or environmental issues. What explains this paradox? Additionally, why do some Americans see climate change as an urgent, immediate danger, while others view it as a gradual, incremental problem, or not a problem at all?

While useful, public opinion polls have limited ability to explain public risk perceptions of global climate change. Most polls use only relatively simple, holistic measures of concern (e.g., "how serious of a threat is global warming?"), which provide little insight into the determinants and components of public risk perception. For example, a critical finding of recent research on risk perception is that public perceptions are influenced not only by scientific and technical descriptions of danger, but also by a variety of psychological and social factors, including personal experience, affect and emotion, imagery, trust, values, and worldviews – dimensions of risk perception that are rarely examined by opinion polls (Slovic, 2000).

Research attention has recently focused on the role of affective imagery in risk perception. *Affect* refers to the specific quality of "goodness" or "badness" experienced as a feeling state (with or without conscious awareness) or the positive or negative quality of a stimulus. *Imagery* refers to all forms of mental representation or cognitive content. "Images" include both perceptual representations (pictures, sounds, smells) and symbolic representations (words, numbers, symbols) (Damasio, 1999: 317–21). In this sense, "image" refers to more than just visually based mental representations. Affective images thus include "sights, sounds, smells, ideas, and words, to which positive and negative affect or feeling states have become attached through learning and experience" (Slovic *et al.*, 1998: 3). The study of affective images in risk perception attempts to identify, describe, and explain those images that carry a strongly positive or negative emotional "charge," and guide risk decision-making. For example, an early study found that many of the images the American public associated with the stimulus "nuclear waste repository" (images such as death, cancer, and the mushroom cloud) evoked strong feelings of dread and judgments that a "nuclear waste repository" was an extremely dangerous risk (Slovic *et al.*, 1991). More broadly, these images also influenced risk perceptions of nuclear energy and were strongly associated with intended voting behavior (*ibid.*) and support (or the lack thereof) for construction of new nuclear power plants (Peters and Slovic, 1996).

To explore the role of affective imagery in American global warming risk perceptions, policy preferences and behaviors, a national survey was conducted from November 2002 to February 2003. The study was implemented with a 16-page mail-out, mail-back survey of a representative sample of the

American public. A total of 673 completed surveys were returned for an overall response rate of 55.4 percent. Compared to population distributions from the 2000 US Census, the sample overrepresented males (65 percent) and persons 55 and older (47 percent). The results were weighted by sex and age to bring them in line with actual population proportions.

Risk perceptions of global warming

This study found that Americans as a whole perceived global climate change as a moderate risk (Figure 2.1). On average, Americans were somewhat concerned about global warming, believed that impacts on worldwide standards of living, water shortages, and rates of serious disease are somewhat likely and that the impacts will be more pronounced on non-human nature. Importantly, however, they were less concerned about local impacts, rating these as somewhat unlikely. The moderate level of public concern about climate change thus appears to be driven primarily by the perception of danger to geographically and temporally distant people, places, and non-human nature.

This conclusion is confirmed by the results of a separate question that asked respondents to indicate which scale of climate change impacts was of greatest concern to them (Table 2.1). The question asked, "Which of the following are you *most* concerned about? The impacts of global warming on ... (1) you and

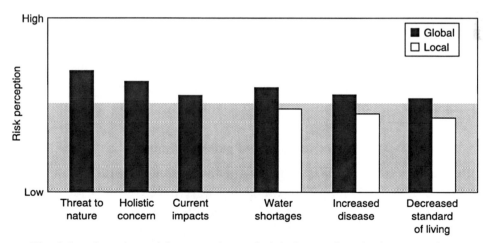

Fig. 2.1. American risk perceptions of global warming. Scales range from 1(low) to 4 (high). Response categories include seriousness of threat to nature and current impacts around the world (not at all to very serious); level of concern about global warming (not at all to very concerned); and the likelihood of specific impacts locally and worldwide (very unlikely to very likely). $N = 590$. Source: A. Leiserowitz (2005).

Table 2.1. *Scale of impacts of greatest concern*

Most concerned about impacts on 4.1.1?	%	Cum. %
You and your family	12	12
Your local community	1	13
The United States as a whole	9	22
People all over the world	50	72
Non-human nature	18	90
Not at all concerned	10	100
Total	100	

Source: A. Leiserowitz (2005). $N = 551$

your family; (2) your local community; (3) the US as a whole; (4) people all over the world; (5) non-human nature; or (6) not at all concerned."

A clear majority of respondents (68 percent) were most concerned about the impacts on people around the world and non-human nature. Only 13 percent were most concerned about the impacts on themselves, their family or their local community. This may help explain why global climate change remains a relatively low priority in issue-ranking surveys (e.g., Dunlap and Saad, 2001). Higher-ranking national issues (e.g., the economy, education, health care) and environmental issues (e.g., clean air, clean water, urban sprawl) are all issues that are more easily understood as having direct local relevance. "Global" climate change, however, is not yet perceived as a significant local concern among the American public. Former Speaker of the US House of Representatives Tip O'Neill once famously stated "all politics are local." To the extent that this is true, climate change is unlikely to become a high-priority national issue until Americans consider themselves personally at risk.

Affective images of global warming

Affective images were measured using a form of word association to the stimulus "global warming." Respondents were asked to provide the first thought or image that came to mind when they heard the words "global warming." Each association was then rated by the respondent using a scale ranging from −5 (very negative) to +5 (very positive). A content analysis of these associations identified a total of 24 distinct thematic categories. The top eight categories, however, represent 97 percent of all respondents (Figure 2.2). Associations to melting glaciers and polar ice were the single largest category of responses, indicating that this current and projected impact

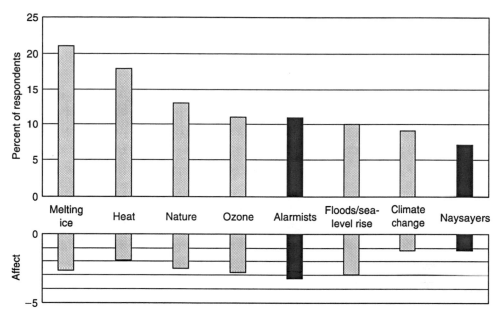

Fig. 2.2. American images of global warming. Affect was rated on a 10-point scale where $+5 = very\ positive$ and $-5 = very\ negative$.
Source: A. Leiserowitz (2005).

of climate change is currently the most salient image of global warming among the American public. This was followed by generic associations to heat and rising temperatures (Heat), impacts on non-human nature (Nature), ozone depletion (Ozone), images of devastation (Alarmists), sea-level rise and the flooding of rivers and coastal areas (Flooding/Sea-Level Rise), references to climate change (Climate Change), and finally associations indicating skepticism or cynicism about the reality of climate change (Naysayers). Mean affect scores (below) indicate that "global warming" had negative connotations for almost all respondents. Alarmist images of disaster produced the strongest negative affect, while naysayers displayed very low negative affect.

Thus, two of the four most dominant images (melting ice and non-human nature), held by 34 percent of all respondents, referred to impacts on places or natural ecosystems distant from the everyday experience of most Americans. Most of the references to "heat" were relatively generic in nature and likely indicated associations with the word "warming" in "global warming." Finally, 11 percent of Americans provided associations to the separate environmental issue of stratospheric ozone depletion, indicating that a substantial proportion of Americans continue to confuse and conflate these two issues (see also Bostrom and Lashof, Chapter 1, this volume). Thus, 61 percent of Americans provided associations to impacts geographically and psychologically distant, generic increases in temperature, or to a different environmental problem.

These results help explain the paradox in public risk perceptions, in which Americans appear concerned about climate change, but do not consider it a high priority relative to other national or environmental issues. This study found that, in aggregate, Americans perceive climate change as a moderate risk, but think the impacts will mostly affect people and places that are geographically distant. Critically, this study found that most Americans lack vivid, concrete, and personally relevant affective images of climate change.

Further, one of the most important findings was what was missing in these results. There were no associations to the impacts of climate change on human health. There were no associations to temperature-related morbidity and mortality (e.g., heatstroke), health effects of extreme weather events (e.g., tornadoes, hurricanes, or precipitation extremes), air-pollution health effects (e.g., asthma and allergies), water and food-borne disease (e.g., cholera, E-coli, giardia), or vector and rodent-borne disease (e.g., malaria, West Nile Virus, Hantavirus Pulmonary Syndrome), all of which are potential health consequences of global climate change (McMichael and Githeko, 2001; National Assessment Synthesis Team, 2001; Patz *et al.*, 2000). Yet human health impacts are likely to be among the greatest dangers of climate change for human societies, especially for the poor and children in developing countries (as well as the poor in developed nations) who lack access to adequate nutrition, clean water, or medical care (IPCC, 2001: 12; Watson and McMichael, 2001).

This finding that Americans do not currently associate global warming with impacts on human health is supported by the results of four questions which asked respondents to estimate the current and future human health effects of global warming (Figure 2.3). On average, Americans said that current deaths and injuries due to global warming number in the hundreds, and in 50 years will number only in the thousands. Perhaps more importantly, 38 percent to 41 percent of respondents selected "don't know" as their answer to these four questions – by far the dominant response. This is another strong indication that Americans do not currently perceive global warming as a grave danger to human health either now or in the future. Further, this research also found that very few Americans associate global warming with extreme weather events, like heat waves, hurricanes, and droughts – all of which may increase in severity due to global warming.

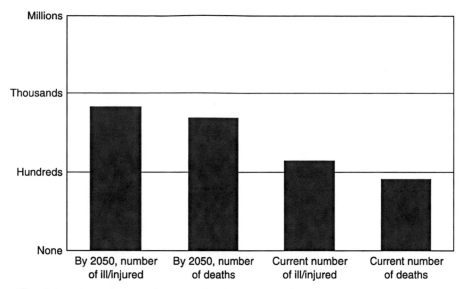

Fig. 2.3. American estimates of the number of deaths, illnesses, and injuries each year because of global warming ($N = 307 - 330$). Source: A. Leiserowitz (2005).

Interpretive communities of climate change

The above aggregate results, however, gloss over substantial variation in risk perceptions within the American public. In particular, this study identified several distinct "interpretive communities" within the American public. An interpretive community is defined as a group of individuals that share mutually compatible risk perceptions, affective imagery, values, and sociodemographic characteristics. Risk perceptions are socially constructed, with different interpretive communities predisposed to attend to, fear, and socially amplify some risks, while ignoring, discounting, or attenuating others. For example, this study found one interpretive community that perceived climate change as a very low or non-existent danger – climate change "naysayers." This group, identified by their affective images, was subsequently found to be predominantly white, male, politically conservative, holding pro-individualism, pro-hierarchism, and anti-egalitarian values, anti-environmental attitudes, distrustful of most institutions, highly religious, and to rely on radio as their main source of news (Leiserowitz, 2003). This interpretive community was significantly different than all other respondents (excluding alarmists) on 13 different risk perception variables (Figure 2.4).

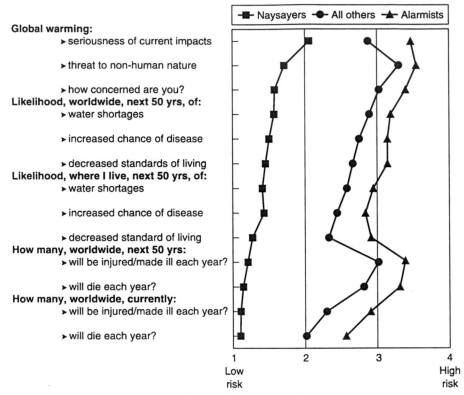

Fig. 2.4.   Mean risk perceptions of global warming.
Source: A. Leiserowitz (2005).

Further, the "naysayer" interpretive community articulated five distinct reasons why they doubted the reality of global climate change:

(1) Belief that global warming is natural ("Normal Earth cycles"; "It is just the natural course of events"; "A natural phenomenon that has been going on for years").
(2) Hype ("It is not as bad as the media portrays"; "The 'problem' is overblown"; "Environmentalist hysteria").
(3) Doubting the science ("There is no proof it exists"; "Around ten years or so ago it was global cooling"; "Junk science").
(4) Flat denials of the problem ("A false theory"; "There is no global warming").
(5) Conspiracy theories ("Hoax"; "Environmentalist propaganda"; "Scientists making up some statistics for their job security").

The diversity of these responses demonstrates that climate change naysayers had different rationales for their disbelief, ranging from acceptance of the reality of climate change (although naturally caused or overblown) to flat denials and outright conspiracy theories. This interpretive community is thus predisposed to discount or flatly reject scientific assessments of climate change. While only 7 percent of the US adult population (or approximately 12 million people), according to these survey results, naysayers are politically active, are significantly more likely to vote, have strong representation in national government and powerful allies in the private sector (see also McCright, Chapter 12, this volume).

This study also identified a contrasting interpretive community with high risk perceptions of climate change – alarmists. Some members of this group provided extreme images of catastrophic climate change, such as: "Bad...bad...bad...like after nuclear war...no vegetation"; "Heat waves, it's gonna kill the world"; "Death of the planet." Alarmists held pro-egalitarian, anti-individualist, and anti-hierarchist values, were politically liberal, strongly supported government policies to mitigate climate change (including raising taxes), and were significantly more likely to have taken personal action to reduce greenhouse gas emissions. This interpretive community was significantly different than all other respondents (excluding naysayers) on every risk perception variable (Figure 2.4). It is also important to note, however, that all other respondents had climate change risk perception levels much closer to alarmists than naysayers. This demonstrates that most Americans are predisposed to view climate change as a significant danger, albeit not as extreme as the alarmists, while climate change naysayers have substantially lower risk perceptions than the rest of American society.

Education and communication strategies

Overall, the findings of this study help explain the paradox in American risk perceptions of global warming. While large majorities of Americans believe global warming is real and consider it a serious problem, global warming remains a low priority relative to other national and environmental issues. In other words, global warming currently lacks a sense of urgency (Moser and Dilling, 2004). Most of the American public considers global warming a moderate risk that is more likely to impact people and places far distant in space and time. These findings suggest that multiple communication strategies are needed.

Strategy 1: Highlight potential local and regional climate change impacts

Local threats are generally perceived as more salient and of greater urgency than global problems. This suggests that efforts to describe the potential national, regional, and local impacts of climate change and communicate these potential impacts to the public are critical. As Rajendra Pachauri, chair of the Intergovernmental Panel on Climate Change recently argued, "there is an opportunity for much political debate when you start to predict the impact of climate change on specific regions. But if you want action you must provide this information" (Schiermeier, 2003: 879). The IPCC and many other scientific teams are currently working to develop and improve regional and local-scale models of climate change impacts. Meanwhile, some potential regional impacts may already be predicted with a significant degree of certainty (e.g., reduced snow pack in the Pacific Northwest; National Assessment Synthesis Team, 2001: 415–18; Mote *et al.*, 2005). Yet other local and regional-scale impacts are still difficult to predict and remain highly uncertain (e.g., climate change impacts on respiratory diseases, see National Assessment Synthesis Team, 2001: 447). Educators and communicators can already draw on studies of the local and regional impacts of climate change in the United States, with appropriate caveats on scientific certainty (see below). Examples include the US National Assessment, *Climate Change Impacts on the United States* (e.g., National Assessment Synthesis Team, 2001); a series of regional reports describing the current and projected impacts of climate change on American ecosystems produced by the Union of Concerned Scientists and the Ecological Society of America (e.g., Field *et al.*, 1999; Twilley *et al.*, 2001; Kling *et al.*, 2003; see also Cole with Watrous, Chapter 11, this volume); state-by-state reports produced by the US Environmental Protection Agency (EPA, 2005); and countless regional case studies appearing in the scientific literature.

Strategy 2: Climate change is happening now

Immediate threats are generally perceived as more salient and of greater urgency than future problems. This suggests that educators and communicators should highlight the *current* impacts of climate change around the world, which in some places are already profound. For example, climate change is already having severe impacts on the Arctic, including the US state of Alaska (Arctic Monitoring and Assessment Program, 2005). Alaska's climate has warmed about 4 °F since the 1950s and 7 °F in the interior during winter. The state experienced a 30 percent average increase in precipitation between 1968 and 1990. The growing season has lengthened by two weeks. Sea ice has retreated by 14 percent since 1978 and thinned by 60 percent since the 1960s, with widespread effects on marine ecosystems, coastal climate, and human settlements. Permafrost melting has caused erosion and landslides, and has damaged infrastructure in central and southern Alaska. Recent warming has been accompanied by "unprecedented increases in forest disturbances, including insect attacks. A sustained infestation of spruce bark beetles, which in the past have been limited by cold, has caused widespread tree deaths over 2.3 million acres on the Kenai Peninsula since 1992, the largest loss to insects ever recorded in North America" (National Assessment Synthesis Team, 2001: 285; see also McNeeley and Huntington, Chapter 8, this volume). Robert Correll, lead scientist of the recent international Arctic Climate Impact Assessment, recently said, "If you want to see what will be happening in the rest of the world 25 years from now, just look at what's happening in the Arctic" (Borenstein, 2003: 21).

Other recent studies have documented warming-related changes in ecosystems and species behavior in the United States (e.g., Clean Air-Cool Planet and Wake, 2005; Parmesan and Galbraith, 2004; Root *et al.*, 2005). These provide convincing evidence that climate change is not just a problem for the future, but one affecting species, ecosystems, and human communities right now.

The affective image results reported above demonstrate that when Americans think of global warming, they are already predisposed to think of melting ice and glaciers in the Arctic with strongly negative affect. Thus the frame for illustrating the severity of climate change in the Arctic is already well established in the American mind. What is needed now are the concrete details, images, and stories of climate impacts – on people, places, economies, cultures, and ecosystems – to fill out the picture, bring the issue to life, and help people understand the potential dangers for the rest of the world. In short, educators and communicators need to make global climate change *local* and to discuss climate change in the *present*, as well as the future, tense.

Strategy 3: Highlight the potential impacts of climate change on human health and extreme weather events

This research found that the American public does not currently associate global warming with any impacts on human health. Communicators need to articulate and emphasize these impacts, which are among the most serious consequences of projected climate change. An emphasis on the projected impacts on human health is also likely to elevate public concerns about global warming, especially compared to the associations currently dominant (melting ice, generalized heat, and impacts on non-human nature). Some of these health implications are relatively well understood (e.g., increased likelihood of heatstroke), while others remain more uncertain (impacts on asthma) (National Assessment Synthesis Team, 2001). Physicians for Social

Responsibility (PSR) is an example of an organization doing this kind of educational outreach. PSR has produced a series of state-by-state reports on the projected impacts of climate change on human health and conducted a public education campaign, including a lecture series by concerned physicians (PSR, 2004).

Likewise, this research found that few Americans associate global warming with extreme events, such as heatwaves, hurricanes, flooding or drought, despite the fact that all are projected to increase in severity due to climate change. Extreme events are vivid, dramatic, and easily understood. Further, extreme events happen each year and thus provide recurring "teachable moments" to explain the potential impacts of future climate change. For example, in 2004 severe to extreme drought continued to grip much of the West, while wildfire incinerated a record 6.3 million acres in Alaska alone (Rippey, 2004; National Climatic Data Center, 2004). While neither of these events can be definitively said to be caused by global warming (causal connections with climate change can only be determined through analysis of long-term trends), these events are consistent with scientists' projections of the future impacts of global warming and may be harbingers of things to come. Educators and communicators can highlight the connections between climate change, human health, and extreme weather events, while being careful to respect current levels of scientific understanding.

Strategy 4: Talk openly about remaining uncertainties

While communicators and educators should highlight those impacts that are scientifically more certain, some potentially important impacts (e.g., human health) are still relatively unpredictable. In the absence of certainty about particular impacts, what should communicators do? Critically, they should never suggest more scientific certainty than actually exists. Definitive claims based on uncertain science are vulnerable to attack, potentially mislead the public, and can irrevocably destroy trust and credibility. What communicators can do, however, is openly describe and discuss the known likelihood and severity of potential impacts, and narrate scenarios that describe possible local and regional futures. To make these possible futures more "imaginable," communicators can use appropriate historical and geographic analogies. When there remains significant uncertainty about a specific impact, communicators should explain why the uncertainty exists, e.g., because the science hasn't been done yet or the systems involved are so complex that science has yet to understand them sufficiently. It is also important to recognize that scientific uncertainty alone is not an adequate justification for inaction or business-as-usual. Rather, it suggests that, at a minimum, it would be prudent to expect the unexpected, develop contingency plans, and adopt adaptive management strategies.

Strategy 5: Tailor messages and messengers for particular interpretive communities

This research also identified several distinct interpretive communities within the American public — groups of individuals who share mutually compatible risk perceptions, affective imagery, values, and sociodemographic characteristics. Each of these interpretive communities is likely to respond to information about climate change in very different ways. Messages about climate change need to be tailored to the needs and predispositions of particular audiences: in some cases to directly challenge fundamental misconceptions; in others to resonate with strongly held values.

For example, educators and communicators must specifically target those members of the public who confuse global warming with the ozone hole. This study found some Americans continue to believe global climate change is caused by stratospheric ozone depletion — a critical misconception also identified by mental models researchers (e.g., Kempton *et al.*, 1995; Bostrom *et al.*, 1994; Read *et al.*, 1994; see also Bostrom and Lashof, Chapter 1, this volume). The ozone hole became a public issue several years before global warming was widely reported in the media. Kempton *et al.* (1995: 67) found that when Americans first learned about global warming and the "greenhouse effect" they assimilated this new information into pre-existing mental models of the ozone hole. This led to several important misconceptions and confusions between the two environmental issues. For example, many people reason that if there is a "hole" in the ozone layer and a global "greenhouse" effect, then there must be a "hole" in the "greenhouse." This "hole" either allows more solar radiation into the biosphere — warming the planet, or alternatively, the "hole" is allowing heat to escape — cooling the planet. This metaphorical reasoning is logical, but of course incorrect. Thus, many Americans still hold inaccurate beliefs about the causes of global warming, which leads them to support inappropriate solutions. These "solutions" can range from choosing not to buy an aerosol spray can, despite the fact that "sprays containing CFCs have been banned in the US for [three] decades" (Cutter, 1993: 53), to the solution proposed by one survey respondent:

> The problem of global warming is the hole in the ozone layer over Antarctica, right? Scientists can produce ozone in labs, right? The solution

is simple. A huge amount of ozone should be created, then a team should fly it to Antarctica. Once there, a plane should fly at a high altitude, meanwhile releasing the ozone into the torn ozone layer. Then, the pollution worldwide should TRY to be regulated. Hey, I'm not saying it'll work, but it's worth a try.

(Leiserowitz, 2003, p. 173)

Educators need to target the source of the confusion – the inappropriate application of knowledge about the ozone hole to the problem of global warming. What makes this complicated, however, is that there do exist links between the two issues, although not in the way the public believes. First, the CFCs that destroy ozone are also highly efficient greenhouse gases that trap heat and thus increase global warming, although their contribution is relatively small compared to the carbon dioxide produced by fossil-fuel burning (IPCC, 2001). Thus, while thinning of the ozone layer itself does not cause climate change, the same culprit (CFCs) is involved in both environmental problems. Another complication is that while the ozone hole does not cause global warming, there is evidence that global warming is exacerbating ozone destruction. As recently reported by Cambridge University, climate change appears to be generating an increase in the number and spatial extent of stratospheric ice clouds, which provide a platform for the chemical reactions that destroy ozone (University of Cambridge, 2005; Brown, 2005). Thus, there may be a causal link between the two issues, but the reverse of what is commonly and mistakenly believed.

While these complications certainly make educational efforts more difficult, they also present an opportunity. If this causal link between climate change and ozone depletion is borne out by further research, educators and communicators do not have to convince people that their current understanding is completely wrong; instead they need to invert the causal sequence – climate change actually contributes to ozone depletion, not the other way around. Further, these recent findings also suggest a complex causal link between climate change and cancer – global warming produces more stratospheric ice clouds, which catalyze more ozone destruction, which allows more ultraviolet light to reach the Earth's surface, which can cause cancer. Relatively rapid international action to ban ozone-depleting chemicals was in part driven by the dread evoked in mass publics by the finding that unprotected exposure to sunlight could now lead to a greater risk of skin cancer or other health implications (e.g., cataracts). This newly discovered causal link – from climate change to stratospheric ozone depletion to increasing cancer risks — has the potential to provide additional scientific and public pressure for action on climate change.

Meanwhile, a different set of strategies would be needed to convince naysayers that global warming is a serious concern. This group will be difficult to reach, however, as many appear to distrust scientists, governments, environmentalists, and the media as sources of information. This may help to explain why increasing scientific certainty, information, and media reports about climate change have not convinced this minority of the public (who are also more likely to vote). Naysayers are also more likely to get their information from conservative talk radio, thus may only acquire information and perspectives that reinforce their disbelief. Further, increased scientific certainty and information may serve only to strengthen some naysayers' conspiracy theories (see McCright, Chapter 12, this volume).

Some may argue that engaging professional climate change naysayers only lends them credibility and gives their views more weight than they deserve; thus climate change scientists, activists, and educators should ignore them. This is a mistake. While a small, but vocal minority of the public (also more likely to vote), this interpretive community includes leading politicians, contrarian scientists, political pundits, Washington think-tanks, and major corporations that have already greatly influenced public opinion and government policy. Left unchallenged, they will continue to cast doubt on climate science and scientists, and impede public policies to mitigate or adapt to climate change. Further, the research reported here suggests that some naysayers could yet be convinced to support policies to reduce fossil fuel use, in some cases with more scientific evidence, in others with framings that resonate with strongly held values, or with messengers that they already trust. For example, some naysayers may respond favorably to a focus on the economic opportunities presented by the global effort to shift energy production toward renewable energy. Entrepreneurial innovation will be critical to this effort and the global marketplace will amply reward those individuals and companies that are best positioned to provide the energy and jobs of the future. Other naysayers may be more persuaded by the national security implications and consequences of continued fossil fuel dependence, which range from the economic, political, and military resources devoted to acquiring and protecting American access to oil in the Middle East (and the geopolitical consequences) to the potentially destabilizing consequences of climate change on governments, economies, and societies around the world. Columnist Thomas Friedman of *The New York Times* has recently argued for a "geo-green" strategy to end American and world dependence on fossil fuels by developing alternative sources of energy, which will both reduce future climate change and further neo-conservative goals to undermine totalitarian regimes and promote freedom and democracy in the Middle East by cutting off the oil revenues that support them (Friedman, 2005a,b). Likewise, Set America Free is an organization that proposes a large-scale investment in alternative energies to reduce American dependence on fossil fuels, and to thus "deny adversaries the wherewithal they use to harm us; protect our quality of life and economy against the effects of cuts in foreign energy supplies and rising costs; and reduce by as much a 50 percent emissions of undesirable pollutants" (Set America Free, 2005). The organization includes prominent members such as Gary Bauer (conservative evangelical leader and 2000 Republican candidate for president), Robert McFarlane (former National Security Advisor under President Reagan), and James Woolsey (former director of the CIA under President Clinton). A parallel movement advocating for action on climate change is also emerging from the religious community, including evangelical Christians. Religious groups, including the National Association of Evangelicals, "argue that global warming is an urgent threat, a cause of poverty and a Christian issue because the Bible mandates stewardship of God's creation" (Goodstein, 2005). Some religious leaders argue that "climate change would have disproportionate effects on the poorest regions of the world," and "caring for the poor by reducing the threat of global warming is caring for Jesus Christ," thus addressing climate change is a moral and ethical requirement (The Rev. Jim Ball quoted in Goodstein, 2005; see also Bingham, Chapter 9, this volume). All of these are examples of organizations, messages, and messengers that are more likely to reach and persuade climate change naysayers than traditional scientific or environmental arguments.

Alarmists, however, already exhibit grave concern regarding the issue. They strongly support policies to mitigate climate change and are already predisposed to be attentive to and believe scientist, government, and environmentalist messages regarding climate change risks. On the other hand, some of these respondents hold extremely negative images that go well beyond scientific assessments of climate change risks. These extreme responses are often apocalyptic, predicting "the end of the world" or the "death of the planet." These are overreactions to an otherwise very serious problem and may lead some to a sense of resigned fatalism. Further, these dystopic visions provide ammunition to naysayers who in turn claim that global warming is a hoax perpetrated by "doomsayers" and "Chicken Littles" (Moser and Dilling, 2004; and Moser, Chapter 3, this volume).

Overall, however, these and other research findings demonstrate that Americans as a whole are already predisposed to view climate change as a significant risk and to support local, national, and international action to reduce the heat-trapping gases that are warming the globe. What is lacking is a sense of public urgency, strong leadership, and political will.

Notes

1. This chapter builds on the findings from a national study (Leiserowitz, 2003) on American risk perceptions of global warming.
2. These basic tenets of energy policy have not been changed in the energy policy discussions in 2005.

References

Arctic Monitoring and Assessment Program (2005). *Impacts of a Warming Climate — Arctic Climate Impact Assessment.* Cambridge, UK: Cambridge University Press.

Borenstein, S. (2003). Alaska is less frozen than it once was, and the effects are devastating. *Milwaukee Journal Sentinel*, August 10, p. 21.

Bostrom, A., Morgan, M. G., Fischhoff, B., *et al.* (1994). What do people know about global climate change? *Risk Analysis*, **14**, 6, 959–70.

Brown, P. (2005). Ozone layer most fragile on record. *The Guardian*, April 27, p. 3.

Clean Air-Cool Planet and Wake, C. (2005). *Indicators of Climate Change in the Northeast, 2005.* Portsmouth, NH: CACP and Durham, NH: University of New Hampshire, Climate Change Research Center.

Cutter, S.L. (1993). *Living with Risk: The Geography of Technological Heizakis.* New York: Edward Arnold.

Damasio, A. (1999). *The Feeling of What Happens.* New York: Harcourt, Inc.

Dunlap, R. E. and Saad, L. (2001). *Only one in four Americans are anxious about the environment.* Available at http://poll.gallup.com/content/default.aspx?ci=1801; accessed January 13, 2006.

Environmental Protection Agency (2005). *Global warming — where you live.* Available at from http://yosemite.epa.gov/OAR/globalwarming.nsf/content/coun-unitedstates.html; accessed January 13, 2006.

Field, C. B., Daily, G. C., Davis, F. W., *et al.* (1999). *Confronting Climate Change in California: Ecological Impacts on the Golden State.* Cambridge, MA and Washington, DC: Union of Concerned Scientists and Ecological Society of America.

Friedman, T. (2005a). The geo-green alternative. *The New York Times*, January 30, p. 17.

Friedman, T. (2005b). Geo-greening by example. *The New York Times*, March 27, p. 11.

Goodstein, L. (2005). Evangelical leaders swing influence behind effort to combat global warming. *New York Times*, March 10, p. 16.

Intergovernmental Panel on Climate Change (IPCC), Working Group II (2001). *Climate Change 2001: Impacts, Adaptation, and Vulnerability. Summary for Policymakers.* Contribution of Working Group II to the Third Assessment report. Available at http://www.ipcc.ch/pub/wg2SPMfinal.pdf; accessed January 13, 2006.

Kempton, W., Boster, J. S., and Hartley, J. A. (1995). *Environmental Values in American Culture*. Cambridge, MA: The MIT Press.

Kling, G. W., Hayhoe, K., Johnson, L. B., *et al.* (2003). *Confronting Climate Change in the Great Lakes Region: Impacts on Our Communities and Ecosystems*. Cambridge, MA and Washington, DC: Union of Concerned Scientists and Ecological Society of America.

Leiserowitz, A. (2003). Global warming in the American mind: The roles of affect, imagery, and worldviews in risk perception, policy preferences and behavior. Unpublished dissertation, University of Oregon, Eugene, OR.

Leiserowitz, A. (2005). American risk perceptions: Is climate change dangerous? *Risk Analysis*, **25**, 6, 1433–1442.

Marland, G., Boden, T., and Andres, B. (2006). *Trends: a compendium of data on global change*. Available at http://cdiac.esd.ornl.gov/trends/emis/top2000.tot; accessed September 22, 2006.

McMichael, A. J. and Githeko, A. (2001). Human Health. In *Climate Change 2001: Impacts, Adaptation, and Vulnerability*, ed. Intergovernmental Panel on Climate Change, Working Group II. Cambridge, UK: Cambridge University Press, pp. 453–85.

Moser, S. and Dilling, L. (2004). Making climate hot: Communicating the urgency and challenge of global climate change. *Environment*, **46**, 10, 32–46.

Mote, P., *et al.* (2005). Declining mountain snowpack in western North America. *Bulletin of the American Meteorological Society*, **86**, 1, 39–49.

National Assessment Synthesis Team (2001). *Climate Change Impacts on the United States: The Potential Consequences of Climate Variability and Change: Foundation Report*. Cambridge, New York: Cambridge University Press.

National Climatic Data Center (2004). *Climate of 2004: Wildfire season summary*. Available at http://www.ncdc.noaa.gov/oa/climate/research/2004/fire04.html; accessed January 13, 2006.

Parmesan, C. and Galbraith, H. (2004). *Observed Impacts of Global Climate Change in the US*. Arlington, VA: Pew Center on Global Climate Change.

Patz, J. A., Engelberg, D., and Last, J. (2000). The effects of changing weather on public health. *Annual Review of Public Health*, **21**, 271–307.

Peters, E. and Slovic, P. (1996). The role of affect and worldviews as orienting dispositions in the perception and acceptance of nuclear power. *Journal of Applied Social Psychology*, **26**, 1427–53.

Physicians for Social Responsibility (2004). *Death by degrees/degrees of danger*. Available at http://www.envirohealthaction.org/degreesofdanger/index.cfm; accessed January 13, 2006.

Pianin, E. and Goldstein, A. (2001). Bush drops a call for emission cuts. *The Washington Post*, March 14, p. A01.

Program on International Policy Attitudes (PIPA). (2005). *Global warming*. Available at http://www.americans-world.org/digest/global_issues/global_warming/gw_summary.cfm; accessed January 13, 2006.

Read, D., Bostrom, A., Morgan, M. G., *et al.* (1994). What do people know about global climate change? 2. Survey results of educated laypeople. *Risk Analysis*, **14**, 6, 971–82.

Revkin, A. (2001). Bush's shift could doom air pact, some say. *The New York Times*, March 17, p. 7.

Rippey, B. (2004). *U.S. Drought Monitor*. Available at http://www.drought.unl.edu/dm/monitor.html; accessed January 13, 2006.

Root, T. L., *et al.* (2005). Human-modified temperatures induce species changes: Joint attribution. *Proceedings of the National Academy of Sciences*, **102**, 21, 7465–9.

Schiermeier, Q. (2003). Climate panel to seize political hot potatoes. *Nature*, **421**, 879.

Set America Free (2005). *An open letter to the American people.* Available at http://www.setamericafree.org/openletter.htm; accessed January 13, 2006.

Slovic, P. (2000). *The Perception of Risk.* London: Earthscan.

Slovic, P., Layman, M., and Flynn, J. H. (1991). Perceived risk, trust, and the politics of nuclear waste. *Science*, **254**, 1603–8.

Slovic, P., MacGregor, D. G., and Peters, E. (1998). *Imagery, Affect, and Decision-Making.* Eugene, OR: Decision Research.

Twilley, R. R., Barron, E. J., Gholz, H. L., *et al.* (2001). *Confronting Climate Change in the Gulf Coast Region: Prospects for Sustaining Our Ecological Heritage.* Cambridge, MA and Washington, DC: Union of Concerned Scientists and Ecological Society of America.

US Census Bureau, Population Division (2005). *US and World Population Clocks.* Available at http://www.census.gov/main/www/popclock.html; accessed January 13, 2006.

US Senate (1997). Senate Resolution 98 (The Byrd–Hagel Resolution). Washington, DC: The US Senate.

United States, National Energy Policy Development Group & The President of the United States (2001). *Reliable, Affordable, and Environmentally Sound Energy for America's Future.* Report of the National Energy Policy Development Group. Washington, DC.

University of Cambridge (2005). *Large ozone losses over the Arctic.* Available at http://www.admin.cam.ac.uk/news/press/dpp/2005042601; accessed January 13, 2006.

Watson, R. T. and McMichael, A. J. (2001). Global climate change – the latest assessment: Does global warming warrant a health warning? *Global Change & Human Health*, **2**, 1, 12.

3

More bad news: the risk of neglecting emotional responses to climate change information

Susanne C. Moser

National Center for Atmospheric Research

> If one does not look into the abyss, one is being wishful by simply not confronting the truth about our time. ... On the other hand, it is imperative that one not get stuck in the abyss.
>
> *Robert Jay Lifton (1986)*[1]

Introduction[2]

Listening to climate change communicators, advocates, and scientists, there is a growing frustration that politicians and the public don't pay more attention to the issue. In their attempts to ring the alarm bells more fiercely, many are tempted either to make the issue scarier or to inundate people with more information, believing that if people only understood the urgency of global warming, they would act or demand more action. When the desired response then fails to materialize, they get disappointed, yet plow ahead undeterred. Surely, if people aren't getting the message, we must give it more loudly! Yet is "not getting the message" really the problem? And is scarier and more information the answer?

Almost every new story about global warming brings more bad news. In 2005 alone, people opened the morning papers to stories that warming could be far worse than previously projected, that our emissions are committing us to warming and sea-level rise for decades to centuries even if we could stop all of them point-blank, today. Increasingly urgent is the news about the rapidly accelerating melting of the Greenland and West Antarctic ice shields.

Esther Elizur (Jerusalem), Sarah Conn (Arlington, MA), Veronica Laveta (Boulder, CO), Neena Rao (Naropa University), Jeffrey Kiehl (NCAR), Hal Shorey (University of Kansas), Lisa Dilling (University of Colorado), and Susan Watrous (Santa Cruz) provided helpful resources, comments and constructive feedback on an earlier draft of this chapter. The final interpretation of the works cited here and any remaining oversights are mine.

Meanwhile, every freak of the weather makes us more uneasy — be it relentless heat in Phoenix or the devastating hurricane season of 2005.

It gets worse when international climate negotiators are unable to commit to significant emissions reductions, and the US Congress can't agree on any binding caps or market-based mechanism to bring emissions down. Then we hear of administration officials editing climate reports to downplay the issue's seriousness, and even the environmental community is internally torn on the issue. The best news yet is of the hundreds of local communities, numerous states, selected businesses and industries, and faith communities around the United States reducing their emissions and usually saving money in the process (see the chapters by Young, Chapter 24, Watrous and Fraley, Chapter 25, Pratt and Rabkin, Chapter 26, Arroyo and Preston, Chapter 21, duVair *et al.*, Chapter 27, Tennis, Chapter 26 and Bingham, Chapter 9, this volume). Most experts agree that these actions alone — while exceedingly important to build precedent, experience, and wider engagement on the issue — won't solve the climate problem (e.g., Betsill, 2001; Victor, House, and Joy, 2005).

For most people, it is challenging to keep listening to such depressing news. Many find it hard to stay optimistic about climate change, about society's ability to significantly reduce its emissions, overcome the technological challenges, and appropriately address the economic, social, and ethical concerns in the process (Agyeman *et al.*, Chapter 7, this volume). Besides, what could we as individuals do about it anyway? The problem is too big, too complicated, too overwhelming — it's hopeless.

Or so it would seem. This chapter speaks to these psychological responses to climate change information and to the effectiveness of emotional appeals used to motivate action. I argue that neglecting the emotional reception of climate-related news makes communication and outreach efforts more likely to fail. Below I describe common feelings people may experience, how communication strategies aimed at increasing the sense of urgency either deliberately or inadvertently play into them, and often produce undesirable outcomes. I conclude with some good news: better alternatives.

Emotional responses to climate change information

> Merely thinking about [the looming realities of our world] is to flirt with despair.
>
> *Paul Rogat Loeb (2004a: 2)*

To some, the introduction to this chapter is merely an enumeration of facts. Others may experience any number of emotional reactions: fear, guilt, anger, defiance, a desire to blame someone, powerlessness, despair, a sense of exhaustion or annoyance at having to hear the litany one more time. After a few sentences into this barrage maybe numbness sets it. Of all these, numbness was probably the least unpleasant. This is precisely the reason why even those who are interested in climate change can "go numb" over time – it's just too hard to stay present to the constant onslaught of bad news.

"Bad news" tells us of a potential danger or of examples in place and time where a threat has materialized in a significant incident. It also informs us – directly or indirectly – about the source or reason for the danger. For example, when we see disasters as "acts of nature" (or God), we might feel we have no influence (see also Bostrom and Lashof, Chapter 1, this volume). If on the other hand we are the culprits, then we may feel implicated and guilty. Or we might feel angry that not more has been done to prevent or contain the problem. We may get worried about who will have to pay. What do we do with all these emotional responses? All too often we keep them to ourselves or suppress them altogether.

The critical importance of emotions

In the western cultural context, we tend to think of a person's cognitive responses to a stimulus in binary ways: rational or irrational. Many interpret "emotional" as synonymous with irrational, and thus as opposite to, and exclusionary of, rational thought. Because of cultural ideals associated with rationality (and related concepts of reason, objectivity, analysis, scientific knowledge, expertise, etc.), we have attached many negative labels to "being emotional." Researchers studying emotions tell us, however, that our cultural suppositions lag significantly behind modern scientific insights (see e.g., Slovic *et al.*, 2004; Wallenius, 2001; Davidson, 2000).

From an evolutionary perspective, emotions can be life-savers. Together with experience and memory, they offer critical "interpretations" of situations around us. In the case of immediate dangers, these basic reactions can induce us to fight, flee, or freeze (e.g., Levine, 1997; Wallenius, 2001).

Insights from cognitive and risk studies further suggest that without emotions (or affect, see Leiserowitz, Chapter 2, this volume), our thinking would be impaired (e.g., Slovic *et al.*, 2004; Davidson, 2000). As Slovic and colleagues (2004: 311) suggest, "analytic reasoning cannot be effective unless it is guided by emotion and affect." This is particularly true in situations where the threat is near and immediate. In situations where the threat is not (yet) directly perceived – as in the case of climate change – we may misleadingly believe that there is no danger at all. Thus, emotionally under- or overreacting without the help of our cognitive facilities will lead to inappropriate responses. According to Davidson (2000: 91), "Cognition would be rudderless without the accompaniment of emotion, just as emotion would be primitive without the participation of cognition."

Finally, emotions play a critical role in health and the course of illness (e.g., Groopman, 2004). Fear, hope, despair, and denial affect a patient's psychological and physiological health (e.g., Groopman, 2004; Snyder *et al.*, 2000; Goldbeck, 1997; Ramachandran and Rogers-Ramachandran, 1996; Morse and Doberneck, 1995; Hall, 1994; Schussler, 1992). Moreover, they are critically important for an individual's ability to persist – physically and psychologically – in extreme adverse situations (e.g., Lifton, 1967).

Together, these scientific insights suggest that emotions are to be seen in a much more positive light than we often do; in fact, they are essential. What is equally true is that unhampered, or unattended to, feelings can also paralyze or mislead us. This just further underscores why we ignore or dismiss them at our peril.

The danger of unhampered and unattended emotions

The psychological purpose of fight, flight, or freeze reactions is to control either the external danger or the internal experience of fear (see reviews in Johnson, 2005; Ruiter *et al.*, 2004; Ruiter *et al.*, 2003; Wallenius, 2001; Witte, 1998). Both responses prove to be positive adaptations if they increase a person's ability to cope with the dangerous situation. An initial focus on fear control, for example, can enable an individual to deal more effectively with reducing the danger (Witte, 1998; Ruiter *et al.*, 2004). On the other hand, if a person's reaction only aims to control the fear or pain without reducing the danger, psychologists consider such a response maladaptive.

Maladaptive or avoidant behaviors (i.e., defense mechanisms) on the individual or collective scale typically include the following (see also Grothmann and Patt, 2005; Ruiter *et al.*, 2004; Osbaldiston and Sheldon, 2002; Cohen, 2001; Opotow and Weiss, 2000):

- the denial of the existence of the threat;
- a belief that the problem won't happen here/to us – a form of exceptionalism;
- the projection of responsibility onto someone else (blaming others for it or transference that experts will fix it);
- wishful thinking or rationalization that the problem will go away on its own, is less severe than believed, or that silver-bullet solutions will be found;
- a traditionalist refusal to do anything different ("we've always done it this way");
- the uncertainty trap ("we don't yet know enough to act");

- displacement of one's attention on other, maybe more immediate issues; and
- feeling trapped, fatalism, or "capitulatory imagination" (Loeb, 2004b: 128) – thoughts that lead to giving up.

Numerous studies have shown that these various forms of denial can be useful in the short run, but tend to be an obstacle over the long term (e.g., Goldbeck, 1997; Ramachandran and Rogers-Ramachandran, 1996; Greer, 1992).

Maybe the leading maladaptive response to threats that are particularly scary, ill understood, difficult to control, overwhelming, and in which we are complicit – such as global climate change – is psychic numbing or apathy.[3] While several writers have suspected that environmental problems may contribute to numbness and apathy (e.g., Clayton and Brook, 2005: 89; Macy and Brown, 1998; Searles, 1972), only a few empirical studies have actually examined the emotional and cognitive responses to climate change, its impacts and solutions (Stoll-Kleemann, O'Riordan, and Jaeger, 2001; Grothmann and Patt, 2005).[4] None of these explicitly examined psychic numbing. Weber's (2006) "finite pool of worry" hypothesis is relevant here: people can only worry so much about currently salient risks (e.g., terrorism), which makes it difficult to worry about another (e.g., climate change). This suggests that people may not be numb to all danger, but that numbness to one issue may simply result from being deeply concerned elsewhere.

Numbing requires at least a minimal amount of engagement with the threat: one first must realize the magnitude of the threat and perceive an inability to affect it before one goes numb. This secondary reaction should not be confused with the apathy many individuals display toward the issue in the first place, which prevents them from even learning about it and having an informed emotional reaction (Cafaro, 2005; Marshall and Lynas, 2003). It is likely that this primary apathy occurs in response to the combination of problems that converge and in their totality overwhelm us. Differently put, climate change does not exist in isolation. The synergistic effect of a drumbeat of news about various overwhelming environmental and societal problems on people's perception of the world may have this numbing effect. Add to this the immediate demands of daily life, and it will take something even worse to break through the walls of apathy (e.g., Gidley, 2005).[5]

Not facing the magnitude of such problems, however, requires conscious (or more likely subconscious) psychological repression: people find it highly unpleasant to feel pain, despair, or guilt, overwhelmed and powerless, to appear morbid or cause distress to others, or to be viewed as unpatriotic and weak (Clayton and Brook, 2005; Macy and Brown, 1998: 26–32). Some have argued that repression of this onslaught of emotions both requires and contributes to commonly observed "cultural ailments" such as alienation from and distrust of oneself and others, isolation, escapist activities, blaming and scapegoating, political passivity, avoidance of information that could reawaken the unpleasant feelings, diminished intellectual performance, dysphoria and depression, and so on (e.g., Gidley, 2005; Macy and Brown, 1998; Searles, 1972).[6] Such repressive mechanisms may even be sanctioned if dire predictions do not (immediately) materialize, allowing listeners to turn away from alarming news (McComas and Shanahan, 1999).

Other maladaptive responses to fear and frustration can include reactance[7] and counterproductive behaviors that may in fact increase one's objective risk to external danger (e.g., Gray and Ropeik, 2002). Survey studies have found, for example, that one common response to information about the threats of climate change is a desire to buy a sports utility vehicle (SUV) as a means of protecting against an unpleasant or unpredictable environment (FrameWorks Institute, 2001). Unfortunately, of course, SUVs at current low levels of fuel efficiency exacerbate the climate problem (Plotkin, 2004; see also Dilling and Farhar, Chapter 23, this volume).

The above discussion illustrates how humans can react to unbridled fear and overwhelming circumstances, or even just the prospect of a potentially overwhelming situation. Clearly, emotions can be powerful motivators as well as de-motivators of action. Thus, playing with emotional appeals to create urgency is like playing with fire.

Emotional appeals to create urgency

> If a red light blinks on in a cockpit, should the pilot ignore it until it speaks in an unexcited tone? . . . Is there any way to say [it] sweetly? Patiently? If one did, would anyone pay attention?
>
> *Donnella Meadows (1996)*

The conundrum so well expressed by the late Donnella Meadows is one that concerned scientists and other communicators face daily. If scientific findings about serious environmental risks presented "sweetly and patiently" (not to speak of their obscure, jargon-heavy technical cousins) cannot capture public or political audiences, then what would move someone to act? The temptation to break through indifference and apathy by using fear appeals is very real, especially as the problem grows more urgent.

If global warming is made scarier, it might become more salient. Scientists and editors of flagship science journals deplore the inattention given to climate change, and on and off the record suggest that "a useful catastrophe or two" and other fear-provoking measures (such as terror alert systems for the state of the climate) are needed to motivate adequate policy response.[8] Similarly, policy advisors and politicians evoke currently resonant public fears, such as weapons of mass destruction, terrorism, and war, and compare the seriousness of climate change to that of more frightening issues (e.g., Blix, 2004; Gorbachev, 2004; King, 2004). But can such fear appeals generate a sustained and constructive engagement with climate change? The answer is: usually not.

Fear appeals have been effective at various times through US history. The recent American response to fear appeals regarding terrorism (including "pre-emptive" war) achieved the desired intent.[9] Further back in history, Franklin D. Roosevelt famously said during the Great Depression, "Let me assert my firm belief that the only thing we have to fear is fear itself — nameless, unreasoning, unjustified terror which paralyzes needed efforts to convert retreat into advance." Acknowledging people's legitimate fears, he went on to say, "only a foolish optimist can deny the dark realities of the moment" and then called for a renewed vision and concrete action plan (cited in Wilson, 2004).

Numerous studies, however, caution us about using fear appeals. Empirical studies show, for example, that fear may change attitudes and verbal expressions of concern but not necessarily increase active engagement or behavior change (e.g., Ruiter, Abraham, and Kok, 2001). More specifically, threat information is more likely to be persuasive, causing persistent attitude change, and motivating constructive responses only when people

- feel personally vulnerable to the risk;
- have useful and very specific information about possible precautionary actions;
- positively appraise their own ability (self-efficacy) to carry out the action;
- feel the suggested action will effectively solve the problem (response efficacy);
- believe the cost associated with taking precautionary action is low or acceptable;
- view the reward for *not* taking the action as unappealing; and
- tend to consciously and carefully process threat information (i.e., engage in central/systematic processing as opposed to peripheral/heuristic information processing).

These findings — while extracted from different empirical studies and based on a variety of theories — appear to be relatively stable and consistent (Johnson, 2005; Ruiter *et al.*, 2004; Das, de Wit, and Stroebe, 2003; Ruiter *et al.*, 2003; Osbaldiston and Sheldon, 2002; Ruiter, Abraham, and

Kok, 2001; Bator and Cialdini, 2000; Floyd, Prentice-Dunn, and Rogers, 2000; Witte, 1998; Bandura, 1997; Baldassar and Katz, 1992; Ajzen, 1991; Block and Keller, 1998; Hine and Gifford, 1991; Leventhal, Safer, and Panagis, 1983; Lynn, 1974). Perceived self-efficacy in responding to a threat, expected response costs, social support, and intention to act are the strongest predictors of concurrent or future behavior (Milne, Sheeran, and Orbell, 2000).[10] If threat information is unspecific, uncertain, perceived as manipulative, or if it comes from little-trusted sources, it may not even evoke fear but resentment, dismissal, or no response at all (Gray and Ropeik, 2002; Osbaldiston and Sheldon, 2002; Slovic, 1993).

Guilt − the emotional response to some self-perceived shortfall with respect to one's own standards of conduct − is another potentially powerful (and sometimes inadvertently used) motivator of individual or social response. People who feel guilty want to make amends or feel a moral responsibility to behave differently (O'Keefe, 2002a). Research suggests that explicit guilt appeals can indeed evoke such feelings, but do not necessarily persuade or induce behavior change because individuals just feel resentful or annoyed with overt manipulation (O'Keefe, 2002a). Yet milder guilt appeals are less persuasive and thus less motivational than overt techniques. To the extent that guilt challenges one's sense of personal integrity, it can initiate the search for self-affirmation. Responses to guilt thus aim primarily at maintaining one's sense of a moral self, and may or may not also motivate behavior that ends or rectifies the guilt-invoking action (O'Keefe, 2002a,b; Nabi, 2002). Take someone's reaction to the criticism that they drive an SUV, by many perceived as "guilt-tripping." The almost invariably resentful reaction is frequently followed with justifications for owning such a vehicle. People may argue they need the vehicle to reach off-road or mountainous locations, transport big items, or protect their children − thus reaffirming the sense of being a reasonable person and responsible parent. The implications are twofold: one, guilt appeals are unreliable as motivators of environmentally benign behavior; and two, people will maintain their sense of self and identity before changing an environmentally damaging behavior, *unless* the new behavior is consistent with who they want to be in the world (see also Clayton and Brooke, 2005; Vasi and Macy, 2003).

In summary, the key to whether or not threat and guilt appeals cause the desired impact is the presence or absence of concurrent supportive, enabling conditions (Eagly and Kulesa, 1997). The recommended alternative behavior must also reinforce (or at least not undermine) one's self-identity. Thus, given the highly complex and uncertain outcomes of such appeals and the limited ability to control people's emotional responses to information, using positive motivations and forms of communication may prove more successful in engaging social actors.

Implications for communication

> One of the penalties of an ecological education is that one lives alone in a world of wounds. Much of the damage inflicted on land is quite invisible to lay [people]. An ecologist must either harden his [or her] shell and make believe that the consequences of science are none of his [her] business, or [s]he must be the doctor who sees the marks of death in a community that believes itself well and does not want to be told otherwise.
>
> *Aldo Leopold (1953, 1993: 165)*

Communicators and change agents can become very frustrated with the indifference and ignorance they encounter in some of their audiences. I have yet to meet one who does not nod sadly to this quote from Aldo Leopold. Some audiences, of course, are interested and eager to learn more, and care deeply about the environment, but it is still hard to prevent them from slipping into hopelessness as they begin to fully grasp the magnitude of the challenge. Other chapters in this volume suggest that people are motivated by different things; here I focus only communication strategies that are cognizant of the insights of the psychological literature discussed above.

Greater self-reflexivity among communicators

Maybe the first insight is for communicators themselves to acknowledge their own emotional responses to environmental degradation and society's responses. Many choose to work on climate change because of deep passions and emotional, identity- and value-driven motivations, and thus are likely to experience strong emotional reactions.

The benefits of such greater self-awareness are manifold. First, unacknowledged feelings among communicators can lead to the impulsive, frustrated, or at least unskilled use of threat and guilt appeals which are unpredictable at best and counterproductive at worst. Second, with some audiences, acknowledging one's own feelings — in subtle if sincere ways — can be an important rapport- and trust-building process. While some want "just the facts, ma'am," especially lay audiences respond well when they feel that they are spoken to as "whole" people. If allowed to bring both head *and* heart to the issue, people can pay attention more fully and engage more deeply. Finally, taking time (the nearly impossible hurdle!) to acknowledge one's emotional responses to climate change and seek and provide emotional support with colleagues and friends helps in the prevention or healing of burn-out (e.g., Moyer *et al.*, 2001, Maslach and Leiter, 2005).

From fear appeals to fostering true hope

Various aspects of climate change, its impacts and even what society may choose to do about it can evoke fear. Many people don't even realize these potentially scary aspects yet, thus lack the emotional motivation to act on climate change. As Block and Keller (1998) and Das, de Wit and Stroebe (2003) among others suggest, at this early stage, people need first to accept that they are vulnerable to the risks of climate change and thus need messages that increase their personal sense of vulnerability. As people move toward contemplating taking action, stronger fear appeals can help form a behavioral intent. However, such fear appeals must be coupled with constructive information and support to reduce the danger (see the discussion in the section on "Emotional appeals to create urgency" above).

Well-designed behavior change efforts seem more effective, as others in this book attest: individual transportation behavior is hard to change when there are limited behavioral alternatives, high response cost, low self-efficacy, and lack of social support and norms (Tribbia, Chapter 15, this volume). Neighborhood and community efforts to reduce emissions (Rabkin with Gershon, Watrous and Fraley, Chapters 19 and 25, this volume) work better with simple instructions, social support, low-cost alternatives, and regular feedback. The literature on hope confirms this. People need a minimum amount of information, a realistic assessment of the threat or diagnosis, a sense of personal control over their circumstances, a clear goal, an understanding of the strategies to reach that goal (including the possible setbacks along the way), a sense of support, and frequent feedback that allows them to see that they are moving in the right direction (e.g., Groopman, 2004; Morse and Doberneck, 1995). Importantly, fostering true hope is not erasing fears or doubts, but facing reality full on, while banking on promising strategies and uncertainty. Precisely because the future is not fully determined by present conditions concerted efforts can now make a positive difference. Communicators must learn to better hold up a positive future, depict an engaging goal and what Snyder *et al.* (1991) have called "the will and the ways," i.e., provide a sense of empowerment, clear instructions on what to do, and helping people see themselves on the path of reaching the goal (Courville and Piper, 2004).

From defense mechanisms to embracing positive values

Defense mechanisms — as mentioned above — are intended to reduce worry and to deflect personal responsibility for a situation, thus making it hard to reach people. Increasing the sense of personal vulnerability, responsibility, and empowerment is critical in tapping into people's positive motivation to pay attention. But as Clayton and Brook (2005: 90) contend, "Most people drastically overestimate the impact of individual motivations and disposi-tions on behavior, discounting the effect of the situational context. ... People's behavior is heavily influenced by the behavior and expectations of other people, especially important others." This statement points toward the immense significance of social values, identities, and support for behavior change (Lane, 2000; Opotow and Weiss, 2000; see also the chapters by Tribbia; Michaelis; and Chess and Johnson, Chapters 15, 16, and 14, this volume).

Ultimately, humans are social beings, whose physical, psychic, and social survival depends on others. Communicators can tap into this deeply anchored need. Social norms give us a compass for "good" behavior. They are maintained by deep values, core beliefs, and relevant institutions (e.g., the courts, faith communities), and are re-enacted daily within our social cohorts (at work, with families, neighbors, and friends). Americans, for example, deeply cherish values such as competitiveness, leadership, ingenuity, innovation, fairness, team play, stewardship, and responsibility for the welfare of others (FrameWorks Institute, 2001). Several of these directly counteract the defense mechanisms mentioned above. Communication strategies can focus on these cherished values to frame needed action as furthering what is "good." Communicators can encourage deeper conversa-tions about the "good life" (see Michaelis, Chapter 16, this volume), about deeply held and sometimes conflicting values (see Regan, Chapter 13, this volume), or personal responsibility in the global context (see Bateson, Chapter 18, this volume).

A recent study tested the appeal of various messages encouraging action on climate change (commissioned by The Advertising Council and Environmental Defense and conducted in 2005 by S. Radoff Associates).[11] Participants responded best to one which suggested that taking action would be "doing good," which appealed to their innate goodness. The second-best response was to a message that appealed to logic and responsibility. Positive psychologists confirm that people have a deep desire to live a "good" or "meaningful" life, in which they derive gratification from exhibiting their strengths, talents, and virtues, and use these skills and strengths to belong to and serve a larger purpose (Seligman, 2004). Reminding people of this larger common good provides meaning beyond self-serving goals, and is essential to counter individuals' sense of isolation and futility vis-à-vis global warming.

Envisioning a future worth fighting for

Finally, building on all the elements of a more positive communication strategy suggested so far, it will be critical to engage people in envisioning a future worth fighting for. While many have spoken to the importance of positive visioning, it is actually difficult to begin. A grand positive vision may well be something that no one creates but eventually emerges out of a myriad of images, stories, and on-the-ground efforts in developing alternatives (lifestyles, technologies, behaviors, environments, communities, institutions, etc.). The lack of such a compelling vision – one that is believable, inclusive, problem-solving, and meaning-giving (Olson, 1995) – has been implicated in the apathy of young people (e.g., Gidley, 2005), and has recently served as a provocative rallying cry to the environmental community (Shellenberger and Nordhaus, 2004). Communicators can serve the emergence of such a vision, first, by ceasing to conjure a doomsday scenario in people's imagination; second, by pointing to the many positive efforts under way; and finally, by providing fora where people can engage in the visioning process. Charismatic leaders have played important roles in this regard in past social movements ("I have a dream ..."). Yet waiting for the Martin Luther King Jr. or Mahatma Gandhi of climate change might waste precious time. Instead, creating a vibrant vision together, drawing up pathways there, and supporting each other in working toward this goal, will get us more than halfway there.

Notes

1. This quote appears here and in other publications, slightly rephrased, numerous times.
2. This chapter is an adapted version of a portion of an article by Moser and Dilling (2004). I am grateful for the permission from Heldref and *Environment* (www.heldref.org/env.php) to expand on it here.
3. Etymologically, the term apathy means the absence of feeling or, more specifically, of suffering (whether due to an inability or a refusal to feel pain). The term "psychic numbing" was first introduced by Lifton (1967). Most work on psychic numbing has focused on war atrocities, the Holocaust, extreme abuse, and nuclear catastrophes (e.g., Lifton, 1967; Cohen, 2001).
4. See also the review in Grothmann and Patt (2005) of a small number of additional studies, conducted mostly in Europe.
5. Witness the world's response to the Indian Ocean tsunami in late 2004. Indirect evidence for people's response to the combination of big problems comes from a study conducted

by the American Geophysical Union (Immerwahr, 1999), which found that focus group participants had very low expectations of society ever being able to solve the climate problem. They related these low prospects to a perception of general moral decay of society.

6. This chapter cannot do justice to these topics. For a longer treatment, see Nicholsen (2002), esp. chapter 5; Greenberg, Solomon, and Pyszczynski (1997); and Ruiter, Abraham, and Kok (2001).

7. Psychological reactance occurs in response to a perceived restriction of choices. For example, an individual who feels manipulated, caught, embarrassed, hurt, frightened, or otherwise restricted in his/her freedom will try to reassert him/herself by attaching negative characteristics on another person, e.g., by denigrating that person as a liar or evil person.

8. The call for a few "useful catastrophes" comes from Tickell (2002). Kennedy (2004) deplored the lack of attention that climate change is receiving in an editorial in *Science*.

9. See a discussion of this issue with UC-Berkeley linguist and framing expert George Lakoff at: http://www.berkeley.edu/news/media/releases/2004/08/25_lakoff.shtml; accessed January 29, 2006.

10. Many of these insights come from research on precautionary behavior related to health and crime – issues that affect a person directly, immediately, and tangibly. Transfer of these findings to environmentally significant behavior (especially on a remote, systemic, and complex issue like climate change) must still be interpreted with caution. While numerous studies have been conducted on the relationship between environmental concern, attitudes, values, and behavior, a careful assessment of the psychodynamics underlying this relationship is still wanting.

11. Thanks to George Perlov from the Ad Council for sharing this information.

References

Ajzen, I. (1991). The theory of planned behavior. *Organizational Behavior and Human Decision*, **50**, 179–211.

Baldassar, M. and Katz, C. (1992). The personal threat of environmental problems as predictor of environmental practices. *Environment and Behavior*, **24**, 5, 602–16.

Bandura, A. (1997). *Self-Efficacy: The Exercise of Control*. New York: Freeman.

Bator, R. J. and Cialdini, R. B. (2000). The application of persuasion theory to the development of effective proenvironmental public service announcements. *Journal of Social Issues*, **56**, 527–41.

Betsill, M. M. (2001). Acting locally, does it matter globally? The contribution of US cities to global climate change mitigation. Paper presented at the Fourth Open Meeting of the Human Dimensions of Global Change Research Community, Rio de Janeiro, October 6–8, 2001.

Blix, H. (2004). Global warming as big a threat as WMD. *New Perspectives Quarterly*, **21**, 32–3.

Block, L. G. and Keller, P. A. (1998). Beyond protection motivation: An integrative theory of health appeals. *Journal of Applied Social Psychology*, **28**, 17, 1584–608.

Cafaro, P. J. (2005). Gluttony, arrogance, greed, and apathy: An exploration of environmental vice. In *Environmental Virtue Ethics*, eds. Sandler, R. and Cafaro, P. Lanham, MD: Rowman and Littlefield, pp. 135–58.

Clayton, S. and Brook, A. (2005). Can psychology help save the world? A model for conservation psychology. *Analyses of Social Issues and Public Policy*, **5**, 1, 87–102.

Cohen, S. (2001). *States of Denial: Knowing About Atrocities and Suffering.* Cambridge and Oxford: Polity Press and Blackwell Publishers.

Courville, S. and Piper, N. (2004). Harnessing hope through NGO activism. *The Annals of the American Academy of Political and Social Science*, **592**, 1, 39−61.

Das, E., de Wit, J., and Stroebe, W. (2003). Fear appeals motivate acceptance of action recommendations: Evidence for a positive bias in the processing of persuasive messages. *Personality & Social Psychology Bulletin*, **29**, 650−64.

Davidson, R. J. (2000). Cognitive neuroscience needs affective neuroscience (and vice versa). *Brain and Cognition*, **42**, 89−92.

Eagly, A. H. and Kulesa, P. (1997). Attitudes, attitude structure and resistance to change. In *Environment, Ethics, and Behavior: The Psychology of Environmental Valuation and Degradation*, eds. Bazerman, M. H., Messick, D. M., Tenbrunsel, A. E., and Wade-Benzoni, K. A. San Francisco, CA: The New Lexington Press, pp. 122−53.

Floyd, D. L., Prentice-Dunn, S., and Rogers, R. W. (2000). A meta-analysis of research on protection motivation theory. *Journal of Applied Social Psychology*, **30**, 2, 407−29.

FrameWorks Institute (2001). *Talking Global Warming* (Summary of Research Findings). Washington, DC: FrameWorks Institute.

Gidley, J. (2005). Giving hope back to our young people: Creating a new spiritual mythology for western culture. *Journal of Futures Studies*, **9**, 3, 17−30.

Goldbeck, R. (1997). Denial in physical illness. *Journal of Psychosomatic Research*, **43**, 6, 575−93.

Gorbachev, M. (2004). Pre-empt global warming. *New Perspectives Quarterly*, **21**, 17−19.

Gray, G. M. and Ropeik, D. P. (2002). Dealing with the dangers of fear: The role of risk communication. *Health Affairs*, **21**, 106−16.

Greenberg, J., Solomon, S., and Pyszczynski, T. (1997). Terror management theory of self-esteem and cultural worldviews: Empirical assessments and conceptual refinements. In *Advances in Experimental Social Psychology*, 29, ed. Zanna, M. P. New York: Academic Press, pp. 61−139.

Greer, S. (1992). The management of denial in cancer patients. *Oncology*, **6**, 12, 33−36.

Groopman, J. (2004). *The Anatomy of Hope: How People Prevail in the Face of Illness.* New York: Random House.

Grothmann, T. and Patt, A. (2005). Adaptive capacity and human cognition: The process of individual adaptation to climate change. *Global Environmental Change*, **15**, 3, 199−213.

Hall, B. A. (1994). Ways of maintaining hope in HIV disease. *Research in Nursing & Health*, **17**, 4, 283−93.

Hine, D. W. and Gifford, R. (1991). Fear appeals, individual differences, and environmental concern. *Journal of Environmental Education*, **23**, 36−41.

Immerwahr, J. (1999). *Waiting for a signal: Public attitudes toward global warming, the environment and geophysical research.* AGU. Available at http://www.agu.org/sci_soc/attitude_study.pdf; accessed February 10, 2006.

Johnson, B. B. (2005). Testing and expanding a model of cognitive processing of risk information. *Risk Analysis*, **25**, 3, 631−50.

Kennedy, D. (2004). Climate change and climate science. *Science*, **304**, 1565.

King, D. A. (2004). Climate change science: Adapt, mitigate, or ignore? *Science*, **303**, 176−7.

Lane, M. (2000). Environmentally responsible behavior: Does it really matter what we believe? *Planning Forum*, **6**, 33–9.

Leopold, A. (1993). *Round River*. New York: Oxford University Press (1953).

Leventhal, H., Safer, M. A., and Panagis, D. M. (1983). The impact of communications on the self-regulation of health beliefs, decisions, and behavior. *Health Education Quarterly*, **10**, 3–29.

Levine, P. A. with Frederick, A. (1997). *Waking the Tiger — Healing Trauma: The Innate Capacity to Transform Overwhelming Experiences*. Berkeley, CA: North Atlantic Books.

Lifton, R. J. (1967). *Death in Life: Survivors of Hiroshima*. New York: Simon and Schuster.

Lifton, R. J. with Markusen, E. (1986). *The Genocidal Mentality: Nazi Holocaust and Nuclear Threat*. New York.

Loeb, P. R. (ed.) (2004a). *The Impossible Will Take a Little While: A Citizen's Guide to Hope in a Time of Fear*. New York: Basic Books.

Loeb, P. R. (2004b). Introduction to Part Four. In *The Impossible Will Take a Little While: A Citizen's Guide to Hope in a Time of Fear*. New York: Basic Books.

Lynn, J. R. (1974). Effects of persuasive appeals in public service advertising. *Journalism Quarterly*, **51**, 622–30.

Macy, J. and Brown, M. Y. (1998). *Coming Back to Life: Practices to Reconnect Our Lives, Our World*. Gabriola Island, BC: New Society Publishers.

Marshall, G. and Lynas, M. (2003). Why we don't give a damn. *New Statesman*, **132**, 4666, 445–7.

Maslach, C. and Leiter, M. P. (2005). Reversing burnout: How to rekindle your passion for your work. *Stanford Social Innovation Review*, 2005, Winter, 42–9; available at: http://www.ssireview.com/pdf/2005WI_Feature_Maslach_Leiter.pdf; accessed January 2, 2006.

Meadows, D. (1996). How environmentalists ought to talk. *Global Citizen*, Donella Meadows Archive. Available at http://www.sustainer.org/dhm_archive/search.php?display_article=vn635environmentalistsed; last accessed May 26, 2004.

Milne, S., Sheeran, P., and Orbell, S. (2000). Prediction and intervention in health-related behavior: A meta-analytic review of protection motivation theory. *Journal of Applied Social Psychology*, **3**, 106–43.

Morse, J. M. and Doberneck, B. (1995). Delineating the concept of hope. *Image — Journal of Nursing Scholarship*, **27**, 4, 277–85.

Moser, S. C. and Dilling, L. (2004). Making climate hot: Communicating the urgency and challenge of global climate change. *Environment*, **46**, 10, 32–46.

Moyer, B., McAllister, J., Finley, M. L., *et al.* (2001). *Doing Democracy: The MAP Model for Organizing Social Movements*. Gabriola Island, BC: New Society Publishers.

Nabi, R. L. (2002). Discrete emotions and persuasion. In *The Persuasion Handbook: Developments in Theory and Practice*, eds. Dillard, J. P. and Pfau, M. Thousand Oaks, CA: Sage Publications, pp. 289–308.

Nicholsen, S. W. (2002). *The Love of Nature and the End of the World*. Cambridge, MA: The MIT Press.

O'Keefe, D. J. (2002a). Guilt as a mechanism of persuasion. In *The Persuasion Handbook: Developments in Theory and Practice*, eds. Dillard, J. P. and Pfau, M. Thousand Oaks, CA: Sage Publications, pp. 329–44.

O'Keefe, D. J. (2002b). Guilt and social influence. In *Communication Yearbook* 23, ed. Roloff, M. E. Thousand Oaks, CA: Sage Publications, pp. 67–101.

Olson, R. L. (1995). Sustainability as a social vision. *Journal of Social Issues*, **51**, 15–35.

Opotow, S. and Weiss, L. (2000). Denial and the process of moral exclusion in environmental conflict. *Journal of Social Issues*, **56**, 3, 475–90.

Osbaldiston, R. and Sheldon, K. M. (2002). Social dilemmas and sustainability: Promoting people's motivation to 'cooperate with the future'. In *Psychology of Sustainable Development*, eds. Schmuck, P. and Schultz, W. P. Amsterdam: Kluwer, pp. 37–57.

Plotkin, S. (2004). Is bigger better? Moving toward a dispassionate view of SUVs. *Environment*, **10**, 8–21.

Ramachandran, V. S. and Rogers-Ramachandran, D. (1996). Denial of disabilities in anosognosia. *Nature*, **382**, 501.

Revkin, A. (2005). Glacial gains in global talks on cleaner air. *The New York Times*, December 11.

Ruiter, R. A. C., Abraham, C., and Kok, G. (2001). Scary warnings and rational precautions: A review of the psychology of fear appeals. *Psychology and Health*, **16**, 613–30.

Ruiter, R. A. C., Verplanken, B., Kok, G., *et al.* (2003). The role of coping appraisal in reactions to fear appeals: Do we need threat information? *Journal of Health Psychology*, **8**, 465–74.

Ruiter, R. A. C., Verplanken, B., De Cremer, D., *et al.* (2004). Danger and fear control in response to fear appeals: The role of need for cognition. *Basic and Applied Social Psychology*, **26**, 13–24.

Schussler, G. (1992). Coping strategies and individual meanings of illness. *Social Science & Medicine*, **34**, 4, 427–32.

Searles, H. F. (1972). Unconscious processes in relation to the environmental crisis. *Psychoanalytical Review*, **59**, 3, 361–74.

Seligman, M. E. P. (2004). Can happiness be taught? *Daedalus*, **133**, 2, 80–7.

Shellenberger, M. and Nordhaus, T. (2004). *The Death of Environmentalism: Global Warming Politics in a Post-Environmental World*. Available at http://www.thebreakthrough.org; accessed February 10, 2006.

Shorey, H. S., Rand, K. L., and Snyder, C. R. (2005). The ethics of hope: A guide to social responsibility in contemporary business. In *Positive Psychology in Business Ethics and Corporate Social Responsibility*, eds. Giacalone, R., Dunn, C., and Jurkiewicz, C. L. Greenwich, CT: Information Age, pp. 249–64.

Slovic, P. (1993). Perceived risk, trust and democracy: A systems perspective. *Risk Analysis*, **13**, 675–82.

Slovic, P., Finucane, M. L., Peters, E., *et al.* (2004). Risk as analysis and risk as feelings: Some thoughts about affect, reason, risk and rationality. *Risk Analysis*, **24**, 2, 311–22.

Snyder, C. R., Ilardi, S. S., Cheavens, J., *et al.* (2000). The role of hope in cognitive-behavioral therapies. *Cognitive Therapy and Research*, **24**, 6, 747–62.

Snyder, C. R., Harris, C., Anderson, J. R., *et al.* (1991). The will and the ways: Development and validation of an individual-differences measure of hope. *Journal of Personal and Social Psychology*, **60**, 4, 570–85.

Stoll-Kleemann, S., O'Riordan, T., and Jaeger, C. C. (2001). The psychology of denial concerning climate mitigation measures: Evidence from Swiss focus groups. *Global Environmental Change*, **11**, 107–17.

Tickell, C. (2002). Communicating climate change. *Science*, **297**, 737.

Vasi, I. B. and Macy, M. (2003). The mobilizer's dilemma: Crisis, empowerment, and collective action. *Social Forces*, **81**, 3, 983–1002.

Vergano, D. (2005). Wolves teach experts about global warming. *USA Today*, May 31.

Victor, D. G., House, J. C., and Joy, S. (2005). A Madisonian approach to climate policy. *Science*, **309**, 1820–1.

Wallenius, C. (2001). Why do people sometimes fail when adapting to danger? A theoretical discussion from a psychological perspective. *International Journal of Mass Emergencies and Disasters*, **19**, 2, 145–80.

Weber, E. (2006). Experience-based and description-based perceptions of long-term risk: Why global warming does not scare us (yet). *Climatic Change*, **77**, 103–120

Wilson, K. (2004). Global warming – Facing our fears. *truthout*, May 6, http://www.truthout.org/cgi-bin/artman/exec/view.cgi/9/4388; accessed February 10, 2006.

Witte, K. (1998). Fear as motivator, fear as inhibitor: Using the extended parallel process model to explain fear appeal successes and failures. In *Handbook of Communication and Emotion: Research, Theory, Application, and Contexts*, eds. Andersen, P. A. and Guerrero, L. K. San Diego, CA: Academic Press, pp. 423–51.

4

Public scares: changing the issue culture

Sheldon Ungar
University of Toronto–Scarborough

The communication of urgency about climate change is a central theme of many chapters in this book. But the selling of a social problem is not done in a vacuum and ultimately depends on wider social phenomena such as issue cultures, bridging metaphors, and cultural whirlwinds. For that matter, simple luck in the timing of fortuitous events can be critical. Success cannot be guaranteed for any issue to get on the radar screen of public attention, but these wider social processes provide a landscape for artful activity that can improve the chances of gaining public and media attention.

Issue cultures

Issue cultures can be defined as cognate sets of social problems that become a commanding concern in society. Perhaps the clearest example is *anything* to do with the security in the United States after the 9/11 terror attacks. Another issue culture has built around the fear of emerging diseases, ranging from Ebola and mad-cow disease though West Nile, SARS, and maybe more recently avian flu. Scientific findings or real-world events related to these problems are immediately selected for coverage by the media and often occasion attention from spokespersons in different public arenas. Social problems that can be linked to and coalesce with extant issue cultures are thus far more likely to attract sustained media and other coverage than problems that are "outliers." These problems are also likely to avoid the "balancing" predicament, where, to take a pertinent example, a handful of climate skeptics (or "contrarians") are given as much media coverage as the vast majority of climate scientists who believe that climate change poses a real threat (Boykoff and Boykoff, 2004; see also the chapters by McCright and Dunwoody, Chapters 12 and 5, this volume). As developed below in the discussion of cultural whirlwinds, the vortex of concern that surrounds such problems tends to yield one-sided coverage, something that can be seen in the still relatively muted criticism in the US media of the Bush Administration's war in Iraq.

Starting around 1980 and continuing through the decade, an issue culture built up around the atmosphere as a number of social problems from this domain rose in quick succession (Ungar, 1998). The popular theory that climatic change caused the extinction of the dinosaurs was followed by a furor over the threat of nuclear winter. But the Cold War began to wind down after 1985, just in time for the discovery of the ozone hole. Here the timing is remarkable. Near the end of the 1980s, with the successful negotiations of the Montreal Protocol, the general public saw a resolution of the ozone problem coming, just in time for its sister issue, climate change, to emerge as a celebrity problem. Prior claims-making about global warming occasioned only sporadic interest in different public arenas. Now the issue was put on the map by the "greenhouse summer" of 1988 with its severe heat and drought over much of North America. The oil industry then took a hit with the *Exxon Valdez* oil spill in early 1989. So strong was the public concern about the environment from the mid-1980s onward that Dunlap and Scarce (1991: 652) speak of a "miracle" of public opinion.

Clearly, proponents of a social problem would prefer to hook up with a current issue culture and thereby garner supportive coverage. However, there are limits on the deliberate marketing of a problem. One clear limit is that problems acquire a trajectory, and claims-makers are constrained by historical, scientific, and practical characteristics that accrue to the problem.[1] From the start of concerted scientific claims-making in the late 1970s, a future orientation became a "sticky" characteristic of global warming (Ungar, 2000). First, the doubling of pre-industrial CO_2 levels was not predicted to occur until about 2060. Doubling can be considered a benchmark measure, a binary that is more intuitively clear than claiming that levels have increased by, say, 40 percent. Doubling was also significant because scientists held that their computer models of the climate system were too primitive to deal with smaller changes on a shorter-term basis. At the time, scientists were only beginning to collect the long-term observations that could be used to document climate changes over time. In order to generate concern, the size of impacts delineated in scientific scenarios had to be sufficiently large or visible on a human scale (e.g., a few feet of sea-level rise, rather than a few centimeters or inches) that they would take decades to occur on a natural scale (e.g., Bernard, 1993). Finally, since computer models are still too coarse in their resolution to predict fine-scale changes, particular extreme weather events cannot be directly attributed to climate change. By implication, efforts to reverse this trajectory and claim that "strange weather" is a sign that climate change is already occurring have largely failed (Bernard, 1993; Ungar, 1999).[2] From the point of view of selling the problem, a future orientation creates a clear discursive liability since concern about the future is discounted in virtually every institutional arena (Cline, 1992).[3]

Bridging metaphors

A second limitation on marketing a social problem derives from the availability of "bridging metaphors" to the popular culture. Scientific ideas and results are encoded in a distinct language and need to be decoded to be accessible to the public. Overall, scientific illiteracy is the norm (Shamos, 1993), and issues that break through the veil of ignorance and gain widespread public acceptance and understanding require explanation. Consider here the success of the ozone hole.

The signal advantage of the ozone hole is that it can be encapsulated in a simple, direct, and widely familiar "penetration" metaphor. Stated succinctly, the hole leads to the increased bombardment of the Earth by lethal rays. The idea of rays penetrating a damaged "shield" meshes nicely with abiding and resonant cultural motifs, including "Hollywood affinities" ranging from the shields on the Starship Enterprise to Star Wars. That the threat can be linked with Darth Vader means that it is encompassed in common-sense understandings that are deeply ingrained and widely shared. The penetration model is ubiquitous in video games and children's television shows. It is also allied with a theory that has captivated the public's imagination: the claim that an asteroid striking Earth caused the disappearance of the dinosaurs (Clemens, 1986).

The key to favorable bridging metaphors is to provide the resources for lay theorizing. If a popular cultural template affords an appropriable theory, an "object to think with" or that can be "played with" — as in Freudian analysis of dreams — it has the capacity to go beyond the scientific domain and to capture the imagination of the public at large (Turkle, 1999). This is underscored by evidence indicating that people learn more from other individuals than from any other source of information (Freudenberg and Pastor, 1991; see also Dunwoody, Chapter 5, this volume). It is conversational presence, encompassing things like talk radio and informal talk related to mundane practices, rather than media coverage per se, that can put an issue in the air and let it acquire a life of it own.

The importance of mundane metaphors that ordinary people are able to think with can be seen in a comparison of the ozone hole with climate change. Weigh up the fundamental metaphor used to frame each problem. It is apparent to anyone that a "hole" is an aberration, something that a protective shield should not have. The greenhouse effect, in contrast, seems like a benign and essential natural phenomenon (see also Bostrom and Lashof, Chapter 1, this volume). Global warming is an extension of this phenomenon, creating the problem of finding the human "fingerprint" amidst highly variable and complex natural processes.[4] More fundamentally, there are apparently no ready-made metaphors in the popular culture – as with genetically modified "Frankenfoods" – that mesh with and provide a simple schematic for understanding the science of climate change.

Cultural whirlwinds

Finally, bridging metaphors and fortuitous events can give rise to cultural whirlwinds – rapidly evolving and progressive sequences of dynamic and often surprising events that create a vortex, hurling through a variety of arenas, creating strong conversational and practical presence. Here it should be underlined how the rapid sequence of events surging through different arenas – boycotts of McDonald's and styrofoam, medical warnings about melanoma accompanied by President Reagan's timely surgery of skin cancer in 1985 and 1987, as well as political gaffes – US Interior Secretary Donald Hodel's advocacy of a "personal protection" plan instead of international action led to the retort that "fish don't wear sunglasses" – all served to unleash a whirlwind around the ozone hole.

More important over the long haul, ozone depletion holds everyday relevance for curbing exposure to the sun. In short, the problem became a fertile source of interest, anxiety, and practical knowledge, talk and action as it centered on the need to protect especially children (Ungar, 1998). It was an issue that people could discuss without feeling overwhelmed or stupefied. Bridging metaphors about "safe sun" were boosted by the growth of a companion industry encompassing sunscreens, lip gloss, sunglasses, UV-safe hats, clothing, umbrellas, awnings, and so on.[5] The issue was medicalized by reports of increased rates of skin cancer plus the personal need to watch for skin changes. At worst, people were (and still are) to avoid the sun between 11:00 a.m. and 4:00 p.m., rendering the outdoors dangerous.

The prospects of global warming as a marketable problem

Issue cultures, bridging metaphors, and cultural whirlwinds cannot be concocted at will. Clearly, global warming has some real liabilities as a marketable social problem. Still, concerted efforts can be undertaken to make the most of opportunities that arise. Three such opportunities are currently on the horizon: linking climate change to extreme events, to security concerns, and to technological solutions such as energy efficiency and renewables. Can these linkages be exploited to create resonant issue cultures? Can they capture the cultural imagination?

The first of these is the most problematic, as there are acrimonious debates among scientists over whether extreme events are actually increasing (historical data are too limited, and more and more people are putting themselves in harm's way). Even if the extreme events are increasing, there is still the problem of whether any increase is in fact attributable to climate change. Security concerns and energy supplies seem to be better candidates for an emergent issue culture, especially if clear links can be drawn between the two in public discourse.

Weingart, Engels, and Pansegrau (2000) observe that in Germany dramatic warnings by a group of scientists drew an extreme picture of an "impending climatic catastrophe" that gained incredible momentum in political discourse.[6] In North America, in contrast, there has been little effort to date to link global warming to extreme weather events (Ungar, 1999; see also Leiserowitz, Chapter 2, this volume). The concatenating series of hurricanes in the summer of 2004 did engender a "blip" of media coverage, but whether or not this issue link will persist beyond one intense hurricane season is yet to be seen.[7] Not surprisingly, because of record-breaking costs from hurricanes, floods, and other unusual extreme weather events, insurance companies have been the first to jump on the global warming bandwagon.[8] For the public, this could mean painful increases in insurance premiums, especially in areas deemed to be high risk. However, there is no evidence to suggest that the public has been informed about, much less grasped, the impending threat of huge increases in insurance. Similarly, there has been no real attempt to use the issue culture that has built up around emerging diseases to draw attention to the disease burden that is likely to ensue from climate change (see also Leiserowitz, Chapter 2, this volume). Clearly, there are numerous cultural images (and frequent reminders from the real world) of extreme events, physical destruction, and harm to humans that would resonate with a wide population. The obvious limitation of building an issue culture around extremes is scientific credibility and the

problem of constructively engaging people, as many feel powerless in the face
of such events (see the chapters by Bostrom and Lashof, and Moser,
Chapters 1 and 3, this volume).

Undoubtedly, the selling of fear in the aftermath of the 9/11 terror attacks
has been an unparalleled success (but in other cases it can fail to motivate; see
Moser and Dilling, 2004). Security concerns have dominated much public
and political discourse since those terror attacks, not just in the United
States, but more recently also in Europe. But the insurgency in Iraq coupled
with rising oil prices and discussions of oil depletion are once again throwing
into relief America's dependence on foreign oil. If oil prices remain high,
energy dependence and (radical) reassessments of the energy economy could
well emerge as a prevailing issue culture. Global warming may then become
part of a larger discourse that may well be of greater interest and concern
to Americans (including politicians in Washington) than climate change on
its own.

Finally, the marketing of energy efficiency and alternative energy sources
could be done without drawing links to global warming (see, e.g., the
chapters by Young; Arroyo and Preston; and Dilling and Farhar, Chapters
24, 21, and 23, this volume); but the latter adds a robust dimension to the
problem and can be aligned with a no-regrets policy and possibly more
compelling bridging metaphors. The potential significance of an issue culture
built around energy efficiency and alternative energy sources is underscored
by the apparent inadequacy of the Kyoto process, as discussed by Taverne
(2005). Beyond increased opposition by economists, neither the United
States, India, nor China — all of which are building many new coal-fired
power stations and together will emit most greenhouse gases — are currently
willing to commit to mandatory limits on their greenhouse gases. Taverne
cites Tony Blair as saying: "If we don't have America, China and India
taking the action necessary to reduce greenhouse gas emissions, then we
don't solve the problem of climate change." Taverne cogently argues that this
will necessitate greater focus on technology and research and development —
rather than targeted emission reductions and attendant penalties for failure
to meet them (the latter is an aspect of Kyoto that garners little notice). The
rapid development of new technologies, combined with the rapid application
of the best available technologies, appears to be the response of choice at this
point. This encompasses everything from using the best available technology
for coal-burning plants to the production and purchase of fuel-efficient
vehicles down to the selection of light bulbs.

Technological fixes already have considerable cultural resonance; the
desire for such solutions reflects deeply held values in Western culture.

Hollywood stars have recently made it trendy to own fuel-efficient hybrid vehicles like the Toyota Prius, a sign that many more opportunities may be there for the taking. Here it is imperative to avoid past mistakes and not sell energy efficiency as a return to the simple life. Instead, energy efficiency can be an involving practice that requires bottom—up campaigns to increase voluntary actions and adoption of more efficient tools and practices, as well as top—down mandatory standard setting and incentive programs (see also Dilling and Farhar, Chapter 23, this volume). With recycling, much of the push to develop and extend programs came from ordinary people, as it afforded them opportunities for concrete involvement and understanding. The curbside recycling box became a common sight and drove home the beneficial impacts of individual actions. The halocline days of the SUV seem to have passed, and the advent of much smaller hybrids in every driveway promises to ease air pollution, greenhouse gas emissions, and even parking problems. Ultimately, however, the trick will not be done by technological fixes but in the emergence of relevant issue cultures and potent bridging metaphors that engage a wide population in the necessary behavior and political change. It would be nice if one could suggest what these might be, but they are best regarded as emergent elements of ongoing efforts to confront the threat of climate change.

Notes

1. Claims-making is central to the sociology of social movements, since in the absence of successful claims about a (deteriorating) condition, the problem will not be recognized and hence acted upon. Claims-makers can be individuals and groups and can include scientists, environmentalists, politicians, policy-makers, and so on. Key issues include the power of claims-makers, the nature of their claims, and their strategies for pressing them.
2. The increased media attention to the severe storm—climate change link during the hurricane series in 2004 may be the first indication of this changing, but whether or not the public discourse shifts to a greater orientation toward the here and now remains to be seen.
3. NIMTOO — a common acronym for Not In My Term Of Office — further illustrates how political actors treat even short-term future considerations with less significance than the more "vivid" and "pressing" problems operating in real time.
4. Scientists use "fingerprint" to capture the difficulty of finding the human signature. People are more likely to notice a footprint, and a large one at that.
5. In addition to the "companion industry," which produced the means to save oneself from harmful UV rays, another industry emerged around the technological innovations that allowed replacement of ozone-destroying substances, albeit somewhat less visible than the safety products industry.
6. Climate catastrophes were termed "Klima-SuperGAU," a term still resonant from the hey-days of the anti-nuclear movement in Germany, which had coined the term "GAU" — greatest imaginable nuclear accident.
7. So strong is the popular culture in North America that it took the rather ridiculous film, *The Day After Tomorrow*, to spawn some public discussion of abrupt climate change (Leiserowitz, 2004). Note also that an increasingly politicized debate was spawned in late 2004 and 2005 over the linkage between hurricanes and climate change as two rather

visible scientists clashed over the issue in public, but the issue has not (yet) gained wider public attention.
8. For example, see http://www.abi.org.uk/climatechange (accessed January 5, 2006) for recent studies and press releases dealing with climate change and the British Insurance industry.

References

Bernard, H. (1993). *Global Warming Unchecked: Signs To Watch For*. Bloomington, IN: Indiana University Press.

Boykoff, M. and Boykoff, J. (2004). Balance as bias: Global warming and the U.S. prestige press. *Global Environmental Change*, **14**, 125−36.

Clemens, E. (1986). Of asteroids and dinosaurs: The role of the press in the shaping of scientific debate. *Social Studies of Science*, **16**, 421−56.

Cline, W. (1992). *The Economics of Global Warming*. Washington, DC: Institute for International Economics.

Cortese, A. (2002). As the earth warms, will companies pay? *The New York Times*, August 18, p. A1.

Dunlap, R. and Scarce, R. (1991). The polls − poll trends: Environmental problems and protection. *Public Opinion Quarterly*, **55**, 651−72.

Freudenberg, W. and Pastor, S. (1991). Public responses to technological risks: Toward a sociological perspective. *Sociological Quarterly*, **33**, 389−412.

Leiserowitz, A. (2004). Before and after *The Day After Tomorrow*: A U.S. study of climate change risk perception. *Environment*, **46**, 9, 22−39.

Moser, S. C. and Dilling, L. (2004). Making climate hot: Communicating the urgency and challenge of global climate change. *Environment*, **46**, 10, 32−46.

Shamos, M. (1993). *The Myth of Scientific Literacy*. New Brunswick, NJ: Rutgers University Press.

Taverne, D. (2005). Political climate. *Prospect*, **113**, August, 28−31.

Turkle, S. (1999). Looking toward cyberspace: Beyond grounded sociology. *Contemporary Sociology*, **28**, 643−8.

Ungar, S. (1998). Bringing the issue back in: Comparing the marketability of the ozone hole and global warming. *Social Problems*, **45**, 510−27.

Ungar, S. (1999). Is strange weather in the air?: A study of US national news coverage of extreme weather events. *Climatic Change*, **41**, 133−50.

Ungar, S. (2000). Knowledge, ignorance and the popular culture: Climate change versus the ozone hole. *Public Understanding of Science*, **9**, 297−312.

Weingart, P., Engels, A., and Pansegrau, P. (2000). Risks of communication: Discourses on climate change in science, politics, and the mass media. *Public Understanding of Science*, **9**, 261−83.

5

The challenge of trying to make a difference using media messages

Sharon Dunwoody
University of Wisconsin—Madison

When social problems erupt, one classic response of governments and organizations is to wade in with an information campaign. From automobile seat belts to AIDS to recycling, policy-makers wage war on our inappropriate behaviors with newspaper stories, brochures, and public service announcements.

The goals are often noble ones, the dollars spent gargantuan, and the outcomes all too predictable: Messages seem to change the behaviors of some people some of the time but have almost no discernible impact on most people most of the time (McGuire, 1986; Hornik, 1989).

The situation has so discouraged policy-makers in the past that the pattern was given its own, dismal label: minimal effects (for perhaps the earliest articulation of this name, see Lazarsfeld, Berelson, and Gaudet, 1944). If media messages have no impact, policy-makers opined, why bother?

Today's communication scholars would agree with yesterday's policy-makers that media messages are often poor catalysts for behavior change, but many would disagree with that "minimal effects" label. Mediated messages can have pronounced effects, they would suggest, just not the ones envisioned by those who design them.

As countries around the globe face the prospect of encouraging massive behavior changes in order to try to prevent or, at the least, cope with climate change, it will be tempting to resort to information campaigns. I would urge us to succumb to that temptation; after all, media campaigns offer dramatic economies of scale by reaching large audiences at relatively low cost with potentially useful information.

But campaigners must become savvy about the roadblocks that they will encounter along the way. In this brief chapter, I will reflect on some of those roadblocks. They will dominate this narrative, as some of the most important barriers to effective use of the mass media for social change are our naïve

89

assumptions about cause and effect in these matters. At the end of this cautionary tale, however, I will suggest some roles that mass media channels could realistically play within the setting of global climate change information campaigns.

Roadblock 1: Existing beliefs will have a stronger impact on your message than vice versa

Beliefs are cobbled together from what we know and how we feel about things. Sometimes that cognitive/affective combination is newly minted or, for other reasons, not intensely maintained, making it amenable to change. For instance, although I have rented a safety deposit box for years from my bank, I don't have strong beliefs about its utility and am open to information that might convince me otherwise.

But if the belief is robust − the outcome of years of pondering, personal experiences, conversations with others, habitual behaviors − then messages that collide with that belief may bounce helplessly off. Even worse, individuals who believe something strongly and passionately may mentally "reconstruct" a contrary message in order to render it consonant with their sense of what is right and true.

For example, I maintain a strong belief that recycling is a good and environmentally important act. That belief has been bolstered by years of recycling behaviors, conversations with other recyclers, and an increasing financial commitment to recycling paraphernalia, among them outdoor compost bins and indoor bins built right into my new kitchen. A few years ago, I ran into a spate of articles by skilled science writers about the complexities of recycling that called into question recycling's economic payoff, its ability to minimize the amount of trash going into landfills and even the assumption that landfill space itself is increasingly scarce. I keenly recall my discomfort as I read these articles, then remember a reaction that I now characterize as the typical robust-believer response: rejection. I threw the articles away and now cannot even remember who wrote them or where they were published. (For an extended discussion of factors underlying beliefs and attitudes, see Eagly and Chaiken, 1993.)

The imperviousness of strong beliefs makes them difficult to influence. Information campaigners need to understand the belief structures of their audiences well enough to turn to audience subsets whose beliefs are either already somewhat consonant with the message or are pliable enough to make change possible. It can be painful to "write off" strong believers whose attitudes collide with one's message, but it is important to put one's resources into campaigns that have a fighting chance of working.

For example, an environmental NGO seeking to increase the number of Americans who value biodiversity began its campaign some years ago by surveying the national public to better understand the variance in beliefs about biodiversity that were already in place. The research revealed sub-groups of Americans whose concerns for the environment were extremely low, making them poor candidates as biodiversity supporters. Among these "least likely supporters," for example, were individuals in whom religious involvement, political conservatism, and lack of engagement in public affairs had converged to create levels of indifference to biodiversity that would be resistant to change, so the NGO reluctantly set them aside (The Biodiversity Project, 1998). But the study also found individuals whose value systems offered fertile ground for campaign messages. Specifically, the survey data showed a strong link between interest in biodiversity and such moral beliefs as feeling a responsibility to future generations and feeling that humans have a responsibility to protect God's creations, nature among them. The organization is currently targeting individuals with these particular beliefs (see, e.g., The Biodiversity Project, 2002).

Roadblock 2: The strongest message effects may be unintended ones

When an information campaign fails to change most individuals' behaviors, policy-makers are prone to declare the effort a failure and glumly begin an adjustment to a "minimal effects" world. But it is possible — even likely — that the campaign did have an effect on the intended audience. It just may not have been the effect policy-makers were hoping to see.

What is that unintended effect? The strengthening of existing beliefs. Paradoxically, even those beliefs that conflict with the message may get stronger. For example, in 1984, Taiwan's state-operated power corporation, Taipower, announced plans for a new nuclear power plant. Concerned that then-recent incidents such as the 1979 nuclear accident at Three Mile Island in Pennsylvania had created a wary and fearful public, the government embarked on an information campaign to convince the Taiwanese citizenry that a new plant would be both useful and benign. Underlying the campaign was the assumption that messages would produce a strong effect, specifically a shift in beliefs about nuclear power that would support construction of the proposed plant.

Well, the campaign indeed had an effect, but not the one intended by the campaigners. Those individuals who were undecided about nuclear power at the beginning of the campaign did become more supportive. But for the bulk of the Taiwanese public who started the campaign with reasonably strong beliefs, either pro or con, the campaign served to intensify – not change – those beliefs (Liu and Smith, 1990).

That intensification process is a robust one in the information effects literature, lending support to the argument that it is critical to understand the nature and strength of audience beliefs before communicating with them. It also raises the important question of the extent to which policy-makers actually want to invite the perverse effect of strengthening rather than dampening inappropriate beliefs, as that is a likely outcome of an information campaign.

For many issues, however, strong beliefs will be in short supply. For example, in the Taiwanese nuclear power case, 27 percent of the sample questioned prior to the campaign expressed no opinion about nuclear power despite the campaign's temporal proximity to the TMI accident, at that time the worst nuclear power plant accident to have ever happened. And one major finding of The Biodiversity Project survey discussed above was that most respondents had no idea what biodiversity was! As I will note later, knowledge gain is an important outcome of most good information campaigns, so a world of minimal or moderate beliefs can be fertile territory for media messages.

Roadblock 3: All information channels are not created equal

While folks invest much time and effort in constructing what they hope will be the right messages in a campaign, they often give scant time to selecting the appropriate channels in which to embed those messages. Channels are the carriers, the work-horses of information dissemination. They can range from human beings (interpersonal channels) to a wide range of mass media, including magazines, radio, television, newspapers, and the World Wide Web (often categorized as mediated channels).

Many campaign designers select channels for political or economic reasons. Television reaches millions of viewers, so public service announcements look like good buys. *The New York Times* or *The Washington Post* will be particularly good at reaching science and environmental decision-makers, so such outlets may be preferred for those campaigns seeking to influence the small subset of Americans that political scientist Jon Miller judges to be scientifically literate and actively seeking science information (Miller, 1998).

While those decisions can be savvy ones, missing from all these calculations is another important metric: We lay individuals seem to employ different communication channels to achieve different purposes. Put another way, this is not a one-channel-fits-all world.

Communication researchers have been exploring this issue in recent years and are finding some interesting, counterintuitive patterns. For example, while it seems that lay audiences readily use mass media messages to learn something about a risk or issue, they appear to resist interpreting those messages as being about themselves personally. That, in turn, is a problem for information campaigners, who prefer the economies of scale provided by mass media dissemination yet typically want their messages to move individuals to action.

Here is an example: A common finding in the many studies of our fear of crime is that, while exposure to mass media crime stories influences a person's perception of the level of crime "out there," it has virtually no impact on the individual's perception of crime in his/her neighborhood. That is, the more crime stories you see on your local TV news, the greater the level of crime you perceive in your city or region. But those stories seem to have little influence on your perception of the level of crime in your immediate vicinity (Tyler and Cook, 1984).

Similarly, one would expect that media coverage of global warming might lead a person to a greater awareness of the prospects of such warming for Earth and for folks "out there" but would not influence that individual's sense either of the possible impacts of the greenhouse effect on her personally or of her impact on the larger process through actions she takes in her daily life. In her view, she would remain strikingly independent of those causal patterns, a kind of free agent whose actions remain impact-neutral (for data documenting this pattern, see the chapter by Leiserowitz, Chapter 2, this volume).

How do we maintain that impact-neutral status in a world full of messages telling us otherwise? When it comes to messages that convey negative information or that ask the recipient to make an effort to change his/her behavior (many risk messages would qualify), we seem to distinguish between "us" and "them" and then greatly restrict the array of channels that we would be willing to interpret as being "about us." In fact, in many studies, the channels that fall into the "us" side of the equation sum to just one: interpersonal. Put another way, the best information predictor of behavior change is not seeing a public service announcement on late-night television but talking to someone (Mutz, 1998).

Typical of the studies of this phenomenon was a survey conducted in Milwaukee, WI, to explore predictors of channel use about an environmental risk (Griffin *et al.*, 2000). The US EPA had determined that lead was a likely contaminant of the drinking water of residents in older homes and asked the city to communicate with those residents about the problem in order to prompt behaviors that might lower their risk of ingesting lead. The city complied by placing a brochure in the next water bill: the brochure, a mediated channel, identified the problem and offered a number of risk-reducing behaviors (letting water run from the tap for a period of time before drinking, buying bottled water, etc.).

When the researchers sought linkages between channel use and behavior change, they found, predictably, that exposure to the brochure had virtually no effect. Most of the residents, in fact, had no memory of even receiving the brochure. But what did predict adopting safer behaviors was a visit, to the home, by a public health person: an interpersonal channel.

Thus, while mediated channels such as television and newspapers may reach millions and provide a cost-effective source of information about global climate change, they may not convince individuals that such changes will influence them *personally* or that they can do something *personally* about the problem. Having an impact on someone's personal beliefs and behaviors may require a different array of information channels. Specifically, the gold standard for behavior change remains interpersonal channels. If you want someone to change his beliefs or, even more dramatically, to change his behaviors in ways that are novel and at least initially inconvenient, the best advice that information campaigners can offer is "talk to him." Another aspect of this interpersonal contact is support and accountability, which is why behavior change approaches in ecoteams work (see the chapters by Tribbia, and Rabkin with Gershon, Chapters 15 and 19, this volume).

Roadblock 4: Experiences matter more than data

Science is evidence-based. Those individuals trained as scientists learn to privilege evidence and to employ "data" when trying to be persuasive. It just seems obvious that a well-done study, or five or six, will carry the day when arrayed against emotion, politics, or ignorance. If someone is skeptical about the contributions of human beings to the greenhouse effect, then let the data speak for themselves.

Ah, if only it were so! But psychologists have been telling us for years that personal experiences will trump data almost every time (see, e.g., Gilovich, 1991; Nisbett *et al.*, 1982; Zillmann and Brosius, 2000). Nisbett and Ross (1980) identify such characteristics as vividness and concreteness as important to the power of the anecdote. A vivid, concrete experience is more easily recalled and seems to carry more weight than "pallid" data with no discernible personality, something also found repeatedly in risk perception studies. It is no accident that, in a society awash in cancer statistics, the NCI cancer hotline lights up when someone famous is publicly diagnosed or that an extreme weather event (the searingly hot summer of 1988 in the United States, for instance, or the 2004 and 2005 US hurricane seasons) produces a discernible spike in both public and media interest in global climate change (Trumbo, 1995; see also Ungar, Chapter 4, this volume).

Psychologists link the power of the anecdote to the larger domain of intuitive judgments and to the power of perceptions. Note Kahneman and Frederick, "The boundary between perception and judgment is fuzzy and permeable: The perception of a stranger as menacing is inseparable from a prediction of future harm" (2002: 50).

One example of this emphasis on the anecdote within the domain of risk communication occurred some years ago in Wisconsin, when the US EPA mandated conversion from regular to reformulated gas at Milwaukee gas stations in order to alleviate air pollution. As consumers pumped the new gas, health complaints burgeoned. Local Milwaukee television stations took notice and covered the disgruntled gas station patrons extensively; the governor even requested that EPA withdraw its reformulated gas requirement. When a later state epidemiological study demonstrated no relationship between pumping reformulated gas and becoming ill, an analysis of television coverage of that study showed that many journalists still believed the sick consumers more than the data (Trumbo *et al.*, 1998).

Vivid, anecdotal experiences can help or hurt a cause. A hot summer may contribute to growing public concern about global warming while a cool spell may detract (see also Bostrom and Lashof, Chapter 1, this volume). The power of the anecdote does suggest, however, that campaigners need to build appropriate, concrete examples into their campaigns to have an effect. The operational definition of "appropriate" here would be anecdotes that are consistent with the best available scientific evidence. For instance, when an international consortium of researchers released a major report on the effect of global warming on the Arctic in late fall 2004 (ACIA, 2004), the team made available a variety of graphics and photographs that could accompany stories about the report. The extensive media coverage that followed took full advantage of the vivid images, a strategy that probably enhanced the impact of the stories on readers and viewers.

Roadblock 5: Audiences who specialize in "fast and frugal" information processing are poor targets for complex science explanations

Although we like to think of ourselves as thoughtful and deliberate decision-makers, we humans are known for our superficial information processing and quick decision-making (Gilovich, Griffin, and Kahneman, 2002; see also Moser, Chapter 3, this volume). Often busy to the point of distraction, we course like greyhounds over the surface of most of the information we see, stopping occasionally for a quick look at a provocative sentence or image but always resuming a pace that permits retention of snippets of information rather than chunks of meaning. We move equally quickly to decisions, whether as trivial as selecting a brand of pasta or as important as making a judgment about a health recommendation.

Even when we feel a need to know, an important precursor to more effortful information gathering and processing, we may resort to the Cliff Notes version of reality by relying on such heuristic aids as expertise ("If the doctor tells me to do something, that's good enough for me"), redundancy ("If the same issue appears in several newscasts or newspapers, it must be important"), and our existing knowledge of an issue, however incomplete (Bostrom, Fischhoff, and Morgan, 1992; Cialdini, 1993; Bostrom et al., 1994; Morgan et al., 2001; Kahlor, Dunwoody, and Griffin, 2004). Research suggests that individuals can be motivated to seek and process information more systematically when they judge a situation to be of significant personal importance (Chaiken, 1980) and by a sense that they know less than they need to know about that situation (Griffin et al., 2004). But it appears that our default mode is heuristic processing, a process of relying on a small information base and using clues as shortcuts through complex information.

German psychologist Gerd Gigerenzer acknowledges this default strategy, but, while many researchers define it as too superficial to get the job done well, he argues that "fast and frugal" processing works just fine in most instances. Heuristic processors indeed tend to rely on a small set of information characteristics, but, Gigerenzer argues, those characteristics are not haphazardly chosen; instead, they are often the product of longstanding social experiences. A reliance on an expert, for instance, means that an individual is behaving heuristically in the sense that she is not seeking a full range of information options to consider before making up her mind or making a decision. But experience has demonstrated that experts are often better equipped than the rest of us to make decisions within their areas of expertise. Thus, relying heavily on an expert means that your final decision is more likely to be a good one than a bad one (Gigerenzer and Goldstein, 1996; Gigerenzer, Czerlinski, and Martignon, 2002).

Fast and frugal processing is not a good recipe, however, for understanding complex concepts and processes. It does not tolerate uncertainty well. A heuristic information processor will want answers, not explanations or hedging. And studies of superficial information seekers and processors suggest that their resulting beliefs and attitudes are more volatile, leaving them vulnerable to new information that contradicts what they may already know (Petty, Priester, and Brinol, 2002). In a media campaign environment, the best way to manage this volatility is through redundancy: sharing consistent messages repeatedly, over a long period of time. That sounds like a simple recipe for success, but the high cost of campaigns looms as a significant limiting factor.

The heuristic predisposition of climate change audiences can work for or against an information campaign manager. Our willingness to believe "experts" is an advantage in a consensus environment. However, dueling experts will confuse a heuristic processor. One study of individuals' reactions to newspaper stories about climate change that contained contrasting viewpoints found that, rather than concluding that one of the experts may be more "right" than the others, the respondents deemed them all to have legitimate claims to truth and concluded that, thus, "no one knows what's true" about climate change (Rogers, 1999). More recent studies (Corbett and Durfee, 2004) support the tendency of contrasting viewpoints in global warming stories to increase public perceptions of the uncertainty of the science (for a more extensive reflection of media coverage of uncertainty, see Friedman, Dunwoody, and Rogers, 1999).

One all-too-frequent policy reaction to this human tendency to default to heuristic information processing is to abandon heuristic seekers and processors and to concentrate, instead, on communicating with policy-makers and other information elites, who are more likely to process systematically and, thus, to construct and maintain stable beliefs systems about global climate change. But giving up on the bulk of the American public ignores the possibility that individuals can become systematic processors under the right circumstances. The trick lies in helping individuals to develop a strong need to know — about global climate change in this instance — and equipping them with a perception that their individual acts may actually ameliorate the problem (see also Tribbia, Chapter 15, this volume).

In most cases, the strength of an individual's need to know will depend on the conviction that a particular problem or issue is relevant to her personally. For example, in the dead of winter, many of us in temperate climates open our monthly utility bills with genuine trepidation. Whether we care or not about global climate change, we have a keen need to know about energy issues in those frigid times. A global climate change information campaign could take advantage of such moments by providing information not only about affordable energy conservation options but also about the long-term impacts of continued use of fossil fuels.

Roadblock 6: The power of journalism to influence perceptions of big, long-term issues such as global climate change is muted in a landscape rich with other influences

It is tempting to reduce cause and effect to a simple duet: persuasive information equals behavior change. As I have tried to point out above, persuasion is difficult under any circumstances, particularly when one is trying to persuade via the mass media. But to make things even more complicated, even the most skillful media messages face formidable competition from everyday life.

Even when we claim to be riveted on a media message, we aren't. Studies of individuals "paying close attention" to a television program, for instance, reveal a distinctly different pattern. Although our eyes occasionally focus on the screen, we spend most of our time doing just about everything else: reading, eating, talking, getting to our feet to wander elsewhere in the house, dozing. Studies of what we learn from the typical television news program, as a result, find that comprehension is poor. Most of us cannot even recall a news item from the previous evening's news fare (Robinson and Levy, 1986).

Researchers who specialize in understanding "informal learning" about science and technology, the kind of learning that takes place in real-world settings such as living rooms and museums, have long understood the need to confront the complexities imposed on process by other factors in an individual's environment. How does one measure the learning that takes place from a single exhibit in the context of a person's day-long visit to a science museum (Falk and Dierking, 2000)? What kind of realistic learning expectations can the producers of a NOVA program have (Crane *et al.*, 1994)?

The take-home message of all this work is that no one snippet of information will make much of a difference. Some scholars suggest that the cumulative weight of many messages over long periods of time may well be greater than the sum of the parts. For example, decades of research have shown that a lifetime of heavy television viewing seems to "cultivate," in individuals, a perception of reality similar to that presented on television as compared to the reality of every day (for a summary of this work, see Gerbner *et al.*, 2002). But even in these longitudinal (and very expensive) studies, the variance in perceptions accounted for by media factors is small. The role played by information, even over the long haul, seems modest indeed compared to other – probably experiential – factors. At least one set of studies of heavy television viewers found that exposure to environmental stories generally left these individuals *less* knowledgeable about environmental issues than were lighter TV viewers, as well as more fearful about specific environmental problems (Shanahan, Morgan, and Stenbjerre, 1997).

So, what's the good news?

In spite of these and other roadblocks, information campaigns abound. The sheer ubiquity of commercial advertising in our culture also suggests that using mediated channels to promote behavioral outcomes must succeed in some cases. So let's turn briefly to a discussion of some relationships between media messages and cognitive, affective, or behavior outcomes that are realistic and achievable.

- Economies of scale will sometimes produce enough behavior change in a "minimal effects" world. Commercial advertising, it turns out, rarely has a strong effect in the sense of influencing the behaviors of most of the individuals who encounter it. But it doesn't need to. Even a small effect, at a national or international scale, will mean that millions of people will behave in the desired way. The proportionately small response to an ad campaign is literally large enough, in a profit-making sense, to justify the high cost of the campaign itself. Other kinds of campaigns can similarly benefit from these economies of scale.
- Influencing a few people may be a precursor to influencing many. When it comes to risk issues, one of the most important predictors of behavior change is an individual's perception that relevant others expect him to care about the issue and to behave appropriately. Called "subjective norms," these expectations seem to matter most in situations that have no obvious personal repercussions, that is, in situations where worry, fear, joy, and other emotions are muted. In the case of a global warming information campaign, then, if a person cannot grasp the personal implications of the greenhouse effect, she may still be responsive to pressure from relevant others who may have been influenced by the campaign.
- Media messages are important signaling devices. One of the most successful research domains for communication scholars is the study of media agenda-setting, the tendency of media news accounts to alert audiences to issues or problems (McCombs and Bell, 1996). Studies of reasons for attending to the mass media find that surveillance – a person's need to keep track of what is happening

in her environment — ranks high on the list of motivations (McLeod and Becker, 1981).

The agenda-setting effect is at its most potent, not surprisingly, when the media communicate about novel phenomena, and much of science certainly qualifies as novel to much of the public. Trumbo's longitudinal analysis of media coverage of global warming from the mid-1980s through the mid-1990s, for example, found that television stories had strong and immediate impacts on the public's perception of global warming as a serious issue (Trumbo, 1995).

- Next to influencing judgments of issue importance, the strongest direct effect of media messages is probably knowledge gain. Time and again, communication scholars have gone looking for attitudinal and behavioral impacts of media messages and have found, instead, modest increases in knowledge. A good example is the extensive literature that has examined the effects of presidential debates. These widely viewed events have a small but discernible impact on our collective understanding of candidate positions on issues but seem ineffective at changing attitudes or behavioral intentions (see, e.g., Sears and Chaffee, 1979; Chaffee, Zhao, and Leshner, 1994). Most studies of message effects find a stronger impact of print messages than of broadcast messages on knowledge gain.

One salutary effect of communicating about global climate change, thus, is that individuals attending to the messages likely will learn something. And the tendency of media accounts to treat global warming as contested terrain may actually enhance learning. Scholars find evidence that controversy increases the salience of issues and prompts individuals to pay closer attention to messages (Tichenor, Donahue, and Olien, 1980). In other words, controversy turns audience members into systematic information processors, albeit for a short period of time.

- Don't underestimate the power of narrative. Media accounts — whether full-page ads or television documentaries — are, at base, narratives. Much recent research (see, e.g., Shanahan and McComas, 1999) suggests that narratives are important arbiters of meaning, at least for those individuals whose belief systems permit them to ingest the messages without transforming them. Message campaigners spend much time designing these narrative threads, and with good reason.

- Effects of media messages may be stronger on policy-makers than on the public. One fascinating outcome of the Trumbo longitudinal study (Trumbo, 1995) is that the strongest agenda-setting effects of media coverage of global warming occurred not with the public but with members of Congress. His time series analyses demonstrated systematic feedback loops between media stories and mentions of global warming in the Congressional Record. That is, creating policy apparently served as a catalyst for media stories, which in turn seemed to galvanize Congress to further action, which produced more media coverage, and so on.

In another study that lends support to the power of agenda-setting among experts, Phillips and colleagues compared the number of citations in the scientific literature to medical studies that had been both published in *The New England*

Journal of Medicine and then covered by *The New York Times* with citations to comparable *NEJM* studies that had not been picked up by *The Times*. Those studies "legitimized" by the newspaper received 73 percent more citations in the peer-reviewed literature than did the comparable studies (Phillips *et al.*, 1991).

It may be the case, in fact, that the *most* attentive audiences for mass media messages are policy-makers and elites. It appears that these individuals turn to the mass media as a way of keeping track of public opinion, as well as to monitor the activities of other elites (Price, 1992).

- Media information campaigns promote a perception of social consequences. The flip side of people's unwillingness to see media messages as being about them personally is that such messages encourage perceptions of a problem as a collective, political one. That, in turn, may lead individuals to seek to take collective action in response. Notes Mutz, "Personal experiences and concerns tend to have very little political impact; instead, concern at the collective level drives the political consequences of social problems" (1998: 103).
- And if you don't want to give up on persuading individuals to change their behaviors? Information campaigns may be most valuable when other components of society add their voices to the call for change and when the campaigners can afford to settle in for the long haul. Anti-smoking campaigns have been under way since the early 1950s and, although no one campaign seems to have had much of an effect (see, e.g., Peterson *et al.*, 2000), the accumulated weight of 40 years of campaigns has helped produce a kind of cultural shift in perceptions of smoking as a social act. Although the cost of cigarettes remains the most powerful predictor of smoking cessation, we now have a public that is well informed about the dangers of smoking and smokers who at least express an interest in quitting (McAlister *et al.*, 1981).

Will it take another 40 years to sensitize Americans to the perils of a rapidly warming climate? Probably not, but a quicker cultural response to the problem will require not just an understanding of the effects of media messages, but also the long-term commitment of both individuals and institutions to attitudinal and behavioral change. In a culture that privileges individual decision-making, that is a substantial commitment indeed.

References

Arctic Climate Impacts Assessment (ACIA) (2004). *Impact of a Warming Arctic: Arctic Climate Impact Assessment*. Cambridge, UK: Cambridge University Press.

Bostrom, A., Fischhoff, B., and Morgan, M. G. (1992). Characterizing mental models of hazardous processes: A methodology and an application to radon. *Journal of Social Issues*, **48**, 4, 85–100.

Bostrom, A., Morgan, M. G., Fischhoff, B., *et al.* (1994). What do people know about global climate change?: 1. Mental models. *Risk Analysis*, **14**, 6, 959–70.

Chaffee, S. H., Zhao, X., and Leshner, G. (1994). Political knowledge and the campaign media of 1992. *Communication Research*, **21**, 3, 305–24.

Chaiken, S. (1980). Heuristic versus systematic information processing and the use of source versus message cues in persuasion. *Journal of Personality and Social Psychology*, **39**, 752–66.

Cialdini, R. B. (1993). *Influence: Science and Practice*, 3rd edn. New York: HarperCollins.

Corbett, J. B. and Durfee, J. L. (2004). Testing public (un)certainty of science: Media representations of global warming. *Science Communication*, **26**, 2, 128–51.

Crane, V., Nicholson, H., Chen, M., *et al.* (1994). *Informal Science Learning: What the Research Says About Television, Science Museums, and Community-Based Projects*. Ephrata, PA: Science Press.

Eagly, A. H. and Chaiken, S. (1993). *The Psychology of Attitudes*. Fort Worth, TX: Harcourt Brace Jovanovich.

Falk, J. H. and Dierking, L. D. (2000). *Learning from Museums: Visitor Experience and the Making of Meaning*. Walnut Creek, CA: AltaMira Press.

Friedman, S. M., Dunwoody, S., and Rogers, C. L. (1999). *Communicating Uncertainty: Media Coverage of New and Controversial Science*. Mahwah, NJ: Erlbaum.

Gerbner, G., Gross, L., Morgan, M., *et al.* (2002). Growing up with television: Cultivation processes. In *Media Effects*, 2nd edn., eds. J. Bryant and D. Zillmann. Mahwah, NJ: Erlbaum, pp. 43–67.

Gigerenzer, G., Czerlinski, J., and Martignon, L. (2002). How good are fast and frugal heuristics? In *Judgment Under Uncertainty: Heuristics and Biases*, eds. D. Kahneman, P. Slovic, and A. Tversky. Cambridge, UK: Cambridge University Press, pp. 559–81.

Gigerenzer, G. and Goldstein, D. G. (1996). Reasoning the fast and frugal way: models of bounded rationality. *Psychological Review*, **103**, 650–69.

Gilovich, T. (1991). *How We Know What Isn't So*. New York: Macmillan.

Gilovich, T., Griffin, D., and Kahneman, D. (eds.) (2002). *Heuristics and Biases: The Psychology of Intuitive Judgment*. Cambridge, UK: Cambridge University Press.

Griffin, R. J., Dunwoody, S., Dybro, T., *et al.* (2000). The relationship of communication to risk perceptions and preventive behavior related to lead in tap water. *Health Communication*, **12**, 1, 81–107.

Griffin, R. J., Neuwirth, K., Dunwoody, S., *et al.* (2004). Information sufficiency and risk communication. *Media Psychology*, **6**, 23–61.

Hornik, R. (1989). The knowledge–behavior gap in public information campaigns: A development communication view. In *Information Campaigns: Balancing Social Values and Social Change*, ed. C. T. Sandman. Thousand Oaks, CA: Sage, pp. 113–38.

Kahlor, L., Dunwoody, S., and Griffin, R. J. (2004). Predicting knowledge complexity in the wake of an environmental risk. *Science Communication*, **26**, 1, 5–30.

Kahneman, D. and Frederick, S. (2002). Representativeness revisited: Attribute substitution in intuitive judgment. In *Heuristics and Biases: The Psychology of Intuitive Judgment*, eds. T. Gilovich, D. Griffin, and D. Kahneman. Cambridge, UK: Cambridge University Press, pp. 49–81.

Lazarsfeld, P. F., Berelson, B., and Gaudet, H. (1944). *The People's Choice: How the Voter Makes Up His Mind in a Presidential Campaign.* New York: Duell, Sloan and Pearce.

Liu, J. T. and Smith, V. K. (1990). Risk communication and attitude change: Taiwan's national debate over nuclear power. *Journal of Risk and Uncertainty*, **3**, 332–49.

McAlister, A., Ramirez, A., Galavotti, C., *et al.* (1981). Antismoking campaigns: Progress in the application of social learning theory. In *Public Communication Campaigns*, 2nd edn., eds. R. E. Rice and C. K. Atkin. Thousand Oaks, CA: Sage, pp. 291–307.

McCombs, M. E. and Bell, T. (1996). The agenda-setting role of mass communication. In *An Integrated Approach to Communication Theory and Research*, eds. M. Salwen and D. Stacks. Mahwah, NJ: Erlbaum, pp. 93–110.

McGuire, W. J. (1986). The myth of massive media impact: Savagings and salvagings. *Public Communication and Behavior*, **1**, 173–257.

McLeod, J. M. and Becker, L. B. (1981). The uses and gratifications approach. In *Handbook of Political Communication*, eds. D. D. Nimmo and K. R. Sanders. Thousand Oaks, CA: Sage, pp. 67–97.

Miller, J. D. (1998). The measurement of civic scientific literacy. *Public Understanding of Science*, **7**, 203–23.

Morgan, M. G., Fischhoff, B., Bostrom, A., *et al.* (2001). *Risk Communication: A Mental Models Approach.* Cambridge, UK: Cambridge University Press.

Mutz, D. C. (1998). *Impersonal Influence: How Perceptions of Mass Collectives Affect Political Attitudes.* Cambridge, UK: Cambridge University Press.

Nisbett, R., Bordiga, E., Crandall, R., and Reed, H. (1982). Popular induction: Information is not necessarily informative. In *Judgment Under Uncertainty: Heuristics and Biases*, eds. D. Kahneman, P. Slovic, and A. Tversky. Cambridge, UK: Cambridge University Press, pp. 102–16.

Nisbett, R. and Ross, L. (1980). *Human Inference: Strategies and Shortcomings of Social Judgment.* Englewood Cliffs, NJ: Prentice-Hall.

Peterson, A. V., Kealey, K. A., Mann, S. L., *et al.* (2000). Hutchinson Smoking Prevention Project: long-term randomized trial in school-based tobacco use prevention – results on smoking. *Journal of the National Cancer Institute*, **92**, 24, 1979–91.

Petty, R. E., Priester, J. R., and Brinol, P. (2002). Mass media attitude change: implications of the Elaboration Likelihood Model of persuasion. In *Media Effects: Advances in Theory and Research*, eds. J. Bryant and D. Zillmann. Mahwah, NJ: Erlbaum, pp. 155–98.

Phillips, D. P., Kanter, E. J., Bednarczyk, B., *et al.* (1991). Importance of the lay press in the transmission of medical knowledge to the scientific community. *The New England Journal of Medicine*, **325**, 1180–83.

Price, V. (1992). *Public Opinion.* Thousand Oaks, CA: Sage.

Robinson, J. P. and Levy, M. R. (1986). *The Main Source: Learning from Television News.* Thousand Oaks, CA: Sage.

Rogers, C. L. (1999). The importance of understanding audiences. In *Communicating Uncertainty*, eds. S. M. Friedman, S. Dunwoody, and C. L. Rogers. Mahwah, NJ: Erlbaum, pp. 179–200.

Sears, D. O. and Chaffee, S. H. (1979). Uses and effects of the 1976 debates: An overview of empirical studies. In *The Great Debates: Carter vs. Ford, 1976,* ed. D. Kraus. Bloomington, IN: Indiana University Press, pp. 223–61.

Shanahan, J., Morgan, M., and Stenbjerre, M. (1997). Green or brown? Television's cultivation of environmental concern. *Journal of Broadcasting & Electronic Media*, **41**, 305–23.

Shanahan, J. and McComas, K. (1999). *Nature Stories: Depictions of the Environment and Their Effects.* Kresskill, NJ: Hampton Press.

The Biodiversity Project (1998). *Engaging the Public on Biodiversity: A Road Map for Education and Communication Strategies.* Madison, WI: The Biodiversity Project. Available at www.biodiversityproject.org/roadmap.htm.

The Biodiversity Project (2002). *Ethics for a Small Planet: A Communications Handbook on the Ethical and Theological Reasons for Protecting Biodiversity.* Madison, WI: Biodiversity Project.

Tichenor, P. J., Donohue, G. A., and Olien, C. N. (1980). *Community Conflict & the Press.* Thousand Oaks: Sage.

Trumbo, C. (1995). Longitudinal modeling of public issues: An application of the agenda-setting process to the issue of global warming. *Journalism & Mass Communication Monographs*, No. 152.

Trumbo, C. W., Dunwoody, S., and Griffin, R. J. (1998). Journalists, cognition, and the presentation of an epidemiologic study. *Science Communication*, **19**, 3, 238–65.

Tyler, T. R. and Cook, F. L. (1984). The mass media and judgments of risk: Distinguishing impact on personal and societal level judgments. *Journal of Personality and Social Psychology*, **47**, 693–708.

Zillmann, D. and Brosius, H.-B. (2000). *Exemplification in Communication: The Influence of Case Reports on the Perception of Issues.* Mahwah, NJ: Erlbaum.

6

Listening to the audience: San Diego hones its communication strategy by soliciting residents' views

Linda Giannelli Pratt

*Office of Environmental Protection and Sustainability, City of San Diego
Environmental Services Department*

Sarah Rabkin

University of California–Santa Cruz

A city synonymous with its climate

In a city that calls its climate "perfect,"[1] where the coastline includes miles of America's most beautiful beaches, where temperatures are high almost all year round and annual total rainfall inches are measured in single digits,[2] concern about the dangers of climate disruption seem out of place. Can laid-back Southern Californians even relate to the threat of global warming, much less take action?

In fact, San Diego's enviable climate makes it a particularly apt setting for civic discussions about climate change. Two of the area's most important economic activities — tourism and agriculture[3] — rely on the exceptional climate to continue flourishing, yet are vulnerable to rising temperatures. A growing population puts increasing pressure on energy and water, resources that are primarily imported from far away. Attention to climate issues will be essential to sustaining the city's celebrated quality of life, including the availability of fresh air, clean water, and affordable utilities. Shifting temperature and rainfall may lead to a significant decrease in agricultural products and may diminish the region's desirability as a tourist destination.

Moreover, the global threat is not remote at all. Climate change at the local level is already well documented. The region's average temperature has been increasing over the past century, and the number of days with temperatures above 100 degrees Fahrenheit (38 °C) has risen even more sharply. Sea level along San Diego's coastline is rising at a rate of three to eight inches (7.5 to 20 cm) per century, and scientific projections put it another 13 to 19 inches (33 to 48 cm) higher by 2100 (California Energy

Commission, 2003; Hayhoe *et al.*, 2004). The implication of this rise in sea level includes a loss of popular beaches and more frequent flooding of shorefront areas.

Increasingly cognizant of these realities, city leaders have begun taking responsibility for San Diego's contributions to global climate change. Over an eight-year period beginning in 1994, they succeeded in reducing greenhouse gas emissions by reducing municipal energy and water consumption, improving fuel efficiency, and expanding the use of methane gas from sewage treatment plants and landfills. In doing so, they saved the city more than US$15 million over a period of 10 years (City of San Diego, 2005).

At the culmination of this period, in January, 2002, then-Mayor Dick Murphy and the City Council approved *San Diego's Greenhouse Gas Emission Reduction Plan*, later re-titled the *Climate Protection Action Plan* (City of San Diego, 2005). The plan establishes new climate protection targets in conjunction with the Cities for Climate Protection program (CCP) of the International Council for Local Environmental Issues (ICLEI) (see also Young, Chapter 24, this volume). Goals include, among others, reducing municipal fuel consumption, shifting to alternative fuels where appropriate, minimizing solid waste generation, planting urban vegetation, improving energy conservation throughout the city, and increasing alternative energy sources.

Success in most of these climate protection endeavors depends upon the involvement of local residents – whose cooperation, in turn, hinges on effective citizen outreach. And in this arena, San Diego has pioneered a simple, surprisingly rare, and potentially powerful strategy. Before attempting to educate and mobilize residents, the city tried to find out how they think.

An innovative strategy born of experience

In February and March of 2004, the Sustainable Community Program of the city's Environmental Services Department (ESD) surveyed residents about what it called "community sustainability indicators": a set of circumstances that affect quality of life in the region. Environmental Services staff wanted to learn what city residents actually knew – and what they thought they knew – about these local economic, social, and environmental conditions. They hoped to design, with survey results in hand, a re-calibrated communication strategy. Understanding how San Diegans perceived their home, city staff reasoned, would help them plan more effectively for the city's future.

The survey was born of experience. More than 20 years of community outreach efforts had endowed department staff with a certain wariness about making assumptions concerning public awareness. They had learned from a few mistakes that before organizing forums and producing educational materials, it's a good idea to do your homework.

For example, when local environmental agencies began engaging city businesses in pollution prevention efforts, their communication strategy initially emphasized altruistic motivations. Outreach materials and presentations focused, reasonably enough, on the benefits of becoming a "good community neighbor." To the surprise of agency staff, this approach failed utterly. Not until they began emphasizing the power of pollution reduction to save money did the business community start climbing aboard. The successful business communication strategy ultimately adopted by the department focused completely on dollars-and-cents arguments, with public-forum titles such as "More Cost-Effective Hazardous Waste Management." Pollution prevention became a secondary aside, slipped quietly into presentations and publications. Sometimes the best communication strategy is simply to cut to the bottom line.

Similar difficulties arose when the department began developing a city-wide community sustainability network. The background and training that staff from local environmental agencies brought to this task gave them a particular and deeply ingrained understanding of the concept of sustainability. These environmentally-oriented professionals were primarily concerned about air and water quality and other elements of the physical environment. For them, improving the sustainability of the region meant influencing the choices that people and institutions make at the residential, commercial, and municipal levels: choices having to do, for example, with environmentally preferable purchasing and energy-efficient transportation alternatives.

Community members who showed up at public forums, however, brought to the table a much broader definition of the term (compare the experience in Santa Monica, Watrous and Fraley, Chapter 25, this volume). For many of these citizens, social and economic issues such as health care and insurance trumped environmental concerns. And the scope of participants' vision ranged from simplistic to sweeping. For some, making the community sustainable meant picking up litter, while for others it involved developing major social service programs.

The city-wide "network" of citizens who care about the sustainability of the region needed a better anchor. Staff realized that they could not successfully bring citizens together to promote the city's "sustainability" without first expanding their own understanding of what that means to the community, and without developing strategies for engaging diverse constituencies.

The ESD stepped forward to develop a clear and detailed portrait of the community — a profile of what residents in various segments of the city knew, valued, feared, desired, and misunderstood about local environmental conditions. In consultation with other city programs, offices, and departments, ESD created a public survey designed to elicit this information. The topics covered included improving air quality, creating affordable housing, reducing pollution of beaches and bays, conserving energy, providing access to transit stops and retail shops from residential areas, and increasing use of public transit (bus, train, trolley). But the survey also included questions on how to reduce local poverty, reduce crime rates, improve local K-12 education, reduce traffic congestion, and conserve water.

The survey, entitled "Quality of Life in San Diego," was conducted via mail and the web and its results ultimately published in a July 2004 report, entitled *Which Way Do We Grow?*[4]

Survey design, distribution, and response

San Diego State University's Social Science Research Laboratory helped ESD design clear, unambiguous, consistently phrased, and non-leading questions that could generate useful responses. The laboratory also tabulated the results of the mailed survey, while an inexpensive commercial service handled online responses.

The survey was mailed to 3000 randomly selected addresses throughout all of the city's zip codes — with each zip-code area receiving a number of mailings proportional to its population. Recipients completed and returned more than 600 of the mailed surveys. The survey was also made available on the city's website, and the URL was e-mailed to various community planning groups and listserves. More than 1,700 respondents replied online, for an overall total of just over 2,300 survey respondents — about half of them male and half female (see Textbox 6.1).

The array of "sustainable community indicators" from which survey questions were generated had emerged from a multi-year process, involving community forums and city staff development. The survey's primary funder, the California Public Utilities Commission, was mostly interested in the four areas that have a direct bearing on local and global climate: energy use, water use, air pollution, and transportation.

In connection with each of the areas of interest in the survey, respondents were asked three basic questions: "How important is this issue to you?"

Textbox 6.1 **Who participated in San Diego's sustainability survey?**

More than 2,300 San Diego residents took part in the sustainability survey. Demographic information was compiled according to variables such as ethnicity, age, income level, and gender. The information was divided into two groups, the mailed survey and the online survey respondents. As illustrated by the figures below, most respondents were white and 36–55 years old.

Ethnicity: Averaging the two survey groups, 69 percent were white, 7 percent were Asian/Pacific Islander, 7 percent were Hispanic/Latino, and 4 percent were Black.

Age: Averaging the two survey groups, 24 percent were 18–35 years old, 41 percent were between 36 and 55 years old, and 28 percent were older than 55 years.

Gender: Survey respondents were evenly split by gender, with 49 percent females and 45 percent males, with 6 percent not responding to the question.

Income: There was a broad representation of income levels from below US$20,000 per year extending to more than $100,000 annually. More than 27 percent of the online respondents indicated that they earn more than $100,000 per year, and 3 percent made less than $20,000 per year. For the mailed survey, 23 percent of the respondents earn between $60,000 and $99,999 per year, and 17 percent made more than $100,000 per year.

(with five possible answers ranging from *very important* through *neutral* to *very unimportant*); "How satisfied are you today with current conditions?" (with five possible responses ranging from *very satisfied* through *neutral* to *very dissatisfied*); and (for respondents in residence for five or more years) "How do you think this aspect of the quality of life has changed in the past five years?" (with five possible answers ranging from *much better* through *no change* to *much worse*).

The survey's results provide a telling snapshot of the ways in which San Diego's residents experience their quality of life. Some responses reveal significant divergences between residents' perceptions and actual circumstances — discrepancies that have already begun to help the city refine its outreach and communication strategies.

Energy

City staff expected that the majority of respondents would view energy conservation as a very important issue, expressing dissatisfaction with the current condition and a significant decline in quality over the past five years. Public outreach had been designed based on those assumptions. However, not even half of those surveyed rated conservation of San Diego's energy resources as very important. Most respondents thought conditions related to energy conservation were unchanged over the previous five years — years that in reality encompassed the dramatic and debilitating energy crisis that traumatized the city in 2000–2001. More than 40 percent of those who responded stated that they had no strong feelings one way or the other about current energy conservation levels — even though another wave of energy shortfalls was predicted for the summer of 2004, i.e., at the time when the survey was being administered.

These responses startled city staff. Many employees still nursed acute memories of crisis-time community meetings, at which poor and elderly residents complained of their inability to power their lights or refrigerators because of skyrocketing utility rates. People were rushed to hospitals for heatstroke during the summertime crisis. Many small businesses went bankrupt due to rising energy costs, and some larger ones left the city. In short, these survey findings forced city personnel to contemplate how residents could appear to have such a stunningly short-term memory for a serious crisis that had plagued the city for months.

During the crisis, when the threat of rolling blackouts loomed and utility charges tripled, the average number of kilowatt hours (kWh) per year used in San Diego homes decreased by 8 percent, from 6,007 kWh to 5,501 kWh. After the perceived end of the crisis, residential energy use resumed an upward climb; in 2003, it was 7 percent higher than in 2001.

San Diegans' surprising post-crisis ignorance and nonchalance about energy, as revealed in the survey, may be the result of a number of extraordinary conditions — such as the manipulation of the energy market by ENRON and other energy providers. It would be comforting to believe that energy is unlimited, that the crisis was the result of human wrong-doing, and thus the need for conservation only temporary. Unfortunately, that is not the case. Power generation capability, transmission line capacity, and an ever-increasing energy appetite throughout the state all impose limits on local energy security — as does the more global problem of continuing fossil-fuel-based energy production and the concomitant release of heat-trapping greenhouse gases. A growing population — with 100 000 new residents added between 2002 and 2004 alone, and more than one million additional residents expected by 2030 — exacerbates local energy demand (San Diego Association of Governments, 2004).

Water

The implications San Diego's need to import nearly 90 percent of its water from outside of the region have been raised many times in public forums. At the time of the survey, analysts assumed that the community would fully understand the need to conserve this precious resource. While the survey revealed that residents believed their city was conserving water with increasing efficiency, most lacked strong feelings one way or the other about water conservation as an important issue.

Water conservation constituted another significant case where public perception did not match the facts. San Diego is an increasingly thirsty city in an arid region. Residential water use rose by four gallons per day between 1995 and 2002, and total municipal water use increased even more, by some 24 percent.[5] Though this may not look significant at first glance, in a population of 1.2 million, simple math suggests that this represents a staggering annual city-wide increase of more than 1.75 *billion* gallons. At the same time, drought threatens unprecedented constrictions in future supply from the Colorado River, San Diego's primary water source. Moreover, rising temperatures cause early melting of Sierra Nevada snow packs, increasing flows during winter when existing reservoirs lack capacity to store it, yet reducing needed supply during the dry Southern California spring and summer months (Miller *et al.*, 2001; Hayhoe *et al.*, 2004).

Apparently, residents make little connection between their faucets and essential water sources hundreds of miles away; nor do they fully understand the value of this precious resource for home and business needs. From the survey report, city staff inferred a need to drive home the conservation message more forcefully. Citizens need to understand that seemingly small individual increases in resource use can translate into enormous city-wide impact, with potentially dire consequences.

The desire to avoid scaring or inconveniencing residents sometimes leads to overly optimistic public communications; this may have held true in the case of San Diego's outreach to water consumers. It is evident that city personnel need to put out stronger messages that applaud the success thus far with water conservation, but make clear the need for further concessions and cooperation.

Air

Public health can be a powerful motivator for serious changes in behavior (see Leiserowitz, Chapter 2, this volume). For that reason, linking issues such as childhood asthma and respiratory illness to air pollution from cars may create a strong incentive to reduce per capita vehicle miles traveled. However, the survey revealed that 50 percent of respondents were satisfied with the region's air quality, and improving the air quality ranked only sixth out of the 11 indicators listed in the survey. In other words, the lack of a perceived air quality problem rendered the link with public health would be harder to illustrate and/or less compelling. Indeed, much of the city's air pollution is hard to see, especially when compared with the infamous brown pall that frequently hunkers in the Los Angeles basin and western Riverside County, San Diego's neighbors to the north.

But the harmful pollution is there — much of it ground-level ozone and particulates generated by motor vehicle exhaust. In 1998, for example, the San Diego Air Basin exceeded the California standard for ground-level ozone on 54 days. The number of days declined every year until 2001, when the standard was exceeded on 13 days. In 2002, the number of days increased to 15, in 2003 it increased even more to 23, and in 2004 it declined to 12 days (County of San Diego Air Pollution Control District, 2005). If the temperature continues to increase, as is projected with climate change models, it will become even more difficult to reduce the ground-level ozone, which can aggravate respiratory conditions, interfere with the ability of plants to produce food, and damage materials such as rubber. About 1.3 million residents in the San Diego region — 43 percent of the total population — are especially vulnerable to ozone pollution. They include those with cardiovascular and chronic respiratory disease, individuals over 65, children under 14, and athletes.

San Diego's air pollution thus poses a special outreach challenge because the problem's severity so significantly outstrips its visibility and public concern with the issue. Efforts to encourage fuel efficiency, vehicle maintenance, and the use of transportation alternatives are unlikely to succeed unless outreach architects can find more effective ways to illustrate the dangers of inaction. The recent increase in fuel prices may bring about changes that ultimately reduce emissions. For example, large gas guzzlers, which also generate more vehicle emissions, are rapidly losing popularity (Wolk, 2005). Once again, the bottom line proves to be a powerful driver.

Transportation

Grousing about urban traffic has become a national pastime, so it comes as no surprise that a majority of San Diego survey residents believe traffic congestion in the city needs improvement. They're right. Both the number of registered vehicles per person and the number of annual vehicle miles traveled per person have increased steadily in San Diego since 1997. By 2004 the city had 0.82 cars per capita – representing a 5 percent increase in four years. In 2001, the average miles traveled per capita (based on the entire city population, not just drivers) was 9,000 – an increase of more than 300 miles from the 1997 level (SANDAG, 2003, 2005a,b).[6] Commute distances stretch as San Diegans live increasingly far from their workplaces.

What did surprise survey analysts was the public perception that use of public transit had been increasing. Perhaps this reflected wishful thinking. In fact, total ridership for the city's trolley, bus, and train systems had decreased in the five years, with a total in 1999 of 216,604 and 208,701 in 2003. Moreover, transportation mode choices varied by income, with wealthier San Diegans more frequently choosing travel in their personal vehicles and lower-income residents tending to choose public transportation more often (SANDAG, 2003). Traditional public transit may not be the best solution at this time for the San Diego region. According to the San Diego Association of Governments, vanpooling has risen consistently over the past four years, from about 600,000 vehicle trips reduced in 2000 to nearly 1.3 million vehicle trips reduced in 2004.[7] Another part of the solution is flexible working hours, coupled with telecommuting. These findings suggest that future communication strategies need to advertise a broader range of transportation alternatives to arouse interest and mobilize changes in travel behavior.

Lessons learned

In 1986, San Diego's Solid Waste Department changed its name to Environmental Services Department in order to embrace a far broader mission than trash collection and disposal. With those responsibilities still at its core, it now also provides a range of services including coordination of city-wide energy efficiency and renewable energy efforts, hazardous waste and asbestos control programs, recycling and reuse of materials, environmentally preferable purchasing, and development of policies to reduce the environmental footprint of city activities. Additionally, it includes the first municipally managed landfill in the country to win rigorous ISO 140001[8] environmental certification. The department's Sustainable Community Program, established in 2000, provides technical assistance and support to the city in its efforts to reduce greenhouse gas emissions.

A significant component of this charge involves reaching out to educate city staff, which numbers close to 10,000, and community members about water and energy conservation, recycling, environmentally preferable purchasing, and other sustainability-related issues. Currently, the outreach comprises four main modes of communication: a series of public forums on planning and environmental issues, some with big-name keynote speakers; a staff training program on fostering sustainable behavior; a website; and a "green schools" program that involves regular presentations at city high schools as well as an annual youth forum.

Sustainable Community Program staff who visit high schools regularly ask their audience whether they have heard of climate change. Usually half the students usually raise their hands. When speakers ask them to define the concept, however, their answers are not always accurate. Previous campaigns that effectively banned the use of chlorofluorocarbons in spray cans to protect the ozone layer are commonly mixed up with the greenhouse effect (see chapters by Bostrom and Lashof; and Leiserowitz, Chapters 1 and 2, this volume). Unfortunately, what the high school students know is typically a reflection of what is the common belief at home. At the same time, the high school program may be a promising avenue to reach much further into the local population, as kids bring what they learn in school home to their families.

This opportunity was underlined in the 2004 and 2005 Youth Forum, where the students were asked to complete the same Sustainability Indicators survey mailed to other San Diego residents. Interestingly, students seemed to be far less satisfied with the state of the environment. However, as might be expected, they were less able to distinguish how conditions had changed over the last five years, with most of them listing "no change." When the students saw the results and were asked for recommendations, they exhibited a distinct "just do it" attitude: if we need to conserve more water and energy, then tell us how much and we will do it; if we need to reduce vehicle emissions, we trust that technology will bring us the answers. This attitude both holds promise and demands caution: outreach efforts directed at younger San Diegans may well prove more effective and multiply by reaching many more older residents indirectly, yet younger people also need much further education about the opportunities and limits of technological fixes.

As city staff attempt to blaze inroads in a community still largely ignorant about climate change and its implications, the *Which Way Do We Grow?* survey report has already offered helpful guideposts — both to fine-tune its outreach strategies and to boost the city's efforts to clean up its own act.

For example, the city presented the findings of the survey to residents through a lengthy newspaper article in the San Diego *Union Tribune*, thus holding up a mirror to San Diegans about their own attitudes. The article also generated interest and quotes from a City Council member and Planning Commissioner, thus raising sustainability issues as a political issue.[9] A follow-up story with a slightly different emphasis ran in the San Diego *Daily Transcript*, a paper targeted toward the business community.[10] In total, more than 500,000 people received this information, and even if only a fraction read the article, awareness was raised among thousands.

The city continues its ongoing outreach efforts using the results of the survey. As in the case of outreach to businesses, the survey revealed that moral or altruistic arguments seemed less effective than arguments related to the "bottom line." Residents also seemed to make little connection between the environment (e.g., water sources) and consumption. In presentations, city staff now make a point of including these issues in public talks and events. Rather than trying to change citizens' fundamental value systems, for example, it may be more fruitful for city leaders to tie environmental and sustainability concerns to matters of personal well-being that residents find more pressing.

Another finding from the 2004 survey related to misperceptions concerning water and energy conservation, air quality, and traffic congestion. Average daily vehicle miles traveled increased and the use of conventional public transit decreased; water conservation was judged moderately important, but it is a critical issue for the city; energy conservation was believed to be unimportant and unchanged, yet the city and state went through a major energy crisis; air quality and public health were found not to be connected in people's minds, yet air pollution severely affects city residents. Rather than hoping residents might come to value energy efficiency and water conservation, city communication efforts now emphasize more strongly the compelling connections between business development and the reliability of energy and water supplies, and increasing home insurance costs associated with floods from extreme storms.

Moreover, the city decided to delve deeper in its 2005 follow-up survey "Water, Power and Public Health" into what people have done specifically to reduce their individual use of energy, water, and the types and numbers of vehicles in their household.[11] First results indicate that the public now views water and energy conservation as important to the future of this region. The 2005 survey asked what degree of responsibility people are actually accepting, and how this sense of responsibility translates into action. An additional new layer in the 2005 survey is a greater emphasis on public health, asking residents about their knowledge and attitudes about the link between air pollution, climate change, and health. While respondents still make few connections between air quality, climate and health, the city sees this as a new educational target for future outreach.

Finally, the results of the *Which Way Do We Grow?* survey have served as justification for a model policy for the vehicle fleet operated by the City of San Diego. This "Community Fuel Efficiency and Reduction Policy" applies to more than 2,000 passenger vehicles, and seeks to reduce fuel use by 5 percent per year. This would be achieved not only through the purchase of more fuel-efficient vehicles, but through reduction of vehicle trips. For example, teleconferencing, e-mails, and other modern communication methods can reduce the need to travel to meetings. If successful, this approach can be mirrored by other public and private entities with vehicle fleets.

Conclusion

More than one environmental protection professional has suggested in recent years that public understanding of, and support for reducing greenhouse gas emissions may be best garnered by sociologists and psychologists, rather than through the insights of chemists or engineers. If city governments are to help shift their citizens' behavior in more sustainable directions, they must learn not only to speak, but to listen, to residents. San Diego's unique survey models one way of initiating such two-way communication — and information obtained through the survey has begun to inform the city's outreach and planning. As cities develop more detailed and realistic visions of their constituents' concerns, they will be better equipped to meet residents where they live.

Notes

1. US Weather Bureau and *Holiday Magazine*, both via http://www.sandiego.gov/ economic-development/glance/quality.shtml; accessed January 9, 2006.
2. See http://www.sandiego.org/nav/Visitors/VisitorInformation/Weather; accessed January 9, 2006.
3. San Diego County ranks as the twentieth largest agriculture producer in the nation, with the value of agricultural production reaching nearly $1.5 billion in 2004 http://www.sdcounty.ca.gov/awm/docs/crop_reports/cropreport2004.pdf; accessed January 9, 2006). Tourism generated more than $5.5 billion in visitor spending for the county in 2004 (San Diego Convention & Visitors Bureau Statistics, *San Diego County 2004 Visitor Industry General Facts*).
4. The survey is available at www.sandiego.gov/environmental-services/sustainable/pdf/ grow.pdf; accessed January 9, 2006.
5. Data are taken from the *Where Do We Grow?* report.

6. Data also cited in the *Where Do We Grow?* report. Figures are compiled from data provided by the San Diego Association of Governments (http://www.sandag.org/resources/demographics_and_other_data/transportation/adtv/index.asp; accessed January 9, 2006) the California Department of Finance (http://www.dof.ca.gov/HTML/DEMOGRAP/repndat.htm; accessed January 9, 2006), and Counting California (http://countingcalifornia.cdlib.org/matrix/c198.html; accessed January 9, 2006).

The *2030 Regional Transportation Plan* states in ch. 3, "Throughout the 1980s, travel (as measured in vehicle miles traveled or VMT) grew about twice as fast as population, primarily because of growth in two-worker households and longer commute distances. During the 1990s, growth in vehicle miles traveled was 50 percent higher than population growth" (SANDAG, 2004).

7. See references in note 6.

8. ISO (the International Organization for Standardization), an independent non-governmental network of the national standards institutes of 148 countries, establishes quality, safety, and environmental standards for a range of products and services, and, on a voluntary basis, certifies businesses and organizations that meet those standards. ISO 14000-family standards focus on minimizing environmental impact and improving environmental performance.

9. Available at: http://www.signonsandiego.com/uniontrib/20040829/news_1h29survey.html; accessed January 9, 2006.

10. Available at: http://www.sandiego.gov/environmental-services/sustainable/pdf/052402.pdf; accessed January 9, 2006.

11. The 2005 survey results are available on the city's website at: http://www.sandiego.gov/environmental-services/sustainable/index.shtml.

References

California Energy Commission (2003). *Climate Change and California*. Sacramento, CA. Available at: http://www.energy.ca.gov/reports/2003-11-26_100-03-017F.PDF.

City of San Diego (2005). *Climate Protection Action Plan*. San Diego, CA. Available at: http://www.sandiego.gov/environmental-services/sustainable/pdf/action_plan_07_05.pdf; accessed January 9, 2006.

County of San Diego Air Pollution Control District (2005). *2004 Annual Report*. Available at: http://www.sdapcd.org/info/reports/ANNUAL.pdf; accessed January 9, 2006.

Hayhoe, K., *et al.* (2004). Emissions pathways, climate change, and impacts on California. *Proceedings of the National Academy of Sciences*, **101**, 34, 12422−7.

Miller, N. L., Bashford, K. E., and Strem, E. (2001). *Climate Change Sensitivity Study of California Hydrology: A Report to the California Energy Commission*. LBNL Technical Report 49110. Available at: http://www-esd.lbl.gov/RCC/outreach/Miller-Bashford-Strem.pdf; accessed January 9, 2006.

San Diego Association of Governments (SANDAG) (2003). *2030 Regional Transportation Plan*. Available at: http://www.sandag.org/index.asp?projectid=197&fuseaction=projects.detail; accessed January 9, 2006.

SANDAG (2004). *2030 Regional Growth Forecast*. San Diego, CA. Available at: http://www.sandag.org/uploads/publicationid/publicationid_1077_3212.pdf; accessed January 9, 2006.

SANDAG (2005a). *State of the Commute 2005*. Available at: http://www.sandag.org/uploads/projectid/projectid_257_3841.pdf; accessed January 9, 2006.

SANDAG (2005b). *Commute Characteristics: San Diego Region, 2004.* Available
    at: http://www.sandag.org/uploads/publicationid/publicationid_1118_3689.pdf;
    accessed January 9, 2006.
Wolk, M. (2005). Gas prices eat into sales of large SUVs. *MSNBC*, March 15.
    Available at: http://www.msnbc.msn.com/id/7181566/; last accessed January 9,
    2006.

7

The climate-justice link: communicating risk with low-income and minority audiences

Julian Agyeman
Tufts University
Bob Doppelt and Kathy Lynn
University of Oregon
Halida Hatic
Tufts University

Introduction

In this chapter, we briefly detail the growth of the US environmental justice movement and one of its offshoots, the international climate-justice movement. This movement is attempting to "put a human face" on climate change. The "Bali Principles of Climate Justice" shift climate change from a scientific–technical issue to one of human rights and environmental justice. We then look at how these issues can be communicated in disadvantaged communities: Roxbury, a predominantly African-American area in Boston, Massachusetts; and in poor, rural communities in the western United States documented in a 2001 study by the University of Oregon Program for Watershed and Community Health (now Resource Innovation), focusing on wildfire management and preparedness.

Environmental justice

Environmental justice *concerns* have been around in North America since the Conquest of Columbus in 1492. Yet, as a social movement, Faber (1998: 1) calls the US environmental justice movement "a new wave of grassroots environmentalism" and Anthony (1998: ix) calls it "the most striking thing to emerge in the US environmental movement." Whether it developed "in" the environmental movement or "from" the civil rights movement is perhaps a moot point. However, the US environmental justice *movement*, as opposed to environmental justice *concerns*, is generally believed to have started around fall 1982, when a large protest erupted in Warren County, North Carolina. The state wanted to dump more than 6,000 truckloads of soil contaminated with polychlorinated biphenyls (PCBs) into what was euphemistically

119

described as "a secure landfill." The protesters came from miles around. They were black and white, people of low income, ordinary, outraged citizens and prominent members of the civil rights movement and the National Black Caucus. Police arrested over 500 protestors in what Geiser and Waneck (1994: 52) describe as "the first time people have gone to jail trying to stop a toxic wastes landfill."

In 1983, the General Accounting Office (GAO, 1983) examined the location of four hazardous waste landfills in Environmental Protection Agency (EPA) Region IV (the South East), where *average* minority populations across the region are 20 percent. However, the four landfills were found to be in communities where racial minorities made up 38 percent, 52 percent, 66 percent and 90 percent of the population, respectively. The GAO concluded that there was enough evidence to be concerned about siting inequity. Later, the landmark 1987 United Church of Christ study, *Toxic Wastes and Race in the United States*, showed that certain communities, predominantly those of color, bear a disproportionate burden from commercial toxic waste sites. It also led Dr. Benjamin Chavis to coin the term that became the rallying cry of many: *environmental racism*.

In October 1991, the "People of Color Environmental Leadership Summit" was held in Washington, DC. Attracting over 600 delegates from 50 states, the main outcome of the Summit was the "Principles of Environmental Justice." These are "a well developed environmental ideological framework that explicitly links ecological concerns with labor and social justice concerns" (Taylor, 2000: 538).

At the federal level, there is an Office of Environmental Justice in the EPA and a National Environmental Justice Advisory Council. Both were developed in 1994, as a result of former President Clinton's Executive Order 12898 on environmental justice, which reinforced the 1964 Civil Rights Act Title (VI) prohibiting discriminatory practices in programs receiving federal funds and directed all federal agencies to begin to develop policies to reduce environmental inequity. Perhaps most importantly, the Order heralded the arrival of the movement's claims into the environmental policy mainstream.

Not only are minorities disproportionately exposed to environmental toxins, but waste sites and brownfields are also cleaned up more slowly by the EPA (Lavelle and Coyle, 1992). This unjust situation has ensured that there is now a fully fledged environmental justice movement, spreading from Alaska to Alabama and from California to Connecticut. It is driven by the grassroots activism of African American, Latino, Asian, Pacific Islander, Native American and poor white communities who are organizing themselves around LULUs (Locally Unwanted Land Uses) such as waste facility siting, lead contamination, pesticides, water and air pollution, workplace safety, and transportation issues. More recently, sprawl and smart growth (Bullard *et al.*, 2000), sustainability (Agyeman *et al.*, 2003), and "climate justice" (International Climate Justice Network, 2002; Environmental Justice and Climate Change Initiative, 2002; CBCF, 2004) have been included in the environmental justice critique.

Framing climate justice

The emerging climate-justice movement shifts the discursive framework of climate change from a scientific–technical debate to one about ethics focused on human rights and justice. Climate-justice advocates argue that this scientific debate has stymied productive global warming discussions and hindered more equitable policy solutions such as equal per capita emissions rights (Baer *et al.*, 2000). The International Climate Justice Network's 2002 *Bali Principles of Climate Justice* (see Textbox 7.1), the Environmental Justice and Climate Change Initiative's (EJCC) 2002 *Ten Principles for Just Climate Change Policies in the U.S.* (see Textbox 7.2), the Congressional Black Caucus Foundation's report, *African Americans and Climate Change: An Unequal Burden* (CBCF, 2004), as well as non-profit organizations such as EcoEquity (http://www.ecoequity.org) and India's Centre for Science and the Environment (http://www.cseindia.org) are all united around one overarching argument: climate change is no longer a theoretical possibility, but rather a certainty that is supported through scientific evidence, a diverse body of research, and countless reports produced by highly credible organizations such as the Intergovernmental Panel on Climate Change (IPCC), The Pew Center on Global Climate Change, and the World Meteorological Organization (WMO).

While accepting the reality of climate change, the scientific evidence and the knowledge of the causes contributing to it, there is still uncertainty as to how climate change will manifest itself locally and at what pace changes will occur. Nevertheless, as the CBCF (2004) report states, and a growing social-science literature documents, those *least* responsible for climate change are also those likely to be *first*, and *most*, impacted by its effects (see also McNeeley and Huntington, Chapter 8, this volume). The catastrophic damage wreaked by Hurricane Katrina in 2005 (whether one sees this as *evidence* of climate change, or as a *harbinger* of climate change impacts) revealed the deep-seated injustice and racism underlying the vulnerability and limited response capacity of poor and minority

Textbox 7.1 *Bali Principles of Climate Justice*, developed by the International Climate Justice Network — August 28, 2002

We, representatives of people's movements together with activist organizations working for social and environmental justice resolve to begin to build an international movement of all peoples for Climate Justice based on the following core principles:

1. Affirming the sacredness of Mother Earth, ecological unity and the interdependence of all species, Climate Justice insists that communities have the right to be free from climate change, its related impacts and other forms of ecological destruction.
2. Climate Justice affirms the need to reduce, with an aim to eliminate, the production of greenhouse gases and associated local pollutants.
3. Climate Justice affirms the rights of indigenous peoples and affected communities to represent and speak for themselves.
4. Climate Justice affirms that governments are responsible for addressing climate change in a manner that is both democratically accountable to their people and in accordance with the principle of common but differentiated responsibilities.
5. Climate Justice demands that communities, particularly affected communities, play a leading role in national and international processes to address climate change.
6. Climate Justice opposes the role of transnational corporations in shaping unsustainable production and consumption patterns and lifestyles, as well as their role in unduly influencing national and international decision-making.
7. Climate Justice calls for the recognition of a principle of ecological debt that industrialized governments and transnational corporations owe to the rest of the world as a result of their appropriation of the planet's capacity to absorb greenhouse gases.
8. Affirming the principle of ecological debt, Climate Justice demands that fossil fuel and extractive industries be held strictly liable for all past and current life-cycle impacts relating to the production of greenhouse gases and associated local pollutants.
9. Affirming the principle of ecological debt, Climate Justice protects the rights of victims of climate change and associated injustices to receive full compensation, restoration, and reparation for loss of land, livelihood, and other damages.
10. Climate Justice calls for a moratorium on all new fossil fuel exploration and exploitation; a moratorium on the construction of new nuclear power worldwide; and a moratorium on the construction of large hydro schemes.

11. Climate Justice calls for clean, renewable, locally controlled, and low-impact energy resources in the interest of a sustainable planet for all living things.

12. Climate Justice affirms the right of all people, including the poor, women, rural and indigenous peoples, to have access to affordable and sustainable energy.

13. Climate Justice affirms that any market-based or technological solution to climate change, such as carbon-trading and carbon sequestration, should be subject to principles of democratic accountability, ecological sustainability, and social justice.

14. Climate Justice affirms the right of all workers employed in extractive, fossil fuel, and other greenhouse-gas producing industries to a safe and healthy work environment without being forced to choose between an unsafe livelihood based on unsustainable production and unemployment.

15. Climate Justice affirms the need for solutions to climate change that do not externalize costs to the environment and communities, and are in line with the principles of a just transition.

16. Climate Justice is committed to preventing the extinction of cultures and biodiversity due to climate change and its associated impacts.

17. Climate Justice affirms the need for socio-economic models that safeguard the fundamental rights to clean air, land, water, food, and healthy ecosystems.

18. Climate Justice affirms the rights of communities dependent on natural resources for their livelihood and cultures to own and manage the same in a sustainable manner, and is opposed to the commodification of nature and its resources.

19. Climate Justice demands that public policy be based on mutual respect and justice for all peoples, free from any form of discrimination or bias.

20. Climate Justice recognizes the right to self-determination of indigenous peoples, and their right to control their lands, including sub-surface land, territories, and resources and the right to the protection against any action or conduct that may result in the destruction or degradation of their territories and cultural way of life.

21. Climate Justice affirms the right of indigenous peoples and local communities to participate effectively at every level of decision-making, including needs assessment, planning, implementation, enforcement, and evaluation, the strict enforcement of principles of prior informed consent, and the right to say "No."

22. Climate Justice affirms the need for solutions that address women's rights.

23. Climate Justice affirms the right of youth as equal partners in the movement to address climate change and its associated impacts.

Textbox 7.1 **(cont.)**

24. Climate Justice opposes military action, occupation, repression, and exploitation of lands, water, oceans, peoples, and cultures, and other life forms, especially as it relates to the fossil fuel industry's role in this respect.
25. Climate Justice calls for the education of present and future generations, emphasizes climate, energy, social and environmental issues, while basing itself on real-life experiences and an appreciation of diverse cultural perspectives.
26. Climate Justice requires that we, as individuals and communities, make personal and consumer choices to consume as little of Mother Earth's resources, conserve our need for energy; and make the conscious decision to challenge and reprioritize our lifestyles, re-thinking our ethics with relation to the environment and the Mother Earth; while utilizing clean, renewable, low-impact energy, and ensuring the health of the natural world for present and future generations.
27. Climate Justice affirms the rights of unborn generations to natural resources, a stable climate, and a healthy planet.

Adopted using the "Environmental Justice Principles" developed at the "1991 People of Color Environmental Justice Leadership Summit," Washington, DC, as a blueprint.

Endorsed by

CorpWatch, United States
Friends of the Earth International
Greenpeace International
groundwork, South Africa
Indigenous Environmental Network, North America
Indigenous Information Network, Kenya
National Alliance of People's Movements, India
National Fishworkers Forum, India
OilWatch Africa
OilWatch International
Southwest Network for Environmental and Economic Justice, United States
Third World Network, Malaysia
World Rainforest Movement, Uruguay

Textbox 7.2 **The Environmental Justice and Climate Change Initiative:**
Ten Principles for Just Climate Change Policies in the US

1. Stop Cooking the Planet

Global climate change will accelerate unless we can slow the release of
greenhouse gases into the atmosphere. To protect vulnerable Americans,
we must find alternatives for those human activities that cause global
climate change.

2. Protect and Empower Vulnerable Individuals and Communities

Low-income workers, people of color, and indigenous peoples will suffer the
most from climate change's impact. We need to provide opportunities to adapt
and thrive in a changing world.

3. Ensure Just Transition for Workers and Communities

No group should have to shoulder alone the burdens caused by the transition
from a fossil-fuel-based economy to a renewable energy-based economy. A just
transition would create opportunities for displaced workers and communities
to participate in the new economic order through compensation for job loss,
loss of tax base, and other negative effects.

4. Require Community Participation

At all levels and in all realms, people must have a say in the decisions that affect
their lives. Decision makers must include communities in the policy process. US
federal and state governments, recognizing their government-to-government
relationship, must work with tribes as well.

5. Global Problems Need Global Solutions

The causes and effects of climate change occur around the world. Individuals,
communities, and nations must work together cooperatively to stop global
climate change.

6. The US Must Lead

Countries that contribute the most to global warming should take the lead in
solving the problem. The U.S. is 4 percent of the world's population but emits
25 percent of the world's greenhouse gases. All people should have equal rights
to the atmosphere.

Textbox 7.2 **(cont.)**

7. Stop Exploration for Fossil Fuels

Presently known fossil fuel reserves will last far into the future. Fossil fuel exploration destroys unique cultures and valuable ecosystems. Exploration should be halted as it is no longer worth the cost. We should instead invest in renewable energy sources.

8. Monitor Domestic and International Carbon Markets

We must ensure that carbon emissions and sinks markets are transparent and accountable, do not concentrate pollution in vulnerable communities, and avoid activities that harm the environment.

9. Caution in the Face of Uncertainty

No amount of action later can make up for lack of action today. Just as we buy insurance to protect against uncertain danger, we must take precautionary measures to minimize harm to the global climate before it occurs.

10. Protect Future Generations

The greatest impacts of climate change will come in the future. We should take into account the impacts on future generations in deciding policy today. Our children should have the opportunity for success through the sustainable use of resources.

The EJCC is an effort by 28 US environmental justice, climate justice, religious, policy, and advocacy groups to call for action on climate change from the Bush Administration and Congress. The 20 organizations or individuals that have joined include:

Black Leadership Forum
Bunyan Bryant (Professor, School of Natural Resources, University of Michigan: individual endorsement, affiliation provided for identification purposes only)
Church Federation of Greater Indianapolis
The Church of the Brethren
Communities for a Better Environment
CorpWatch
Corporation for Enterprise Development
Council of Athabascan Tribal Government
Deep South Center for Environmental Justice; Xavier University

EcoEquity
Environmental Justice Resource Center; Clark Atlanta University
Georgia Coalition for a Peoples' Agenda
Indigenous Environmental Network
Intertribal Council on Utility Policy
Just Transition Alliance
National Black Environmental Justice Network
Kids Against Pollution
Native Village of Unalakleet
New York PIRG
North Baton Rouge Environmental Association
Redefining Progress
Southern Organizing Committee
Southwest Network for Economic and Environmental Justice
Southwest Workers' Union
United Church of Christ Justice and Witness Ministries
United Methodist Church
West County Toxics Coalition
West Harlem Environmental Action (WE ACT)

populations – much as the CBCF (2004) report foresaw. Importantly, it revealed the differential vulnerability of people *today* to present climate variability and extremes, thus making climate justice not just a future issue, but one that should concern us today. Wealthy and predominantly white New Orleans residents fled in droves in their cars and SUVs, while the sick, the poor, the elderly, African Americans, and undocumented immigrants were left to fend for themselves. This hurricane may well help shift the climate change and present-day disaster risk debate in many ways, not least from the science and engineering arena into the territory of ethics, human rights, and social justice.

A number of other key themes have emerged along with the development of the climate-justice movement. First, low-income and minority populations are faced with a disproportionate burden from the impacts of climate change on both health and economic well-being. While the impacts of climate change will occur on a global scale, the CBCF report emphasizes the fact that the effects will not be spread evenly over the population (CBCF, 2004: 15). Rather, climate change is likely to have different impacts on people of different socio-economic and racial groups (*ibid.*). The Preamble to the

Bali Principles, likewise, identifies a similar notion in its statement: "the impacts [of climate change] will be most devastating to the vast majority of the people in the South, as well as the 'South' within the North."

Second, there is a growing body of literature exploring issues of environmental injustice and, in effect, demonstrating that in the United States, African Americans are significantly more likely to live and work in locations where they are exposed to higher levels of pollution than the remainder of the general public (CBCF, 2004: 36). The factors contributing to this disparity are related to both socio-economic status and race. The CBCF report highlights data from an earlier study by Keating and Davis indicating "in 2002 an estimated 71 percent of African Americans lived in counties in violation of federal air pollution standards as compared to 58 percent of the white population" (*ibid.*). In addition, "78 percent of African Americans are located within 30 miles of a coal fired power plant, where the environmental and health impacts of the smokestack plumes are most acute, as compared to 56 percent of whites" (*ibid.*; see also Bingham, Chapter 9, this volume).

Third, there is an increasing amount of data supporting the notion that African Americans are simply less responsible for the US contribution to climate change than whites. The CBCF report reveals a significant disparity between African Americans and white Americans with regard to carbon emissions and issues of responsibility. Quantitative data indicate that on average blacks emit 20 percent less carbon dioxide in total than whites on both a per household and a per capita basis. In addition, on average, white Americans generate 14 percent more carbon directly through fuel use and are indirectly responsible for emitting 36 percent more carbon through other purchases (CBCF, 2004: 68).

Fourth, there are, as we saw in the Hurricane Katrina and Rita incidents, differential coping and adaptive capacities between middle-class and lower-income and minority populations, with cars and insurance abundant in the former, but not the latter. Thus, telling people to get in their cars and evacuate may only be effective if everyone has a vehicle, and one is prepared to ignore road capacity issues. Such class differences also suggest why lower-income populations have greater attachment to their limited possessions (homes and other property). It is because they are predominantly uninsured that they may be more reluctant to leave, even if they had a vehicle to escape in.

Finally, the climate-justice movement emphasizes the fact that the structure of policies aimed at addressing the issue of climate change will come at either great cost or great benefit to vulnerable populations. The CBCF report reviews a number of policy strategies and the positive and negative effects of each on the African-American community. For instance,

> poorly designed climate policies will most directly harm African American families. Such policies include those that suddenly increase the price of energy but do not raise revenue and recycle it in a progressive manner, or fail to promote clean energy technologies. In contrast, properly designed energy policies can create large net benefits for African Americans. When the revenue from carbon charges is used to offset distortionary taxes, such as payroll taxes, dramatic employment benefits can be reaped across the nation. Several studies find net job creation from climate policies on the order of 800,000 to 1,400,000 jobs. Based on historic hiring patterns, this increase in employment will disproportionately profit African Americans.
>
> *CBCF (2004: 4)*

It is therefore imperative that any climate mitigation strategy take into consideration issues of ethics, human rights, and justice.

It is precisely the unethical and unjust lack of connection between *responsibility* and *burden* that has given rise to the climate-justice movement. It fosters the observed shift in the framing of climate change to reflect issues of equity, justice, and accountability. In planning for a climate treaty and appropriate policy measures, both domestically and internationally, the climate-justice movement focuses on an evaluation of the disproportionate health and economic burdens with which many low-income and minority communities are faced. In effect, the framing of climate change must shift in order to ensure that principles of equity are incorporated into any climate policy or international treaty. Reframing climate change as a matter of social justice will also help mobilize portions of the population heretofore alienated from the science-dominated discourse. Both EcoEquity and the Centre for Science and the Environment lay out a vision of "fairness" that in their words is "equal per capita environmental rights to the atmosphere" (http://www.ecoequity.org). This marks a shift from the usual principle of "grandfathering" (i.e., allowing the continuation of old practices even as new policies are being introduced), based on historical or established use. Internationally, this vision of justice is captured in the proposed "contraction and convergence" approach, which reduces emissions from developed, high-emission countries, and over time, comes to a worldwide equal but much-reduced per capita emission standard (Global Commons Institute, 2001; Meyer, 2000).

Lack of environmental concern?

Contrary to popular belief, African Americans, as Mohai (2004) has recently shown, *do* place a high value on environmental sustainability (see also Tribbia, Chapter 15, this volume). They are *as* concerned about global environmental and nature preservation issues as whites, but are *more* concerned about pollution issues, especially in local neighborhoods, than whites. While Mohai's research didn't talk specifically about climate change but air pollution more generally, we can assume that his findings hold good for all manifestations of air pollution, including climate change. This heightened concern is attributed by Mohai and Bryant (1998) to the *"environmental deprivation"* effect: the result of African Americans living in more polluted neighborhoods. As we will show through the two cases below, we have a long way to go in figuring out how to tap into that environmental concern.

Risk communication research over the past two decades – despite the findings of Mohai (2004) and Mohai and Bryant (1998) – still only provides rather limited insight into the specific communication needs of poor and minority populations. That research has argued, however, that such populations have to be addressed in unique and specific ways to better connect information about (potential) risks to their concerns, circumstances, and cultural contexts. As Elaine Vaughan (1995: 170) points out, for example, "Framing risk communication issues from a socio-cultural perspective is uncommon in the environmental policy literature," as environmental policy professionals are overwhelmingly more comfortable with framing issues from a technical perspective. Furthermore, Vaughan notes, "the fact that environmental risks are not uniformly distributed across society provides an additional reason to examine the implications of socio-cultural variability for the risk communication process" (*ibid.*: 171).

Conventional risk communication strategies have focused on analyzing factors, often technical, that influence the form and effectiveness of communications without including "the social contexts within which individuals adapt to risk and information exchanges occur" (*ibid.*). The environmental justice frame and the principles upon which it was founded have introduced issues of injustice into the broader understanding of risk communication. New approaches to risk communication are considering the fact that more factors shape the perspective and understanding of risk, and thus, how one interprets and manages it (*ibid.*: 171–2). The two cases presented below begin to point out what else matters in effective communication with poor and minority populations.

Case 1: Roxbury, MA

One poor minority neighborhood is that of Roxbury, MA. The Roxbury community, just south-east of downtown Boston, is 5 percent white, 63 percent black, 24 percent Hispanic, 1 percent Asian or Pacific Islander, less than 1 percent Native American, 3 percent other, and 4 percent multi-racial, according to the United States Census (2000). The corresponding figures for the City of Boston are 50 percent white, 24 percent black, 14 percent Hispanic, 8 percent Asian or Pacific Islander, less than 1 percent Native American, 1 percent other, and 3 percent multi-racial (United States Census, 2000). Alongside this demographic difference, Roxbury has higher percentages of people speaking community languages, particularly Spanish, French, French Creole, Portuguese, Portuguese Creole, and African languages, than does the City of Boston. In 1999, 73.2 percent of its residents were classified as low to moderate income, as compared to 56.2 percent for the City of Boston. These statistics reveal one of metro Boston's poorest neighborhoods, and an inner urban "Environmental Justice Population," according to the Commonwealth's criteria:

- The median annual household income is at or below 65 percent of the statewide median income for Massachusetts; or
- 25 percent of the residents are minority; or
- 25 percent of the residents are foreign born; or
- 25 percent of the residents are lacking English language proficiency.

Commonwealth of Massachusetts (2002: 5)

A Roxbury-based environmental justice organization, Alternatives for Community and Environment (ACE), is dedicated to working primarily within the local community to promote local empowerment in decision-making for environmental, social, and economic issues. During interviews to understand the relationship between the environmental justice and sustainability movements, Agyeman (2005), like Vaughan (1995), found that there is evidence that if an organization wants to put out communications in this community, it may not be as simple as putting out a "relevant" message, no matter how worthy. While the specific framing of the message of climate change is important, it needs to account for *injustice*, namely the fact, as the CBCF (2004) report shows, that African Americans emit 20 percent less carbon dioxide in total than whites on both a per-household and a per-capita basis. Using the injustice frame, with its links to Roxbury's record asthma levels (even though CO_2 is not responsible), would be a way to frame a conversation on air pollution more generally. As Mohai (2004) has shown, this is a key concern. From there, climate change could be broached as a global air pollution issue affecting predominantly the poor and communities of color, such as Roxbury.

There are, however, two caveats. First, as Kollmuss and Agyeman (2002) have shown, merely providing information to people or communities does not guarantee behavior change (see also chapters by Tribbia; and Chess and Johnson, Chapters 15 and 14, this volume). Second, Agyeman (2005) concluded that even more than the message, of supreme importance is the messenger, the person bringing the message. An implication of this is worth pondering. In some communities, especially those of color and low income such as Roxbury, trust of "outsiders" or people labeled "experts" is low (Kasperson *et al.*, 1992). Messages will get lost or go unheard without a trusted messenger.

This has hiring and board-level recruitment implications for organizations looking to communicate with such audiences, but also speaks to the need for partnerships between organizations looking to communicate climate change with organizations based already within the community such as ACE. There are organizations like ACE in most large cities and towns.

Case 2: Economically distressed communities in rural Southwest Oregon

How poverty puts low-income communities at risk from the impacts of climate change and other natural hazard events was illustrated in a 2001 study by Resource Innovations at the University of Oregon (UO RI) (formerly called the Program for Watershed and Community Health). It compiled the financial and social costs from wildfires to low-income communities in the western United States. The study, which focused on wildfires and rural poverty, examined the pervasive and disproportionately negative impacts that wildfires have on poor households, communities, and local governments. The report suggested that approximately 3–5 million of the 10–15 million residents in the wildland–urban interface in the West lack incomes high enough to meet basic economic needs with enough left over to cover the expense of wildfire protection.

In the study area, low-income citizens and poor communities often live in districts with the greatest exposure to natural areas, on properties and in structures that are most susceptible to natural disasters, and they have the most limited resources available to take measures to protect their homes and properties.

To provide resources to rural low-income and minority communities at risk from disasters, public agencies, decision-makers, and community-based organizations must understand the problems and take steps to assist the low-income communities in securing the funds and resources needed for reducing risk and responding to disasters. Working to build capacity among the low-income and underserved communities can be achieved by helping public and private organizations increase the access that poor and isolated communities have to disaster management programs and resources.

As a result of the study, the UO RI initiated a multi-year project to assist rural and underserved communities in building their capacity to reduce wildfire risk. One facet of this program has been focused on developing and implementing community wildfire protection plans. The UO RI focused initial efforts in Southwestern Oregon, and Josephine County in particular, an area with some of the highest rates of rural poverty in Oregon. This area, because of its extensive forest cover, has a high risk of wildfires — risks that are likely to intensify in the West as global warming proceeds. This was illustrated in the Biscuit fire, one of the nation's largest in 2002, which burned large swaths of Southern Oregon. UO RI worked through Josephine County government to develop the fire plans, organized a steering committee composed of local government officials and citizens, and partnered with numerous local organizations, such as rural fire districts and other citizen groups, to develop the fire plans.

Efforts to build local capacity focused extensive outreach on educating the rural poor (and social service providers) about the risks of wildfires, how they can protect their homes and properties, and the resources available to assist them. While the project does not specifically focus on climate change, and quantitative studies have not been undertaken to analyze the effectiveness of the communication mechanisms used, it is possible to offer some observations about communicating environmental risks to these poor, rural audiences. We summarize the lessons learned from both case studies — Roxbury and Southern Oregon — in the section below.

Lessons learned

Based on these two cases, there are many lessons that organizations communicating risk to low-income and minority communities may want to bear in mind.

Framing the problem

Citizens in poor, rural areas and disinvested inner urban cores are likely to be skeptical of programs that originate outside of their community.

Traditionally defined environmental issues (e.g., wilderness protection, wildlife conservation) may not be a priority in rural or urban core households struggling to put food on the table and take care of basic needs. However, Taylor (2000: 508) suggests that "the environmental justice 'frame' is a master frame that uses discourse about injustice as an effective mobilizing tool." Frames can mobilize people. To do this, the issue or problem frame should tap into people's dreams and fears in their homes and local communities and address their concerns. For instance, "Clean Buses for Boston," a partnership led by ACE to bring 350 compressed natural gas (CNG) buses to Boston was not just about clean air, which the Roxbury community wants (it has five times the state asthma rate), but it also desperately needs a reliable, affordable bus service. This combination of sustainability and justice issues is what Agyeman (2005) calls "just sustainability."

Trust and credibility

It is necessary for outside groups wanting to communicate risk to go the extra mile to establish an open two-way dialogue with stakeholders and community residents to build trust and credibility. This can be done in two ways: first, by using actionable issues, i.e., issues that are both relevant and framed in a way that makes sense to a local community, and second, by partnering with a credible, local organization. When trust between stakeholders and community residents, and groups wanting to communicate risk, is initially low as it usually is, *how* you deliver the information (the framing) is as important as, or more important than, *what* is shared. In Southern Oregon, over 12 meetings were held across the region in the first 12 months, and many additional meetings have been held since then. The UO RI structured each meeting as a dialogue and learning session, not as an event simply to distribute data.

Empowering messages

Although environmental risks such as wildfires are very tangible in areas such as Southern Oregon, as is asthma in Roxbury, today's atmosphere is rife with fears related to the risks of terrorism, West Nile virus, avian flu, "natural" disasters and other potential calamities. It is therefore vital to deliver any information and engage in dialogue in a direct, calm, and reassuring manner, to be honest about the situation and avoid speculation, omit jargon, watch non-verbal communication, and avoid words that may increase fear and concern. The point is to avoid unintentionally creating overwhelming fears that can cause people to turn away and ignore the issues (see also Moser, Chapter 3, this volume). In Southern Oregon, this was accomplished by organizing an advisory group that included community-based and social service-agencies to help plan and execute community meetings, by extensive pre-meeting work to carefully developing the materials and presentations, and by carefully monitoring each meeting and evaluating how things went afterwards.

Careful choice of messengers

Based on the case studies described above, we made the fundamental observation that the selection of spokespersons is very important. Just because a government official or scientist knows "the issues" and can articulate them does not mean they can communicate effectively to a concerned public – especially if those individuals have low trust in the first place. The wrong framing, word, or phrase in a public meeting can generate the opposite effect that you are seeking and lead to hostility directed at you or your organization. Spokespersons should understand the socio-cultural values of the stakeholders, be able to communicate sensitive information effectively, not take criticism personally or push back when challenged, and in general be able to engage in an open and equal dialogue with stakeholders.

An asset-based approach

Finally, perhaps the most important lesson learned through the effort in Southern Oregon and research in Roxbury, MA is to adopt an *asset-based*, rather than a *problem-based*, approach. This approach sees that minority and low-income community members have skills and other assets which make them part of the solution. One-way communication does not solve a problem, nor can it usually elicit the involvement of others. The outcome from community fire-planning efforts in Southern Oregon has resulted in increased capacity among fire districts and community-based organizations that not only assists them in reducing wildfire risk, but these organizations have also changed the perspective of their own community members (for the better). Citizens are more likely to trust their fire district or community group to bring them good information and services.

While the 2001 UO RI study and project is rural, and focuses on wildfire issues, and the ACE study is urban and focused on interactions between environmental justice and sustainability groups, the lessons learned may be applicable to other environmental risks such as climate change, especially rapid climate change because poverty will continue to exacerbate the risk that low-income and minority communities face in dealing with issues related to global warming and other natural disasters.

Conclusions

Communicating risk in the context of climate change requires that we rethink our entire approach. We need to recognize the real concerns, constraints, and strengths in many communities and use them to promote climate justice and mobilize communities heretofore disenfranchised from the climate conversation. We also see the need for partnerships with trusted and credible organizations in such communities, such that the messenger(s) come(s) from within the community. We also see, as Vaughan (1995: 175) does, that "Individuals are not passive receivers of risk information, rather, communications are actively filtered through the 'lens' of *a priori* belief and value systems. Cultural experiences and norms, which may vary among diverse communities, contribute to the structure and content of these systems."

The experience with Hurricanes Katrina and Rita in 2005 teaches us that responding to, and preparing for, climate change needs a radical rethink. The impacts of weather- and climate-related disasters will be experienced according to the underlying racism and social injustices that are present already. To the affluent and more powerful, these may be latent or hidden until a disaster reveals them, but they are felt by poor communities already today. Moreover, these communities are less responsible for causing climate changes than more affluent ones, and they are less able to deal with their consequences. All these points, along with what we know from risk communication (see discussion above), lead to the clear conclusion that communicating disaster- and climate-change-related messages to culturally and socio-economically diverse audiences needs to change fundamentally along the lines of our suggestions above.

This leads to a powerful final thought. Many disenfranchised low-income and minority communities such as those in our cases (but also in developing nations) have the attitude, perfectly understandable, that they should be allowed to consume, drive, or pollute to the level that members of the more affluent classes do before they begin to cut back. They see this as their "right," their argument being that, once they've consumed as much as the middle classes, then they'll begin to think of sustainability, of cutting back, but not before. Because we have not yet approached communication as a two-way process, we as academics and environmental communicators may not have heard that message, much less begun to address it.

References

Agyeman, J. (2005). *Sustainable Communities and the Challenge of Environmental Justice*. New York: New York University Press.

Agyeman, J., Bullard, R., and Evans, B. (2003). *Just Sustainabilities: Development in an Unequal World*. London: Earthscan/The MIT Press.

Anthony, C. (1998). Foreword. In *The Struggle for Ecological Democracy: Environmental Justice Movements in the United States*, ed. Faber, D. New York and London: The Guilford Press. pp. ix–xi.

Baer, P., Harte, J., Haya, B., *et al.* (2000). Equity and greenhouse gas responsibility. *Science*, **289**, 2287.

Bullard, R., Johnson, G., and Torres, A. (eds.) (2000). *Sprawl City: Race, Politics and Planning in Atlanta*. Washington, DC: Island Press.

Commonwealth of Massachusetts (2002). *Environmental Justice Policy*. Boston, MA: State House.

Congressional Black Caucus Foundation (CBCF) (2004). *African Americans and Climate Change: An Unequal Burden*. Oakland, CA: Redefining Progress.

Environmental Justice and Climate Change Initiative (2002). *Ten Principles for Just Climate Change*. Oakland, CA: Redefining Progress.

Faber, D. (ed.) (1998). *The Struggle for Ecological Democracy: Environmental Justice Movements in the United States*. New York and London: The Guilford Press.

Geiser, K. and Waneck, G. (1994). PCBs and Warren County. In *Unequal Protection: Environmental Justice and Communities of Color*, ed. Bullard, R. San Francisco, CA: Sierra Club Books. pp. 43–52.

General Accounting Office (1983). *Siting of Hazardous Waste Landfills and Their Correlation with Racial and Economic Status of Surrounding Communities*. Washington, DC: Government Printing Office.

Global Commons Institute (2001). *Contraction and Convergence: Equity and Survival*. London: GCI; available at: http://www.gci.org.uk/refs/ C&CUNEPIIIg.pdf, accessed November 9, 2005.

International Climate Justice Network (2002). *Bali Principles of Climate Justice*. Press Release of August 29, Johannesburg, South Africa.

Kasperson, R. E., Golding, D., and Tuler, S. (1992). Social distrust as a factor in siting hazardous facilities and communicating risks. *Journal of Social Issues*, **48**, 4, 161–87.

Kollmuss, A. and Agyeman, J. (2002). Mind the gap: Why do people act environmentally and what are the barriers to pro-environmental behavior? *Environmental Education Research*, **8**, 239–60.

Lavelle, M. and Coyle, M. (eds.) (1992). Unequal protection: The racial divide in environmental law. *National Law Journal*, **15** (Special Supplement), 52–4.

Meyer, A. (2000). *Contraction & Convergence: The Global Solution to Climate Change*. Schumacher Briefing 5. Devon, UK: Green Books.

Mohai, P. (2004). Dispelling old myths: African American concern for the environment. *Environment*, **45**, 5, 10–26.

Mohai, P. and Bryant B. (1998). Is there a "race" effect on concern for environmental quality? *Public Opinion Quarterly*, **62**, 4, 475–505.

Program for Watershed and Community Health (PWCH). (2001). *Wildfire and Poverty: An Overview of the Interactions Among Wildfires, Fire-related Programs, and Poverty in the Western States*. Eugene, OR: ECONorthwest.

Taylor, D. (2000). The rise of the environmental justice paradigm. *American Behavioral Scientist*, **43**, 4, 508–80.

United Church of Christ Commission for Racial Justice (1987). *Toxic Wastes and Race in the United States*. New York: United Church of Christ Commission for Racial Justice.

United States Census (2000). Roxbury Data Profile; available at: http://www.cityofboston.gov/DND/PDFs/Profiles/Roxbury_PD_Profile.pdf, accessed June 12, 2004.

Vaughan, E. (1995). The significance of socioeconomic and ethnic diversity for the risk communication process. *Risk Communication*, **15**, 2, 169–80.

8

Postcards from the (not so) frozen North: talking about climate change in Alaska

Shannon McNeeley
University of Alaska—Fairbanks, National Center for Atmospheric Research

Orville Huntington
Alaska Native Science Commission

In May 2003 and again in January of 2004 a small group of climate change researchers and educators, we among them, met with Alaska Natives in the remote Koyukon Athabascan village of Huslia. We gathered in the log community hall to discuss local and regional impacts of climate change and wildfire. The Huslia Tribal Council hosted the workshop, called in response to concerns heard from Elders from the region about the recent changes seen in the weather, the animals, and the landscape. These were not the expected variations in the climate — something these people had been adapting to as long as they inhabited the region. Rather, these were changes never seen before or told of in their traditional stories passed down for generations. The weather and its spirit, the Elders said, had become different, less predictable.

> A long time ago our Elders used to be able to tell the weather, but nowadays they can't even tell weather. My grandfather and my grandfather before that, they said once man starts fooling around with the moon, the weather's going to change. Well, the weather did change ....We're not having the weather we usually have ... and it affects the wildlife too ... I'm sure a lot of the Elders here have the same question that I'm asking right now — what's happening to our Earth's atmosphere? What's going on out there with this crazy weather? We had a warm winter — no 80 below in the last 3 years, never got more than 40 below. They used to tell the weather by the ring around the moon — that used to be a storm coming, now they don't know because it might rain in the middle of winter like it did this year. There's a lot of changes.
>
> *Jack Wholecheese, Huslia (May 2003)*

Now, the Elders observed how the rivers were freezing up later in the fall and breaking up earlier in the spring. The lakes all over the Alaskan Interior were drying and causing problems in travel and acquiring food and resources, such

139

as firewood and house logs. Many animal and bird species were either significantly declining or behaving very differently. They spoke about many other environmental changes they were witnessing — changes that were echoed by indigenous people throughout Alaska and the entire Arctic/ sub-Arctic region (Krupnik and Jolly, 2002). The people of Huslia invited the scientists to help them understand why these changes were happening.

The scientists came to Huslia to hear how the rapid warming in Alaska was affecting the people and their environment on a local level — something that observational instruments and global climate models cannot "tell" as well as people (Georgi and Hewitson, 2001). This meeting served to exchange knowledge and observations between scientists and Native experts with the common goal of expanding each others' understanding of climate change impacts through the marriage of different, yet complementary, ways of knowing. Collectively, they are sending an early warning to the rest of the world about the changes that are happening as a result of recent warming trends.

This early warning, combined with experiences and stories about coping with climate change in real time, is changing the dialogue about climate change in Alaska. The complexities involved force us to rethink how to talk about climate change in Alaska and to the rest of the world. It is no small challenge, but with careful thought about how to frame the issue, create institutions to help communicate a common vision, and build capacity for coping with the change, a move toward political depolarization of the issue and a collective vision of solutions can emerge.

Early warning

For thousands of years the Native peoples of Alaska have been living off the land and have fine-tuned survival skills that allowed them to adapt to change and survive the extremes that nature in the northernmost latitudes presented them. Today in rural Alaska, life is increasingly modernized, but survival still largely depends on acquiring food through subsistence hunting, trapping, fishing, and gathering. This way of life requires detailed knowledge and wisdom about the dynamics of one's environment as well as an ability to predict the weather and identify the variability and severity in daily and seasonal weather patterns. Quite different than modern life in the United States or European mid-latitudes, where people are more insulated from the whims of weather and climate, rural and Native Alaskans' close connection to the land through subsistence activities means that they receive direct cues from nature that provide insights about ecosystem dynamics and environmental changes. This relationship entails receptivity to these cues for success (and in some cases survival), and creates an ability to observe, learn, and adapt to the changes accordingly. Because subsistence fishers, trappers, hunters, and gatherers are regularly out on the land, they feel the effects of climate variability and change directly and must continually negotiate and adapt to a rapidly changing milieu. A long-standing presence of these tribes and their ancestral clans in this region also means a deep, collective history and knowledge about the environment, and appropriate responses to environmental changes passed down through traditional stories and teaching of older generations to new ones (McIntosh, Tainter, and McIntosh, 2000; Raboff, 2001).[1]

There is a growing body of research documenting Arctic and sub-Arctic Natives' intimate and intricate knowledge about their natural environment and the observations of environmental changes that appear to be out of the normal variability regime (Krupnik and Jolly, 2002; Krause, 2000). Change is nothing new to these people as they have been adapting to a changing environment for millennia. However, the concern expressed across the region is that the *rapid rate* of change is something new, and that the weather has become unpredictable, which is extremely dangerous for subsistence activities and living in Arctic and sub-Arctic rural areas (Krupnik and Jolly, 2002). In the past, the Elders say, the changes were slower and the weather much more predictable. This allowed for more time to adapt while traditional ecological knowledge (TEK) continued to accumulate and get stored in social–ecological memory. Fikret Berkes (1999: 8) defines TEK as a "cumulative body of knowledge, practice, and belief, evolving by adaptive processes and handed down through generations by cultural transmission, about the relationships of living beings (including humans) with one another and with their environment." Accumulated TEK about climate–ecosystem–land interconnections has historically made northern Natives resilient to environmental change. However, a new set of circumstances that combines rapid environmental change in a context of rapid social, economic, and cultural changes has brought about a new era of vulnerability to climate change (Nuttall, 2005). This vulnerability could make adapting to climate change in the future increasingly challenging.

Should these direct personal observations of rapid climate-related environmental change, this early warning from Alaska Natives, be of concern to the rest of the world? The answer is undoubtedly yes. Scientists have predicted for some time now that the poles will see the earliest signs of planetary warming and that it will be only a matter of time before the lower latitudes follow. In addition to being an early warning, the Arctic climate plays an important role in the mechanisms of global climate change through various processes, and will cause additional change through various feedbacks. First, melting snow and ice that diminish reflectivity (i.e., albedo) of radiation from the sun cause additional warming as land surfaces turn from white to dark and absorb more heat in a continuous feedback cycle. Second, as Arctic land-based ice melts, the meltwater contributes to sea-level rise; it also adds freshwater to the oceans, potentially altering ocean/atmosphere cycles that move warm and cool masses throughout the planet. And third, changing landscape processes in the North altered by warming (e.g., melting of permanently frozen ground or increasing incidence of forest fires) will add more greenhouse gases to the atmosphere (ACIA, 2004).

Many environmental groups and the media thrive on fear-provoking messages and imagery to make global warming seem an attention-worthy threat (Moser and Dilling, 2004; see also Moser, Chapter 3, this volume). "Dying" cultures of the Arctic are a favorite image/symbol of the media to convey the urgency. The tendency to portray dying indigenous cultures at the hands of the dominant imperialistic culture is not a new phenomenon; however, indigenous cultures throughout the North have proven themselves resilient in the face of ongoing outside pressures as they have maintained many of the attributes that characterize their unique cultures (Sahlins, 1999). This is not to downplay the issue of climate change impacting northern cultures. Rather it is to show how Alaska Natives along with other Arctic indigenous peoples have become, perhaps unwillingly, a symbol for the human impacts of a changing climate. Northern indigenous people are frequently referred to by scientists and the media as the "canaries in the coal mine" for global warming. While this metaphor is useful to convey meaning to the public about early warnings, it also obscures more meaningful messages coming from northern indigenous peoples. Legitimate concerns about increasing vulnerabilities notwithstanding, as an alternative to this public "discourse of vulnerability" (Thomas and Twyman, 2005) in which Native people are portrayed as vulnerable and helpless victims, we emphasize the need to move past this image and consider what other important messages northern indigenous people are relaying to people elsewhere about their real-time experiences with climate change impacts on a local level, and how their worldviews and value systems determine the ways in which they cope with those changes.

Beyond persuasion: coping with climate change

While natural climate variability plays a role, the evidence is mounting that rapidly rising temperatures in Alaska are related to global climatic changes and are in fact causing the ecological and physical changes observed on the ground (ACIA, 2004; Krupnik and Jolly, 2002). Because Arctic Natives have no choice but to be attentive to these changes and adapt as quickly as possible, they are increasingly being looked to as local-level experts who can help understand current and ongoing adaptation strategies and coping mechanisms as well as the interconnections between changing climate and ecosystem responses (Fox, 2002; Riedlinger and Berkes, 2000). In Huslia, Elder Tony Sam gave the example of how Alaska Natives have had to adapt to the warming by changing the timing of the fall hunt:

> The old timers would tell us that the weather is gonna get old,[2] and it's gonna stay warm all the time in years to come, and that's what's happening now. Right now it's happening. Like in the fall time it stays warm till way in November sometimes. We hardly go out anymore. Long time ago it used to get cold right away, and it would freeze up and then we'd go out with the dog team. Say around the 1960s, that's when it used to be pretty good yet [the weather]. But around the 1980s it started to get a little warmer. In the 1990s it was getting worse. These last four to five years I don't go out in the end of October anymore – hardly go out because it don't freeze up – it stays warm the whole time. How I know is we used to hunt in the fall time – we'd hunt bears in the den in the fall time. That's how I know it. Right now we don't hardly go out – in November that's when we go out. But even at that time it don't really freeze up good. Like these last few years it just stays warm all winter.

> *Tony Sam, Huslia (May 2004)*

In Alaska climate change is not some futuristic notion; rather it is an extant reality. Thus, with most people, there is no longer a need to convince them of that reality; nor does talk of climate projections in 50 to 100 years from now (while important) make the issue any more real. It also does not help people cope with and adapt to the changes that are happening now. The dominant framing elsewhere has primarily been one of climate change as a future issue, reported as global trends in mean annual temperature. Frequently, projections are given for large regions or the entire globe. Since the establishment of the Intergovernmental Panel on Climate Change (IPCC) in 1988, the climate change discussion has in large part been framed by the global scientific establishment. The IPCC became the nexus of international science and policy on climate change, and thus has focused the discourse about the science and impacts on national and larger scales, while local impacts and adaptation strategies – at least until recently – were largely left on the periphery (Thomas and Twyman, 2005). Better higher-resolution climate models, increasingly sophisticated downscaling methods, and empirical place-based work on impacts and adaptation strategies in recent years are beginning to change this focus and, consequently, the framing of climate change.

Because the discourse in the media and political realm has been framed by the dominant knowledge-producing community – i.e., the international network of credible scientists with authority and a shared understanding of climate change (Forsyth, 2003) – the focus has been primarily on two issues: First, is the planet warming, yes or no? and second, if so, how much of the warming is caused by humans (as opposed to resulting from natural cycles)? As the science continues to improve, scientists are providing answers to these questions with increasing confidence. In Alaska, where the mean annual temperature since 1949 has increased by 3.5 degrees Fahrenheit, and where wintertime average temperatures in the Interior have risen by as much as 9.4 degrees Fahrenheit[3] (Alaska Climate Research Center, 2005), the conversation has moved beyond these questions. Instead, the focus in Alaska is shifting to questions of real-time coping with the impacts of warming that are ongoing today and projected to accelerate in the future.

Craig Fleener, director of the Council of Athabascan Tribal Governments,[4] spoke to the Alaska Forum on the Environment in 2005 about how climate change is affecting interior Native people on a local level and how they respond to these impacts:

> The message is that the climate is changing. These things are happening all over the place. We're experiencing these changes here every year and they're drastic, they're dramatic, and they're often painful. And we live with them everyday and we think to ourselves here in rural Alaska, how are we going to make it, how are we going to make it tomorrow? I wasn't able to get my moose this season, so what am I going to eat? Hopefully, I can get some salmon if I can't get a moose, or hopefully I can get a caribou. And if I can't get those three, well then I'm going to have to get some rabbits or some porcupine. But eventually, if the climate continues to change and the wildlife populations continue to be impacted like they have in our region, and the numbers of those populations continue to decline like they have been over the past 20 or 30 years, then it's going to continue having a dramatic impact on us. So, the message that you need to take home is – I can do something about it, and I will do something about it.
>
> *Craig Fleener, Anchorage (February 2005)*

The need for Alaska Native communities to cope with changes in the environment and the impacts on subsistence activities overshadows the issue of human versus natural causes that preoccupies scientists and the media. In the Athabascan villages of the Alaskan Interior residents discuss climate change in terms of specific experiences related to higher temperatures, changes in wind patterns, snow and rainfall amounts and timing, animal behavior, population dynamics, and taste/consistency of fish and game when eaten, weather patterns, water levels and ice in rivers and lakes, riverbank erosion, permafrost melt, and so on. People are observing and experiencing these very real and tangible changes, and they make for a different conversation than scientific descriptions of statistical averages. Linking indigenous observations, knowledge, and experiences to western scientific observations and explanations is where the real power lies in understanding place-based climate change. From a communications perspective, however, this convergence of scientific observation and lived experience is not sufficient to catalyze broad-based support for real solutions to the problem, neither in Alaska nor elsewhere. In other words, while the discussion of climate change in the North has moved beyond persuading people of the reality of the issue, individual or collective action has not followed automatically. We — in Alaska and in the rest of the world — still need to find other ways of framing the issue that stop the polarization of climate change politics and move toward more enlightened approaches to collective solutions.

Reframing climate change in Alaska

There is a shared responsibility for addressing the issue of human-caused climate change implicit in the fact that all of Alaska (which includes Alaska Natives) is heavily dependent on the oil industry. Understanding the human fingerprint on climate and how this connects to village Alaska is important to point out but difficult to reconcile. Until rural villages can transition to alternative sources of energy, the use and access to cheap fossil fuels is a must. Alaska Natives are a case in point: they use fossil fuels to heat and light their homes, to cook, and to travel by boats, snow machines and all-terrain vehicles to pursue their traditional subsistence activities. Without a viable alternative to cheap fossil fuels they cannot get to the fish and game they depend on. They are also dependent on the oil industry for the revenue it generates and returns to them through the dividend checks, jobs created, and state-sponsored services. Thus, local measures to reduce greenhouse gas emissions are not currently a focal point in the native discourse on climate change. This must also be understood within the context of other more pressing social and economic issues, such as the mounting cost of living in rural Alaska due to rising costs of energy and basic services amidst high unemployment and poverty levels.

In the media and in academia it is common to underscore that the majority of the emissions responsible for the recent warming in the Arctic region came from fossil fuel use in the lower latitudes. While this is true in absolute terms because the Arctic/sub-Arctic region is thinly populated compared to the warmer mid-latitudes, Alaska is one of the United States' biggest oil-producing states, providing an economic base that sustains the economy of Alaska for all residents — Native and non-Native alike. So, is it fair to ask people in "the Lower 48," for example, to minimize their consumption while those in the North do not? Or is it fair to ask the people in the North to minimize consumption as a matter of principle? Neither is realistic in isolation from the other, and the logic behind both notions is based on a dichotomous "us versus them" mentality that divides people and their rights based on geography and ethnicity. When it comes to the debate about climate change, the divisions are murky when all consume fossil fuels and contribute emissions. This calls for a new way of framing climate change that facilitates meaningful dialogue and collective solutions (see also Regan, Chapter 13, this volume). It is thus imperative that alternative energy sources are developed in the Arctic to help solve a variety of local and global problems. Implementing alternative energy sources and efficient vehicle technology would both lower emissions and be more economically and environmentally sustainable. Talking about ways to take local actions to deal with climate change impacts shifts the discourse toward one of empowerment and unification (see also the chapters by Young; Agyeman *et al.*; Tribbia; and Moser, Chapters 24, 7, 15, and 3, this volume).

> We need to ask ourselves "what can I do?" We can't just blame everyone else for our problems. In the Yukon Flats we had to take matters into our own hands. We can't just wait for some big government agency or multi-million dollar organization to pick the Flats to help with our problems.
> *Craig Fleener, Council of Athabascan Tribal Governments, Anchorage*
> *(February 2005)*

It is common when discussing climate change in Alaska to fall into a dichotomous debate cast as a matter of justice. The Inuit Circumpolar Conference (ICC),[5] for example, chooses to frame climate change as a human rights issue. ICC representative Sheila Watt-Cloutier was quoted in the media as saying "The U.S. and others feel they can continue business as usual, when our entire way of life as we know it may end in my grandson's lifetime" (Johnson, 2005). Similarly, the small coastal Inupiaq Eskimo village of Shishmaref is considered by many to be among the world's first "climate refugees" as the community has chosen to move their village further inland as sea ice rapidly retreats and increasing storm surges and coastal erosion eat away their land. Justice is one way to frame the issue, and perhaps the right one for the Inuit, based on their own history of interactions with the dominant culture and political system. Given the Bush administration's stance on climate change with its reluctance to move toward mandatory emissions reductions and its insistence on the negative impacts of emissions restrictions on the US economy, this might be a necessary response, but it is not sufficient. It is a polarizing relationship that inevitably perpetuates divisions along ethnic lines and hinders collective solutions. In the meantime new ways to frame the issue that reach others in positions to make real changes, such as more progressive-minded industry leaders, for example, should be developed concurrently (see also Arroyo and Preston, Chapter 21, this volume).

Blame or guilt tactics to force action tend to shut down the dialogue by alienating those who would be potential change agents (e.g., Moser and Dilling, 2004). This does not mean we cannot speak of responsibility, differential impacts, and equity and justice. Arctic and sub-Arctic indigenous people bear a disproportionate burden of the impacts of human-caused climate change relative to their contributions of emissions. And history has shown that native peoples have suffered when outside paradigms are imposed on them that strip their power to decide their own fate (Smith, 1999). However, the lines are blurred in a modern context where all native communities participate in the fossil-fuel-based market economy. The need to collectively find solutions cross all scales of place and time − i.e., local, national, and global for both current and future generations − is evidenced by the experiences and stories of indigenous people throughout the northern latitudes.

Toward unification and empowerment

We need new ways of talking about climate change in Alaska that can be heard within and beyond this state and begin to affect conversations elsewhere. These new ways need to go beyond persuasion and employ a non-polarizing framing of the issue. Institutions can help in the transformation of this dialogue. Institutions commonly mediate the relationship between governments and communities, and they may also play an important role in voicing people's concern with climate change across levels of government. As such, they can serve as a unifying voice for people dispersed across a wide region and concerned with a range of issues. Who in Alaska could play this important role?

Some organizations that represent Arctic/sub-Arctic indigenous people have taken on climate change as an important issue, such as the Inuit Circumpolar Conference and the Gwich'in Council International, acting as representatives to the intergovernmental Arctic Council for the Arctic Climate Impacts Assessment process (see Textbox 8.1). In Alaska there are

Textbox 8.1 **The Arctic Climate Impact Assessment (ACIA)**

In recent decades Alaska has felt more warming than almost any other place on Earth (Whitfield, 2003). Since 1949 Alaska's mean annual temperature has warmed about 3.5 degrees Fahrenheit (Alaska Climate Research Center, 2005).[6] Temperatures in parts of Alaska are rising faster than the rest of the world due to the *albedo effect* that occurs when melting snow and ice change the landscape from white to dark and the Earth absorbs more heat (ACIA, 2004). Climate modelers have been predicting for some time that the poles would warm the fastest from increases in greenhouse gases, and observations are proving that to be true.

In 2004 the intergovernmental Arctic Council released a report, the Arctic Climate Impact Assessment (ACIA), which states that there is a consensus among hundreds of Arctic researchers and indigenous populations on the basis of research and practical experience that the climate of the North is warming at an alarming rate and that this warming is having many impacts on the landscape and people across the Arctic (ACIA, 2004).[7] In November of 2004 the Arctic Council released the overview document of the four-year ACIA effort called *Impacts of a Warming Arctic*, which concluded:

> The Arctic is now experiencing some of the most rapid and severe climate change on Earth. Over the next 100 years climate change is expected to accelerate, contributing to major physical, ecological, social and economic changes, many of which have already begun. Changes in arctic climate will also affect the rest of the world through increased global warming and rising sea levels.

On November 16, 2004, Senator John McCain (R-AZ) held a federal hearing before the Senate Committee on Commerce, Science and Transportation in response to the results of the assessment (Senate Committee on Commerce, Science, and Transportation, 2004). McCain, who has called the Bush administration's performance on the issue of climate change "disgraceful" (Cockerham, 2004), has been actively attempting to push the US government to implement an emissions reduction program. While there were no Native representatives present at the hearing, Senator McCain pointed out the important role of Natives in the ACIA and the role of indigenous knowledge in understanding climate change, stating, "I think this approach can and should be used as a model for future assessments" (Cockerham, 2004). McCain and several of the scientists who testified at the hearing expressed the importance of incorporating indigenous people into the study and understanding of global climate change.

other institutions – such as the Council of Athabascan Tribal Governments or the Alaska Native Science Commission – that voice concerns of local people on a variety of issues, including climate change. However, representation is either on a small, regional scale or not entirely representative of the rural perspective when organizations are located in urban centers, somewhat removed from village life. Moreover, many other Native organizations have not yet taken on climate change. For example, the Alaska Federation of Natives (AFN), a statewide organization that is often the voice for Native people in governmental negotiations, has yet to make any resolution for mitigation and/or adaptation in the face of climate change. More recently, one statewide representative organization, the Alaska Inter-Tribal Council (AITC) – a tribal advocacy group based in Anchorage – passed a resolution in December 2005 urging the government to take action on global warming (Rasheed, 2005). The AITC action is significant, but perhaps a new institution is required that provides a mechanism to unite all indigenous people in Alaska across Aleut, Eskimo, and Indian cultures, living in both rural and urban areas, and specifically focusing on climate change in its efforts to influence government, the media, and the public at large. Having such a focused mission (even if embedded in broader ambitions for comprehensive sustainability) and a unifying voice could help reframe an issue that clearly requires fundamental new thinking and policy change.

One example of an effective inter-tribal organization is the Yukon River Inter-Tribal Watershed Council, which works with Tribes in Alaska and First Nations in Canada to protect the Yukon River watershed. The Watershed Council is a grassroots service organization that was created by and for the indigenous communities in the watershed, and addresses issues on local, national, and international levels as well as interfacing with local, state, and federal governments to represent local concerns. The Watershed Council has a broad mission and an issue-specific, but large, vision – to someday be able to drink straight from the Yukon River – combined with practical and accomplishable goals in areas of monitoring, clean-up and capacity building for the communities in the watershed. The Watershed Council provides one possible organizational model to unite tribes and empower local voices for a shared goal of coping with the impacts of climate change. While the Council does not focus on climate change *per se*, it already promotes alternative energy sources by providing technical assistance to communities for renewable energy development, energy conservation, and evaluating energy options. In one instance, the Watershed Council helped with the installation of solar energy systems in the Native villages of Venetie and Arctic village, which will help them rely less on expensive and polluting fossil fuel energy sources.

Conclusion

To continue a "discourse of vulnerability" that characterizes poor, vulnerable Natives of the Arctic/Sub-Arctic as passive bystanders to the impacts of climate change is to disempower those already affected by global warming, while undermining their efforts that aim at building adaptive capacity from the bottom up (Thomas and Twyman, 2005). Of course, the total share of rural and native Alaskans in causing the problem is far lower than that of mid-latitude residents and industries. It is also obvious that the people of the North are already experiencing greater impacts of climate change than most other places on the planet. People in the North need little more convincing that change is under way; they need help adapting to it, and fora to discuss these very real and increasingly pressing needs.

The conclusion from these findings, however, is not to pursue a framing of climate change as a matter of villains and victims. What's needed instead is a discourse of empowerment, a conversation that offers practicable alternatives and new opportunities to Alaska Natives. It would acknowledge the complexities of their modern lives while moving them forward, away from fossil fuel dependencies toward true twenty-first-century energy solutions. It would also unify their voices into one that is loud and clear enough to attract the attention of the media and distant decision-makers and empower them to create a desirable future for the rapidly changing North.

Notes

1. With the increasing influence of television, Western lifestyle patterns, and the cash economy on contemporary youth in Native communities, the transfer of traditions has changed. While the older generations still retain a good deal of knowledge about the local environment, there is growing concern that this socio-cultural trend will result in a diminishing of this knowledge if new ways of transferring traditional knowledge are not systematized.
2. In Koyukon culture the statement about the weather getting "old" refers to change and the loss of its strength or intensity (similar to when people get old). When the Elders said "the weather was going to get old" they meant that it was not going to be the extreme cold that they were used to.
3. Record for the town of Big Delta, AK.
4. The Council of Athabascan Governments (CATG) is a tribal consortium in the Yukon Flats region of Interior Alaska that builds capacity and provides services to local Native villages on matters of health, education, and natural resource management.
5. The ICC is an international organization that represents Inuit/Eskimo peoples throughout the Arctic, including Alaska.
6. This compares to a warming of 1 °F across the globe during the last half of the twentieth century. The increase of 3.5 °F represents an annual average across the entire state. Some places in Alaska, such as Talkeetna, have seen winter mean temperatures increase of 9.4 °F.
7. The Arctic Climate Impact Assessment was a four-year project comprising over 300 scientists and indigenous participants to analyze the impacts of climate change across the Arctic region. The project was commissioned and coordinated by the Arctic

Council and the International Arctic Science Committee. The Arctic Council is a high-level intergovernmental organization comprised of eight member states (Canada, Denmark, Finland, Iceland, Norway, Russia, Sweden, and the United States) and six indigenous Permanent Participant groups (Aleut International Association, Arctic Athabaskan Council, Gwich'in Council International, Inuit Circumpolar Conference, Saami Council, and Russian Association of Indigenous Peoples of the North).

References

Arctic Climate Impact Assessment (ACIA) (2004). *Impacts of a Warming Arctic: Arctic Climate Impact Assessment*. Arctic Council. New York: Cambridge University Press.

Alaska Climate Research Center (2005). Temperature Change in Alaska 1949–2003. Available online at http://climate.gi.alaska.edu/ClimTrends/30year/index.html; accessed January 14, 2006.

Berkes, F. (1999). *Sacred Ecology: Traditional Ecological Knowledge and Resource Management*. Philadelphia, PA: Taylor and Francis.

Cockerham, S. (2004). Climate change divides – Arctic warming: Stevens, McCain at odds over fuels' effect. *Anchorage Daily News*, November 17. Refer to http://www.adn.com; accessed January 31, 2006.

Forsyth, T. (2003). *Critical Political Ecology*. London and New York: Routledge, Taylor and Francis Group.

Fox, S. (2002). These are things that are really happening: Inuit perspectives on the evidence and impacts of climate change in Nunavut. In *The Earth is Faster Now: Indigenous Observations of Arctic Environmental Change*, eds. Krupnik, I. and Jolly, D. Fairbanks, AK: Arctic Research Consortium of the United States, pp. 13–53.

Georgi, F. and Hewitson, B. (lead authors) (2001). Regional climate information: Evaluation and projections. In *Climate Change 2001: The Scientific Basis*. Contribution of Working Group I to the Third Assessment Report of the IPCC. Cambridge, UK: Cambridge University Press, pp. 585–638.

Johnson, J. (2005). Global Warming is a Human Rights Issue. *Indian Country Times*. http://www.indiancountry.com; accessed August 22, 2005.

Krause, L. (2000). Global Warming Melts Inuit's Arctic Lifestyle. *National Geographic News*. http://news.nationalgeographic.com/news/2000/12/122900inuits.html; accessed January 31, 2006.

Krupnik, I. and Jolly, D. (eds.) (2002). *The Earth is Faster Now: Indigenous Observations of Arctic Environmental Change*. Fairbanks, AK: Arctic Research Consortium of the United States.

McIntosh, R. J., Tainter, J. A., and McIntosh, S. K. (eds.) (2000). *The Way the Wind Blows: Climate, History, and Human Action*. New York: Columbia University Press.

Moser, S. C. and Dilling, L. (2004). Making climate hot: Communicating the urgency and challenge of global climate change. *Environment*, **46**, 10, 32–46.

Nuttall, M. (2005). Hunting, herding, fishing, and gathering: Indigenous peoples and renewable resource use in the Arctic. In *Arctic Climate Impact Assessment*, ed. Arctic Council. New York: Cambridge University Press, pp. 649–90.

Raboff, A. P. (2001). *Iñuksuk: Northern Koyukon, Gwich'in & Lower Tanana 1800–1901*. Fairbanks, AK: Alaska Native Knowledge Network.

Rasheed, N. (2005). Tribal council wants government to recognize impact of global warming. *Channel 2 Broadcasting, Inc.* Available at: http://www.ktuu.com; accessed January 3, 2006.

Riedlinger, D. and Berkes, F. (2000). Contributions of traditional knowledge to understanding climate change in the Canadian Arctic. *Polar Record*, **37**, 203, 315–28.

Sahlins, M. (1999). What is anthropological enlightenment? Some lessons of the Twentieth Century. *Annual Review of Anthropology*, **28**, i–xxiii.

Senate Committee on Commerce, Science and Transportation (2004). *Global Climate Change*. United States Senate, November 16.

Smith, L. T. (1999). *Decolonizing Methodologies: Research and Indigenous Peoples*. London: Zed Books, Ltd.

Thomas, D. S. G. and Twyman, C. (2005). Equity and justice in climate change adaptation amongst natural-resource-dependent societies. *Global Environmental Change*, **15**, 115–24.

Whitfield, J. (2003). Alaska's climate: Too hot to handle. *Nature*, **425**, 338–9.

9

Climate change: a moral issue

The Rev. Sally Bingham
The Regeneration Project

Introduction

Talking about climate change is difficult at best, but when the subject is raised from the pulpit, it becomes not just difficult, but very complicated and sensitive. Because climate change is often seen as a political issue or even as part of a liberal conspiracy, clergy have to be extremely careful. On the other hand, religious leaders carry an important moral voice in the community, thus they are in a strong position to raise awareness and influence public opinion. This moral voice is increasingly being heard in broad, denomination-wide statements and open letters, some directed at political leaders, others at congregations; it is also being heard at the very local level, where priests and rabbis, ministers, imams, and monks speak to their congregations and communities (see Textbox 9.1). According to polls conducted after the 2004 elections in the United States, moral values played a major role in voters' decisions, thus giving rise to even more important reasons for the moral dimensions of issues such as climate change to be brought up front and center.

Clearly, choosing to speak out on climate change places religious leaders at risk of alienating their parishioners if they are seen as unprofessional or as using the pulpit as a soapbox. There is a thin line between engaging and mobilizing a community around an issue that has important social justice, environmental, and ethical dimensions (undoubtedly a justifiable religious matter) and being dismissed as a political agitator. The subject is difficult to address in the secular world, but even more so in the faith community. While small, there still is some scientific controversy and debate hovering around climate change. Non-scientific speakers have to step lightly to maintain credibility and integrity. We thus must be primarily grounded in theology. Even when there is overwhelming evidence on an issue as a result of much study and discernment, it would be insensitive and alienating to dismiss other

---

Textbox 9.1 **Religious networks, interfaith coalitions, faith-based campaigns on climate change, and selected resources**

Faith-based communities, networks, and activities for the protection of our climate are growing every day. Thus this selection of links is necessarily incomplete. But it offers some important gateways to statements, activities, and information, as well as to communities who have taken on climate change and the protection of the environment as a central moral issue.

Religious networks

National Council of Churches: Eco-Justice Working Group
http://www.webofcreation.org/ncc/; accessed January 5, 2006.

Interfaith Climate Change Network (with link to an interfaith study guide to global warming) http://www.protectingcreation.org/; accessed January 5, 2006.

Interfaith Climate Change Campaigns (by state)
http://www.webofcreation.org/ncc/climate.html accessed January 5, 2006.

The National Religious Partnership for the Environment (NRPE)
http://www.nrpe.org/; accessed January 5, 2006.

Evangelical Environmental Network (with access to the "What Would Jesus Drive" Campaign)
http://www.creationcare.org/; accessed January 5, 2006.

Coalition on the Environment and Jewish Life
http://www.coejl.org/; accessed January 5, 2006.

Interfaith Center on Corporate Responsibility
http://www.iccr.org; accessed January 5, 2006.

Denominational statements, resolutions, and open letters

Clearinghouse of Faith Statements & Resolutions.
http://www.webofcreation.org/ncc/statements/faithstatements.html; accessed January 5, 2006.

Letter of the US Conference of Catholic Bishops to the Senate, July 2004
http://www.usccb.org/sdwp/ejp/news/climatechangeltr.htm; accessed January 5, 2006.

Oxford Declaration on Global Warming, July 2002
http://www.climateforum2002.org/statement.cfm; accessed January 5, 2006.

Other resources

Bassett, L., Brinkman, J. T. and Pedersen, K. P. (eds.) (no year). *Earth and Faith: A Book of Reflection for Action.* New York: Interfaith Partnership for the Environment and UNEP.

Center for a New American Dream (2002). *Responsible Purchasing for Faith Communities.* Available at: http://www.newdream.org/publications/purchguide.pdf; accessed January 5, 2006.

Cooper, D. E. and Palmer, J. A. (eds.) (1998). *Spirit of the Environment: Religion, Value and Environmental Concern.* Oxford: Routledge.

Earth Ministry and the Interfaith Climate & Energy Campaign (no year). *The Cry of Creation: A Call for Climate Justice.* An Interfaith Study Guide on Global Warming. Available at: http://www.protectingcreation.org/about/documents/Cry%20of%20Creation.pdf; accessed January 5, 2006.

ENERGY STAR for Congregations (2000). *Putting Energy Into Stewardship* (guide). EPA, Washington, DC. Available at: http://www.energystar.gov/congregations/; accessed January 5, 2006.

Golliher, J. and Logan, W. B. (eds.) (1995). *Crisis and the Renewal of Creation: World and Church in the Age of Ecology.* London, New York: Continuum International Publishing Group.

Gottlieb, R. S. (ed.) (2003). *The Sacred Earth: Religion, Nature, Environment.* Oxford: Routledge.

Green Faith Initiative (2004). *The Green Faith Guide.* Washington, DC. Available at: http://www.dcenergy.org/Green%20Faith%20Guide.pdf; accessed January 5, 2006.

opinions in the pews as irrelevant. Parish leaders must and do respect the dignity of every human being no matter what opinions or beliefs they maintain. It is encouraging to witness over the past several years, however, that parishioners across denominations are coming to realize that ecological issues are spiritual issues and thus belong in congregational life. The sections below speak in more detail about how to approach climate change theologically. Drawing on my experience as executive director of The Regeneration Project (www.theregenerationproject.org; see also Bingham, 2002, 2004) in San Francisco, California, and as an Episcopal priest, I will give examples for how clergy, to be credible and persuasive in a congregation, must use the teachings of their denominations to explain the moral opinions they preach. In the Judaeo-Christian tradition, for example, there are mandates throughout scripture and biblical history calling people of faith to be good stewards of God's Creation, and resources are readily available to support that claim. Similar teachings and resources exist in other religious traditions.

Moral pitfalls: A friendly warning

To begin this exploration of how to talk about climate change in a religious context, I offer this anecdote, an example from my own early – and difficult – beginnings of bringing climate change into my parish. I preached a sermon in which I said that people who drive around in big monster cars are only thinking about themselves and their own protection. It was selfish behavior and did not show love of neighbor. It was one of my first attempts at trying to lead people to realize that every one of their behaviors counts and affects others. We matter and our behaviors matter. I went on to argue that in the event of a collision with a smaller car, the driver of the big car is more likely to kill the passengers in the other car. If we all took that attitude the cars would get bigger and bigger and we'd all soon be behind the wheels of Hummers (see Plotkin, 2004). A gentleman walked out in the middle of my sermon when I made the bold statement that knowing the bad effects of SUVs on society and driving them anyway was a sin. I felt badly that he was offended, but six months later, that same man approached me and confessed that he had been thinking about what I said. He had bought a VW bug that he uses around town and for work. He kept his "big monster car" though, and drives it to the mountains. We had a hug and have become friends since.

I, too, learned my lessons from that sermon: First, no one likes to feel guilty (or worse, to be publicly shamed) (see also Moser, Chapter 3, this volume) and, second, over time the voice of clergy does have a powerful impact. For me, the experience was humbling and empowering at the same time. In order to maintain one's integrity as a member of the clergy, careful communication and great sensitivity to the feelings of our congregants are critical. We must show respect for our parishioners and never be so bold and controversial that a parishioner feels the need to walk out. A good pastor does not offend his or her parishioners. We may not get the chance for reconciliation.

Dialing into the moral compass

The facts and scientific evidence on climate change and our human role in it are convincing for many (see Leiserowitz, Chapter 2, this volume), but few of us, especially lay people, completely understand the way the climate system works. What we do believe, however, is that the climate has already shifted, storms and their impacts are becoming more severe, and nature is already responding to these shifts in various ways. Moreover, scientific evidence shows that there is more carbon dioxide in the atmosphere now than at any other time in the last 420,000 years, and maybe as much as 750,000 years. If humans are contributing to this problem, as science indicates we are, then humans can also influence a change in the other direction. Research on public attitudes and my own experience suggest that people want to know that there is something they can do to help solve the problem (FrameWorks Institute, 2001). Most people, religious or not, have an inner moral compass which suggests that if we created the problem, it is up to us to fix it. People want to know that we can and respond far better to a positive approach than to a doomsday one (see Moser and Dilling, 2004; see also Moser, Chapter 3, this volume). But even within these broad parameters of current public opinion, the issue still is how to talk about climate change in the religious context. The strategies offered below stem from my own mistakes and the lessons I have learned over the years about how to frame the moral argument for protecting our climate and our future.

Relying on credible information sources

Most clergy have little scientific background, and thus must rely on other credible sources to understand and explain climate change. In the past the media have created a sense that climate science is highly contested. More recently, however, most newspapers are beginning to reflect more adequately the overwhelming scientific consensus that the issue is real and under way (see Boykoff and Boykoff, 2004). The most useful, credible, and up-to-date information sources I have found include the Union of Concerned Scientists (http://www.ucsusa.org/global_environment/global_warming/index.cfm), Environmental Defense (http://www.undoit.org/), and the Intergovern- mental Panel on Climate Change (http://www.ipcc.ch/), representing the findings of some 2,500 scientists from 55 countries[1] (impressive numbers for parishioners). But even relying on expert knowledge does not mean religious leaders all of a sudden should pretend they have become the experts themselves. Admitting to not know the answer to a specific question and acknowledging that not all scientists agree helps maintain credibility. Even being familiar with some of the names, backgrounds, and arguments of those who do not agree can be helpful (see McCright, Chapter 12, this volume). Fred Singer, for example, is one who disagrees with the scientific consensus; he has no recent peer-reviewed publications, and his work is funded by the fossil fuel industry (Environmental Defense, 2004). The strategies of "climate contrarians" like him are actually recycled from previous anti-environmental and public health campaigns (Monbiot, 2006). For example, doctors and scientists were telling people that smoking caused cancer 20 years before the tobacco industry would admit to it. As trials later brought to light, the tobacco industry meanwhile paid some doctors and scientists to dispute what most other experts were saying. The fossil fuel industry has taken much the same attitude about the link between burning fossil fuels and climate change. From a moral standpoint, however, the question becomes whether we want to take the chance that if scientists are correct about global warming and we do not take heed, it may soon be too late to prevent the worst impacts from this growing global problem.

Acknowledging doubts and making room for the doubters

No matter how solid the scientific evidence, there will always be those who are not (yet) convinced (see the discussion on "naysayers" in Leiserowitz, Chapter 2, this volume). There is not one major social issue where that is not the case. Expecting doubt from the audience and having to deal with vocal "non-believers" should be part of any clergy's repertoire. Another anecdote from my own experience makes that point: About six years ago, when I was just beginning to give talks and teach about a religious response to climate change, I made a lot of mistakes. I used to talk about our dependence on fossil fuel for energy in this country and how wrong it was. On one of those occasions, a man in the audience came prepared with paper and pencil. I saw him furiously taking notes while I was talking, nodding his head, and it actually made me glad to see someone so intent on learning that he wrote things down. After 30 minutes, I came to my bottom line that we had the technology to use renewable, clean energy, which does not produce heat-trapping gas emissions and does not harm Creation. I added that as religious people and as a major moral voice in this country it was our responsibility to lead the way whether it cost more or not. During the Q&A, the first hand to go up was that of the man with the notepaper. Even though other hands were up, too, I called on him (don't make this mistake either). The next ten minutes he spent taking my presentation apart point by point. It turned out that he was a retired nuclear engineer, he disputed most of my arguments for renewable energy and his bottom line was if the congregation listened to this "little lady," the women would be back using scrub boards to wash clothes, there would be no electric toasters or TVs, and most of us would be using horses for transportation, reading by candle light — essentially a return to the Middle Ages. He went on to pick apart the science with an article from *The Wall Street Journal* and poked holes in my moral argument with the dagger of the only thing that really mattered — economics.

Unprepared to discuss nuclear issues and long before the terrorist attacks of 9/11, I had little to come back with. This incident taught me much humility. I learned not to present myself as one with all the answers, as the one who holds "the Truth." It also taught me to make a better effort about knowing my audience: I begin every talk now with "who's out there?" I thank people for their interest and questions about the subject. I expect that many may know more than I do about climate change, and simply come looking for companionship and like-minded thinkers. I have also learned never to call on the first hand that goes up. When someone challenges me, I acknowledge his or her point, but no longer feel obligated to answer. In my role as presenter I can also facilitate the discussion, passing the question or comment on to the audience and engaging them in the debate. Moreover, I often begin by asking, "How many of you doubt that the climate is warming?" And then, "How many believe that human behavior is playing a major role in climate change?" Thus, right up front the audience identifies itself and the naysayers know they are in the minority — which is usually the case.

Drawing strength from history

Even if we succeed in convincing our audience of the science of climate change, the problem may appear so new and daunting to them that they cannot see their way through it. It can be useful to draw a lesson (pointed out to me by David Orr, Professor of Environmental Studies at Oberlin College) from an enormous case of social injustice that we Americans sought to fix and eventually did: slavery. When slavery was still a common practice, southern slave owners and businessmen argued that their economy depended on slaves and that everything, including the lives of the slaves, would come apart if slavery were abolished. Some slave owners even claimed — outrageously — that the slaves liked being slaves. It rings of Fred Singer's claim that everyone likes a warmer climate. In retrospect, it is hard to believe that such claims and falsehoods ever were part of the public debate, but the parallels to the current public debate about fossil fuel use are unavoidable. History may well prove how foolish our dependence and insistence on the use of fossil fuels are. We currently are dependent on coal, oil, and gas, but there already are, and we will create additional, options to wean ourselves off that dependence, so that we do not destroy Creation and human health or have to send our young to war to secure those politically unstable parts of the world from which we import much of our oil.

Faith into action: cleaning up our own house — what are we called to do?

After I have made the case for climate change and the human responsibility for much of the warming already observed, and after I have shown how we have solved seemingly impossible problems before, parishioners want to hear about the alternatives. What can be done? What can we do? It is important to illustrate what is being done around the world, in the United States, and even in one's state or community. It is particularly encouraging and empowering to tell congregations what other parishes do — in other words, what they themselves could do, right in the place where they come to worship. Wind is the fastest-growing electricity industry in the world (e.g., AWEA, 2004a). Other countries are already using wind far more than America does, yet we, too, have the technology. We are sending it overseas while we could employ it here more frequently, and create jobs in the process (AWEA, 2004b). I explain to my congregation that in order to encourage the further development of renewable energy sources such as wind, solar, geothermal energy, and biomass, it would only be fair to increase the limited tax incentives and subsidies they currently get to make them competitive with fossil fuels (which are heavily subsidized). This resonates with Christians and many other faith traditions, given their commitment to justice. Pointing to local or regional examples of successful wind farms and solar houses allows people to see that those alternatives work (e.g., AWEA, 2004c; NREL, 2004). Telling people of examples from other parishes where energy efficiency measures are being taken, or where renewable energy has been installed, will bring practical solutions even more "down to Earth." For example, there is a congregation near San Jose, California that is getting 70 percent of its electricity from solar panels on its roof (Shir Haddash, 2002). The congregation raised the money among its own members and invested in the future. Similar steps were taken at the Roman Catholic Cathedral in Los Angeles (Torres, 2002). An Episcopal church in the center of Boston installed a geothermal system that takes care of all its heating and cooling needs (Trinity Church of Boston, 2003). The Interfaith Power and Light Campaign, currently has over one thousand congregations who have agreed to cut CO_2 emissions and practice conservation in order to fulfill the covenant agreed upon when they joined our program. The main focus of The Regeneration Project is to encourage Interfaith Power and Light programs in all 50 states. The project currently has affiliates in 20 states (access state affiliates via: http://www.theregenerationproject.org/ipl/).

Becoming Earth stewards

Between showing people what can be done and getting them actually to do it often lies one big barrier: denial. Denial makes us sound a bit like the apologists (who rejected the existence of the historical Jesus of Nazareth), but in some cases denial can be a necessity to retain our sanity. On the other hand, denial is often simply the reaction that comes out of fear and a sense of disempowerment (Moser and Dilling 2004; Moser, Chapter 3, this volume). Few of us are politicians with a lot of power. We are simply citizens concerned about our future and the state of the planet that we are leaving for our children and grandchildren. As people of faith who sit in pews and profess a love for God and God's Creation, we have a responsibility to be stewards of that Creation. So, while we may not be the experts with knowledge, the politicians with power, or the economists who can calculate the most cost-effective solutions, we are called to set an example of what is righteous and just. As the biblical scripture says: "What does the Lord require of you but to do justice, and to love kindness and to walk humbly with your God?" (Micah 6:8).

Lighten up on guilt

People do act on climate change for many reasons. But using guilt to motivate action rarely works − not even in faith communities where congregations are used to lessons based on guilt. The subject of climate change is still too new and too abstract and untouchable to lay blame on individuals or communities. Most of us do not know we are destroying Creation or polluting our neighbor's air when we turn on the lights or drive our cars (see Bostrom and Lashof, Chapter 1, this volume, on the difficulty of understanding the causal mechanisms behind global warming). Rather than making people feel bad, it is more important to point out that we now know that every single one of our actions matters and that it is time to become more mindful of our behaviors (a concept that is already quite familiar to Buddhists). What is powerful about this argument is that if everything you do matters, *you* matter. People like hearing that. We all want to matter, and we need to be reminded. By contrast, "guilt-tripping" people for polluting the air of poor communities when they turn on an appliance is typically not received well by parishioners. I tend to lose my audience if I make them feel bad. But the same information can be passed on in less direct ways, yet ironically in ways that can be heard more easily. "Did you know that the highest rates of asthma, lung disease, and respiratory problems are found in those communities where the coal and gas-powered plants are located?" Poor communities certainly are aware. To them, the best link between those coal plants and global warming is air quality. They may at first not care about the global issue, but they can see the link via something that they do care about deeply.

This suggests that when we try to find ways to talk about global warming, we must be aware of the needs and concerns of the community to which we are speaking. Once I was speaking at a breakfast gathering in the African-American Baptist church in Bay View Hunter's Point, a poor neighborhood in San Francisco. The rector, a friend, warned me ahead of time not to emphasize global warming but rather to talk about energy savings to his congregation or I would lose them immediately. "If you want them to cut down on energy use, talk to them about saving money, not carbon dioxide emissions." So I held up a utility bill from a neighboring congregation and showed them the decrease in cost over a year due to the energy efficiency measures they had taken. The group "got it" instantly and thus saw the link between saving energy, saving money, and saving the planet.

Telling stories of interconnectedness

Helping people see the connections between their actions, their neighbors, and the global environment in their terms is not only necessary to engage them; it can also be inspiring. For all of human history, telling stories has served to inspire us and to remind us of how we are connected to nature. Take the example of collecting driftwood, shells, and stones, and taking them home in our pockets. Why do we do that? (And there is hardly a person who hasn't.) We do it because they are part of who we are. We feel connected to those sticks and stones and shells and they call to us. No one can deny picking up some little piece of nature and carrying it around — a token, a symbol, a reminder of a special time or place. But the connection goes even deeper. We share the DNA with an oak tree. It is our brother. This way of putting it serves as a reminder that we are connected and that everything is related and interdependent upon everything else. Therefore, each one of our behaviors affects the things that we are connected to. We cannot emit carbon dioxide into the air and not affect something somewhere. It makes sense to people once they have identified with the metaphor of being related to oak trees and shells. They begin to see that everything is intrinsically interdependent. This means that, if we poison the air, it is not only our neighbor who has to breathe it, but we ourselves, too. The same goes for the water, oceans, and rivers.

Loving our neighbors

Then there is our interconnectedness to each other. Typically, those who are affected first and most severely by the use of coal for electricity are those who work in those power plants and those who live nearby. As mentioned above, the highest rates of lung disease, respiratory problems, and asthma are in the poor communities where these facilities are located. No one likes to think that they are poisoning their neighbor's air every time they turn on the lights, particularly if they are called to "love their neighbor." (Again, this point is best made as information, not by directly appealing to guilt.) Thus, unhealthy air and power plants can be linked to a particularly sensitive social justice issue. Using less dirty fossil fuel power sources thus not only cleans the air for us and our neighbors and benefits the climate, but it alleviates social injustice, too.

How this message of interconnectedness and social justice works was nicely demonstrated during the California "energy crisis." The California Interfaith Power and Light leaders were preaching and teaching that what we turned off, our neighbors could keep on. As Californians were threatened with rolling blackouts, the religious community responded with the charge that if you love your neighbor (who needs electricity to live), then turn off everything except what you absolutely need for your own survival. The basis for our call was the knowledge that most of us want to help, not hurt, our neighbors. It contributed to the astounding result that Californians reduced their electrical use by 15 percent during that time (California Public Utilities Commission, 2001). The experience confirmed for me that people will respond positively when they understand a situation and are empowered with making a choice that will either help or harm others. We all have seen communities helping each other after disasters. That moral ethic is a vast reserve to tap into when looking for positive behavioral changes related to climate change.

Making the far-away tangible, meaningful

Research and experience show that as soon as climate change becomes something that people can feel, see, or experience close to home, it becomes persuasive and meaningful (FrameWorks Institute, 2001). But one can also tap into the "love thy neighbor" ethic to bring compassion to places and people in far-away regions already experiencing the early impacts of climate change (see also McNeeley and Huntington, Chapter 8, this volume). Thus, with or without visuals, concrete examples of how Bangladesh and islands in the South Pacific are already impacted by the rising seas can be very powerful. The inhabitants of Tuvalu have asked to be taken in by Australia because their island is rapidly disappearing. A total of 15,000 environmental refugees are searching for new homes (e.g., Paddock, 2002). This example offers an opportunity to ask parishioners how they would feel if everything they knew and cared about were about to vanish forever beneath the ocean. It is an opportunity to describe the patterns of increasing disasters worldwide – entire seasons of crops being lost, homes destroyed, entire communities wiped out and many, many lives lost (Simms, 2003). These are the kinds of situations that make people think about the implications of continuing with business-as-usual, i.e., about not changing their behavior. "What if it were me?" It is essential that we give evidence of the harm we are causing to others with our unconscious and irresponsible behavior – without actually calling it that. When people feel that they are to blame and the problem is insurmountably large, they stop listening, turn off, go under-ground, and pretend it's really not a problem (Macy and Brown, 1998). While people won't always do the right thing every time, with gentle persuasion and a consistent message, they will try.

Conclusion

As the scientific consensus on climate change solidifies, the moral obliga-tion to act on this knowledge becomes more pressing. While scientists refine their understanding, politicians debate policies, and engineers develop more answers to deal with this global problem, the people who come into our churches, temples, and mosques need to be acquainted with climate change and become engaged in creating and supporting solutions.

The strategies offered above lay out the enormous task before clergy and religious leaders who believe that climate change is a topic worthy of moral and spiritual attention. Our task is to educate gently, but with facts, and to tell stories that inspire and remind us of our interconnectedness with nature and our neighbors, nearby and far away. Our task is to create a safe and welcoming place for our doubts and fears as well as for our desires to help and leave our children a livable world. Our teachings must remain theological in nature and include not only the bad stories, but the good news, too. In that sense, there is nothing particularly different or new about climate change. If we communicate well, people of faith will want to be part of the solution rather than part of the problem, and intentions and motivations to change, and eventually the behaviors, too, will change. How priests and ministers, rabbis, and imams will combine the arguments of science,

economics, jobs, technological innovation, environmental stewardship, and
social justice will differ from individual to individual and from audience
to audience. The subtleties and emphases will differ by situation. But there
hasn't been a story yet − if told from the heart and with a good deal of
humility − that couldn't move a mountain.

Note

1. See http://www.un.org/News/facts/; accessed January 16, 2006.

References

American Wind Energy Association (AWEA) (2004a). *Global Wind Power Growth
Continues to Strengthen*. Washington, DC: AWEA. Available at http://www.
awea.org/news/news040310glo.html; accessed January 5, 2006.
AWEA (2004b). *Wind Energy and Economic Development: Building Sustainable Jobs
and Communities*. Washington, DC: AWEA. Available at: http://www.
awea.org/pubs/factsheets/EconDev.PDF; accessed January 5, 2006.
AWEA (2004c). *Wind Energy Projects Throughout the United States of America*.
Washington, DC: AWEA. Available at http://www.awea.org/projects/; accessed
January 5, 2006.
Bingham, S. (2002). The regeneration project. *Grist Magazine*, March 25. Available
at: http://www.grist.org/comments/dispatches/2002/03/25/bingham-rp/
index.html; accessed January 5, 2006.
Bingham, S. (2004). The energy dilemma. *The Witness Magazine*. Available at:
http://www.thewitness.org/agw/bingham042204.html; accessed January 5,
2006.
Boykoff, M. T. and Boykoff, J. M. (2004). Balance as bias: Global warming and the
US prestige press. *Global Environmental Change*, **14**, 125−36.
California Public Utilities Commission (2001). *CPUS 2001 Energy Efficiency and
Conservation Programs*. Report to the Legislature. San Francisco, CA.
Available at: http://www.cpuc.ca.gov/static/industry/electric/energy+efficiency/
01energyconservationrep.htm; accessed January 5, 2006.
Environmental Defense (2004). *Global Warming Skeptics: A Primer*. Available at:
http://www.environmentaldefense.org/article.cfm?contentid=3804; accessed
January 5, 2006.
FrameWorks Institute (2001). *Talking Global Warming* (Summary of Research
Findings). Washington, DC: FrameWorks Institute.
Macy, J. and Brown, M. Y. (1998). *Coming Back to Life: Practices to Reconnect Our
Lives, Our World*. Gabriola Island, BC: New Society Publishers.
Monbiot, G. (2006). *Heat: How to Stop the Planet Burning*. London: Allen Lane.
Moser, S. and Dilling, L. (2004). Making climate hot: Communicating the urgency
and challenge of global climate change. *Environment*, **46**, 10, 32−46.
National Renewable Energy Laboratory (NREL) (2004). *Innovation for Our Energy
Future*. Golden, CO: NREL. Available at: http://www.nrel.gov/; accessed
January 5, 2006.
Paddock, R. C. (2002). Tuvalu's sinking feeling. *The Los Angeles Times*,
October 4, A1.

Plotkin, S. (2004). Is bigger better? Moving toward a dispassionate view of SUVs. *Environment*, **46**, 9, 8–21.

Shir Haddash (2002). *Congregation Shir Hadash Goes Solar*. Available at: http:// www.shirhadash.org/solar/solar-020919.html; accessed January 5, 2006.

Simms, A. (2003). Unnatural disasters. *The Guardian*, October 15. Available at: http://www.guardian.co.uk/climatechange/story/0,12374,1063181,00.html; accessed January 5, 2006.

Torres, M. L. (2002). Merging spirit and sunlight in downtown L. A. *Tidings Online*. Available at: http://www.the-tidings.com/2002/0823/solar.htm; accessed January 5, 2006.

Trinity Church of Boston (2003). *Trinity's Geothermal Heating and Cooling System*. Available at: http://www.trinityboston.org/arc_bld_upd-events.asp; accessed January 5, 2006.

10

Einstein, Roosevelt, and the atom bomb: lessons learned for scientists communicating climate change

Lucy Warner

University Corporation for Atmospheric Research

There has arguably never been a more critical moment in the history of communication between scientists and those in power than 1939. Physicists had demonstrated nuclear fission. Some speculated that uranium could be harnessed to trigger a chain reaction, unleashing the vast stores of energy locked inside the atom, and that this energy could be channeled to build a bomb with unprecedented destructive force.

The story of how a few prescient scientists convinced President Roosevelt of the danger and potential of the atom bomb is a fascinating case study, riddled with delays, uncertainties, and miscalculations that could be a cautionary tale for climate communicators. Many of the lessons to be drawn from that iconic episode almost 70 years ago echo themes familiar to climate change communicators.

What follows is an attempt to illuminate today's very different scientific urgency within this highly contrasting framework. The hope is that both the differences in context and substance and the sometimes-surprising similarities may provoke reflection on the difficulties of communicating the threats inherent in world-altering risks – the atom bomb on the one hand and the more slowly ticking time-bomb of climate change on the other.

In the narrow context of the story below, the communications campaign was a success – the scientists persuaded the government to build the bomb and, some argue, shortened the Second World War as a consequence. Beyond this context, of course, the bomb has led to an ongoing nuclear proliferation that threatens the stability of the world, is producing growing numbers of nuclear states, and most recently brought a war justified on the grounds of a nuclear threat.

The communication of climate change is a work in progress, arguably a partial success depending on whether the frame of reference is regional, national, international, governmental, or non-governmental. Despite the obviously limited analogy, the story of the early precursors to the building of the atom bomb brings up issues relevant to climate change communicators: the ambiguities of the risk and the difficulty of crafting messages, identifying the appropriate messengers, translating the concerns into action, and, ultimately, changing the world for better or worse.

The story

The story[1] begins with Leo Szilard, the Hungarian physicist who first conceived of the nuclear chain reaction and grasped its potential dangers. In 1939, his experiments at Columbia University demonstrated that uranium, when split, would emit more neutrons than it captured. Those experiments verified the possibility of a chain reaction, and Szilard and his colleague Enrico Fermi began work on the design of a nuclear reactor. Aware of the potential military significance of their work, they and a small group of other European émigrés to America, including Edward Teller, Eugene Wigner, and Isidor Rabi, debated whether to publish or repress work in this potentially world-altering field. Their hand was forced by the publication of two scientific papers from a French team in the early weeks of 1939 concluding that uranium would most probably chain-react under the right conditions (Joliot, von Halban, and Kowarski, 1939a,b). Szilard, Wigner, and Teller decided that no time should be lost in warning the Roosevelt administration before the Germans gained the upper hand in nuclear experimentation. The group knew little about what their German colleagues were doing in this area, but knowing the lethal potential of a uranium chain reaction, they assumed they were in a race to build the world's first nuclear weapon.

Getting the word out

Fermi had briefed a group of naval officers in mid-March of 1939 on recent advances in chain-reaction research and was met with cool condescension. His listeners were reportedly ignorant of his scientific credentials and suspicious of his motives (Herken, 2000: 7). Following this less-than-enthusiastic reception, the group came up with a plan to involve a scientist everyone had heard of: Albert Einstein. Their goal was to prevent the Belgian Congo, which had the world's largest reserve of high-grade uranium, from providing the element to the Germans. Since Einstein knew the queen of Belgium, the scientists decided to ask him to write to the queen asking her to keep the Congo's supplies out of German hands.

Szilard set out with Wigner as chauffeur one summer morning to find Einstein's vacation home on Long Island. Einstein had surprisingly never heard of the work on possible chain reactions, but he immediately grasped its significance and offered to help. As Szilard commented, "The one thing most scientists are really afraid of is to make fools of themselves. Einstein was free from such a fear and this above all is what made his position unique on this occasion" (Rhodes, 1986: 305). He would help them craft a letter.

Proper channels

In lieu of the queen, Einstein suggested they get in touch with an acquaintance in the Belgian cabinet, but all agreed that before contacting another government, the US State Department should be notified. Einstein drafted a letter to his Belgian acquaintance, in German, which Wigner took away to be translated and typed, and to which Szilard added a cover letter to the State Department.

Meanwhile, Szilard had located a messenger to Washington: Dr. Alexander Sachs, a Russian-born economist and biologist, who had no background in nuclear physics but had an advantage that turned out to be critical – he understood the political process. As it happened, he knew Roosevelt personally. When Sachs and Szilard met, the economist came up with what struck Szilard as a radical suggestion – he would take a statement to Roosevelt in person. There were, at the time, few direct routes for communication between the scientific community and the White House.[2] Szilard was reportedly stunned by the boldness of Sachs's proposal, but the group of physicists agreed it was worth a try (Rhodes, 1986: 306).

The message and the messenger

In late July 1939, Szilard went for tea to Einstein's house, this time with Teller as chauffeur, to finalize the letter to the president. Szilard and Sachs had suggested some additional text, and Einstein ultimately signed two versions – one long, one short. "How many pages does the fission of uranium rate?" Szilard is quoted as musing (Herken, 2000: 9). There was also much discussion of the proper emissary for what was now a letter from the world-famous Einstein to the president of the United States. Names proposed were as diverse as Charles Lindberg, financier Bernard Baruch, and Karl Compton, president of the Massachusetts Institute of Technology. In the end, the job went to Sachs.

What to ask for

The final draft is neither long nor technical by modern standards,[3] but its recommendations are curiously mild: one, to appoint a liaison with the physicists working on chain reactions in the United States, and two, to speed up the research through private funds and use of industrial laboratory equipment.[4] In an addendum, Szilard included a long discussion of the possibilities and dangers of fission and sent it to Sachs on August 15, a month after he had first visited Einstein. In the end, the deliberations over how to word the message didn't matter, because Sachs was to use none of them.

Delivering the message

Roosevelt's calendar was understandably full. A week after Sachs received the letter, on August 23, 1939, Hitler invaded Poland; war was declared on September 1. Sachs thought it essential to see Roosevelt in person, but by October 3, he had failed to get onto Roosevelt's calendar and the physicists were about to give up on him. Finally, on October 11, Sachs met with an aide, who immediately realized the information was worth the president's time and escorted him into the Oval Office.

Over brandy, Sachs attempted to impress on Roosevelt the importance of the message he had been consigned to deliver. Despite the carefully prepared draft, Sachs abandoned Einstein's letter and Szilard's memorandum in favor of his own written paraphrase. As he explained it, "I am an economist, not a scientist, but I had a prior relationship with the President ... No scientist could sell it to him" (Rhodes, 1986: 313). Before warning of the new weapons of mass destruction, he emphasized the positive – nuclear power, nuclear medicine. His explanation to Roosevelt has been described as initially "leaving the president baffled and uncomprehending until Sachs pointed out that it was likely others would get the bomb unless Americans built it first" (Herken, 2000: 10). When Sachs finally got to the point, Roosevelt interrupted to say, "Alex, what you are after is to see that the Nazis don't blow us up." He promptly summoned his aide with the words, "This requires action."

The long road to implementation

True to his reputation, Roosevelt was the skilled communicator here – rescuing the well-intentioned but rambling attempts of Sachs through his own abilities to listen and to frame the issue in immediate and relevant terms. But despite Roosevelt's call to action, there was a long way to go before implementing an effective atomic bomb program.

A committee including the nuclear physicists and military personnel met and issued a report within three weeks. It recommended a crash program of experiments but requested only a few thousand dollars. Members of the committee worried that if an expensive program undertaken on the advice of foreign scientists were to fail, they would be the target of a congressional investigation (Herken, 2000: 12) — and there was no scientific consensus that a bomb could be built. The following summer, the fission project was put under Vannevar Bush and his newly formed National Defense Research Council. (An electrical engineer and inventor, Bush was then president of the Carnegie Institution.) It was not until November of 1941 that Bush, initially skeptical, delivered to Roosevelt a report from the National Academy of Sciences affirming that an atom bomb was feasible and two months after that, in January 1942, that Roosevelt approved an atom bomb program.

Three years after Szilard and Einstein had penned their letter to Roosevelt, the world's first nuclear reactor was born under the bleachers at the University of Chicago's Stagg Field. Within a year, the Manhattan Project was launched. The first atom bomb was detonated on July 16, 1945, exactly six years to the day from Szilard's first meeting with Einstein on Long Island.

For better or worse?

The scientists themselves had begun with a utopian vision of the outcome of their work. Early in his life, Szilard had been influenced by a tract by H. G. Wells in which scientists, industrialists, and financiers form a small band to establish a world republic — the *Bund*, as he called them. He subsequently read H. G. Wells's prophetic *The World Set Free*, a 1914 novel in which a world war featuring an atom bomb leads ultimately to a peaceful world government. Although saving the world from fascism was clearly an overriding motive in these scientists' minds, Szilard's idealism was quickly dampened in the event. After watching the debut of the reactor under Stagg Field, he is said to have remarked to Fermi that "this day would go down as a black day in the history of mankind" (Herken, 2000: 14). And right he was, given the sequelae of Hiroshima, Nagasaki, 50 years of Cold War, and ongoing nuclear proliferation.

The climate change analogy

What relevance does this almost 70-year-old account have to the issue of communicating a modern urgency – the potential impacts of climate change? Clearly, the world is a much different place. Few American scientists require chauffeurs and most write coherent English. Few would dream of having the opportunity to explain an Earth-shattering scientific discovery over a glass of brandy in the Oval Office. And now there exists an enormous, government-supported science complex that simply didn't exist in the United States before the Second World War.

The outlines of the 1939 scenario and the current one have little in common at first glance. In 1939, the issue was relatively clear and simple. A small group of scientists, in close personal touch, had a consensus message to deliver to a single person with the power to take direct action. It was wartime, and the future of the nation was at stake. The scenario of climate change is fundamentally different. A large group (in the thousands) of scientists from around the world is trying to deliver a message whose broad outline is becoming clear but whose details are not. They need to approach a diverse group of policy-makers in different regions and countries whose futures will be affected in varying ways, many still uncertain. At the national level, even an activist head of state with deep understanding of the scientific problems can only address some climate issues in isolation from the international community, local and regional interests, economic sectors, and social/consumer issues. In fact, unlike with the top-secret discussions of 1939, the diverse groups who make up the general public are front and center in efforts to win hearts and minds and influence behaviors.

Climate change communication is now arguably more a matter of policy, politics, and values than it is of science. The big-picture science is out there, in the media, in the popular imagination, and even in international protocols, national policies elsewhere in the world, and regional programs and efforts in the United States. But the critical issue of communicating the certainties, the uncertainties, the risks, the costs, and the consequences (both scientific and societal) continues to be of vital importance to scientists, policy-makers, and communicators. Several of the issues posed by the nuclear fission example offer insights into issues we face today.

Getting the word out

There are two overriding themes that tie these stories together: the sense of urgency implied by the risks, and the political context that overshadows the science. In both cases, there is also a good deal of scientific uncertainty. The threat of a wartime enemy obtaining a weapon of mass destruction, of course, is much more obvious than the slowly creeping threat of climate change, which is likely to take place over decades to centuries and to have multiple, complex effects (in some cases positive). Much like the potential peaceful uses of nuclear power, the apparent potential benefits of a warmer climate can undermine the perception of the larger underlying threat.

The trap for the climate change communicator is the temptation to find the climate equivalent of the nuclear bomb (see also Moser, Chapter 3, this volume). Every weather catastrophe or natural disaster poses the potential for an erroneous or overstated connection to the climate story – a year of drought, a season of heavy rain or snowfall. Most recently the connection between the devastating hurricanes of 2005 and global warming has been much debated in the scientific literature and in the press. "Hurricanes Intensify Global Warming Debate" was *Discover* magazine's top story of 2005, and eminent hurricane researchers have gone back to their data to issue new analyses in *Nature* and *Science*. Opinion seems to be shifting toward a connection between hurricane intensity and global warming, but as *Discover*'s headline emphasizes, it's still a debate. When the dust settles the complex web of causalities can't ever be reduced to global warming "causing" any specific weather or climate event. Sea-level rise from warmer oceans will cause widespread coastal flooding, but at a much slower pace and, it is hoped, with better warning mechanisms than the tsunami that killed over a hundred thousand in December 2004.

Naysayers notwithstanding, scientists generally deliver a coherent message that global warming is real but have no clear pathway to get the message out to the public or policy-makers. They need help in deciding when to issue a press release, when to hold a press conference, when to schedule a congressional briefing (see Cole with Watrous, Chapter 11, this volume, on one example of a collaboration between scientists and advocacy groups that helps do just that). Reporters and politicians may not know which experts to consult, let alone how to weigh conflicting evidence. By the same token, many scientists hesitate to speak out on directly policy-relevant issues, either because the scientific culture discourages self-promotion or because they feel that the policy arena is not objective in the scientific sense. Szilard's comment that "The one thing most scientists are really afraid of is to make fools of themselves" is hardly unique to the research community, but a culture that values accuracy and precision does not take kindly to over-statement or premature conclusions. Take, for example, James Hansen's controversial statement before a congressional committee in 1988 that global

warming had begun; it garnered a lot of press attention but was viewed by
many in the scientific community as overstatement or even grandstanding. It
is often an act of courage within the scientific culture to issue a call to action,
especially if that call is based on incomplete data, preliminary results, or (as
with Szilard's early intuition) a hunch. The scientific method screams against
these kinds of premature judgments and wars with a scientist's role as a
conscientious citizen.[5]

Proper channels

Within their academic circles at Columbia, the University of Chicago, etc.,
the nuclear physicists had departmental channels, and to a certain extent they
followed them in deciding what should be published and what kept
confidential. Today, climate change researchers disseminate their work
primarily through the peer-reviewed literature, and − in an attempt to get
wider media coverage and public attention − through the varying structures
of a university department or school, an agency, or a non-governmental
organization. The publication of peer-reviewed literature has little institutional oversight, but that is not always the case for press releases and other
forms of public information. In the realm of public information, institutions
tend to become more directly involved and the communications process is
complicated by the fact that the subject has become increasingly politicized.

A recent *New York Times* article documents incidents at National Oceanic
and Atmospheric Administration headquarters and at NASA's Goddard
Institute for Space Studies, where news releases on new global warming
studies were revised by political appointees to play down definitiveness or
risks (Revkin, 2004). It also describes how analyses from the Department of
Energy and the Environmental Protection Agency were ignored in the weeks
before the Bush administration reversed its election promise to curb power
plant emissions of carbon dioxide. A broader pattern of politicization of the
scientific process, including climate change, is alleged in a recent Union of
Concerned Scientists' report (UCS, 2004). The issue is not one of political
party − on regional and local levels, Republicans can be some of the most
active in enacting forward-looking climate policies, and Democratic
administrations are often guilty of using scientific information for political
ends.

In this politicized climate, the professional societies are taking on a more
direct role. The American Geophysical Union and the American
Meteorological Society have taken consensus positions on climate change

to supplement bodies such as the Intergovernmental Panel on Climate Change.

Institutions — just like individual scientists — that honor the culture of academic freedom, such as my own, universities, and NCAR's sponsoring agency, the National Science Foundation, must continually guard against the tendency to politicize results in either direction — by either sensationalizing or burying new climate findings. And, of course, within the context of academic freedom individual scientists must be careful to express their opinions responsibly in a highly charged atmosphere where comments can easily be misconstrued or taken out of context.

Most institutions have an infrastructure of media relations and government relations staff to help in both crafting messages and finding the appropriate channels for communication with policy-makers and the media. Scientists often do not take advantage of these resources, which lie outside their normal frame of reference. While few empirical data exist explaining this under-use, we can speculate that scientists may not know that they exist, how to use them effectively, or they may not feel comfortable with being in the media spotlight. Just as media activities can seem uncomfortably self-promoting, government affairs can seem tainted as lobbying. However, these professionals are often working to educate policy-makers to make better-informed decisions. There are also useful written guides and programs to train scientists as communicators (see, e.g., Wells, 1992; Kendall, 2000).[6]

The message and the messenger

Crafting the message may have seemed a laborious task to the harbingers of atomic power in 1939 — Einstein and Szilard labored over multiple drafts of two versions of their communiqué and Sachs abandoned these and created a third. While the physics they were describing was complicated, it was arguably nothing compared to the intricacies and uncertainties of climate change in all its potential scenarios across multiple natural systems, regions, cultures, and economic sectors. Today, there are primers on how to craft scientific communications — using the active voice, incorporating captivating metaphors, telling a story (see, e.g., Shortland and Gregory, 1991).

The answer to Szilard's question about how many pages the fission of uranium takes: only as many words as are needed to understand the metaphorical and actual bombshell — the atomic one. You don't need to know how to construct equations in a climate model to explain the model's results. By the same token, scientists shouldn't be expected and shouldn't expect themselves to understand the full consequences of new science or technology before seeking the ear of policy-makers or the public. For example, much as the Szilard *Bund* had inklings of both patriotic and ominous repercussions of their nuclear experiments, neither they nor Roosevelt could possibly have envisioned the consequences of detonating nuclear bombs, let alone the world that the nuclear era would usher in.

In Sachs's (somewhat biased) judgment, a scientist could not possibly have carried the information to Roosevelt. Of course he may simply have meant that none of them knew the president as he did, but scientists often still deputize economists or journalists or policy-makers or NGO advocates to carry their messages. If the message itself is an economic one or a policy one, this may be wise, but it is also a sign of the cultural reluctance to speak out.

The social or political entrée is as important as it has ever been in gaining an audience with a congressional committee staffer or member of the administration. Few people can aspire to a personal relationship with the president, but the name recognition of a reputable institution, coalition, or professional society will open doors. Public–private coalitions with collective interests can be particularly persuasive, as can coalitions of groups with multiple diverse interests.

What to ask for

Having expended tremendous energy to gain the ear of President Roosevelt, the atomic scientists ended up asking for very little − in both financial and organizational terms. Einstein's letter asked for a liaison to keep the government informed of ongoing developments in nuclear research and a fast track for funding. These were presumably well-thought-out requests, but stunningly modest, and it's tempting in hindsight to wonder why they weren't bolder. When you have the ear of those in power, what is it you really want done? Whether it's funding for additional research, equipment, or support for a policy initiative, give those in authority the information they need to support a request and ask realistically and forcefully for what is needed.[7]

Delivering the message

Roosevelt is the hero in this part of the narrative. He listened patiently to an economist summarizing the respectfully phrased recommendations of a group of physicists who had no experience in communicating to a non-scientist. ("Certain aspects of the situation which has arisen seem to call for watchfulness and, if necessary, quick action ...") Sachs even changed what today we would call the "spin" to try to make the message more palatable. Nuclear power posed good and evil potential. When Roosevelt finally grasped the story he responded spontaneously with the core point or punchline, what today we would call the sound-bite: "What you are after is to see that the Nazis don't blow us up." That's what Sachs should have said to Roosevelt. Craft the message to be brief and to the point; include the impact, the emotion, the relevance. There are lots of those messages in climate change, even given its complexity and nuance.

Sachs chose to emphasize the positive, watering down the destructive power of the atom by beginning with the potentially beneficial capabilities of nuclear power. It is a very human tendency to want to find the rosy side of potentially frightening or dangerous possibilities, and Sachs was trying to give a balanced view, perhaps in order not to frighten the president into paralysis. In the case of climate change, the drive to look on the bright side is used by the naysayers, who try to persuade the public and policy-makers that a warmer world will be like an endless summer vacation – where plants will grow better and New England will feel like Florida (see also McCright, Chapter 12, this volume). As mentioned above, most scientists are appropriately reluctant to oversimplify the course, causes, and implications of climate change. It is their business to deal in uncertainties and nuance – for example, that warming will not be uniform across the globe and that there will be "winners" and "losers" in global warming. There should be no contradiction between expressing these caveats and delivering a clear message or even a call to action, where appropriate.

The long road to policy change

It took exactly six years from Szilard and Einstein's first meeting to the detonation of the first atom bomb. Wigner described their efforts to prod the government into action as "swimming in syrup" (Herken, 2000: 11). Some of the physicists involved in warning Roosevelt reflected years after the war that their efforts to engage the government might have been more of a hindrance than a help to the cause. They speculated that research in Britain would have led eventually to the Allies' developing the bomb first. They even wondered whether the formation of Roosevelt's committee led to a false sense that the nation's interests were being looked after (Herken, 2000). Those involved in the IPCC document preparations, the US National Assessment, and similar efforts to deliver policy-relevant science may identify with that sentiment. National and international studies, committees, and protocols may in hindsight turn out to be less effective than economic and social forces in driving changes in energy consumption and technology (see chapters by Meyer and Arroyo/Preston, Chapters 28 and 21, this volume). We don't yet have the benefit of hindsight.

For better or worse?

The Szilard *Bund* was driven by idealism, patriotism, and dread at the same time. For much of the post-war era, the arms race was seen as a deterrent to war as well as a threat. Within the climate change debate, there are similar, if less extreme, undercurrents. Among some environmentalists, there is both a utopian flavor to depictions of simpler pre- and post-industrial civilizations and an almost apocalyptic vision of a world in which rapid climate change disrupts life as we know it (see, e.g., McKibben, 1990). No armageddon comparable to nuclear annihilation accompanies the curbing of greenhouse gases, although extreme right-wing economists and policy-makers paint dire pictures of the crippling of modern industrial society should the use of petroleum resources be curtailed.

Given the politically-charged implications of greenhouse gas controls, and the discomfort many scientists feel in extending their quantitative and analytic expertise into such speculative areas, the result is often that even clear communicators with responsible personal opinions prefer to keep their thoughts to themselves. From this professional communicator's perspective, it is important to remember that there are important and scientifically valid stories to tell while remaining squarely in the domain of science. The melting of Arctic sea ice, sea-level rise, and species migrations are all climate impacts with policy implications that can be communicated from within the realm of scientific observation and modeling.

For better or worse, the scientists who first conceived of a nuclear bomb were successful in getting the ear of power because all parties were motivated to defeat an enemy who threatened the world order. One wonders if there is anything that can or will lead to a comparable sense of urgency and unanimity in the context of climate change.

Notes

1. Much of this narrative is drawn from (Rhodes, 1986; Herken, 2000).
2. In 1939, Roosevelt had put in place a Science Advisory Board, and the National Research Council had been charged since 1916 with providing scientific expertise to the government on request. However, the former was focused chiefly on advances in agriculture and weather prediction, and the latter received few requests for technical advice (Herken, 2000: 8).
3. A reproduction of the final letter is available at http://www.dannen.com/ae-fdr.html; accessed January 5, 2006.

4. The letter also points out that Germany has taken over uranium supplies from Czechoslovakia but makes no mention of the Belgian Congo or American sources of uranium.
5. For a discussion of the old and emerging scientific paradigms, see Gibbons (1999).
6. The Aldo Leopold Leadership Program also offers an intensive training program to teach environmental scientists to communicate their work effectively to a variety of lay audiences (see http://www.leopoldleadership.org/content/; accessed January 11, 2006).
7. The issue of advocacy is controversial within the scientific community; many scientists are not comfortable in this role. Organizations can serve as mediators when scientists are uncomfortable making policy recommendations themselves (see chapter by Cole with Watrous, Chapter 11, this volume).

References

Gibbons, M. (1999). Science's new social contract with society. *Nature*, **402**, C81.

Herken, G. (2000). *Cardinal Choices*. Stanford, CA: Stanford University Press.

Joliot, F., von Halban, H., and Kowarski, L. (1939a). Liberation of neutrons in the nuclear explosion of uranium. *Nature*, **143**, 470.

Joliot, F., von Halban, H., and Kowarski, L. (1939b). Number of neutrons liberated in the nuclear fission of uranium. *Nature*, **143**, 680.

Kendall, H. W. (2000). *A Distant Light: Scientists and Public Policy*. New York: Springer-Verlag.

McKibben, B. (1990). *The End of Nature*. New York: Anchor Books.

Revkin, A. (2004). Bush vs. the laureates: how science became a partisan issue. *The New York Times*, October 19, p. D1.

Rhodes, R. (1986). *The Making of the Atomic Bomb*. New York: Simon & Schuster.

Shortland, M. and Gregory, J. (1991). *Communicating Science: A Handbook*. New York: John Wiley & Sons.

Union of Concerned Scientists (2004). Scientific integrity in policymaking: An investigation into the Bush administration's misuse of science. Available at: http://www.ucsusa.org/scientific_integrity/interference/reports-scientific-integrity-in-policy-making.html; accessed January 11, 2006.

Wells, W. G., Jr. (1992). *Working with Congress*. Washington, DC: American Association for the Advancement of Science.

11

Across the great divide: supporting scientists as effective messengers in the public sphere

Nancy Cole
Union of Concerned Scientists

Susan Watrous

Is it enough for a scientist simply to publish a paper? Isn't it a responsibility of scientists, if you believe that you have found something that can affect the environment, isn't it your responsibility to actually do something about it, enough so that action actually takes place?

Mario Molina (2000)[1]

Most breaking science stories have a news lifespan of a few days before they retire, relegated to the pages of arcane journals. But there was one science story in 2004 with a far longer – and wider – news cycle than many. In the world of climate science, this story broke ground in several areas: it delivered a range of the latest climate model simulations, compared higher- and lower-emissions scenarios, and calculated the resulting impacts for a specific region.[2]

On August 16, 2004, the *Proceedings of the National Academy of Sciences* (*PNAS*) published a report on the possible future of California's climate (Hayhoe *et al.*, 2004). Authored by a team of 19 scientists,[3] the report highlighted two sets of scenarios representing higher and lower pathways for future emissions. The study's authors found that, under the higher-emissions scenario, California's average summer temperatures could warm by between 9 and 18 °F – twice as much as in the lower scenario – by the end of the twenty-first century. In this higher-emissions scenario, Californians could expect up to eight times as many heatwave days, with heat-related mortality rising significantly. Projections for Los Angeles, for example, included between 7,900 and 11,820 heat-related deaths over the decade of the 2090s, without accounting for likely increases in population over that time. The Sierra Nevada snowpack would be reduced by 73–90 percent, severely compromising California's water supply. The study's lead author Katharine Hayhoe put the temperature rise into a frame certain to get the attention of media-saturated Californians: during the summer, the state's "inland cities," she told the *San Francisco Chronicle*, "would feel like Death Valley does today" (Hall, 2004).

But what was truly novel about this study was that it showed that aiming for a lower-emissions future globally would have significant payback. "The magnitude of future climate change," the *PNAS* study noted, "depends substantially on the greenhouse gas emission pathways we choose." This message offered Californians a role in the outcome: step up to the policy plate and implement measures to reduce the emissions of heat-trapping gases, and help ensure that your children and grandchildren inherit a healthy world.

The report earned headlines in more than 50 print and online newspapers the week of its release. All the major wires carried the story; the authors appeared on 13 radio interviews; and 19 television stations aired the story.[4] Even India's *Hindustan Times* and Australia's *Sydney Morning Herald* ran the California climate news, as did the BBC News Online. In addition, the study's authors answered speaking invitations from Los Angeles to Washington, DC, and as far afield as Mexico City and Beijing.[5]

Only a month later, the Union of Concerned Scientists (UCS), in collaboration with the *PNAS* authors, published four summaries of the same findings interpreted for the public and policy-makers. Photographs and graphics accompanied the highly accessible text. The main summary, *Climate Change in California: Choosing Our Future*, was supported by others focused on climate change and water supplies, rising heat and human health, and solutions.[6] This release also earned extensive media coverage.

If, as serious organizers say, "Good press coverage opens doors," this story illustrated the point. During the late summer and fall, several of the study's authors, together with outreach and policy staff from UCS, met with more than 75 individuals at more than 30 California state and federal agencies and industry groups to brief them on the findings related to their sectors. The briefings included meetings with three California cabinet secretaries, members of the state legislature, the chair of the California Air Resources Board, representatives from electric utilities, agriculture and water groups, the investment and insurance industries, the faith community, and others.[7]

After all the media coverage and policy briefings, did this science report make a difference? Without a doubt, the California study informed important policy initiatives in California and beyond. The West Coast Governors' report on addressing climate change referenced the *PNAS* study, as did the California Air Resources Board's regulatory document that set new tailpipe emissions standards in the state.[8] In December 2004, the California Public Utilities Commission (CPUC) announced that electric utilities must incorporate the costs of climate change into their choice of energy sources. The new policy mandates that utilities focus on boosting energy efficiency and conservation, as well as cultivate sources of renewable energy.[9]

Less than a year after the report's release, Governor Schwarzenegger announced aggressive new emission-reduction targets for California, arguing, "I say the debate is over. We know the science. We see the threat. And we know the time for action is now."[10] Several of the authors of the *PNAS* study joined the governor at his public announcement and were asked to work with the state on follow-up research modeled on the *PNAS* study (Figure 11.1). The governor also used the occasion to launch the next wave of climate policy discussions, which includes strengthening existing policies like efficiency and renewable energy requirements and pursuing new

Figure 11.1.   Governor Schwarzenegger on June 1, 2005 signing Executive Order S-3-05, which sets stringent emission reduction targets for the state. UCS's Peter Frumhoff (first on left) as well as several PNAS study authors (not shown here) attended the ceremony.
Source: Jason Mark.

strategies such as mandatory market-based programs for emission reductions.[11] Further news of California climate change policy made headlines throughout 2005.[12]

What made this science story more compelling than others that air every day? To begin with, both the *PNAS* study and UCS's public report concerned the future of the world's fifth largest economy, a state that's home to more than 34.5 million people and a $30 billion agricultural industry that feeds the nation. Furthermore, the news, backed by groundbreaking science, was sobering. All these earmarked the communication as a headliner. But perhaps more importantly, the publications had a team of credible messengers − an interdisciplinary group of scientists, many of them from respected in-state institutions and at the forefront of their fields − and a well-planned outreach strategy. In the presentations, the scientists came from their strengths, laying out their scientific findings, while a savvy science-based advocacy group provided the necessary support to bridge the gap from laboratory and lecture hall to the media newsrooms and halls of government and industry.

This chapter tells the story of scientists stepping beyond their traditional roles, across the great divide and into the public limelight, and of a non-profit organization − the Union of Concerned Scientists − whose modus operandi is to bring credible science into public discourse and the policy process in an accessible manner.

A union of "citizens and scientists for environmental solutions"[13]

Sound science guides our efforts to secure changes in government policy, corporate practices and consumer choices that will protect and improve the health of our environment globally, nationally, and throughout the United States. In short, UCS seeks a great change in humanity's stewardship of the Earth.[14]

Founded in 1969 by a group of faculty and students at the Massachusetts Institute of Technology who "joined together to protest the misuse of science and technology in society,"[15] the Union of Concerned Scientists (UCS) is a non-profit with more than 150,000 members and activists. More than 40 percent of the members are scientists. UCS's mission is to combine "rigorous scientific analysis, innovative policy development and effective citizen advocacy to achieve practical environmental solutions."[16]

As Nancy Cole, UCS's Global Environment Program Deputy Director, says, "We try to make it easy for busy scientists to participate effectively in the policy process." UCS's programs focus on global security, food and the environment, biodiversity loss, and scientific integrity in policy-making, as well as climate change science, impacts, and solutions.

In 1990, UCS began addressing the issue of climate change by educating the public about potential hazards, and working to transform policy related to energy and transportation with the "Appeal by American Scientists to Prevent Global Warming."[17] Seven hundred members of the National Academy of Sciences added their names to that document. Since then, UCS has led a number of strategically important initiatives related to the issue. In 1997, upwards of 1,500 scientists from around the globe – with 105 science Nobel laureates topping the list – signed on to the UCS-sponsored "World Scientists' Call for Action at the Kyoto Climate Summit." These letters played an important role in establishing the reality of global warming in the minds of the majority of the public and helped counter an industry-funded campaign to confuse the public and policy-makers on the issue (see McCright Chapter 12 this volume). In 1999, UCS, in collaboration with the Ecological Society of America and a team of leading independent scientists, published its first regional climate change impacts report, *Confronting Climate Change in California* (Field *et al.*, 1999). In an August 2002 letter to UCS, Mary Nichols, California's Secretary for Resources at the time, credited that report with being "influential" in "galvanizing attention in our state around the need to act on climate impacts and solutions" (Nichols, 2002). The report, she noted, gave policy-makers what they needed – "solid science" – to confirm the existence of climate change and establish its relevance to the state.

The 1999 California report, as well as other strategic activities, helped foster political good will for California's Pavley bill (A.B. 1493), which limits emissions of heat-trapping gases from cars and light trucks.[18] It also encouraged California politicians in 2002 to set the strongest renewable energy standard in the United States (20 percent by 2017).[19] Certainly, no lone piece of evidence, nor the isolated action of one advocacy group, pushes through any legislation. However, Nichols noted in her 2002 letter to UCS,

> I also believe *Confronting Climate Change in California* contributed to winning support for the California Climate Change Registry and Inventory bill in 2000, and laid the groundwork for California's historic passage, and signing by [then] Governor Gray Davis, of the greenhouse gas bill, making California the first state in the nation to order automakers to lower global warming emissions from passenger vehicles.
>
> *Nichols (2002)*

In all of its work, UCS encourages scientists who are willing to speak out, and identifies media and policy opportunities where scientists' voices could make a difference, thereby facilitating their more public role. However, the group also makes a considerable effort to help scientists maintain their credibility in these activities. In a professional culture where objective perspectives are everything, promoting a particular viewpoint, much less a policy solution, can be highly problematic. Indeed, some within the scientific community chastise colleagues for simplifying information for a popular audience, out of concern that complexities and uncertainties in an issue will be lost (see also Warner, Chapter 10, this volume). Christopher Field, an ecologist at the Carnegie Institution who was the lead author on the 1999 California report and contributed to the *PNAS* study, notes, "Scientists are trained to be skeptical of overgeneralization, and one of the aspects of broader [public] communication that's difficult for a lot of people to understand is why the sort of headline style that broader communication requires isn't overgeneralization. I think the two are compatible, but many people have trouble understanding that."

Others in the scientific community view scientists' public outreach or participation in "advocacy" as problematic, if not downright unethical (Wagner, 1999). To address these concerns, UCS sticks to a well-honed formula for scientific messengers: Bring what you know into the public debate; make your work policy-relevant, but never policy-prescriptive or policy-driven. Says Jason Mark, UCS's Clean Vehicles Program Director and California Director, "Our internal motto is 'credibility is our currency,' and that's one thing that differentiates UCS in the policy world."

To help scientists hold the credibility line in reports, briefings, media encounters, and public talks, UCS asks them to present only their findings — for example, *if* emissions remain the same, then *this* scenario and *these* impacts are likely to happen. But when the discussion turns to the implications for policy, UCS staff or collaborating advocacy groups make the request for a specific action. It's a method that allows scientists to present scientific data and maintain their unbiased integrity, while at the same time giving policy-makers what they need — a clear policy request that reflects the best scientific information (Warner, Chapter 10, this volume).

With its internal scientific staff, media and outreach experts, and policy analysts, UCS employs a multi-pronged approach to climate change communication and advocacy work. The organization facilitates regional studies in areas where policy is at a major crossroads, organizes media events, sets up meetings for scientist-authors to brief stakeholders on the relevant impacts, and sustains the outreach effort with assistance from the members of UCS's Sound Science Initiative. Each of these is discussed in more detail below.

Thinking globally, reporting regionally

By the time UCS published the second California climate impacts report in 2004, it had a handful of regional reports under its belt. In 1999, UCS and the Ecological Society of America (ESA) had co-sponsored the first study on California. The innovative ESA/UCS partnership, the brainchild of UCS's Peter Frumhoff, Director of the Global Environment Program, was a great success, and continued through two more reports: one, in 2001, focused on climate change in the Gulf Coast/Gulf of Mexico region, and a Great Lakes report followed in 2003. Later in 2003, UCS released a climate impacts report on Iowa, and began preparations for the 2004 California report.

In every case, the choice of the region and the scientific collaborators resulted from an informal but complex equation. Together with media, policy, and outreach specialists, the UCS project manager for the climate impacts assessments answered two essential questions: First, where can UCS contribute something scientifically new? Second, where can the assessment best inform and motivate sound climate policy decisions?

UCS staff assess these educational and policy opportunities. This process has evolved as UCS has come to better understand what state policy-makers need and what might be effective in advancing the issue in the region. For example, historically UCS focused on officials in air quality and environmental protection agencies. But early explorations of stakeholder concerns in the Great Lakes region, and input from regional scientists, revealed that water managers – an important constituency in the region – viewed climate change as an air issue, not a water one; they didn't see the connection between climate change and water resources. For UCS, this highlighted an opportunity to educate a politically important constituency that is usually silent on the climate issue, and then to engage them in support of measures to reduce heat-trapping emissions.

However, Nancy Cole distinguishes UCS's strategic planning process from that of other organizations that focus mainly on education: while "we recognize that people need to be informed and educated in order to make certain decisions," she says, "for us it's always a power analysis. In deciding where to focus our efforts, we always try to determine who the decision-makers are and who influences them."

UCS's strategic assessment is complemented by a search for potential scientific contributors. For the first three reports, a scientific steering committee composed of renowned ecologists and climatologists helped select the region, identify novel scientific contributions, and recruit regional authors. The committee also assisted in shaping each report's focus, reviewed early versions of the document, and – after the report had gone through several layers of peer review – approved the final document.[20]

The geography of what matters: regional impacts, regional experts

Why has UCS found it both important and expedient to work regionally using mainly authors from the region? When it comes to a global problem with impacts that are hard to see, geography matters; regional information addresses impacts that are tangible to residents. "People don't make decisions based on *global* temperatures increasing by a few degrees," says Katharine Hayhoe. "But if you say that you will not be able to grow Merlot grapes in Napa–Sonoma after a certain time, that means something. Or you will not be able to go skiing at Lake Tahoe at Christmas anymore. You have to use really specific examples to reach people and communicate urgency – that's the only way to do it."

Enlisting regional researchers as messengers amplifies the effect. Former UCS staff scientist and project manager for the Gulf Coast and Great Lakes reports Susanne Moser says that in the United States politicians often find regional scientists "more credible." She points out that such scientists are not only experts in the field; they're typically also affiliated with institutions with which the policy-makers are familiar, they live in the region and often have roots there, and they're constituents. "It's the difference between getting information from the credible but distant Intergovernmental Panel on Climate Change and getting it from Professor So-and-So at the university in your backyard," says Moser. "It's what people use as a shortcut for trust."

Having local experts also makes it easier for the scientists to attend events after the report's release. "Over time, we got smarter," says Moser. "Would participants be willing to stand up and speak locally and repeatedly, or come to communication trainings, or attend meetings with policy-makers?" If authors didn't have far to travel, the outreach work would be much easier, which allowed UCS to make better use of briefing opportunities.

Between 1999 and 2005, nearly 50 scientists have served as authors on the regional climate impacts reports; some have returned for a second tour of duty. The ones interviewed here report the experience as mainly positive, while UCS notes the results in the policy landscape as highly effective. California provides one of the best examples of the process.

Why California? A case study

The Union of Concerned Scientists headed out west for its first regional climate change impact report in 1999 for a constellation of reasons. To begin with, says Nancy Cole, "California has a very strong and active scientific community, and policy-makers in the state pay attention to that community. Plus, California has some preeminent scientific institutions, and the scientific community is an important part of the social and political fabric of the state, which is not true in many other places." In the closing years of the 1990s, UCS judged that, as Cole says, "The policy landscape was ripe for some action." Even without a specific policy target, then, it was a chance to test a new approach – the regional report – on promising ground. The policy opportunity in 2004 was far more concrete.

The "power analysis" that drove UCS's return to California five years later had its origin in critical policy developments in the first years of the twenty-first century – in particular, the implementation of the vehicle emissions-cutting Pavley bill. Once the state legislature approved A.B. 1493 in 2002, the California Air Resources Board (CARB) had until September 2004 to hammer out explicit regulations regarding tailpipe emissions for cars and light trucks. UCS understood the enormous pressure that automakers were putting on CARB,[21] and wanted to provide support for policy-makers to hold steady against industry pressure and deliver regulations with real teeth. The hope was that these regulations would significantly reduce local emissions, and also set standards that would ultimately lead to national and global emission reductions.

UCS's strategic equation for California included several other powerful variables: First, the state has historically been a bellwether for national environmental, public health, and safety standards, setting, for example, in 1961, the first motor vehicle pollution standards in the United States.[22] A number of states have followed suit with emissions standards of their own. Second, Golden Staters pride themselves on their trend-setting prowess and environmental consciousness. In July 2004, the non-partisan Public Policy Institute of California (PPIC) found that 81 percent of residents would support a law – the first of its kind in the nation – "requiring automakers to reduce the emission of greenhouse gases from new cars by 2009."[23] Moreover, California is the largest new car market in the United States, with residents buying about 12 percent of all new vehicles sold annually; changes in the state proliferate far beyond its borders. For example, almost one-quarter of the new vehicles purchased in the country meet California's emission standards.[24] No car-maker wants to be barred from the Golden State goose.

Just as importantly for UCS, it was a chance to work with top experts to contribute something scientifically new – and the 2004 California study was ground-breaking in several aspects. First, explains Hayhoe, "We wanted to base our impact analysis on the very latest results from climate models – and we wanted those models to be the very latest versions of all the knowledge accumulated over the last couple of decades." Thus, the study employed the most recent models from the Hadley Center in the UK and the National Center for Atmospheric Research in Colorado. Climate impacts research generally lags anywhere from a few years to decades behind climate modeling results. However, in this case, the team of scientists managed to complete analysis of model runs, downscaling, impacts, and the translation into policy-relevant terms in just eight months. In addition, the team demonstrated the difference in climate impacts between higher- and lower-emissions scenarios that the impact community had called for, but which no one had yet accomplished at the regional level in the United States.

For all these reasons, the 2004 California report made quite a big splash. Jason Mark notes, "We have found ample evidence, especially from policy-makers, but also from some key stakeholders, that the work has really resonated."

For example, after the CARB voted on the Pavley regulations in September 2004, Charlotte Pera, Senior Program Officer for US Transportation at the Energy Foundation, thanked UCS for its work in an email to Mark:[25] "I think this study was probably the single most effective element of a very effective campaign ..."[26,27]

Even the power industry seemed to be responding constructively. Mark reports that, after meetings with UCS staff and several *PNAS*/UCS report authors in late summer 2004, "at least one electric utility might be considering internal follow-on analysis based on the climate study to explore the impacts for their hydropower generating resources if water flows substantially drop, as the scenarios project."[28] For the water and power sectors in California, Mark adds, "It is increasingly the case that the industry recognizes that there's an issue here and there's some value in trying to understand it."

For other sectors likely to be impacted by climate change – the ski, wine, and insurance industries, for example – "the strength of this work," says Mark, "is to be able to demonstrate for a company who's never thought about why they should get involved in climate policy, that they have a stake." The briefings were also effective in that regard, he says, noting that pairing the information about long-term impacts for a specific industry with a positive solutions approach "can be powerful dual messages for people on the receiving end [of climate change]."

In spite of its obvious efficacy, it is still relatively uncommon for scientists to have — or take — such direct opportunities to bring their research findings before affected stakeholders. What did this form of public outreach look like from their end of the briefing table?

Crossing the great divide: scientists speak out

Many scientists whose work focuses on issues affecting the global and human environment face a tough personal choice: When do the benefits of speaking to policy-makers and the media, crossing the great divide between laboratory and press conference or Congressional briefing, outweigh the risks? (see also Warner, Chapter 10, this volume).

Most of the scientist-authors interviewed for this chapter were quick to point out the advantages of having a well-connected and highly skilled organization to help with the outreach, the satisfaction in making their work relevant to a wider audience, and the benefits to their professional careers and collegial relationships. However, they also noted that some scientists may be hesitant to work with an advocacy organization.

Christopher Field identifies several reasons why scientists shy away from what he calls the "broader communication of science." Sometimes it's personal. "Some scientists are fundamentally uncomfortable," he says, "with the kind of headline style that is required for broader communication." In other cases, a conservative academic department may have a stereotyped definition of what a successful academic does that does not include public communication. Still other scientists may be "interested in principle but, for a variety of reasons, haven't overcome the activation energy to be involved." These people, he says, may feel they haven't figured out the right message, or might not be good at public presentations, or perhaps they worry about the time and energy commitment. There are also, notes Field, "people who are concerned they would be labeled as an advocate rather than an independent evaluator."

"In the long run, the most important feature of the scientific process and the integration of the scientific process in broader societal issues," says Field, "is that the science needs to be viewed as fundamentally unbiased." One of the foremost concerns that most scientists have is that their work not be seen as objective, or that the projects, in the words of another PNAS study author, "might become a political football."

None of the scientists reported that they had ever had that kind of negative experience with UCS. In fact, quite the opposite.

"What has always struck me about UCS compared to some organizations is that UCS has always been most interested in the truth," says Don Wuebbles, Professor and Head of the Atmospheric Sciences Department at the University of Illinois and one of the authors of UCS's Great Lakes report. "They're not willing to back away from that even though they might have a bias toward doing something about a given issue – and that I've admired .... I really appreciate what they're trying to do in providing the correct information to policy-makers and politicians."

Hayhoe, who contributed to the Great Lakes and Iowa studies and led the 2004 California report, says, "UCS stands out – head and shoulders – above many others in really allowing scientists to say exactly what they want to say. They allow us to craft the message – they provide assistance, but we are the ones who actually craft it – and agree to it, and make it our own. That's important. You can't work with scientists in any way if you try to alter their results, or put a spin on them. They won't do that again." It is this vulnerable terrain of credibility that UCS understands and preserves; hence the organization's remarkable success in working with scientists.

Moreover, while UCS is relatively small as national advocacy groups go – 150,000 members, compared, for example, to Natural Resources Defense Council's one million members, or the Sierra Club's 700,000 – it cultivates some mighty connections. "It's UCS that has the contacts," says Hayhoe. The scientist-authors do the research, but UCS convenes the briefings with stakeholders, plans the schedules, and provides the forum. "All we scientists have to do," she adds, "is step onto that platform and deliver our message. Without UCS, very few scientists ... have the connections to request such briefings and set them up in order to disseminate the information."

Hayhoe also notes that the naysayers of climate change have well-oiled and -funded publicity machines that shift into gear as soon as these skeptics publish a paper, "and their papers are very few and far between compared to the total body of literature," she adds. "So how much more do the scientists that make up for the 99.99 percent of publications ... need someone to help us do that for our most significant results?" (see also McCright, Chapter 12, this volume).

The chance to put those "significant results" in the hands of decision-makers is clearly a motivating factor for those who have worked with UCS. "The most obvious reward and the original motivation for doing this is that you feel like you are doing something that is both intellectually stimulating – it's very challenging – but at the same time it's practical," maintains Hayhoe.

"[Knowing] that what you're doing will actually matter to people and that decisions will be made based on the work we're doing has enormous significance."

Kim Nicholas Cahill, a PhD student in the Interdisciplinary Graduate Program in Environment and Resources at Stanford University and one of the 2004 California report authors, also mentions this. "I would encourage scientists of all levels to participate in [this kind of study as] an exercise to make their research more politically and socially relevant." She adds, "While [this work] isn't necessarily *the* answer ... it can be contributing to a process that is trying to make improvements. Science can be the first step in terms of providing good information and clarifying what the problem might be, what the impact might be."

Dan Cayan, another of the 2004 study authors and director of the Climate Research Division at the Scripps Institution of Oceanography, at the University of California—San Diego, mentions another value that scientists add to the larger process: they can provide a reference point for decision-makers, who frequently must operate in the landscape of uncertainty, particularly in the area of climate change with its numerous variables and long time lags. "In dealing with problems like this, it's very important that the science community take the high road and focus on providing the best possible information," says Cayan, "along with trying to provide a gauge of confidence and uncertainty."

But some might argue that information is, after all, simply information; why can't politicians and policy-makers just read the journals? Translating raw scientific data into an understandable language is an important aspect of this communication process, as is figuring out which data are pertinent to the decisions at hand. UCS plays a critical role in both these processes. Cayan notes, "UCS is able to translate what scientists are saying into [accessible] language or an actionable agenda." UCS also initiated partnership with credible practitioners, such as water managers or public health workers, who help guide the research. "[The practitioners] told us, 'we don't really want to know about mean temperature, we need to know about temperature extremes in order to understand heat-related mortality.' That input actually sharpens the focus," Cayan says. "It's a very valuable component."

To help with the translation of science into plain English, UCS also offers media workshops to its report authors. At these trainings, media experts teach scientists about "how the media world works and what reporters need," says UCS's Cole. In the workshops, scientists have "an opportunity to identify the most important and relevant messages they wish to communicate, then refine them in ways that people can easily understand."

The training also gives scientists a chance to practice the lay-language delivery they will need for interviews, public talks, or briefings — often a far cry from conference presentations to colleagues. "We assume that scientists, like CEOs or anyone else, need practice doing something they wouldn't ordinarily do," says Cole.

Don Wuebbles, who attended such a media training in 2003, described it as "interesting and valuable ... I still struggle at times with packaging things in short, grabby statements, and making your three or four points ... but I certainly think about it a lot more than I used to because of that kind of training."

Several report authors also mentioned the benefits of networking in the UCS studies. "One of the rewards was that I got acquainted with a science team, some of whom I hadn't worked with before, who I imagine I'll be rubbing shoulders with for the rest of my career," says Cayan. "That was certainly a benefit."

In the briefings, the teams also offer collegial support, especially since some members are "senior statesmen," as Hayhoe dubs them, in the climate science community. Two authors on the 2004 California team, Christopher Field and Stephen Schneider, have long contributed to the IPCC reports, as has Don Wuebbles. The presence of distinguished colleagues and peers helps enormously, says Hayhoe. "They have a wealth of experience and knowledge to share that cannot be duplicated by anyone who's not a scientist and has not done this." Cayan adds, "It helps to have strength in numbers in addressing issues like this ... [and] it helps to have people who have credibility talking with somewhat of a single voice."

By publishing their findings first in the peer-reviewed *PNAS* journal and by their careful delivery to the public, the scientist-authors retain the respect of their peers. Because of that, Hayhoe notes, other scientists are intrigued. "Everyone wants the work they do to have significance, so communicating your results to more than just the few people who read the obscure journal that you publish it in is very meaningful to a lot of people .... Most [people who work in this area] would be very interested in collaborating in such assessments, as long as their concerns regarding integrity were laid to rest."

Naturally, the model and its effectiveness also contain a cautionary tale. If successful translation and release of scientific findings to the media, the public, and decision-makers opens doors, responding to all these opportunities can also burn out the very scientists who have become so skilled at speaking out — scientists who, after all, still have full-time research and teaching duties elsewhere. UCS has planned for that as well.

Giving the reports "legs": SSI and climate science education days

To boost the effectiveness of its regional reports, UCS enlists support from members of its Sound Science Initiative (SSI), an email-based group of scientists interested in speaking on scientific issues in the media or with policy-makers. In 2005, SSI had about 4,000 members from every US state and 56 countries; California leads the list with nearly 600 members.[29] "SSI gives legs to the impact reports," says Cole, "by mobilizing SSI members in the region to support the report authors who may only have a limited amount of time to do outreach, education, and media activities."

Typically, in addition to informational updates, UCS sends SSI members action alerts when the organization notes a media or policy opportunity where scientists could have an impact. Action alerts may go to the whole list, or only regional members, and usually recommend specific activities, such as letters to news editors or state officials, or comments on legislative initiatives to government agencies. "With SSI, we have a range of outreach tools that are available to us," says Nancy Cole, who directs the program."[30]

In California, UCS extended the impact of the 2004 regional study with several activities. First, the organization drafted a sign-on letter thanking Governor Schwarzenegger and the state legislature for the steps they'd already taken to implement emission-reduction policies, and urging them to do more and take a leadership stance on this issue.[31] The letter, with 486 signatures, was delivered to the legislature in April 2005, and also appeared as an ad in the *Sacramento Bee* (Figure 11.2). In addition, UCS organized a Climate Science Education Day in the state capitol, bringing 40 scientists from key districts throughout the state to talk to their representatives about climate change.[32]

Science in the public interest

The strategic model that the Union of Concerned Scientists has finessed seems promising for other groups as well. First, there's the assistance that UCS provides to busy scientists, helping them craft a message that reflects accurately and well on what they do. Second, the organization works hard to ensure valuable scientific findings don't quietly retire to the library stacks. It does this by cultivating connections to people who need to hear about the developments — be they in government, the media, or business. Already this is a big improvement on what often happens to new scientific findings. "Too many scientists don't understand how their published research connects to policy," says Susanne Moser. "Of course, after umpteen studies are

Figure 11.2.   Thank you ad by California scientists in the *Sacramento Bee* in April 2005.
Source: Union of Concerned Scientists, reprinted with permission by Kevin Kirchner from MacWilliams, Robinson and Partners and designMind.

published, there is a cumulative process that works through these dark channels to have an effect – and make a difference." But UCS helps amplify scientific findings that are relevant to policy much more quickly and effectively.

As for the deeper questions of why, for both the organization and the scientists – why do this work? why take the risk of public outreach? and why is it important? – the words of one UCS staff member stand out. "When our first child was born," says Jason Mark, "UCS's work took on new meaning. It literally hit home, if you will." It was an appreciation that came, says Mark, from living with a new, and much longer, time horizon than his own biological life. "A predictable response, but no less powerful," says Mark. "At the end of the day, we're about using the best science and information to help protect the world that my children and grandchildren will inherit." The words reflect the organizational tenets of UCS, which was founded on a dedication to seeing science take a vital role in all the places where it can make a positive difference to the well-being of humans and the planet.

What makes the collaboration between UCS and its scientist partners so successful is a commitment to developing communication strategies that are both effective with the public and policy-makers, and respectful of the need for scientific integrity. Bringing together the latest research, delivered by credible messengers to the right audiences at the right time, the Union of Concerned Scientists helps move the work from the labs and journals onto the radar screens of legislators and stakeholders, from the field to the top of the news hour.

Notes

1. Nobel laureate chemist Mario Molina speaking at the first National Conference on Science, Policy and the Environment, December 2000; quoted in Blockstein (2002:92).
2. This chapter is based on individual interviews conducted with UCS staff and former staff, and scientists who participated as authors in UCS's regional impacts studies. Natasha Fraley contributed research assistance to this chapter.
3. Of the 19 scientists, two (Peter Frumhoff and Julia Verville) were employed at the Union of Concerned Scientists (UCS), two (lead author Katharine Hayhoe and Lawrence Kalkstein) were funded for their work on this project by UCS, and one (Susanne Moser) had previously worked as a UCS staff scientist, but was not funded for her contributions to the *PNAS* study. Frumhoff, Director of UCS's Global Environment Program, conceived of and initiated the study; Hayhoe led the science team. In this project, the UCS and UCS-funded scientists collaborated with the rest of the team, who are independent.
4. Media and outreach information supplied by the Union of Concerned Scientists.
5. In November 2004, UCS's Peter Frumhoff spoke at the Dialogue on Future International Actions to Address Global Climate Change in Mexico City. Frumhoff also addressed the CALSTART conference on the state's Transportation Energy Future in Pasadena, California in December 2004. Two study authors, Dan Cayan and Michael Hanemann,

spoke at the Fall 2004 meeting of the Association of California Water Agencies
in Palm Springs, California. And Katharine Hayhoe addressed the International
Symposium on Key Vulnerable Regions and Adaptation in Relation to Article 2 of the
UNFCCC on "Assessing Regional Vulnerabilities to Climate Change" in Beijing, China
in October 2004. These are just a few of the many talks the authors gave based on the
study.

6. All the documents are available on http://www.climatechoices.org/; accessed January
   19, 2006.
7. Outreach information supplied by UCS.
8. The *PNAS* study is cited as Appendix F in the http://egov.oregon.gov/ENERGY/
   GBLWRM/Regional_Intro.shtml; and one of a list of "Additional Supporting
   Documents and Information" in http://www.arb.ca.gov/regact/grnhsgas/2ndattach.pdf;
   accessed January 19, 2006.
9. To read more about the policy, see http://www.ucsusa.org/news/press_release/california-
   puc-requires-utilities-to-factor-in-costs-of-global-warming.html; accessed January
   19, 2006.
10. Governor's Remarks at World Environment Day Conference, San Francisco, June 1,
    2005. For the complete transcript, see http://www.climatechange.ca.gov/; accessed
    January 19, 2006.
11. In September 2005, the CPUC announced it was funding a $2 billion initiative for energy
    efficiency and conservation. The goal: Reduce greenhouse gas emissions by 3.4 million
    tons of CO_2 by 2008, and save consumers $2.7 billion in energy fees. According to the
    CPUC, this was the "single largest authorization for energy efficiency in US history."
    For the complete press release, see http://www.cpuc.ca.gov/static/hottopics/1energy/
    a0506004.htm; accessed January 19, 2006.
12. For a detailed list of the state's policies, see the California Climate Change Portal, http://
    www.climatechange.ca.gov/; accessed January 19, 2006.
13. The quote is UCS's tagline; see http://www.ucsusa.org/; accessed January 19, 2006.
14. http://www.ucsusa.org/ucs/about/mission.html; accessed January 19, 2006.
15. http://www.ucsusa.org/ucs/about/history.html; accessed January 19, 2006.
16. http://www.ucsusa.org/ucs/about/mission.html; accessed January 19,2006.
17. To view a broader list of UCS's accomplishments, see http://www.ucsusa.org/ucs/about/
    history.html; accessed January 19, 2006.
18. For more information on California A.B. 1493, view the "Report to the Legis-
    lature and the Governor on Regulations to Control Greenhouse Gas Emissions
    From Motor Vehicles" at http://arb.ca.gov/cc/cc.htm; accessed January 19, 2006.
19. To read the full text of the California Renewables Portfolio Standard bill, see http://
    www.leginfo.ca.gov/pub/01−02/bill/sen/sb_1051−1100/sb_1078_bill_20020912_
    chaptered.html; accessed January 19, 2006.
20. UCS reports undergo a level of peer review far exceeding that of many journal articles.
    The Gulf Coast and Great Lakes reports, for example, each had at least 20 external
    scientific peer reviewers, while a typical journal article has two or three. The joint ESA/
    UCS reports did not aim to contribute new science, but rather to synthesize existing
    scientific understanding. By contrast, the 2004 California study reported new scientific
    findings, and the information was vetted in the prestigious, peer-reviewed *Proceedings of
    the National Academy of Sciences.*
21. To read more about the arguments the automobile industry and its associated groups
    employed, see UCS's "How Automakers Threaten and Mislead Consumers;" available at:
    http://www.ucsusa.org/clean_vehicles/vehicles_health/automakers-spin-misleads-
    consumers.html; accessed January 19, 2006.
22. To read a history of California's regulations in this area, see http://www.arb.ca.gov/html/
    brochure/history.htm; accessed January 19, 2006.
23. To read the complete Public Policy Institute of California's "Statewide Survey: Special
    Survey on Californians and the Environment," July 2004, see http://www.ppic.org/
    content/pubs/S_704MBS.pdf; accessed January 19, 2006.

24. Figures quoted here are from http://www.ucsusa.org/clean_vehicles/vehicles_health/
    automakers-spin-misleads-consumers.html; accessed January 19, 2006. According
    to Jason Mark, UCS based these calculations on information from Ward's Auto Online
    2003 (http://www.wardsauto.com).
25. Charlotte Pera, email dated 9/28/04, to Jason Mark, UCS's Clean Vehicles Program
    Director and California Director, Union of Concerned Scientists.
26. At a meeting in October 2004, James D. Boyd, Commissioner on the California Energy
    Resources Conservation and Development Commission, said, "I know [the *PNAS* study]
    will add to the body of knowledge on the subject of climate change impacts in California,
    and it adds to the work that I've always referenced as one of the major benchmarks,
    the Green Book I've called it [for the color of its cover], the report of the Union of
    Concerned Scientists *et al.* of a few years back ... , because that certainly stimulated a lot
    of activity." Boyd was referring to the 1999 UCS/ESA California report. Jason Mark,
    who is a member of the climate advisory panel of the California Energy Commission,
    related this anecdote in a February 17, 2005 interview. The exact quote is from the
    transcript of the October 7, 2004 meeting of the California Energy Resources
    Conservation and Development Commission, California Climate Change Advisory
    Committee, http://www.energy.ca.gov/global_climate_change/04-CCAC-1_advisory_
    committee/documents/2004−10−07_meeting/2004−10−07_TRANSCRIPT.PDF;
    accessed January 19, 2006.
27. In another measure of success, other groups have used the *PNAS* study as the basis for
    scientific follow-up. In 2005, the non-profit advocacy group, Redefining Progress,
    produced *Climate in California: Health, Economic and Equity Impacts*, an environmental-
    justice-focused review explicitly built on the heat-related mortality data from the *PNAS*
    study; two of the *PNAS* authors also contributed to this study (see http://www.
    rprogress.org/newpubs/2004/executive_summary_040922.pdf for the executive summary
    of the study; accessed January 19, 2006).
28. Utilities will be substantially impacted by climate change, not just in reduced hydropower-
    generating capacity and increased demand for electricity during hotter summers, but also
    by increased water-pumping requirements, which is the largest electricity demand in the
    state.
29. SSI members range from graduate students to senior scientists working in the Earth,
    physical, life, and social sciences. For more information, see http://www.ucsusa.org/
    global_warming/sound-science-initiative.html; accessed January 19, 2006.
30. For more information on SSI, and the activities its members engage in, see http://
    www.ucsusa.org/ssi/; accessed January 19, 2006.
31. The letter is the second of this kind in the state, where, following the 1999 report, a similar
    UCS letter was signed by 125 scientists, presented to policy-makers and run as an ad in
    the *Sacramento Bee*.
32. As is typical of such SSI events, UCS provided an issue briefing − in the form of
    a conference call and other information materials − for the participants prior to the
    Education Day, in which participants learned relevant information from California
    report authors. UCS has organized several such days in Washington, DC with members
    of Congress − events that have varied in focus and types of participants. At one,
    scientists from a particular state teamed with religious leaders, economists, and
    business leaders from the same state to argue the scientific, moral, and economic
    aspects of the issue.

References

Blockstein, D. E. (2002). How to lose your political virginity while keeping your
    scientific credibility. *BioScience*, **52**, 1, 91−6.

Field, C. B., Daily, G. C., Davis, F. W., *et al.* (1999). *Confronting Climate Change in California: Ecological Impacts on the Golden State.* Cambridge, MA: Union of Concerned Scientists and Washington, DC: Ecological Society of America. Available online at: http://www.ucsusa.org/; accessed January 19, 2006.

Hall, C. T. (2004). CALIFORNIA – Global warming clouds the future but experts say it's not too late to cut harmful emissions. *San Francisco Chronicle*, August 17, available at: http://www.sfgate.com/cgi-bin/article.cgi?f=/c/a/2004/08/17/BAGTI896B91.DTL&hw=climate+change&sn=001&sc=1000; accessed January 19, 2006.

Hayhoe, K., Cayan, D., Field, C. B., *et al.* (2004). Emissions pathways, climate change, and impacts on California. *Proceedings of the National Academy of Sciences*, **101**, 34, 12422–7.

Nichols, M. (2002). Letter of August 15 to Kevin Knobloch, then UCS's Executive Director (now the organization's President), Union of Concerned Scientists.

Redefining Progress (2004). *Climate in California: Health, Economic and Equity Impacts.* Oakland, CA: Redefining Progress.

Wagner, F. H. (1999). Analysis and/or advocacy: What role(s) for ecologists? EcoEssay Series No. 3. Santa Barbara, CA: National Center for Ecological Analysis and Synthesis. Available at: http://www.nceas.ucsb.edu/nceas-web/resources/ecoessay/wagner/; accessed January 19, 2006.

12

Dealing with climate change contrarians

Aaron M. McCright

Michigan State University

Introduction

Industrial societies have developed through the extraction and consumption of fossil fuels. Many of the most powerful industries in the United States, both historically and at present, depend upon the use of fossil fuels. Also, the fortunes of many of the wealthiest and most powerful families in the United States are founded within the fossil fuels industry. Further, a very large proportion of the manufacturing jobs in the United States are connected with the use of these fuels. Thus it is not surprising that proposals to substantially reduce their use, or even just to reduce emissions of heat-trapping gases from their use, to mitigate the effects of climate change encounter stunning obstacles and are seen as deeply threatening by powerful economic and political actors.

A coordinated anti-environmental countermovement (see, e.g., Austin, 2002; Beder, 1997; Helvarg, 1994; Switzer, 1997) has mobilized in the United States since the late 1980s to challenge the legitimacy of climate change as a problem on which society should act. This response includes both massive lobbying efforts by the American fossil fuels industry[1] (e.g., Gelbspan, 1997; Levy and Egan, 1998; Newell, 2000) and concerted efforts by American conservative think tanks to question the necessity of dealing with climate change (e.g., Luke, 2000; McCright and Dunlap, 2000, 2003). Integral to these efforts has been the promotion of approximately a dozen scientists collectively known as climate change "contrarians"[2] (or sometimes "skeptics"). Climate change contrarians publicly challenge what they perceive as the false consensus of "mainstream" climate science — the reality of anthropogenic

Thanks are due to Lori Baralt, Tom Dietz, Lisa Dilling, Riley Dunlap, Linda Kalof, Susi Moser, Chris Oliver, Scott Sawyer, Rachael Shwom, and Chenyang Xiao for their insightful critiques and suggestions on earlier drafts. All mistakes are the sole responsibility of the author.

climate change. They proclaim their strong and vocal dissent from this growing consensus by criticizing mainstream climate science in general and pre-eminent climate scientists more specifically, often with considerable financial support from American fossil fuels industry organizations and conservative think tanks.

The contrarians' activities pose a significant barrier to substantive communication among the scientific community, policy-makers, and the general public (see, e.g., Brown, 1997; Gelbspan, 1997; McCright and Dunlap, 2003; Ozone Action, 1996a,b,c). Indeed, a major role of the contrarians has been to distort communication efforts of the scientific community with policy-makers and the general public regarding climate change. This chapter will first summarize existing research on the claims, organizational affiliations, tactics, and effectiveness of the contrarians in the American national policy context, and then identify several interrelated strategies for dealing with contrarians – to protect the integrity of the open communication necessary between the scientific community and public policy-makers.

Climate change contrarians: tactics and effectiveness[3]

Institutional inertia, the entrenchment of vested interests, and the relative disempowerment, disengagement, and apathy of many members of the general public all conspire against solving most social problems. Yet, many Western intellectuals in the Enlightenment tradition believe that open communication will force us to seriously confront problems as the weight of mounting scientific evidence and appeals to justice and fairness become overwhelming (e.g., Habermas, 1984, 1987). But this is a slow and difficult process, and substantial effort is needed to bring a social problem to the public agenda, identify politically acceptable but effective solutions, and implement them. For those concerned with climate change, the problem of dealing with such social inertia is compounded by the efforts of the American anti-environmental countermovement.

This countermovement has mounted a sustained assault on scientific communication – attempting to confuse both policy-makers and the general public about climate change. Conventional scientists try to validate new knowledge claims about complex climate phenomena, which challenge the dominant social paradigm about how humans interact with the environment (Dunlap, 2002). On the other hand, the fossil fuels industry, conservative think tanks, and the contrarians they promote advance their objective of maintaining the status quo merely by obstructing communication of these new knowledge claims. Only a minimal amount of confusion about climate change may be necessary to reinforce the social inertia that perpetuates the status quo, even in the face of considerable scientific evidence otherwise. Thus the goals of the contrarians are achieved more easily than are the goals of conventional climate scientists.

Some of the contrarians publish in the peer-reviewed climate science literature, where they oversimplify and even misinterpret existing research while selectively presenting data supporting their own counterclaims (see, e.g., Soon and Baliunas, 2003). Of greater significance, most contrarians challenge climate change knowledge claims largely through activities outside of the scientific community. For presenting their most dubious assertions, they have chosen venues that are free from the constraints of traditional scientific standards. This withdrawal from the institutions and processes that define modern science provides the contrarians with great latitude in making their arguments. For instance, most contrarians present claims that consistently exceed the content of their peer-reviewed work in publications, public appearances, and websites supported by fossil fuels organizations and conservative think tanks. And the contrarians make their assertions to lay audiences who may not detect the technical flaws in their arguments (for a discussion of related vulnerabilities to which lay audiences can fall prey, see Dunwoody, Chapter 5, this volume; Moser and Dilling, 2004). Since their credentials inspire perceptions of expertise and trustworthiness among non-experts, the lack of accountability outside the scientific community makes the contrarians especially dangerous to scientific communication efforts. They can present assertions that do not withstand scientific peer review to an audience that often assumes, because of the contrarians' credentials, that those arguments are sound and constitute scientific evidence.

Not only do most contrarians operate largely outside of the scientific community, but most also benefit substantially from affiliations with fossil fuels industry associations and conservative think tanks. Their relationships with the latter arguably are more crucial since conservative think tanks can enter the public discussion about climate change with a patina of intellectual legitimacy and credibility that fossil fuels industry organizations cannot claim. Furthermore, conservative think tanks have made such efficient use of money, ideas, personnel, and the media that they comprise the most successful policy-planning network in the United States since the mid-1970s.[4]

Several contrarians joined influential conservative think tanks during the 1990s. The think tanks utilized these credentialed scientists to provide scientific legitimacy to their counterclaims about climate change.[5] The contrarians wrote many brief documents (e.g., policy briefs and backgrounders) on climate change for the think tanks, served as expert sources regularly cited in dozens of other think-tank documents, made public appearances at press conferences sponsored by the think tanks, and represented conservative think tanks on radio and television programs to further challenge mainstream scientific knowledge claims about climate change. During this regular interaction with policy-makers and the general public, the contrarians forcefully dismissed the results of hundreds of scientific publications by making overly simplistic and dubious counterclaims — which were frequently embedded within a broader anti-environmental and anti-regulatory discourse (see McCright and Dunlap, 2000, 2003). These activities violate basic ethical norms established to protect the integrity of science.

The anti-environmental countermovement benefited from the shift in control of Congress during the 1990s. The contrarians took advantage of the Republican majority in Congress since the mid-term elections in 1994, after which they achieved a much enhanced visibility in Congressional hearings on climate change. The conservative Congress tended to minimize discussion of climate change overall. For instance, there were more Congressional hearings on climate change in 1992 alone than there were between the 1994 Republican takeover and the December 1997 Kyoto Conference. In the hearings that did take place during this time period, five contrarians testified approximately as often as did thousands of mainstream climate scientists publishing in the scientific literature. The testimony of the contrarians provided ranking Republican politicians with the arguments that allowed them to shift the nature of the debate in Congress away from the question of "What do we need to do to address global warming?" toward the more preliminary question of "Is global warming really a problem?" (McCright and Dunlap, 2003).

Finally, contrarians were able to translate this heightened visibility into increased media presence by successfully exploiting the media's balancing norm — which equates "objectivity" with presenting "both sides of the story" (see Boykoff and Boykoff, 2004). For example, between 1994 and 1997, five contrarians were cited as often in the nation's seven largest circulating newspapers as were the most respected climate scientists of the time (McCright and Dunlap, 2003). By publicly challenging the claims of the mainstream scientific community, contrarians created a sustained drama that journalists have been socialized to consider newsworthy and integral to a "good story." Such a contrived storyline breeds confusion within the general public regarding what is widely accepted knowledge and what is a highly speculative claim. This kind of coverage also confuses the distinction between what are scientific judgments and what are value judgments

(Sarewitz, 2004). This confusion facilitates political inaction and policy gridlock — disproportionately favoring the efforts of the anti-environmental countermovement to challenge the legitimacy of climate change.

The anti-environmental countermovement has successfully distorted climate change communication between the scientific community on the one hand and policy-makers and the general public on the other. Climate change contrarians have been major figures in the creation of our current situation of public confusion about climate change and a climate policy based largely on voluntary action and more research. Since the mid-1990s those conservative policy-makers attempting to prevent the creation and implementation of any substantial climate policy have been able to utilize the counterclaims of contrarians to legitimize their inaction (Brown, 1997; McCright and Dunlap, 2003). They regularly have argued that the science is too uncertain to justify the social change that would come with effective climate change policy. Indeed, the George W. Bush Administration represents the institutionalization of the anti-environmental countermovement, since contrarians' counterclaims have figured prominently in several White House documents about climate change (e.g., United States White House, 2002a,b).

Countering climate change contrarians

Given that climate change may be the most complex scientific problem faced by modern society, it is crucial for policy-makers to have a basic understanding of the best scientific evidence available. Also, the successful implementation of any climate policy may depend upon its public legitimacy, which is facilitated when citizens have a basic understanding of elementary climate science principles or at least a healthy trust in the scientific community. Thus, it is essential to neutralize or contain the activities of the contrarians and their allies in the anti-environmental countermovement who attempt to use non-scientific forums to make ideological claims shrouded in scientific discourse. Since this may be controversial, some clarification is necessary.

The contrarians clearly have had a level of political access and influence that far exceeds what would be expected given the veracity and significance of their scientific contribution. Thus, the issue is *not* how to silence the contrarians but rather how to deal with this demonstrable imbalance of power in a way that upholds the norms and protects the integrity of the scientific enterprise. Certainly, all scientists have the right to voice their personal views as American citizens. Yet membership in a scientific profession tempers this right. That is, when a scientist speaks *as a scientist in the name of science*, he or she becomes a spokesperson for a community of professionals (see also the chapters by Warner, and Cole with Watrous, Chapters 10 and 11, this volume). As such, this scientist has a responsibility to obey the norms of the scientific community while clearly and effectively communicating the state of scientific knowledge on the issue at hand.[6] It is because the few contrarians have violated these expectations that their activities should be marginalized. Minimizing their contribution will allow the consensus position about climate change to be transmitted more clearly among the scientific community, policy-makers, and the general public.

Several factors would enhance such effective communication:

1. greater scientific literacy among policy-makers and the general public, especially how science works as a social process (e.g., the role of peer review as an underpinning for scientific claims, the importance of scientific consensus even in the face of a few dissenters);
2. greater transparency in science-related policy-making processes;
3. better funding for communication efforts by governmental agencies and research organizations;
4. more appropriate norms within journalism for scientific reporting; and
5. better training for environmental scientists in how to communicate effectively with policy-makers and the general public.

The combination of low levels of scientific literacy, opaque policy-making, limited resources earmarked for scientific communication, and the existing norms within scientific journalism will continue to pose a substantial challenge to effective climate science communication for years to come. Nevertheless, in the face of these enduring barriers to effective communication, several suggestions may be made that speak to the challenges contrarians pose. These suggestions, which reinforce those ideas advanced by Moser and Dilling (2004), are directed not only to members of the climate science community, but also to concerned citizens, social justice advocacy organizations, ecological sustainability organizations, and progressive business organizations.

First, *scientists and communicators would benefit greatly from a deeper and more comprehensive awareness and understanding of climate change contrarians as individual actors*. Since the early 1990s, contrarians have taken the initiative in defining the terms of the public discussion of climate change. Thus, an obvious first step is to identify the primary contrarians and better anticipate their counterclaims, rhetorical techniques, and public activities discussed above. The Union of Concerned Scientists has worked on this for nearly a decade through its Sound Science Initiative (see Cole with Watrous,

Chapter 11, this volume). Ultimately, we must come to terms with the motivations and intentions of contrarians. While genuine criticism of various climate science knowledge claims is a valid and common process to advance the science, contrarians demonstrate ulterior questionable motives when they ally themselves consistently with fossil fuels organizations and conservative think tanks to convey their counterclaims outside of the scientific community's normal outlets.

Even further, *scientists and communicators would gain strategic advantage if they fully recognized the long-term involvement of contrarians within the anti-environmental countermovement.* Contrarians perform a key role for this countermovement by lending pseudo-scientific legitimacy to attempts at obfuscating scientific communication for the narrow material and ideological interests of fossil fuels organizations and conservative think tanks. Indeed, employing sympathetic scientists to debunk scientific claims that challenge vested interests has been an enduring pattern within the anti-environmental countermovement over the past decade and beyond on other environmental and public health and safety issues. For instance, anti-environmental operatives, such as the attorney Michael Fumento, JunkScience.com founder Steven Milloy, and the political scientist Michael Sanera, regularly attempt to debunk the knowledge claims of the peer-reviewed scientific community on such issues as pesticide exposure, environmental carcinogens, and ozone depletion. These anti-environmental scientists routinely deny the evidence of environmental problems by exploiting scientific uncertainties, misinterpreting peer-reviewed research, and selectively presenting data that support their own counterclaims (Fumento, 1996; Milloy, 1995; Sanera, 1999).

Third, *scientists and other communicators would provide a valuable service to policy-makers and the general public if they exposed the tactics and goals of contrarians and their relationships with the anti-environmental movement.* In other words, the credentials, expertise, funding, and tactics of contrarians must become problematized within public discourse. The general public should know more about: (1) the marginalized status of contrarians within the scientific community (because they operate largely outside mainstream scientific institutions); (2) the routine violations of scientific standards by contrarians (i.e., misinterpretation of peer-reviewed research, use of suspect methods, selective presentation of those results that support their counterclaims); (3) the enduring conflicts of interests routinely ignored when contrarians accept money from fossil fuels organizations to make negative pronouncements about climate science; and (4) the pseudo-scientific legitimacy contrarians provide to assist the anti-environmental countermovement in preventing new environmental regulations and weakening existing environmental regulations. In essence, we should cultivate in the general public a healthy skepticism of contrarians' credibility, motives, and tactics.

Taking the previous suggestion even further, *scientists and communicators could help move the debate forward if they publicly acknowledged that the crux of the climate change debate at this time is not conflict over science but over very different values.*[7] Clearly, there will always be more to know, and there are legitimate scientific issues that need to be resolved. Scientists would do themselves and the public a great service if they made this point about the nature of science, and that of the scientific debates explicitly. To allow contrarians to abuse science to carry out value debates undermines the scientific enterprise in general and the legitimacy of the climate change problem in particular.

While members of the anti-environmental countermovement promote economic growth and business dominance above most other values, members of the scientific community value open inquiry, rigorous and systematic analyses, and peer review, and members of the environmental community (e.g., the environmental movement and pro-environmental policy-makers) prioritize values of ecological sustainability and social justice. By putting the conflicting values directly in the public eye, we are able to have more honest discussions about the larger political, cultural, social, and economic context of climate change (see also Regan, Chapter 13, this volume). Also, by making the value conflicts more obvious, value-driven motives may be less likely to covertly lay waste to science and the scientific process of working through uncertainties. This entails maintaining vigilance in the face of veiled attempts by anti-environmental groups to continually assert their values to distort the public understanding of climate science. The most frequently recurring example is the use of the terms "sound science" and "junk science" by anti-environmental actors. Past research (e.g., Herrick and Jamieson, 2001) and common sense reveal that these terms are merely rhetorical tools, devoid of any substantive meaning. Yet these buzzwords continue to confuse policy-makers and the general public in discussions about varied scientific matters.

Fifth, *scientists and communicators can help proactively frame the public discussion about climate change rather than only respond to how contrarians frame the debate.* This will help to further marginalize the contrarians, who have been fairly successful at portraying themselves as an oppressed minority of "round-earthers" in the midst of the "flat-earth" climate science establishment. Claiming the disenfranchised underdog role has gained contrarians access to an arsenal of provocative imagery within American culture. However, this is not a very accurate description of reality. Thus, we would do well to convey to policy-makers and the general public a sense of the sheer amount of empirical research and theoretical modeling, years of person-hours, and number of climate scientists that have led to our current understanding of climate processes. So, scientists and communicators can do more to convey this perception of events to non-scientists and shift the burden of proof to contrarians. We may accomplish this effectively by making sure that key dialogue with policy-makers and the general public is guided by established risk communication principles (National Research Council, 1989).

Also, we should promote an internally reinforcing package of basic climate change knowledge claims to policy-makers and the general public. This package would help the public understand the causal link between everyday activities, emissions, climate change, potential impacts, and solutions (see also Bostrom and Lashof, Chapter 1, this volume). It should also present some of the most consensual claims of recent publications from the most prestigious "arbiters" of scientific evidence – the Intergovernmental Panel on Climate Change (2001; and the forthcoming *Fourth Assessment Report* in 2007) and the National Academy of Sciences (National Research Council, 2001) – and other ongoing research. These should be conveyed effectively in language and – whenever possible – with graphics that lay citizens can understand. In essence, these basic claims should seem obvious and straightforward to non-scientists. This is achieved partly through the use of repetition in all possible venues, including some (e.g., editorials in major newspapers) favored by contrarians.

Finally, *these basic knowledge claims should be embedded within a broader discourse that simultaneously is cautious and optimistic – drawing upon ideas having great resonance with central elements in American culture* (see the chapters by Bostrom and Lashof; and Ungar, Chapters 1 and 4, this volume). For instance, preparing for the future is embedded deeply within American culture. Parents attempt to plan a better future for their children, varied organizations create long-term plans, millions of people purchase all varieties of insurance, and even the Boy Scouts tell children to "be prepared." Moreover, the "preponderance of evidence" is a widely known standard of proof from the US civil court system, and it is used often in our personal lives with matters of great concern. We do not wait for knowledge beyond a reasonable doubt before getting treatment on potential diseases and illnesses. If the weight of evidence reaches some compelling threshold, then we take action.

Most of the anti-environmental countermovement has criticized climate science for promoting gloom-and-doom scenarios (e.g., Baden, 1994;

Bailey, 1993). Climate change knowledge easily can provoke pessimism and anxiety, and this just points to the necessity for framing present and future options in ways that encourage action and not paralysis (see Moser, Chapter 3, this volume). We may utilize ideas about the preponderance of evidence and future preparation to reframe climate change in terms of problem-solving, entrepreneurial opportunities, and jobs – each of which is a central aspect of the American identity. For example, a long-term, wholesale conversion from fossil fuels to renewable energy sources provides many opportunities for the development of innovative technologies, the emergence of a new generation of entrepreneurs, and the creation of a significant amount of jobs (see also Young, Chapter 24, this volume). Indeed, it provides us with a chance to create the world anew (see Jamieson, Chapter 30, this volume). Even more generally, we can convey the idea that we are threatened on our current trajectory of almost insatiable dependence on fossil fuels. By contrast, our country's sovereignty and future may be "saved" by renewable energies and other environmentally friendly technologies.

Coda

In a deep sense, more is at stake here than *just* climate change. Climate change represents a case that stirs deeply in issues of scientific literacy, public understanding of science, public trust in science, and the politicization of science. The suggestions offered for dealing with contrarians may indirectly help protect the integrity and relative autonomy of the institution of science. What is ultimately at stake here is the future relevance of science vis-à-vis public policy. This chapter discusses some challenges that climate science encounters when attempting to inform public policy. Policy-making on science-related issues is lacking at best or seriously flawed at worst when productive input from the scientific community is distorted or rejected on the basis of economic or ideological interests. Thus, we must learn to effectively counter contrarians in many areas of science if we are to have the potential to realize the mandate of the National Academies – that our public policies on serious problems of the day are to be guided by the best technical knowledge and scientific advice available.

Notes

1. However, since the December 1997 Kyoto Conference, a significant part of the multinational business community has publicly acknowledged the reality of climate change and has begun to take the lead toward greater energy efficiency and renewable energy (see, e.g., Carpenter, 2001; Levy and Egan, 1998; Newell, 2000; see also the

chapters by James, Smith, and Doppelt; and Arroyo and Preston, Chapters 20 and 21, this volume).

2. The following are the most active and outspoken of the American climate change contrarians (listed alphabetically with primary professional affiliation): Bruce Ames, Professor of Biochemistry and Molecular Biology at the University of California, Berkeley; Sallie Baliunas, Staff Astrophysicist at the Harvard–Smithsonian Astrophysical Observatory; Robert Balling, Jr., Associate Professor of Geography and Director of the Office of Climatology at Arizona State University; John Christy, Professor and Director of the Atmospheric Science Department at the University of Alabama at Huntsville; Hugh Ellsaesser, retired meteorologist from Lawrence Livermore National Laboratory; Sherwood Idso, President of the Center for the Study of Carbon Dioxide and Global Change; David R. Legates, Associate Professor of Geography at the University of Delaware; Richard Lindzen, Professor of Meteorology at the Massachusetts Institute of Technology; Patrick Michaels, Research Professor of Environmental Sciences at the University of Virginia; Frederick Seitz, Chairman of the Science and Environmental Policy Project; S. Fred Singer, President of the Science and Environmental Policy Project; and Willie Soon, Research Physicist at the Harvard Smithsonian Center for Astrophysics.

3. This section draws largely upon earlier research (McCright and Dunlap, 2003) on a sample of five of the best-known American contrarians: Sallie Baliunas; Robert Balling; Richard Lindzen; Patrick Michaels; and Fred Singer.

4. Key works that document the enduring influence of conservative think tanks within the United States are: Allen (1992); Blumenthal (1986); Burch (1997a,b); Clawson and Clawson (1987); Diamond (1995); Himmelstein (1990); Jenkins and Eckert (1989); National Committee for Responsive Philanthropy (1997); People for the American Way (1997); Ricci (1993); Saloma (1984); Stefancic and Delgado (1996).

5. McCright and Dunlap (2003) systematically analyzed the content of approximately 224 documents produced and distributed by some of the more influential American conservative think tanks. They identified three interrelated counterclaims about global warming promoted by many influential conservative think tanks. First, the conservative movement claimed that the evidentiary basis of global warming is weak, if not wrong. Second, conservatives argued that the net effect of global warming would be beneficial should it occur. Third, conservatives argued that the policies proposed to ameliorate the alleged problem of global warming would do more harm than good. These three counterclaims comprised the conservative movement's response to the environmental community's call for ameliorative action on global warming.

6. An illustration might be warranted. An epidemiologist called to Congress to testify about AIDS would be expected to competently discuss what passes as widely accepted knowledge within the scientific community. If that epidemiologist were to use this venue as an opportunity to promote an ideological position – while distorting scientific knowledge – *and* if like-minded members of Congress were to use this to establish policy (or, more to the point, a policy of no action), then a serious public health problem may ensue that puts Americans at greater risk than before.

7. Indeed, this is a common theme in the literature on scientific controversies (see, e.g., Mazur, 1981; Nelkin, 1984).

References

Allen, M. P. (1992). Elite social movement organizations and the state. *Research in Politics and Society*, **4**, 87–109.

Austin, A. (2002). Advancing accumulation and managing its discontents. *Sociological Spectrum*, **2**, 71–105.

Baden, J. A. (1994). *Environmental Gore*. San Francisco, CA: Pacific Research Institute for Public Policy.

Bailey, R. (1993). *Eco-Scam*. New York: St. Martin's Press.

Beder, S. (1997). *Global Spin*. White River Junction, VT: Chelsea Green Publishing Company.

Blumenthal, S. (1986). *The Rise of the Counter-Establishment*. New York: Times Books.

Boykoff, M. T., and Boykoff, J. M. (2004). Balance as bias: Global warming and the U.S. prestige press. *Global Environmental Change*, **14**, 125–36.

Brown, Jr., G. E. (1997). Environmental science under siege in the U.S. Congress. *Environment*, **39**, 2, 12–31.

Burch, P. H. (1997a). *Reagan, Bush, and Right-Wing Politics, Volume I*. Greenwich, CT: JAI Press.

Burch, P. H. (1997b). *Reagan, Bush, and Right-Wing Politics, Volume II*. Greenwich, CT: JAI Press.

Carpenter, C. (2001). Businesses, green groups and the media. *International Affairs*, **77**, 313–28.

Clawson, D. and Clawson, M. A. (1987). Reagan or business? In *The Structure of Power in America*, ed. Schwartz, M. New York: Holmes and Meier, pp. 201–17.

Diamond, S. (1995). *Roads to Dominion*. New York: Guilford Press.

Dunlap, R. A. (2002). Paradigms, theories, and environmental sociology. In *Sociological Theory and the Environment*, eds. Dunlap, R. E., Buttel, F. H., Dickens, P., *et al*. CO: Rowman & Littlefield, pp. 329–50.

Fumento, M. (1996). *Science Under Siege*. New York: William Morrow.

Gelbspan, R. (1997). *The Heat Is On*. Reading, MA: Addison-Wesley Publishing.

Habermas, J. (1984). *Theory of Communicative Action: Volume 1*. Translated by T. McCarthy. Boston, MA: Beacon.

Habermas, J. (1987). *Theory of Communicative Action: Volume 2*. Translated by T. McCarthy. Boston, MA: Beacon.

Helvarg, D. (1994). *The War Against the Greens*. San Francisco, CA: Sierra Club Books.

Herrick, C. N. and Jamieson, D. (2001). Junk science and environmental policy. *Philosophy and Public Policy Quarterly*, **21**, 2/3, 11–16.

Himmelstein, J. L. (1990). *To the Right*. Berkeley: University of California Press.

Intergovernmental Panel on Climate Change. (2001). *IPCC Third Assessment Report: Contributions of IPCC Working Groups*. Geneva: IPCC.

Jenkins, J. C. and Eckert, C. M. (1989). The corporate elite, the new conservative policy network, and reaganomics. *Critical Sociology*, **16**, 121–44.

Levy, D. L. and Egan, D. (1998). Capital contests. *Politics and Society*, **26**, 337–61.

Luke, T. W. (2000). A rough road out of Rio. In *Consuming Cities*, eds. Low, N., Gleeson, B., Elander, I., *et al*. London: Routledge, pp. 54–69.

Mazur, A. (1981). *The Dynamics of Technical Controversy*. Washington, DC: Communications Press.

McCright, A. M. and Dunlap, R. E. (2000). Challenging global warming as a social problem. *Social Problems*, **47**, 4, 499–522.

McCright, A. M. and Dunlap, R. E. (2003). Defeating Kyoto. *Social Problems*, **50**, 3, 348–73.

Milloy, S. (1995). *Science Without Sense*. Washington, DC: Cato Institute Press.

Moser, S. C. and Dilling, L. (2004). Making climate hot: Communicating the urgency and challenge of global climate change. *Environment*, **46**, 10, 32–46.

National Committee for Responsive Philanthropy (1997). *Moving a Policy Agenda*, ed. Covington, S. Washington, DC: National Committee for Responsive Philanthropy.

National Research Council (1989). *Improving Risk Communication*. Washington, DC: National Academy Press.

National Research Council (2001). *Climate Change Science*. Washington, DC: National Academy Press.

Nelkin, D. (1984). *Controversy*, 2nd edn. Newbury Park, CA: Sage.

Newell, P. (2000). *Climate for Change*. Cambridge, UK: Cambridge University Press.

Ozone Action (1996a). Ties that blind I. *Ozone Action Report*. Washington, DC.

Ozone Action (1996b). Ties that blind II. *Ozone Action Report*. Washington, DC.

Ozone Action (1996c). Ties that blind III. *Ozone Action Report*. Washington, DC.

People for the American Way (1997). *Buying a Movement*. Washington, DC: People for the American Way.

Ricci, D. M. (1993). *The Transformation of American Politics*. New Haven, CT: Yale University Press.

Saloma III, J. S. (1984). *Ominous Politics*. New York: Hill and Wang.

Sanera, M. (1999). *Facts Not Fear*. Washington, DC: Regnery Publishing.

Sarewitz, D. (2004). How science makes environmental controversies worse. *Environmental Science and Policy*, **7**, 5, 385–404.

Soon, W. and Baliunas, S. (2003). Proxy climatic and environmental changes of the past 1,000 years. *Climate Research*, **23**, 89–110.

Stefancic, J. and Delgado, R. (1996). *No Mercy*. Philadelphia, PA: Temple University Press.

Switzer, J. V. (1997). *Green Backlash*. Boulder, CO: Lynne Rienner Publishers.

United States White House (2002a). *Climate Change Review*. Washington, DC: White House Staff. Available at http://www.whitehouse.gov/news/releases/2001/06/climatechange.pdf; accessed January 9, 2006.

United States White House (2002b). *Global Climate Change Policy Book*. Washington, DC: White House Staff. Available at: http://www.whitehouse.gov/news/releases/2002/02/climatechange.html; accessed January 9, 2006.

13

A role for dialogue in communication about climate change

Kathleen Regan
Tufts University
(previously at the Public Conversations Project)

Introduction

While many controversies are resolved through analysis, argument, compromise, and resolution, some become defined instead by opposing views that cluster around seemingly irreconcilable poles. This is the case, for example, in debates currently taking place in many communities around the dilemma of climate change and proposed responses. Under these conditions, approaches that work successfully in other contexts often become perversely counterproductive. Argument degenerates into diatribe, discussions deteriorate into shouting matches. Compromise becomes seen as surrender and a widely acceptable solution becomes hard to imagine (Becker *et al.*, 1995).

Given the challenges current conversations about climate change present, how do we begin a new and different conversation, or re-engage without rancor? A communication technique appropriately enough called *dialogue* offers one approach. Although dialogue is not conflict resolution, neither is it "just talk." It invites a conversation rooted in participants' personal experience. It can help to ease strained relationships and establish new ones. The new connections can themselves lead to new ways of understanding contentious issues like climate change and to discussing and developing possible solutions.

Dialogue is a genuinely creative and generative act. Successful dialogue is sometimes simply a willingness to meet again to continue talking, but once people begin to speak honestly about their concerns, and about their own uncertainties with regard to their deeply held positions, ideas for next steps emerge that could not have been imagined before the dialogue. Success can also be a decision to take an action that was previously not imagined or thought possible.

The need for deliberative dialogue

How might dialogue fit in to the public conversation on climate change? At first glance, conversation about climate change may not seem to be as polarized as some other social issues — most Americans view themselves as favoring a safe environment, for example. So why is a dialogical approach needed? People tend to have elaborate and differing opinions on matters that bear directly on environmental problems and solutions, such as private property rights vs. public goods, the roles of religion and science in their lives, and responsibilities to today's children vs. those to future generations. These differences, combined with the potential for expert/non-expert misunderstanding about highly technical issues, have effectively created polarized conversations. Dialogue, or facilitated and structured conversation, offers significant opportunities for improving communication in circumstances such as these (Kloor, 2005).

To get a feel for how conversations on climate change were currently perceived between scientists and non-scientists, I interviewed 14 people. They included both experts on climate change science and policy and non-expert citizens who are on the receiving end of what scientists and other knowledgeable speakers say. My goal was threefold: first, to learn what speakers on climate change thought they were saying or intended to say; second, to learn what they thought listeners heard; and third, to learn what listeners *did* hear. How did listeners evaluate and prioritize the information they heard? Who did listeners say they trusted when it came to technical information? All interviewees were asked the same questions.[1]

The findings were surprising and disturbing at once. Especially troubling was how well intentioned all interviewees were both to speak clearly and to listen openly, but how their efforts clearly failed. The various reactions highlighted just a few of the many challenges to creating an authentic dialogue: lack of shared language, lack of mutual respect and understanding, mistrust of motives, and cultural expectations about science. For example, non-scientists complained of being "talked down to" by experts who dismiss their perspectives and concerns. In addition, conversations on climate change have tended to devolve into debates about some scientific aspects of climate science such as uncertainty. Most of the public is not technically expert in those issues, and so, faced with "dueling" technical experts on opposing sides of the issue, non-scientists often feel as though it is hard to know "whom to trust"[2] (see also the chapters by McCright; and Warner, Chapters 12 and 10, this volume).

The interviews also showed that despite all good intentions, the conversation so far may be having unintended consequences. In fact, as the voice of alarm has grown louder, it sometimes has had the opposite of its intended effect on listeners. One interviewee characterized it as "compassion fatigue." Another likened it to the well-known doctor/patient dilemma in which the patient becomes more resistant to medical counsel as the gravity of his or her illness increases. Consequently, as advocates have increased the urgency of their message, many listeners have become more resistant to it. Speakers and listeners often find themselves caught in an "INSTRUCT/RESIST" or "CRITICIZE/DEFEND" sequence (Madsen, 1992; Moser and Dilling, 2004; see also chapters by Moser; and Dunwoody, Chapters 3 and 5, this volume). The inability of so many to speak to and listen to one another meaningfully signifies the need to look outside our current conversations about climate change for a different approach.

Beginning a new conversation

Dialogue offers a new approach. The differences in how we understand our relationship to the physical world relate not only to our degree of scientific knowledge but also are rooted in our personal values, identities, and world-views, and are legitimated in our religious traditions, our cultural practices, our passion to preserve what we love (see chapters by Tribbia; Bingham; and Jamieson, Chapters 15, 9, and 30, this volume). Each of us has a great deal at stake. The challenge of addressing the scientific complexities of climate change together with the web of underlying values is both daunting and pressing. It compels us to look at ourselves in ways we may never have before. Seeing ourselves differently is a first step.

Talking with others whom we may regard as different from ourselves is the second step. In response to a question about "unmet needs" with regard to climate change communication, one interviewee asked, "Who are the primary caretakers of beauty and the things you love? What are some things you love? ... What are the obstacles to your being more engaged?" Emotional connections to our world can be a much stronger force for action than technical sophistication (see also chapters by Dunwoody; and Moser, Chapters 5 and 3, this volume).

One hallmark of dialogue that distinguishes it from other kinds of talk is its ability to identify an "old conversation" that can act as a discursive trap and its techniques for providing opportunities for participants to step out of that conversation and into a new one. Recognizing what's going on in the old conversation is essential before participants can move forward. Characteristics of an "old conversation" include strong opinions coupled with an urge to convince or persuade others. Under these conditions, listening to other points of view becomes nearly impossible. But if we *can* learn to talk to one another across our current divisions, we have a chance to re-imagine our shared future.

In the context of climate change, the "old conversation" appears to be dominated and carried out within the frame of scientific knowledge, certainty or uncertainty, and predictions of the future. My interviews suggested that citizens get frustrated with scientists' perceived arrogance and dismissal of non-scientific points of view. Some felt that scientists "are unwilling to listen to our real concerns." As a consequence, meaningful conversation, in this context, becomes almost impossible. There is also the added challenge that experts and non-experts, scientists and citizens, bring different kinds of knowledge to the table and they are not generally regarded as equal to one another. Technological information is often perceived as superior to lay knowledge about issues and also as "value-free" because it is scientifically generated, evidence-based data. However, the scientific process also requires subjective judgments, such as which questions to ask, how technical problems are framed, which data to include or exclude, and other culturally-based assumptions that all scientists make in carrying out their work. Broadening the conversation beyond science and trusting the dialogic process enough for scientists to share the conversation with non-scientists is a key ingredient to creating a "new conversation" on climate change.

The nuts and bolts of dialogue

How then can we create this new kind of conversation? What would support our ability to remember that everyone is a stakeholder in the future of the climate? The Public Conversations Project (PCP) is an action-research project that has developed approaches to dialogue facilitation on divisive political issues. PCP began as a group of family therapists who observed similarities between polarized public conversations and "stuck" family conversations. In conflicted families, each person retreats behind generalizations and builds a case against the others. As clinicians, PCP founders observed how families move from impasse to dialogue, from closed to more open conversations. They observed also that relationships initially characterized by fear and anger could be transformed, and that people with deeply differing experiences and ideas could find new ways of talking and listening that did not involve self-silencing or shouting. They wondered if the approaches they used with families could fruitfully be applied to highly contentious public debates (Becker *et al.*, 1995).[3] According to PCP Senior Associate Maggie Herzig, "To offer a dialogue program is to invite people out of or away from conversations they find unsatisfying and into a fresh anddesired conversation." The goal of the dialogues PCP has facilitated is not to broker settlements. Rather, it is to promote constructive ways of communicating across differences, to stimulate fresh thinking about the ideas, beliefs and commitments of the "other" and of oneself, and to open up possibilities for action and change that were either obscured or unimaginable in the prior polarized debate.

A good dialogue offers those who take part the opportunity to:

- Listen and be listened to so that all speakers can be heard
- Speak and be spoken to in a respectful manner
- Develop or deepen mutual understanding
- Learn about the perspectives of others and reflect on one's own worldview (see Public Conversations Project, 2003).

Otto Scharmer at MIT's Global Leadership Initiative characterizes the goals of dialogue somewhat differently, but aptly with respect to conversations about climate change:

- To broaden the perspectives of those whose knowledge or expertise is confined to one or another area of specialization
- To provide opportunities for non-specialists to gain knowledge to which they would otherwise not have access
- To deepen each participant's understanding of how a particular problem or its proposed solutions may affect them and others
- To expand the range of imagined possibilities beyond one's own experience or prejudice.

His colleague, corporate strategist Adam Kahane, adds that dialogue can "build a shared understanding of what is happening (current reality), might happen (alternative scenarios), and should happen (the vision) (Kahane, 2002). Dialogue, so understood, involves an exchange of perspectives, beliefs, and experiences in which people listen openly and respectfully.

In short, dialogue is an alternative approach that holds some hope for new and more generative conversations about climate change. What does it take to get it started, given the strong feelings that exist on all sides?[4]

- A group is identified or identifies itself as wanting to engage in a dialogue, usually in response to a dilemma or crisis. Because people don't usually seek out dialogue except when they are under some stress about an event or occurrence, past or future, there are always strong emotions involved.
- An invitation is composed, one that clearly lays out the purpose of the dialogue but which also takes care not to state that purpose in value-laden language. The idea is to provide an invitation that *is open to multiple perspectives* on the subject at hand while at the same time remaining clear about the issues to be discussed.

For example, the issue might be the siting of a wind farm. Acknowledging the depth of people's concern is a necessary part of the invitation. Implying that one perspective is more acceptable than another is to be avoided.

- As individuals are invited, they are also asked some questions about their own perspectives, concerns, and deeply held beliefs about the topic. This allows the conveners of the dialogue to gain insight into what the core issues are for attendees, or where the emotional and political "rough spots" might be. These "mapping" interviews, as they are known, give conveners an opportunity to model for participants the respectfulness, openness, and curiosity they want participants to demonstrate in the dialogue itself. If those taking part feel respected and supported in their concerns at the outset, they are more likely to risk opening themselves up and being open to others in the dialogue. Mapping interviews also give conveners the opportunity to explain what participants can expect to encounter in the dialogue, to talk to them about proposed ground rules, and most important, give them an opportunity to decline to take part. Not everyone is ready for dialogue on issues of deep concern to them, and offering invitees the opportunity to decline the invitation to dialogue respects them, honors their deeply held beliefs around challenging or contentious issues, and acknowledges the difficulty of dialogue. It also assures that those who choose to take part are doing so with knowledge and understanding about the commitment they are about to make.

The steps within a dialogue can vary depending on the group and its hoped-for outcomes. Preparation is crucial, however, no matter what the goals of the dialogue itself may be. The more work done to prepare participants before they engage in a dialogue, the more likely will be its success. Participants and facilitators jointly create a structure for conversation which is designed to support the stated purposes of the dialogue. This conversational structure is always guided by ground rules created by participants and facilitators together. It is in this act of preparing together that the groundwork for successful dialogue is laid. All are asked to notice the assumptions they make as they listen to others, and to test these assumptions by asking questions. They also are asked to refrain from attempts to persuade or to refute what others say. Adhering to these agreements makes it possible even for people with fundamental and passionate differences to support the spirit and goals of the conversation (Public Conversations Project, 2003).

While ground rules can and do vary from one group to another, some examples of basic ground rules include the following:

1. Participants speak from personal experience, not as representatives of a group.
2. Participants shift from expressing positions to expressing underlying hopes, dreams, concerns, and fears.

3. Participants shift from emphasizing what they are sure about to acknowledging uncertainties, complexities, and gray areas in their own thinking about the issue.
4. Participants share stories about ways in which their views, hopes, and concerns have been shaped by their life experience – that is, they situate their views in their own experience.
5. Participants unpack the meaning of "buzzwords," either technical jargon or perhaps value-laden "hot-buttons."
6. Participants encourage one another to examine the premises, assumptions, and world-views that underlie their own thinking (M. Herzig).

When people speak as complex individuals, they begin to listen more fully to each other. They also may begin to reflect in new ways on their own perspectives and to understand themselves as well as others, better. As mutual understanding increases, trust usually increases as well.

The guidelines offered above are not just applicable to clearly divided or antagonistic groups. They can also be helpful in situations where participants share a general perspective related to the controversy, but differ in the nuances of their beliefs, feelings, and uncertainties. For example, for some, the issue of climate change is part of a larger question, that of sustainability. Yet even among those who share a common concern about sustainability as an underlying question, deep divisions exist. Polarized debates occur about emerging technologies that many believe are as good as or better than what is currently in use. For example, families, technical experts, and whole towns are deeply divided over a proposed wind farm off the coast of Cape Cod, Massachusetts.[5] Most residents support the idea of alternative energy sources, but some oppose it on aesthetic grounds while others object to use of a public resource for private gain. For some, the heart of the matter is that they have felt excluded from the process; for others, it has to do with the role and responsibilities of individual local communities in responding to broader societal dilemmas. Some are concerned about the visual impact on their viewscape, while others see the project as creating jobs and reducing climate impacts. But most residents feel some combination of all the above. Dialogue is uniquely suited to address questions of fairness, participation, roles, and values.

Conclusions

While such conversations can be difficult, they are also urgent. For even if people could agree on some technological responses to climate change, the currently available solutions (Pacala and Socolow, 2004; Hoffert *et al.*, 2002) may not be enough to actually halt the increase in atmospheric

CO_2 concentrations. Many believe that more than technological solutions are needed, more than replacement of one resource with another. They argue that changes in citizens' actions, behavior, and consumption choices are also imperative (see also Michaelis, Chapter 16, this volume). The challenges of climate change, public policy, and private actions engender some critical questions for all of us:

- How can community and policy leaders help shift conversations about our varied perspectives on this issue in such a way that they represent hope for all participants while reflecting and respecting their underlying values?
- What kinds of conversations might invite participants to imagine and envision creative, new ideas for the future?
- Who is currently not part of such conversations, but should be included?
- What could be gained by listening to the voices of others?
- What might be lost by not listening to the voices of others?

Dialogue invites a way to begin to address these questions. It is, however, perhaps only the beginning of a long conversation that has not yet been imagined. Questions like the ones stated above are very big and will not be answered overnight. They present enormous challenges, and we need as many visions as we can gather in order to find our way. Engaging in a "new conversation" on climate change, inviting a full range of voices, preferences, and creative thinking, thus affords a critical opportunity. These questions require deliberate, thoughtful, engaged citizens in all communities to reach across longstanding divisions. Such conversations across the biggest divides in our communities are not yet occurring. Where some progress has been made in understanding and taking action, it has been in communities with shared values, and with individuals willing to champion actions (see chapters by Watrous and Fraley; Rabkin and Gershon; Pratt and Rabkin; and Arroyo and Preston, Chapters 25, 19, 6, 21, this volume). Some would suggest, however, that challenging as it may be, this is a rich opportunity: climate change may offer a chance to mend the divisiveness of our society as a whole (see also Jamieson, Chapter 30, this volume).

> One way or another, the world we inhabit will not long continue on its historical trajectory. At this extraordinary moment, the accelerating pace of global warming is driving us to a unique point in our evolution. Not coincidentally, the antidotes to global warming have the potential to mend a profoundly fractured world.
>
> *Gelbspan (2004: 11)*

If we choose to pursue an ecologically sustainable future, however we define or understand it, radical shifts in the way we express our values, lifestyle choices, and consumption patterns will be required. We will need then to engage in dialogue about what those shifts should be and how we make them if we are to pass on a sustainable/livable world to those who will come after us. Dialogue offers a path out of our deepest fears and a way to realize our most fragile hopes.

Notes

1. List of interview questions. See *Appendix at the end of this chapter*.
2. Quotes are excerpted from my taped interviews. Details of the tapes are available on request.
3. Also, case studies of some environmental dialogues facilitated by PCP such as the Northern Forests Dialogue Project can be found on their website: http://www.Publicconversations.org; accessed January 12, 2006.
4. The following is adapted from PCP's *Guide to Community Dialogue*, available on their website: www.publicconversations.org. The original Guide contains detailed descriptions and plans for convening a dialogue.
5. Cape Cod Times Special Report. Available at: http://www.capecodonline.com/special/windfarm; last accessed January 9, 2006.

References

Becker, C., Chasin, L., Chasin, R., Herzig, M., and Roth, S. (1995). From stuck debate to new conversations on controversial issues: A report from the public conversations project. In *Cultural Resistance: Challenging Beliefs about Men, Women and Therapy*, ed. K. Weingarten. Binghampton, NY: Haworth Press, pp. 143–63.

Gelbspan, R. (2004). Cool the rage: To take the heat out of terrorism, solve the climate crisis. *Orion*, July–August, 11.

Hoffert, M., Caldeira, K., Benford, G., *et al.* (2002). Advanced technology paths to global climate stability: Energy for a greenhouse planet. *Science*, **298**, 981–7.

Kahane, A. (2002). *Changing the World by How We Talk and Listen*. Generon. Available at: http://pioneersofchange.net/communities/foresight/articles/Kahane%20on%20talking%20and%20listening.pdf; accessed January 19, 2006.

Kloor, K. (2005). The holy & the hawks. Longtime foes join forces to combat a common enemy: Global warming. *Audubon Magazine*, September–October.

Madsen, W. C. (1992). Problematic treatment: Interaction of patient, spouse, and physician beliefs in medical noncompliance. *Family Systems Medicine*, **10**, 368.

Moser, S. and Dilling, L. (2004). Making climate hot: Communicating the urgency and challenge of global climate change. *Environment*, **46**, 10, 32–46.

Pacala, S. and Socolow, R. (2004). Stabilization wedges: Solving the climate problem for the next 50 years with current technologies. *Science*, **305**, 968–72.

Public Conversations Project (2003). *Constructive Conversations about Challenging Times: A Guide to Community Dialogue*. Version 3.0. Available at http://www.publicconversation s.org; accessed January 9, 2006.

APPENDIX: Questions for Science Communicators/Listeners

What is your experience in talking to non-scientists about scientific information?

Do you have opportunities to talk with non-scientists specifically about scientific issues related to global climate change?

If not climate change in particular, what are the subjects of your scientific or technical communication with non-scientists?

What do these conversations 'look like' when they work? What do they look like when you sense that they are not working?

Have you been in situations talking about science or the evironment in which you felt as though you had not been heard?

Have you been in communications situations that you think might have been better served by a more structured or facilitated conversation?

What do you think are the speaking challenges for you as a specialist in a field?

What do you think are the listening challenges for you as a nonspecialist?

What are some questions your audiences have raised that have resonated with you? Or given you the opportunity to reflect on your won beliefs?

What are some questions you have raised in conversations with specialists?

How do you decide whom to trust in talking with people who have technical knowledge that you do not share?

Do you – or how do you – change your choice of words/language when faced with people whose views differ strongly from your own?

What questions have you – or might you – pose to a group not in agreement with you?

Are you aware of any "unmet needs" with regard to communication around climate change in particular, or environmental issues more generally?

In what contexts have you seen these kinds of communications succeed? In what contexts have you seen these communications fail?

What would be your advice for scientists in talking with non-scientists about technical information?

What have I not thought to ask you that you think I should know – or – what would you like for me to ask?

14

Information is not enough

Caron Chess
Rutgers University

Branden B. Johnson
New Jersey Department of Environmental Protection

Introduction

Communication about climate change is as complex as the science. Among the most difficult communication challenges is persuading people to act on a problem such as climate change, which is not immediately relevant or easily solved. The challenge becomes even more difficult when you ask people to change routine behavior, such as how they get to work or set their home thermostat.

Not surprisingly, after years of schooling, many readers of this volume probably come to it believing that information changes attitudes and behaviors. Some may go so far as to think information is a cure. Filling eyes and ears with sufficient data may seem a means of cleansing the brain of misguided opinions and incorrect facts, and replacing them with "appropriate" beliefs. Many informational efforts to promote environmentally responsible behavior lean on an implicit theory of behavior (Costanzo *et al.*, 1986) that "right" behavior naturally follows from "right" thinking.

However, more knowledge does not necessarily lead to more appropriate behavior, or even behavior deemed more appropriate by the educator or communicator (Johnson, 1993; Fischhoff, 1995). Social science theory and much empirical research show that links between information and behavior can be tenuous at best. Information is not entirely inconsequential, but it is much overrated as a change agent (see a discussion of the so-called "deficit model" in Sturgis and Allum, 2004; see also Tribbia, Chapter 15, this volume).

Advocates for precautionary action may find it difficult to accept the fragile link between information and behavior. After all, many advocates have devoted much time to crafting educational materials that are clearly important. But we do question whether the focus of educational materials should largely be transmitting basic factual information, or that education is the most effective method of prompting climate change mitigation.

We argue that increasing the popularity of, say, energy-efficient cars will not be simply the result of more people appreciating the potential impacts of climate change. Neither will reduction of home energy usage simply increase with knowledge of the link between use of fossil fuels and global change. Instead, people are more likely to be motivated by prior beliefs and values (see also Dunwoody, Chapter 5, this volume). We then note which kind of information may be most helpful. We also include research showing that individuals will be more attentive to those advocates who seem "like me" or are seen as credible for other reasons. Finally, this chapter suggests that providing information about the threat of global warming needs to be coupled with information about the efficacy of various remedial actions.

Two stories of behavior change

Advocates for precautionary action may find it easier to appreciate the fragile link between information and behavior by considering two examples that have nothing to do with climate but everything to do with changing behavior.

A communication failure

After two years of intensive effort by a local health worker to encourage boiling of water for consumption, only 11 families in a Peruvian village of 200 families boiled their drinking water. This failure, described in one of the classics of the communication field (Rogers, 1962), was in part due to the public health worker being seen by the poor villagers as a "snooper" sent to Los Molinos "to pry for dirt and to press already harassed housewives into keeping cleaner homes" (p. 11).

Another reason for the failure was that the health worker spent a great deal of time promoting water boiling with those who did not have much status in the village. For example, one early adopter had moved from the Andes Mountains to this coastal town. Her highland haircut and poor Spanish marked her as an outsider and she was unlikely to ever gain social acceptance. Because she had nothing to lose socially, she had no problem accepting the health worker's advice. Another early adopter was ill. Her behavior reflected the norms of the village: sick people drink "cooked" water and healthy people don't. Their belief system defined foods as inherently hot (e.g., brandy) and cold (e.g., pork and "raw" water), regardless of their temperature. Because ill people must avoid extremes, cooking the water moderated its fundamental coldness.

In short, the health worker had little credibility with most of the villagers, who distrusted her motives, and her early adopters were marginalized even before they began boiling water. With such role models, the practice of boiling water would never gain acceptance, let alone become routine. In addition, the health worker's transmission of information about germ theory was at odds with the prevailing beliefs about the causes of illness.

A communication success

Local members of Womankind Kenya, a grassroots group opposing genital mutilation of women, tried many times to convince one of the country's most esteemed practitioners to stop the practice, according to *The New York Times* (Lacey, 2004). However, their arguments, including those about women's health, did nothing to change the cutter's perspective, which was deeply rooted in her culture. "I thought my mother would curse me from the grave if I did not carry on the tradition." She said she had no use for people who came around denouncing her way of life. But the advocates knew that "She was respected, and we wanted her on our side." So they tried a very different approach. When the activists brought influential clerics who had become convinced the practice was harmful, this religious woman felt obliged to listen. According to this report, the imams told her that her practices were not consistent with the Koran. She not only had to stop the practice, she had to ask forgiveness from each girl who was maimed. She has since become a leading activist, using her credibility to convince other practitioners to stop cutting (Lacey, 2004).

Motivating precautionary behavior

These stories, although completely removed from the issue of climate change, provide clues for those hoping to encourage individuals to reduce fossil fuel use or heat-trapping gas emissions.

None of these strategies can guarantee success, either alone or collectively. But they can be useful tools for enhancing persuasiveness. Whether the focus is beliefs, values, trust, emotions, or economics, "[t]he greatest enemy of persuasion is assumption;" when motivating changes in behavior, "[y]ou must know what *does* matter to [your audience], not what *should* matter to them" (Reardon, 1991: 10). Hunches about motivation are not enough; professionals should directly assess the motivations of target populations for the particular behavior in mind (see the chapters by Tribbia; Moser; Young; Watrous and Fraley; Michaelis; Dilling and Farhar; and Agyeman *et al.*, Chapters 15, 3, 24, 25, 16, 23, and 7, this volume, all of which touch on the question of motivation). Just as important is appreciating that a motivator might be offensive in some circumstances, or worse in others. For example, tax rebates might encourage some American consumers to consider a hybrid car. But giving money for votes is bribery. Research is also essential to determine what target populations perceive as barriers to specific behaviors. For example, if biking to the store means risking one's life, few people will respond to inducements to bike. Barriers are tied to the specific behavior: obstacles to a suburbanite's car-pooling are different than those that impede conserving energy at home (McKenzie-Mohr, 2000).

Maybe the most important thing to know for motivating behavior is your audience's beliefs and values. The health worker in Los Molinos could provide extensive information in the residents' own language on germ theory and illness, and on links between behavior and disease. But this information did not conform to villagers' views of the world, so it had limited salience. In addition, understanding of germ theory may not have been necessary for them to change their behavior. Boiling water need not depend on accepting that water contains microscopic life that can threaten one's health.

Factual information was even less pertinent to the Kenyan 'cutter,' who found little relevance in the science linking such cutting to physical harm to women when this behavior seemed to her to clearly provide social and cultural benefits. The persuasive clerics did not preach about the rate of infection or the status of women. Instead they argued that a clitoris was as important to God as an eye or an arm. Cutting was a sin. In other words, the cutter did not initially stop performing this rite for the reasons articulated by activists. In turn, the activists did not insist that cutters stop for the "right reasons."

If environmental advocates only attend to those who do the "right" things for the "right" reason, they are unlikely to change substantially the use of fossil fuels. Several studies related to climate change (Bostrom *et al.*, 1994b; Read *et al.*, 1994; Kempton, 1997) also stress the importance of knowing prevailing attitudes and values of the audience.

As Kempton (1997: 20) put it, "anyone trying to communicate with the public about global environmental change has to address preexisting cultural models and concepts and not simply assume that he or she is writing on a blank slate." Presenting scientists' causal models of climate change is not likely to encourage energy conservation if audiences have strong countervailing values or attitudes on the issue. As in Los Molinos, provision of better information may not be sufficient.

For example, according to one recent study, Americans see fossil fuels as the basis of a comfortable and familiar life (CESA and GNP, 2003). By contrast, "clean energy" (the term they preferred to describe electricity from wind, solar, and water sources) was seen as unproven to be able to provide reliably for this same comfortable, familiar lifestyle. Appealing to self-sufficiency was seen as more motivating than a "nagging" appeal from environmentalists (see also Michaelis, Chapter 16, this volume). SUV manufacturers have tapped into such motivating desires, portraying SUVs as a way to survive bad weather or get away from the crowd.

The opportunity to demonstrate high status combined with virtue, particularly if it requires little sacrifice of amenities, appears to motivate early Hollywood and well-to-do suburban purchasers of the Toyota Prius hybrid car (Gross, 2004). People do not have to exhibit high moral standards or honorable motives to demonstrate to others the benefits of behavior change. Demanding too much virtue, perceived as environmentalist "nagging," might even get in the way of persuasion.

Climate change activists might also motivate behavior change by tapping into our desire to conform by convincingly representing reduction in fossil fuel use as an emerging norm, something that "everyone" does, or at least people the audience respects, such as celebrities (Bator and Cialdini, 2000; Cialdini, 2001). Ajzen and Fishbein (1980) posited behavioral intentions as due in part to subjective norms (e.g., beliefs about what others close to the subject think s/he should do). For example, research participants recycled more frequently and greater amounts when they were given feedback with either injunctive social norms (what others think *should* be done) or descriptive social norms (what other people *are* doing) (Bator and Cialdini, 2000). Saying nothing or simply asking people to recycle had less impact (Schultz, 1999; Schultz and Tyra, 2000). As long as claims about buying hybrid cars or "clean energy" electricity are credible — for example, they are highly visible in daily life or mass media coverage, or come from a trusted source — these communications could help increase adoption of such behaviors.

As mentioned earlier, another potential motivator is financial incentives to adopt the new behavior and disincentives to stay with the old (e.g., increased gasoline prices, new carbon taxes, or rebates and tax write-offs for renewable energy) (see also Dilling and Farhar, Chapter 23, this volume). These incentives do not depend upon accurate knowledge about climate change or desire to avoid its effects. This may make financial incentives and disincentives potentially effective tools.

On the other hand, fiscal approaches must be carefully designed to avoid both unintended effects and barriers to this approach. For example, Kenyan cultural norms might have condemned such payments as a rationale for genital cutters to abandon their profession, while religion was an acceptable reason.

Appropriate information for appropriate action

Again, these strategies don't obviate providing information. We simply argue that factual information is usually not sufficient to *motivate* behavior. The usefulness of specific information and the effectiveness of these motivators might vary by the stage in behavior change that a person has achieved. We are all familiar with the example of New Year's resolutions for behavior change (e.g., to lose weight) that are never put into action or soon abandoned. Several theories posit that the process by which people come to adopt a precaution (such as a particular behavior that will reduce fossil fuel use) comprises a number of different stages. For example, stages typically include unawareness of the issue, contemplation, preparation, action, and maintenance of the newly adopted behavior (for a review of behavior change theories, including the stage model, see Halpern *et al.*, 2004; Glanz, Rimer, and Lewis, 2002). There are common barriers to change facing people in the same stage and different barriers facing people in different stages, among which effective communications must distinguish (Prochaska and DiClemente, 1983; Weinstein, 1988; Weinstein and Sandman, 1992; Weinstein *et al.*, 1998a,b).

For example, research is needed to determine how to make an aware person realize climate change is personally salient. We also need to know more about ways to prompt an intention to act in those who already find the issue salient by and to help translate the intention into action.

Information may be important for those *already motivated* to then take the next step. Research would be useful to determine what information reinforces those already motivated, such as those looking to do the "right thing" by buying fuel-efficient cars. This audience may also be interested in how the cars work. Information may also remove barriers to action by others. For example, research might find that people who are interested in buying hybrid cars need to know that maintenance is not difficult.

In addition, unlike the Kenyan choice (cut or not), anyone desiring to take action on climate change has many possible behaviors to consider, not all of them appropriate. Information may be needed to encourage appropriate behavior (Bord, Fisher, and O'Connor, 1997). For example, research shows that citizens hold several myths that can lead to inappropriate behavior (Löfstedt, 1991, 1992, 1993; Bostrom *et al.*, 1994a; Read *et al.*, 1994; Kempton, 1997; Dunlap, 1998: 482; Bord *et al.*, 2000; Bulkeley, 2000; Henry, 2000; Leiserowitz, 2003). For example, many Americans confuse climate change with stratospheric ozone losses, and thus mistakenly believe that aerosol sprays are a major cause of global warming (see also chapters by Bostrom and Lashof; and Leiserowitz Chapters 1 and 2, this volume). This misunderstanding leads people to think they are protecting the environment by buying a hair gel or roll-on deodorant rather than a spray. They are recalling former use of ozone-destroying chlorofluorocarbons as spray propellants, without recognizing that these are now banned from spray cans and most other uses in the United States. This confusion may be amplified by the largely irrelevant fact that CFCs are a small-volume, if powerful, greenhouse gas.

Another myth is attribution of climate change largely to traditional industrial air pollution, even though industry provides a relatively small portion of greenhouse gases. This attribution is partly due to viewing carbon dioxide as a "pollutant" (like sulfates and other toxic substances), allegedly due to frequent use of the phrase "fossil fuel *burning*" in climate change educational materials (Kempton, 1997: 20; emphasis added). The attribution also comforts people that they don't really have to change their own behavior, interfering with their ability to see that their own energy use contributes to climate change even more than industry (Stamm *et al.*, 2000).

The "mental models" method applied by Bostrom *et al.* (1994b) and Read *et al.* (1994) to climate change (also see Morgan *et al.*, 2002) explicitly aims to identify and correct causal myths about risks (see also Bostrom and Lashof, Chapter 1, this volume). When providing information, it is important to acknowledge the seeming validity of such an erroneous belief and explain why it may be wrong *before* showing why another explanation is superior (Rowan, 1991: 376). Otherwise information about climate change has little chance of making it past the prior-belief barrier.

The credibility of the messengers

The background and outsider status of the health worker in Los Molinos made it difficult, if not impossible, for her to motivate people to change their behavior. She was suspect, and by extension, so was her story of how disease was caused. A messenger who seemed more like the other villagers would have been more persuasive.

Once low-status individuals were early adopters, the likelihood of others boiling water became even more remote. If she had persuaded the most popular woman in the village to boil water, she would have elicited more response from other members of the village. People are persuaded by "PLUs" (People Like Us), or those messengers perceived as even "better than us" (Julian Agyeman, personal communication).

Conversely, the Muslim clerics in Kenya had credibility for a cutter who felt she "was doing the right thing in the eyes of God." She could not dismiss them as she could the local activists. Others were then encouraged to change their behavior once they had a visible role model of a respectable converted cutter.

The importance of credible communicators and early adopters has not yet been empirically examined in the climate change field, but an example from energy conservation illustrates the concept. Craig and McCann (1978) found that heavy residential energy users receiving a letter purporting to be from the state utility commission reduced energy use much more than those who received the same letter purporting to be from the company.

Conclusions

As noted earlier, none of the strategies discussed in this chapter offers a guarantee of behavior change. Clearly, information is not unimportant. However, "just the facts, ma'am" is a strategy that on its own is unlikely to encourage Americans to change their energy use or other behaviors that cause greenhouse gas emissions. Activists and communicators should continue to supply information, but they must also understand what else matters to their audiences and adjust their approaches accordingly.

Unfortunately, if theories about the limited impact of information are correct, readers already strongly committed to providing "just the facts" about global climate change may not find this chapter convincing. Those dissatisfied with their limited success to date may be more open to a different approach, and may even find this chapter persuasive.

References

Ajzen, I. and Fishbein, M. (1980). *Understanding Attitudes and Predicting Social Behavior*. Englewood Cliffs, NJ: Prentice-Hall.

Alhakami, A. S. and Slovic, P. (1994). A psychological study of the inverse relationship between perceived risk and perceived benefit. *Risk Analysis*, **14**, 1085–96.

Bator, R. J. and Cialdini, R. B. (2000). The application of persuasion theory to the development of effective proenvironmental public service announcements. *Journal of Social Issues*, **56**, 527–41.

Bord, R. J., Fisher, A., and O'Connor, R. E. (1997). Is accurate understanding of global warming necessary to promote willingness to sacrifice? *Risk: Health, Safety & Environment*, **8**, 339–54.

Bord, R. J., O'Connor, R. E., and Fisher, A. (2000). In what sense does the public need to understand global climate change? *Public Understanding of Science*, **9**, 205–18.

Bostrom, A., Atman, C. J., Fischhoff, B., *et al.* (1994a). Evaluating risk communications: Completing and correcting mental models of hazardous processes, part II. *Risk Analysis*, **14**, 789–98.

Bostrom, A., Morgan, M. G., Fischhoff, B., *et al.* (1994b). What do people know about global climate change? 1. Mental models. *Risk Analysis*, **14**, 959–70.

Bulkeley, H. (2000). Common knowledge? Public understanding of climate change in Newcastle, Australia. *Public Understanding of Science*, **9**, 313–33.

Cialdini, R. B. (2001). The science of persuasion. *Scientific American*, **284**, 76–81.

Clean Energy States Alliance (CESA) and Gardner Nelson and Partners (GNP) (2003). *Message Development Focus Groups: Qualitative Analysis*. New York, NY: CESA, GNP.

Costanzo, M., Archer, D., Aronson, E., and Pettigrew, T. (1986). Energy conservation behavior: the difficult path from information to action. *American Psychologist*, **41**, 521–8.

Craig, C. S. and McCann, J. M. (1978). Assessing communication effects on energy conservation. *Journal of Consumer Research*, **5**, 82–8.

Dunlap, R. E. (1998). Lay perceptions of global risk: Public views of global warming in cross-national context. *International Sociology*, **13**, 473–98.

Fischhoff, B. (1995). Risk perception and communication unplugged: Twenty years of process. *Risk Analysis*, **15**, 137–45.

Glanz K., Rimer, B. K., and Lewis, F. M. (eds.) (2002). *Health Behavior and Education: Theory, Research, and Practice*, 3rd edn. San Francisco, CA: Jossey-Bass.

Gross, J. (2004). From guilt trip to hot wheels. *The New York Times*, June 13, ST-1, p. 14.

Halpern, D., Bates, C., Beales, G., and Heathfield, A. (2004). *Personal Responsibility and Changing Behaviour: The State of Knowledge and Its Implications for Public Policy.* London, UK: Cabinet Office, Prime Minister's Strategy Unit. Available at: http://www.pm.gov.uk/files/pdf/pr.pdf; accessed January 9, 2006.

Henry, A. D. (2000). Public perceptions of global warming. *Human Ecology Review,* **7,** 25–30.

Johnson, B. B. (1993). Advancing understanding of knowledge's role in lay risk perception. *Risk: Issues in Health & Safety,* **4,** 189–212.

Kempton, W. (1997). How the public views climate change. *Environment,* **39,** 12–21.

Lacey, M. (2004). Genital cutting shows signs of losing favor in Africa. *The New York Times,* June 8, p. 3.

Leiserowitz, A. A. (2003). Affective Imagery and Risk Perceptions of Global Climate Change. Presentation given at the Annual Meeting of the Society for Risk Analysis, Baltimore, December 7–10.

Löfstedt, R. E. (1991). Climate change perceptions and energy-use decisions in northern Sweden. *Global Environmental Change,* **1,** 321–4.

Löfstedt, R. E. (1992). Lay perspectives concerning global climate change in Sweden. *Energy and Environment,* **3,** 161–75.

Löfstedt, R. E. (1993). Lay perspectives concerning global climate change in Vienna, Austria. *Energy and Environment,* **4,** 140–54.

McKenzie-Mohr, D. (2000). Promoting sustainable behavior: An introduction to community based social marketing. *Journal of Social Issues,* **56,** 543–54.

Morgan, M. G., Fischhoff, B., Bostrom, A., and Atman, C. J. (2002). *Risk Communication: A Mental Models Approach.* Cambridge, UK: Cambridge University Press.

Prochaska, J. O. and Di Clemente, C. C. (1983). Stages and processes of self-change of smoking: Toward an integrative model of change. *Journal of Consulting and Clinical Psychology,* **51,** 390–5.

Read, D., Bostrom, A., Morgan, M. G., *et al.* (1994). What do people know about global climate change? 2. Survey studies of educated laypeople. *Risk Analysis,* **14,** 971–82.

Reardon, K. E. (1991). *Persuasion in Practice.* Newbury Park, CA: Sage.

Rogers, E. M. (1962). *Diffusion of Innovations.* New York, NY: Free Press.

Rowan, K. E. (1991). When simple language fails: Presenting difficult science to the public. *Journal of Technical Writing and Communication,* **21,** 369–82.

Schultz, P. W. (1999). Changing behavior with normative feedback interventions: A field experiment of curbside recycling. *Basic and Applied Social Psychology,* **21,** 25–36.

Schultz, P. W. and Tyra, A. (2000). A Field Study of Normative Beliefs and Environmental Behavior. Presentation given at the Western Psychological Association, Portland, OR, April.

Stamm, K. R., Clark, F., and Eblascas, P. R. (2000). Mass communication and public understanding of environmental problems: The case of global warming. *Public Understanding of Science,* **9,** 219–37.

Sturgis, P. and Allum, N. (2004). Science in society: Re-evaluating the deficit model of public attitudes. *Public Understanding of Science,* **13,** 55–74.

Weinstein, N. D. (1988). The precaution adoption process. *Health Psychology,* **7,** 355–86.

Weinstein, N. D. and Sandman, P. M. (1992). A model of the precaution adoption process: Evidence from home radon testing. *Health Psychology*, **11**, 170–80.

Weinstein, N. D., Lyon, J. E., Sandman, P. M., *et al.* (1998a). Experimental evidence for stages of health behavior change: The precaution adoption process model applied to home radon testing. *Health Psychology*, **17**, 445–53.

Weinstein, N. D., Rothman, A. J., and Sutton, S. R. (1998b). Stage theories of health behavior: Conceptual and methodological issues. *Health Psychology*, **17**, 290–9.

PART TWO

Facilitating social change

15

Stuck in the slow lane of behavior change? A not-so-superhuman perspective on getting out of our cars

John Tribbia

University of Colorado—Boulder

Introduction

While riding my bike home from work during rush-hour traffic, I noticed once again that I am able to travel from Boulder's downtown to my home on the southern outskirts faster than a personal automobile. This is not because I am a superhuman on a bicycle, but because I do not have to walk far for parking and sit idly in traffic as drivers do. On this particular day, riding at a leisurely pace down the bike path, I passed the usual scene of standstill traffic and waved to several drivers. I wondered "What compels these folks to sit in traffic, especially those that also appear unhappy about their present situation?" I know my personal motives for riding a bike, including saving money on gas and parking, preservation of the environment, avoiding greenhouse gas emissions, and my physical well-being, to name a few. But why are streets backing up with cars if the highly educated people of the most "bicycle friendly and best commuting" community of Colorado — Boulder — are unhappy waiting at the stop light? And maybe more importantly, what would get them out of their stalled cars and join me on the bike path or get into public transportation?

Clearly, my daily observations are not unusual. Traffic congestion is a rapidly worsening problem in many US urban areas. And many people who already act in environmentally responsible ways wonder why others don't do so. My observations also reflect the findings in the sociological and psychological literature that high levels of education, knowledge about environmental impacts of personal actions, or information alone do not guarantee environmentally responsible behavior (ERB). The contrast between "knowing" and "doing" could hardly be greater than it is in my home town.

Situated at the foothills of Colorado's Rocky Mountains, the city of Boulder has become a national hub for well-educated, liberal, outdoor, and environment-conscious individuals who have access to alternative media and a wide variety of relevant information sources. According to the 2000 US census, more than 55 percent of Boulder's residents have either an Associate's, Bachelor's and/or graduate degree, compared to 40 percent among residents of Colorado and 31 percent nationwide (US Bureau of the Census, 2004). In addition to the city's large contingent of educated people, many of its grassroots environmental programs initiated in the 1960s and early 1970s have made Boulder a reputable "green" city – thus attracting visitors and new residents from all over the nation. Many of these grassroots organizations, like EcoCycle – the community's recycling program, Wild Oats – a now national grocer feeding the organic food movement – and GO Boulder – the alternative transportation program of the city – have helped Boulder become a city with a widely recognized green reputation.

Residents are proud, and visitors agree, that the city is one of the "Top Green and Clean Cities" (*The Green Guide*, 2005; see also *Modern Maternity*, 2000 [from City of Boulder, 2005]). Boulder has also been recognized for its pioneering environmental efforts as one of the "Five Most Impressive Cities Making a Difference in the Environment" (*Delicious Living Magazine*, 2004 [quoted by City of Boulder, 2005]). Living up to this reputation, Boulder has great accommodations for bicyclists – it was voted a "Bicycle Friendly Community" by the American League of Cyclists in 2004 (City of Boulder, 2005) and offers the best commuting access for Colorado workers (US Environmental Protection Agency, 2004 [quoted by City of Boulder, 2005]). Boulder also participates in the Cities for Climate Protection Campaign (see Young, Chapter 24, this volume). Despite having these and numerous other accolades for being environmentally friendly, it is not clear that we Boulderites are actually acting out our green reputation any more than other US cities. In fact, there seems to be a dissonance between what Boulderites think, believe, and politically support, and what many of us actually do – notably the cars we drive, the amount of material goods we purchase, what we do *not* teach in schools, and how many of us "jet-set" around the globe with little awareness of the environmental consequences of our behaviors. Consequently, a closer look at what guides individual behavior unveils why Boulder, of all places – despite great efforts by the city – has yet to provide alternative forms of transportation that will make it more convenient and cost-effective in terms of time and money for people to get where they need and want to go without their cars.

The following sections look at the factors that affect an individual's behavior and the barriers that limit ERB or hinder behavior change. Neither of these factors alone appears to be sufficient to explain people's behavior or any change therein, yet if one is missing, change toward ERB seems less likely. In the latter part of the chapter, I will build on these findings and explore the implications for communication and behavior change strategies.

Factors affecting environmentally responsible behavior

The literature on changing human behavior, especially toward environmentally responsible behavior, is multifarious. Researchers from a variety of disciplines have examined the internal and external forces that foster and constrain an individual's actions. I summarize this literature here under five rubrics and discuss them along with potential barriers to behavior change: an individual's inclination to engage in a certain behavior; the motivation to continue or change that behavior; the cognitive and emotional information processing that inform behavioral choices; a person's ability and skill to engage in ERB; and the presence or absence of external support through peers, norms, and infrastructure.

Inclination

The concept of inclination is not uniformly defined or understood. I use it here – as many researchers do – to mean those characteristics of an individual that make him or her more or less likely to engage in ERB. In particular, I focus on an individual's disposition as determined by a combination of socio-economic status, race, gender, age, values, attitudes, and beliefs (Corbett, 2005).

Socio-economic status can influence inclination in several ways. For example, personal income can determine how well a person is able to meet their needs and desires, such as what kind of automobile(s) they may purchase. Indeed, when passing one of the area high schools in Boulder, one notices large numbers of cars driven by area teenagers. This elucidates how affluence in a community can negatively impact the environment, since most of the cars driven by teenagers are second, third, or even fourth vehicles in the household.

People with larger disposable incomes are more likely to own and drive a vehicle than a person who lacks the financial means to even purchase one. Still, socio-economic status also appears to influence inclination to engage in ERB through indirect ways, affecting people's personally pressing concerns, the education they enjoy, the access to information they might have, or maybe their ability and desire to buy hybrid vehicles.

Similarly, age can influence one's inclination toward ERB. "Since the 1980s, the demographic factor most strongly and consistently associated with environmental concern has been age, with younger generation adults more concerned than their counterparts ([...], Corbett, 2005: 369). For example, people who have acquired environmental information and knowledge starting around 1970 "are far more aware of and concerned with environmental issues than prior cohorts" (National Research Council, 2002: 8). Researchers have also found environmental concern among these adults as a result of more years of education (Gardner and Stern, 2002). For example, 31 percent of Boulder's population falls into a 25–44 year-old age range (US Bureau of the Census, 2004), suggesting that there is a vast audience, with a high level of environmental concern, to potentially approach with climate communication and behavior change strategies.

Inclination is also influenced by "race." Ethnic minorities in the United States present attitudinal differences compared to Anglo-Americans regarding political participation, social welfare, environmental justice, and education (Parker and McDonough, 1999; see also the chapters by Agyeman *et al.*; and McNeeley and Huntington, Chapters 7 and 8, this volume). For example, Parker and McDonough (1999) found African Americans were more aware and concerned than Euro-Americans on issues regarding noise pollution, litter, water supply, and other environmental issues frequently of concern in urban areas, as well as endangered wildlife; while Euro-Americans displayed greater concern for wilderness and species protection, land conservation, and overpopulation. Thus, "environmentalism is present in African Americans and Euro-Americans but is exhibited in different ways by these groups" (*ibid.*: 172). Environmentalism among Asian American and US-born Latino populations is similar to environmental action among Anglo-Americans (Johnson, Bowker, and Cordell, 2004).

Gender also influences a person's inclination to engage in ERB. Typically, women show greater proclivity to act in environmentally friendly ways (Zelezny, Chua, and Aldrich, 2000); many women are deeply concerned with environmental problems that either already do, or in the future may, affect their children.

Studies that examine the relationships between demographic variables such as those mentioned above and environmentally significant behavior essentially reveal correlations, but provide no causal explanations. So what does explain the link between demographic characteristics and behavior? Researchers typically find that link in individuals' values, attitudes, and beliefs, which are strongly influenced by people's position in society, as are their concerns and interests (Gardner and Stern, 2002). Social position is also deeply formative of people's values. These value orientations play a key role in guiding individuals' decisions regarding ecological issues (Axelrod, 1994).

Stern and Dietz (1994) suggest three distinct value orientations, which provide the basis for an individual creating their own "belief system." Individuals with an *egoist* value orientation appreciate the environment for what they can get from it or for how it affects them personally. For example, a person may choose to commute to work on a bicycle because they value the positive health benefits of regular exercise. By contrast, a *social altruist* wants to preserve the environment for others' enjoyment, sustained use, and well-being. Such a person may be willing to reduce their greenhouse gas emissions by driving less because they are concerned for their children's future. Finally, a person with a *biospheric* value orientation is concerned about ecosystem health and the biosphere for its own sake. Such an individual may frequently use public transportation because they are concerned with how climate change may impact other species.

These different value orientations toward the environment, described above, form early in a person's life, deeply affect an individual's belief system and attitudes (or inclination) toward certain types of ERB, and also change little over a person's lifetime (see also chapters by Bateson; and Grotzer and Lincoln, Chapters 18 and 17, this volume). As Stern (2000: 411) and others have stated in their "value—belief—norm—activation theory," values influence beliefs, which in turn affect personal norms of appropriate behavior, and those norms in turn activate a "propensity to take actions with pro-environmental intent."

Motivation

The next factor affecting ERB is motivation. Assuming an individual is generally inclined to act in pro-environmental ways, what could motivate that individual to actually do so? Common motivators include knowledge and information, feelings and emotions, identity fulfillment, desires and aspirations, and personal needs.

Knowing that there is an environmental problem such as climate change would seem to be a logical precondition for action. Differently put: if we don't know that there is a problem, why go through the hassle of changing our ways? This is precisely the rationale that has motivated countless education and outreach campaigns. The goal of such educational efforts is to provide the information that will increase knowledge and thus change people's attitudes.

Individuals need to know not only about the problem, however, but also about solutions such as alternative methods of transportation to be motivated to behave environmentally responsible (Gardner and Stern, 2002). For example, a driver of a diesel truck may be conscious that his/her automobile is a contributor to the present pollution problems, but may not be fully aware that biodiesel – a cleaner-burning alternative made from natural, renewable sources such as vegetable oils – can be used in the truck and reduce some of the pollution.[1]

In addition, Kaplan (2000) observes that individuals capable of obtaining and making sense of information tend to be more strongly motivated to change their behavior. For example, a person who understands that using mass transit to travel is one way to reduce the environmental impact of automobiles and also knows the routes and is able to read bus schedules is likely to be more motivated to take the bus to work than a neighbor who does not.

However, merely obtaining knowledge and information can have a counterintuitive effect, one that advocates should be aware of. Studies have shown that individuals can view "getting more information" about a problem as having "acted on it" (Flinger, 1994; see also Rabkin with Gershon, Chapter 19, this volume). In other words, getting more informed can actually terminate any further "active" environmental behavior, a tendency that points to the need for a more comprehensive approach to behavior change than typically taken.

Another challenge with information campaigns is that individuals may not resonate with the environmental information (see also chapters by Leiserowitz; and Agyeman *et al.*, Chapters 2 and 7, this volume). "We see that education can change attitudes and beliefs, but that many barriers, both within individuals and in their social and economic environments, can keep proenvironmental attitudes from being expressed in action" (Gardner and Stern, 2002: 73; see also chapters by Grotzer and Lincoln; and Bateson, Chapters 17 and 18, this volume). Therefore knowledge and information are important to an extent, but motivation through education should combine with other approaches like tapping into an individual's emotional disposition (Gardner and Stern, 2002).

The motivation to act in environmentally responsible ways is strongly influenced by the emotions and feelings that a person has in response to information (see Moser, Chapter 3, this volume) or as a result of various environmental experiences (Flinger, 1994). ERB can be hindered when these emotions result in intimidation and decreased sense of self-efficacy, a sense of helplessness, or feelings of having to sacrifice rather than enhancing one's quality of life (Kaplan, 2000; see also Chess and Johnson, Chapter 14, this volume). In particular, a person might view their contribution to alleviate traffic and reduce pollution (by driving less) insignificant when they observe other people continuing to drive their cars. Similarly, the sense of having to sacrifice personal freedom offered by cars or having to alter one's habits — like allowing a bit of extra time for the bus, biking, or having to bring a change of clothes after biking — can also hinder a person's motivation to choose an alternative mode.

On the other hand, using the motivational component of personal environmental experiences or of emotional responses to information can be vital when promoting ERB, because these experiences can stimulate strong emotions, convictions, aestheticism, or the determination to act. Flinger (1994) cites nature experiences, environmental catastrophes, and participation in environmental activism as stimulants or reinforcing experiences that can motivate ERB. Such experiences can also trigger a new-found "awareness" or "awakening" that presents a source of motivation to fulfill desires and aspirations, which is closely linked to a person's sense of identity. If an individual aspires to be seen as an environmentally responsible person by his or her peers, superiors, or family, ERB will play an essential part in one's behavioral repertoire (Axelrod, 1994; Kaplan, 2000). Such an individual would be compelled to buy less and use fewer resources, seek acceptance from like-minded people, and build his or her sense of competence from acting in environmentally responsible ways. This person would also be more motivated to participate in social altruistic ERB (Stern, 2000), such as voluntarily marketing car-pooling arrangements, participating in an environmental activist group, or donating money to "drive less" campaigns. Ultimately, satisfaction is derived from striving for behavioral competence where individuals enjoy being able to solve problems and complete tasks for themselves or at the benefit of others (DeYoung, 2000). As a result, Kempton and Holland (2003) showed how the development of an individual's identity as an environmentally responsible person is an important factor that fosters ERB.

A final source of ERB motivation is personal need. Will ERB enable somebody to get what they need? Individuals may engage in ERB by biking to work ten miles each direction to increase personal fitness gains and better overall health. Conversely, people may not be able to use bikes or buses because they have to pick up children from daycare or run errands on their way home from work.

Information processing and behavioral intent

Environmentally responsible behavior is also influenced by how individuals perceive their surroundings, the actions of others, and the impact of their own. Differently put, how a person processes information about the environment and possible actions is contingent upon her prior beliefs and mental models (see Bostrom and Lashof, Chapter 1, this volume), effects what kinds of behavioral intents she forms, and what public commitment she may be ready to make to engage in certain behaviors.

Bamberg (2003) claims the intent of a certain behavior is formed when individuals assess the likely consequences of action, how this action will fit within a social norm, and if there are any external factors or controls that may limit or discourage their action. For example, a person may decide to drive to work instead of riding his bike, because there is plenty of free parking and everybody else does the same; besides, he can sleep in a bit later. A personal sacrifice to do without the car may make little sense when everyone else continues to drive. Or people may resist using public transportation because they may be exposed to "odd" people or because doing so might be considered "lower class" by their peers.

Indeed, this example demonstrates that certain beliefs may cause people to reject an ERB, because engaging in such behavior would limit personal lifestyle or quality of life (DeYoung, 2000; see also Dilling and Farhar, Chapter 23, this volume). People like their cars, they feel in control of their environment (radio, solitude, ability to travel whenever and wherever, space to unwind, etc.), and driving for them may manifest freedom and self-determination. Yet for others, lifestyle goals can be met by using alternative transportation: taking the bus to work gives ample time to read the morning paper, chat with neighbors, and avoid the nuisance of traffic jams and searching for a parking spot.

People can derive substantial motivation from the belief or expectation that a certain outcome will be realized from a specific behavior as well as the positive value placed upon that outcome (Axelrod and Lehman, 1993). "One car less" may not be sufficient to make a dent in global warming (a demotivator for many), but it may satisfy someone's desire to be an example and role model, or for simply being part of the solution. Thus, what we believe affects how we process information about a problem and its solutions and what behavioral intent we form. An individual's intent to act in a certain way is arguably the strongest predictor of what a person will actually do (Meinhold and Malkus, 2005; Corbett, 2005; Stern, 2000), especially if that intent is made publicly known and if the individual makes a public commitment to the action. Such public statements create a sense of accountability and tie back to a person's need for social acceptance, identity, and sense of competence (DeYoung, 2000; McKenzie-Mohr, 2000).

Ability and skill

Ability refers to whether or not an individual has the appropriate skill, a sense of self-efficacy or perceived influence, and access to financial, technical, and informational resources to carry out a behavioral intent. Specifically, an individual will not use alternative modes of transportation if he/she lacks the ability to ride a bike, cannot afford the time to travel on the bus, or cannot walk all the way to the next commuter rail station. People enjoy being able to solve problems and complete tasks; we all prefer being seen as individually competent. Thus, those with the necessary skills and abilities are more likely to be engaged in ERB than others (DeYoung, 2000; Stern, 2000).

In addition to actual skills, a sense of self-efficacy − i.e., the perception of being able to meet a given challenge − is an important determinant of a person's likelihood of participating in ERB (Kaplan, 2000; Stern, 2000). The Responsible Person Model introduced by Kaplan (2000: 498) acknowledges that humans are actively curious and problem-solving; they avoid environments and situations that are difficult to navigate or where they feel they cannot make a difference. People want to participate, learn, play a role, and ultimately avoid contexts that leave them feeling helpless.

One might assume that "if people know what they should do and why they should do it, they will know how to do it" (DeYoung, 2000: 521). However, this assumption is misleading, because even if we imagine that an individual will act environmentally responsibly because he/she wants to feel competent, individuals must receive specific procedural information to adopt a certain behavior (DeYoung, 2000). In summary, changing to ERB is possible only when programs are designed to overcome barriers impeding a desired behavior. Lack of skill or sense of personal competence, missing information, and resources can all be such barriers (McKenzie-Mohr, 2000).

External support

External support for ERB comes from several sources: peer groups, social norms and institutions, and enabling infrastructure. It affords an individual with a sense of accountability, positive encouragement and reinforcement to overcome engrained habits of behavior, and the enabling environment in which to enact good intentions. It reduces the possibility of people feeling alone and helpless (DeYoung, 2000). It also provides guidance for certain behaviors through modeling by peers, social norms, laws and regulations, and sometimes reprimand or punishment for non-ERB.

Peer and community participation can take place at work, church, school, in social clubs, the family and with friends, in neighborhood organizations, and local/community gatherings. Participation is elemental in external support systems, because individuals derive satisfaction from being involved in community activities and value opportunities to take action (Kaplan, 2000). Peer and community groups foster commitment and accountability, prosocial inclination, caring about the welfare of other human beings, finding pleasure from working with others toward a common goal, acting with people and helping them, being needed, and sharing of news, skills, and knowledge (DeYoung, 2000).

These interpersonal influences also build and reinforce pro-environmental social norms (Stern, 2000). Such norms guide individual behavior by setting standards for socially acceptable behavior (*ibid.*). As a participating role player, the individual is strongly influenced by community expectations or wider social expectations, for example, about what type of car to own, whether driving to work is acceptable in a community, or what is "cool" (*ibid.*; see also Dilling and Farhar, Chapter 23, this volume).

Laws and regulations – the institutionalized form of social norms – also provide external support for behavior. For example, seatbelt use increased markedly once there was a law mandating it; similarly, legal standards for vehicle fuel efficiency or speed limits offer compelling external support for ERB. Kaplan (2000) suggests that such laws and regulations are more effective at providing external support for ERB if they come about through a process of participatory problem-solving. Civic participation in identifying solutions increases the likelihood that people will accept decisions that are made, find the solutions valuable, and actually implement the new rules.

Finally, external support for enacting behavioral intent comes through the infrastructure available to switch to ERB. The best intent to use alternative transportation will fail if there is no (convenient) public bus system, if there are no safe bike paths, or if driving remains the cheapest way to get around. All but the most internally motivated cannot be expected to change behaviors if environmentally friendly alternatives are not provided or are so inconvenient and costly that people leading full, busy, and complex lives cannot enact them.

Communication and behavior change strategies

The five factors that enable ERB and behavior change toward more environmentally responsible actions — inclination, motivation, information processing and forming a behavioral intent, ability, and external support — suggest that behavior change campaigns must rely on similarly diverse and multifaceted strategies to address and overcome inhibiting barriers. Above all, behavior change strategies must be communicated effectively to individuals, especially to those who can influence others currently disengaged from ERB and address the obstacles individuals face or may perceive head on.

Interestingly, social scientists have made considerable progress in understanding human behavior and how it changes. Yet, all too often, this knowledge is not made available to those who could use it in their campaigns. McKenzie-Mohr (2000), for example, asserts that "there is mounting pressure for professionals to publish in journals, which has resulted in few attempts to actively make behavior change knowledge accessible to those that benefit from it most." ERB may be spread more widely if such strategic knowledge is communicated by credible messengers to designers of behavior change campaigns, or to the individuals who hold influential positions among their peers or in the larger population.

This idea was emphasized in Malcom Gladwell's recent bestseller, *The Tipping Point* (2002), which examines how little ideas, if communicated effectively through appropriate channels, can spread widely and make a big difference. Gladwell describes how innovative ideas, products, or social behaviors can cross a certain threshold — a "tipping point" — to become a ubiquitous social fad, behavior, or paradigm. Three principles help such ideas or behaviors get to the tipping point. First, the "Law of the Few" suggests that ideas are disseminated by a few key individuals who are widely connected to others (the "connectors"); by people who are experts or knowledgeable enthusiasts in a particular field (the "mavens"); and by individuals that are socially influential and persuasive (the "salespeople").

The second, which Gladwell calls the "Stickiness Factor," states that "there are specific ways of making a contagious message memorable [through] relatively simple changes in the presentation and structuring of information" (Gladwell, 2002: 25). And finally, the "Power of Context" refers to the principle that getting people to change their behavior sometimes is contingent on little details of their immediate environment. So how do we apply Gladwell's principles and the insights gleaned from the literature discussed above to the development of effective communication and behavior change strategies?

Applying Gladwell's principles sequentially, ERB can be cultivated by, first, identifying the *few* individuals that do or could engage in environmentally responsible behaviors, i.e., the connectors, mavens, and salespeople; then, creating *sticky* messages to promote ERB; followed by modifying the *context*, to better facilitate ERB. Maybe the most important initial step is to define precisely what is to be achieved with what population/target group. With a goal and target population in mind, screening and selection of key individuals can take place more efficiently. To identify these key people, an ERB advocate would consider these individuals' social position, background, what they do, and how their interactions could positively reinforce ERB among the target population. In sum, they must ensure the selected individuals can be credible, trusted, and legitimate leaders in the target population.

The next step is sorting out which of these individuals are the mavens, i.e., the knowledge hubs that can communicate their knowledge cogently; the most persuasive "salesmen" of the group; and those people who are highly connected, especially across social divides.

Once these key individuals have been identified, most of the communication and advocacy efforts can be focused on them: to increase their motivation using knowledge and education campaigns, to tap into proenvironmental, prosocial, and personal motivations to foster the formation of ERB intent. Thus, advocates enable and support these individuals to carry the message to their social networks.

ERB advocates must also collaborate with these key individuals to appropriately tailor behavior change message to become sticky in the specific socio-cultural context they report back to. The collaboration should include specific skill-building for the messengers, provide appropriate technological, informational, and financial support to communicate the message, and identify where other sources of support are necessary. For example, it may be necessary to boost a sense of self-efficacy where it is low through empowering messages and education; in others it may be more important to focus on obtaining public commitments; in yet others the focus may need to be on removing actual barriers to provide a context that enables, rather than hinders, behavior change.

Although this rough strategy leaves out many details, it depicts a process that differs from many traditional outreach campaigns aimed at undifferentiated masses of individuals. Instead, it seeks out key individuals as entry points into specific target groups and suggests that behavior change is not a one-by-one persuasion task, but a social challenge.

Conclusion

Human behavior releases heat-trapping gases to the atmosphere, thus contributing to increases in global temperatures and altering climate patterns. Global warming solutions will and must be found — as this book suggests — at many levels. To the extent we want to engage individuals in these solutions — through their own behaviors or by way of support of policy solutions — behavior change campaigns have to improve. In particular, we must pay greater attention to the internal and external forces that foster and constrain individual actions. As the highly educated, environmentally conscious, and well-informed Boulderites demonstrate, knowledge about environmental impacts of personal actions does not guarantee environmentally responsible behavior.

Many might have thought that the socio-economics and politics of a city like Boulder would promise widespread ERB, yet the reality is different. We operate out of habit, pursue our high-consumption lifestyles, and insist on a convenient, liberating, and seemingly cheap form of transportation, because no better alternative presently exists. In this small city, a trip across town might take 15 minutes by car or close to an hour and a half by a complicated mix of public transportation and long walks. That is where the rubber of the city's "green mystique" meets the road of reality. It is not until the context of individual behavior is addressed that highly educated, strongly motivated, and environmentally inclined individuals — much less their less inclined neighbors — will get out of their cars and go places in new ways.

Note

1. Contrary to common belief, however, biodiesel is not "fossil-fuel free," at least not from a full product production and delivery perspective. Significant amounts of fossil fuel are used to produce biodiesel and to deliver it to the places of consumption.

References

Axelrod, L. J. (1994). Balancing personal needs with environmental preservation: Identifying the values that guide decisions in ecological dilemmas. *Journal of Social Issues*, **50**, 85–104.

Axelrod, L. J. and Lehman, D. R. (1993). Responding to environmental concerns: What factors guide individual action? *Journal of Environmental Psychology*, **13**, 149–59.

Bamberg, S. (2003). How does environmental concern influence specific environmentally related behaviors? A new answer to an old question. *Journal of Environmental Psychology*, **23**, 21–32.

City of Boulder (2005). *The Best of Boulder*. Available at: http://www.ci.boulder. co.us/comm/Public%20Affairs%20Internet/Best.htm; last accessed on January 12, 2006.

Clark, M. E. (1995). Changes in Euro-American values needed for sustainability. *Journal of Social Issues*, **51**, 63–77.

Corbett, J. B. (2005). Altruism, self-interest, and the reasonable person model of environmentally responsible behavior. *Science Communication*, **26**, 368–89.

DeYoung, R. (2000). Expanding and evaluating motive for environmentally responsible behavior (ERB). *Journal of Social Issues*, **56**, 509–26.

Flinger, M. (1994). From knowledge to action? Exploring the relationships between environmental experiences, learning, and behavior. *Journal of Social Issues*, **50**, 141–60.

Gardner, G. T. and Stern, P. C. (2002). *Environmental Problems and Human Behavior*, 2nd edn. Boston, MA: Pearson Custom Publishing.

Gladwell, M. (2002). *The Tipping Point: How Little Things Make a Big Difference*. New York, NY: Back Bay Books.

Johnson, C. Y., Bowker, J. M., and Cordell, H. K. (2004). Ethnic variation in environmental belief and behavior: An examination of the new ecological paradigm in a social psychological context. *Environment and Behavior*, **36**, 157–86.

Kaplan, S. (2000). Human nature and environmentally responsible behavior. *Journal of Social Issues*, **56**, 491–508.

Kempton, W. and Holland, D. C. (2004). Identity and the sustained environmental practice. In *Identity and the Natural Environment: The Psychological Significance of Nature*, eds. Clayton, S. and Opotow, S. Cambridge, MA: The MIT Press, pp. 317–41.

McKenzie-Mohr, D. (2000). Promoting sustainable behavior: An introduction to community-based social marketing. *Journal of Social Issues*, **56**, 543–54.

Meinhold, J. L. and Malkus, A. J. (2005). Adolescent environmental behaviors: Can knowledge, attitudes, and self-efficacy make a difference? *Environment and Behavior*, **47**, 511–32.

National Research Council (2002). *New Tools for Environmental Protection: Education, Information, and Voluntary Measures*. Committee on the Human Dimensions of Global Change, eds. Dietz, T. and Stern, P. C., Division of Behavioral and Social Sciences and Education. Washington, DC: National Academy Press.

Parker, J. D. and McDonough, M. H. (1999). Environmentalism of African Americans: An analysis of the subculture and barriers theories. *Environment and Behavior*, **31**, 155–77.

Stern, P. and Dietz, T. (1994). The value basis of environmental concern. *Journal of Social Issues*, **50**, 65–84.

Stern, P. C. (2000). Toward a coherent theory of environmentally significant behavior. *Journal of Social Issues*, **56**, 407–24.

US Bureau of the Census. (2004). *Educational Attainment*. Washington, DC: US Government Printing Office.

Zelezny, L., Chua, P., and Aldrich, C. (2000). Elaborating on gender differences in environmentalism. *Journal of Social Issues*, **56**, 443–57.

16

Consumption behavior and narratives about the good life

Laurie Michaelis
Living Witness Project

Introduction

Numerous studies in Europe and North America have pointed to the importance of consumption patterns in shaping greenhouse gas (GHG) emissions (Michaelis and Lorek, 2004; Brower and Leon, 1999). Around 75–80 percent of economy-wide emissions are associated with three areas of consumption: food, transport, and shelter (including household energy use). The location of emissions in the supply chain is shifting away from industry, which is becoming more efficient, and shifting towards lower-carbon fuels, and towards households and institutions (Price *et al.*, 1998).

Emissions of GHG can be reduced *either* by changing consumption patterns and lifestyles, *or* by reducing emissions per unit of consumption — e.g., through improved energy and resource efficiency or switching to renewable sources (see also Dilling and Farhar, Chapter 23, this volume). Environmentalists in the 1970s gained a "hair shirt" image, connecting values of community living, getting back to the land, and simple lifestyles. But after political commentators derided US President Jimmy Carter's appearance on TV wearing a cardigan during the 1979 oil crisis, governments have been careful to focus on supply-side and technological responses to climate change. During the 1990s, environmental groups adopted a similar focus as they were increasingly included in policy advice circles and business partnerships. The mainstreaming of environmental politics has involved the abandonment of its countercultural edge.

The consumption debate in the industrialized world was rekindled in large part by pressure from the developing country delegations at the Rio Earth Summit in 1992. Chapter 4 of the summit's blueprint for sustainable development, *Agenda 21*, calls for "new concepts of wealth and prosperity which allow higher standards of living through changed lifestyles and are less dependent on the Earth's finite resources." The late 1990s saw increasing dialogue in research and government circles about the ethics and practicalities of encouraging more sustainable lifestyles (e.g., Stern *et al.*, 1997; Crocker and Linden, 1998).

Public surveys in Europe (e.g., Kasemir *et al.*, 2000; Stoll-Kleemann *et al.*, 2001) and the United States (Kempton *et al.*, 1995) find a growing sense among citizens that a change in lifestyles will be needed to address climate change. However, most people do not believe that such a change will happen voluntarily and find it hard to see how they can change their own patterns of consumption.

This chapter reviews some of the efforts that have been made in recent years to understand consumption, challenging the assumptions that dominate climate policy discussions. It notes the diversity of cultures of consumption in consumer societies and identifies some of the policies, measures, and communication strategies that might be effective in addressing people with different consumption styles. It focuses in particular on the role of dialogue and discourse as a foundation and mediator of consumption cultures (see also Regan, Chapter 13, this volume). Finally, it reviews some recent community-based efforts to develop new consumption ethics, placing more emphasis on sustainability.

Understanding consumption

In explaining behavior, Westerners tend to focus on individual characteristics and circumstances (Liebes, 1988). This is in profound contrast to other cultures, where more emphasis is placed on social roles and contexts. The economic theory on which most climate policy analysis is based starts from a fundamental assumption that the consumer is a rational, independent individual (Jaeger *et al.*, 1998).

Policy analysis, grounded in the economic theory of utility, often views consumption as the principal way in which people achieve individual and collective well-being. Consumers are assumed to act rationally in the market according to their personal preferences, to maximize their own satisfaction or utility.

The economic approach is certainly useful for tackling some policy questions, but avoids the question of how preferences arise. The central policy goal is to maximize consumers' welfare, which is usually assumed to be achieved through satisfying innate preferences or desires "revealed" in their current behavior. This approach leads to policies that support consumption of more of the same.

Research by psychologists, sociologists, and economists into satisfaction and well-being calls the mainstream assumptions into question. First, consumers do not behave according to the economic concept of rational optimization (Dietz and Stern, 1993; Jaeger *et al.*, 1998). Even when making considered choices, we are only able to consider a small number of variables, leading to behavior better described as "satisficing" (Simon, 1955). Much of our behavior is habitual. Choices are often better understood as a process of internal debate among alternative narratives or behavioral scripts, rather than an internal optimizing algorithm. Hence, we are guided by the scripts available to us — to a large extent supplied by the people around us and through the media (Dennett, 1993; Dickinson, 1998). In recent years, people's models of consumption behavior have increasingly derived from TV soaps and celebrities rather than their immediate communities (Schor, 1998).

Second, we do not necessarily use these decision processes to choose behaviors or options that will increase our personal satisfaction (see also Jamiesou, Chapter 30, this volume). Although we do appear to be strongly motivated to pursue wealth, Argyle (1987) and Inglehart (1997) find that increasing income and consumption are correlated with satisfaction only up to a point. Once national average income exceeds about US$10,000 per capita, further GDP growth does not appear to make people happier. Argyle finds that some of the most important determinants of happiness are health, family relationships, friendships, and having meaningful work. Those with a high *relative* income are more satisfied than those with a low relative income, but the absolute level of income is unimportant provided basic needs are met and individuals are able to function normally in society.

Consumption patterns have been the subject of a great deal of research (reviewed in Jackson and Michaelis, 2003). Consumption can be explained in a wide variety of ways:

- Some kinds of consumption play a role in meeting physiological needs such as nutrition and warmth/shelter.
- Much of our consumption is shaped by habits, norms, and routines.
- Our consumption choices help us to divide the world into what is "allowed," "good," and "desirable," and what is not. We may make that division in terms of the sacred and the profane, clean and unclean, known and unknown.
- Consumption plays a role in mediating power, status, and social competition.
- It may also be the key to social participation and acceptance, sense of belonging, and role definition.
- Consumption can be a displacement activity — used to avoid experiencing negative feelings or a lack of purpose in life. It can be part of fantasy, dreaming, desire, and the pursuit of happiness. It can help to symbolize our pursuit of the Good Life.

- Consumption can play a role in political and social discourse — forming an expression of our individual and collective values, ethics, and worldviews.
- And it can be a way of establishing and expressing our individual and collective identities — our understandings of who we are and our purpose in life.

One of the strongest messages emerging from the research is that consumption and lifestyles are shaped more by people collectively than individually. Consumers often feel they have no choice: they are "locked in" to their consumption patterns (Jackson and Michaelis, 2003).

Social science research is often used to show why behavior is so hard to change, but recognizing the many influences on consumption opens up both the possibility, and the legitimacy, of policies to influence it. Indeed, the research shows that, despite paying lip-service to consumer sovereignty, governments already influence consumption through a wide range of policies that bolster the status quo.

Jackson and Michaelis (2003) point to the need for comprehensive government strategies to encourage behavioral change, including the measures most widely discussed such as fiscal and regulatory policies, information, and education. But strategies are unlikely to work unless politicians and civil servants lead by example (see also Tennis, Chapter 26, this volume). There is also tremendous potential for developing strategies for social and behavioral innovation, in much the same way that governments develop national frameworks for technological innovation (Michaelis, 1997; Sathaye et al., 2001). Such strategies may involve fiscal and regulatory support for local initiatives, but they also entail measures to develop better communication — engaging all kinds of communities and linking them to those with expert and specialist information, skills and resources, as well as to government agencies and others in positions of power.

The most important implication of the research for communication strategies is that messages directed at individuals have little effect. The most effective strategies are those that engage people in groups, and give them opportunities to develop their understanding and their narratives about consumption in dialogue together.

Consumption cultures and narratives

Research also points to the need to recognize the diversity of cultures of consumption, each with a characteristic set of values, worldviews, and consumption styles. Thompson, Ellis, and Wildavsky (1990) seek to characterize the diversity of world views and discourses with "cultural theory" (CT). CT analyzes social groups and cultural orientations along two axes: "group" and "grid." The higher the scores on the "group" dimension, the stronger are the links between people. A high score on the "grid" dimension implies a high level of social differentiation. Strong group, strong grid gives a hierarchical organization, typical of the military or institutionalized religion; weak group, weak grid gives an individualist community such as a stock exchange; strong group, weak grid gives an egalitarian community, such as is found in some monastic orders. Weak group, strong grid gives a fatalist culture – usually coexisting with other groups that hold social, economic, or political power. Dake and Thompson (1999) find from a study of households in northwest England that (a) cultural attitudes can be characterized effectively at the household level using the CT framework and (b) households' CT types are correlated with their consumption patterns and lifestyles. Evidence from Nepal suggests that the framework is valid across wide differences of economic circumstance and social tradition.

Dake and Thompson (1999) identify five distinct cultures of consumption that can be characterized as follows:

- **Hierarchist** consumers place emphasis on values such as duty and order. They eat a conventional diet and have homes that contain old or antique furniture reflecting continuity with the past. Their consumption patterns demonstrate their established place in the community.
- **Individualist** consumers emphasize values such as liberty and innovation. They keep up with fashions in food, home furnishing, and personal transport. These are the conspicuous and competitive consumers.
- **Egalitarian** consumers are politically engaged, concerned with environmental and social values. In their consumption they tend to avoid anything artificial and distinguish "real" needs, which can be met by nature, from "false" needs that we cannot meet except by depriving future generations. Their consumption choices are a moral statement.
- **Fatalist** consumers do not make active choices, but muddle through, perhaps being constrained by budget or unable to take control of their consumption patterns for other reasons.
- **Hermits** are not socially isolated (as fatalists are) but make a positive choice to be independent of social expectations. They live simply without making a political or moral point of it.

Each of these cultural types or "solidarities" (as Dake and Thompson call them) has the potential to be concerned about, and take action for, environmental and social issues, but it has its own narrative about human nature, the environment, and appropriate solutions for sustainability (Thompson *et al.*, 1990; Thompson and Rayner, 1998; see also Regan,

Chapter 13, this volume). Mainstream political debate, including that on consumerism and sustainability, focuses mainly on the tensions between individualist and hierarchist values and narratives. However, both cultures have important elements that support consumerism. The individualist narrative encourages material consumption by emphasizing:

- An understanding of human beings as autonomous, rational individuals whose highest potential is to be achieved through their work and demonstrated in increasing levels of material wealth and consumption.
- An understanding of society as nothing more than a collection of individuals, whose collective purpose is to enable individual members to meet their needs and aspirations.
- An understanding of the natural world as a resource base for meeting human needs.
- An understanding of the Good Life largely in terms of material prosperity, individual freedom and choice.

The hierarchist narrative seeks to reintroduce social solidarity. It calls for a return to traditional gender roles and social structures, with limitations on autonomy justified by external threats or scripture. Because hierarchists tend to support maintenance of existing hierarchies, they are sympathetic to the interests of the wealthy and powerful, especially large corporations. Hence, they argue for protection of economic interests and promotion of economic growth.

Changing behavior

Anthropologist Gregory Bateson (1972) described two broad ways in which behavior can change: "proto-learning" involves adjustments to practices within the existing set of rules and routines; "deutero-learning" involves developing new rules. Argyris and Schon (1996) have popularized Bateson's framework within management studies, naming the two modes of behavior change respectively "single-" and "double-loop" learning (see also Bateson, Chapter 18, this volume). Changes in circumstances or stimuli can lead people to shift their behavior without changing their cultural norms or assumptions; or they can challenge those norms and assumptions, leading to a change in underlying culture.

People within different consumption cultures are likely to respond differently to changes in circumstances, including government policies. We are more likely to succeed in bringing about more climate-friendly and otherwise sustainable behaviors if we tailor interventions to particular cultural groups, for example as suggested in Table 16.1.

Table 16.1. *Designing sustainability policy interventions and communication strategies for cultural types*

Cultural type	Effective sustainability policies	Emphasis in communication strategies
Fatalist	Planning, standards, regulation, and policing.	Cost savings, e.g., through energy conservation.
Hierarchist	Promote new social norms through regulation, fiscal policies, public procurement in schools and hospitals, etc.	Duty, moral obligation, preserving environment as heritage, taking care of children, grandchildren. Support for development of civic processes.
Individualist	Pricing, clear responsibilities backed up by law.	Objective information on financial, social, and economic implications of choices.
Egalitarian	Encourage opportunities to develop green consumption (regulatory and fiscal support for niche markets and products).	Support to develop stakeholder processes and community reflection and dialogue about consumption and lifestyles.
Hermit	Unpredictable: hermits tend to sidestep social norms and coercion unless there is no choice (well-designed combinations of measures are most likely to be effective).	Objective information on environmental, financial, social, and economic implications of choices.

Note: Only those most characteristic of each type are noted here; many policies are effective for several cultural types.

The various cultural groups are likely to vary in their responsiveness to policies. They will also vary in the way they receive, interpret, and act on messages from government, the media, campaigning organizations, and others.

Hierarchists and egalitarians both, for example, have a sense of social solidarity and obligation. They can be expected to take on board and respond to concerns about collective welfare. Hierarchists might be expected to respond well to messages from government, but to reject information from environmental campaigners. Conversely, egalitarians may distrust messages from those in power, whether in government, business, or the media, but trust information from campaigning groups and informal email lists or personal contacts. Both hierarchists and egalitarians may also be influenced by discussions and decisions in their social groups — bearing in mind that for hierarchists, the pronouncements of religious or other leaders may be the key, whereas for egalitarians shared values must be established by mutual agreement.

Individualists and fatalists are less concerned about social solidarity, and both policies and messages might be more effective if they focus on the personal implications of behavior. Individualists might be more influenced by messages that connect environmentally responsible behavior with success, status, and positive affect. They may respond to media messages that connect celebrities with sustainable lifestyles. Fatalists might be most effectively influenced by policies that make environmentally benign choices cheaper and easier, and messages that make them aware of what is available. Hermits reject social conformity and so may be hard to influence, but they may be persuaded to change their behavior both by environmental concern and by the availability of cheaper and easier lifestyle choices.

Ultimately, the greatest potential for a shift towards sustainable lifestyles might be through a change in culture — that is, a shift in assumptions about human nature, our relationship with the world around us, the nature of human society, and our aspirations for the Good Life (see also the chapters by Jamieson; and Bateson, Chapters 30 and 18, this volume). In CT terms, this implies a shift of people from one cultural group to another. Historically, however, efforts to engineer cultural change have not worked well and have often had unintended consequences. Yet, culture does evolve. One of the most important stimuli for cultural change has been the emergence of new technology, especially transport and communication technology. Eisenstein (1983) points to the way the development of printing in Europe in the fifteenth century led to a transformation of understandings of human nature and the world, helping to bring about both the Enlightenment and Romanticism, which are the foundations of modern culture and consumerism (Michaelis, 2000). Castells (1997) similarly points to some of the shifts now occurring as a result of new information technology.

Cultural change may also be occurring in response to a growing sense of risk in individual lives and to society as a whole, associated with technology, environmental threats such as climate change, and the growing dominance of our daily lives by market-based institutions (Beck, 1988; Giddens, 1994). The shift described by Ulrich Beck and Anthony Giddens involves an increase in "reflexivity" — explicit collective examination of goals, worldviews, and ethics. This process could be understood as the emergence of a more egalitarian culture to challenge the dominance of individualist culture. It is also described by Don Beck and Christopher Cowan (1996) as part of a continuing process of cultural evolution.

When challenges emerge to the status quo or the dominant culture, the initial instinct is to deny that there is a problem or that change is warranted.

Once the challenge is acknowledged, Beck and Cowan describe several ways in which a society may respond:

• We may seek to define the challenge within the current paradigm and seek solutions accordingly. This is a "single-loop learning" response. In our climate responses, it entails policies and measures that focus on market instruments and technological innovation.
• If the current paradigm does not seem able to provide a solution, there may be strong pressures to revert to an earlier cultural form — in this case, to the restrictive, protectionist, and regulatory approaches of a more hierarchical society, perhaps including rationing of fossil fuels. Again, this is a single-loop learning response, working within the repertoire of familiar worldviews, values, and practices.
• Society may move forward to a new cultural form, entailing double-loop learning to reflect on and revise worldviews, goals, values, and fundamental practices.

Michaelis and Lorek (2004) sketch out three scenarios of future consumption in Europe following these three possible responses. Each scenario seeks to offer an internally consistent view of the way the world might evolve in terms of economic development, technology, politics, social structure, and culture. The scenarios correspond to the three "active voices" in CT (fatalists and hermits are not socially or politically proactive). The scenarios could easily be generalized to other consumer societies such as the United States. Table 16.2 links these scenarios to those in the IPCC *Special Report on Emission Scenarios* (SRES) (Nakicenovic and Swart, 2000), although the correspondence is not exact, especially in the case of Hierarchical Europe/SRES A2.

It is possible that the real world will include some combination of the responses outlined in the three scenarios, and any region is likely to include groups falling in different cultural types. Achieving significant GHG mitigation may *require* such a combination. However, while the potential for the first two lies clearly in current norms and political discourses, the *egalitarian* response is only beginning to emerge in small groups, some of which will be discussed in the next section.

One study of British consumers by Market and Opinion Research International (MORI) (Hines and Ames, 2000) found five clusters, based on attitudes and approaches to ethical consumption. The clusters do not correspond exactly to Dake and Thompson's CT categories as shown in Figure 16.1, but there are some interesting overlaps, with the largest groups corresponding roughly to hierarchist, individualist, and fatalist consumers, and a small group (5 percent of the sample) which Hines and

Table 16.2. *Climate responses – three scenarios for consumption*

Michaelis and Lorek	Key features	Analogies in SRES
Hierarchical	Political reactions to declining oil reserves and the "war on terror" result in high regulation and centralization. Government and business organizations increase their power. Economic growth and technological innovation are fragmented and slow. GHG mitigation is achieved through reduced economic activity.	In A2, the underlying theme is self-reliance and preservation of local identities. In CT terms the scenario has hierarchist, individualist, and fatalist aspects. An emphasis on local traditions results in reduced diversity of consumption patterns at any location, but increased diversity among locations. However, economic and population growth result in rapidly increasing GHG emissions.
Individualist	Market values result in rapid technological innovation and economic growth. Growing environmental concerns lead to government commitment to internalization of environmental costs, and business innovation focused on GHG mitigation, which is successful.	Corresponds best to A1 (individualist–hierarchist) and B1 (individualist–egalitarian). Both have high rates of economic growth and technological change. In A1 the focus on innovation (fossil fuels, nuclear power, renewables) is the major influence on levels of GHG emissions. These are both scenarios of continuing globalization (with convergence in consumption patterns among locations).
Egalitarian	Environmental disasters and the failure of government to get to grips with the causes of climate change leads to a growth in local activism and direct democracy. Development of stronger communities with less emphasis on economic growth. High rates of social innovation. Increasing local diversity and localization of consumption patterns.	Corresponds best to B2 (egalitarian), which combines local self-reliance with strong environmental concern. GHG grow at lower rates than other scenarios due to a combination of technological and lifestyle change.

Ames called "Global Watchdogs" corresponding roughly to Dake and Thompson's egalitarian consumers. These committed egalitarians are the cultural innovators who demonstrate the attractiveness of a more sustainable way of life.

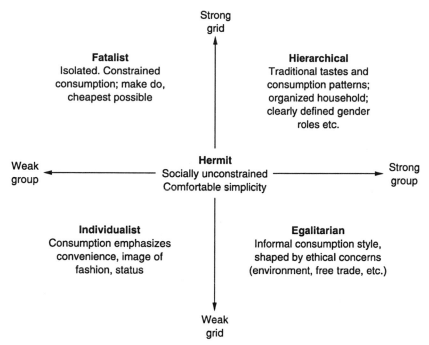

Fig. 16.1.   Cultural theory: Household consumption styles.
Source: Based on Dake and Thompson (1999).

Sustainability pioneers

A growing number of local groups and networks in Europe and the United States are exploring ways of living more sustainably. Michaelis (2004) reviews three national groups in Britain actively exploring sustainable consumption: the non-governmental Global Action Plan (GAP) UK; the Permaculture Association; and the Quakers. All three of these use reflexive processes – group reflection and dialogue through which they develop shared goals, values, and practices. They are committed to the goals and values of sustainability, but their practices differ considerably. The reflexivity in these organizations takes widely differing forms, appealing to people with different cultural backgrounds and interests.

GAP UK is part of an international network of organizations, originating from GAP in the United States, working to encourage people to live sustainably. The overall strategy, values, and principles are developed by the central organization, and center on the idea that people are most likely to change their lifestyles if they are supported to do so within a group (see also Tribbia, Chapter 15, this volume). On the whole, it adopts a hierarchist approach – advising people how to change, measuring the results and providing feedback. GAP UK supports local groups in neighborhoods, schools, churches, or workplaces to learn about the environmental impacts of their lifestyles and to set targets for improvement. The groups or "EcoTeams" follow a process first developed in the United States in the late 1980s, with a series of meetings based on the EcoTeam Handbook (see also Rabkin with Gershon, Chapter 19, this volume). EcoTeam participants in Nottingham, England have achieved on average 50 percent savings in waste, 27 percent savings in gas and electricity, and 17 percent savings in water (personal communication, GAP UK). The GAP approach is designed to reach a mass audience and may be the most promising of the three reviewed here in reaching the general public. GAP receives funding for some of its work from the UK government.

Permaculture has grown from a practical approach to sustainable land management ("permanent agriculture") based on ecological principles, to become a philosophy and way of life (Holmgren, 2002). Originating in Australia, the movement has spread internationally and is rapidly gaining members in Britain and the United States. It is experimenting with different practices, but the UK movement has adopted an approach known as "action learning guilds," in which small groups reflect together on what they are trying to achieve in "changing the world."[1] The action learning approach offers a particularly strong model for innovation by individuals and the small groups. In CT terms, Permaculture has strong individualist and egalitarian elements. It is a movement rather than an organization. It pioneers new approaches but does not have strong mechanisms for having them widely adopted apart from the charisma of prominent advocates.

Although Quaker values are closest to egalitarian culture in CT terms, there are also strong elements of individualism (with each individual free to follow their own spiritual path and theology) and hierarchy (with maintenance of traditional disciplines). Quakers have a well-developed system of institutions and practices for decision-making which seeks to ensure that everyone can be heard and yet a collective conclusion – the "sense of the meeting" – can be reached (Morley, 1993).

Quakers in Britain and in the United States have been slow to take up sustainability as a priority for collective action, but they are beginning to do so with a developing network of groups exploring sustainable living (see, e.g., Quaker Earthcare Witness, 2005). Quaker approaches seem to offer opportunities for the development of collective approaches to sustainable living, with the potential for resolving of some of the tensions and difficulties posed by the conflicts between the approaches of different cultural types.

Conclusions

Consumer culture is often talked about as if it were a single, monolithic body of narratives, practices, and symbolism supporting mass consumption. However, ethnographic studies have found diverse cultures of consumption, among groups with varying levels and types of concern about environmental and social issues, and with varying propensities to act on their concern. Government policies and communication strategies need to be tailored to these different groups. Discussions of communication about environmental issues, including climate change, have tended to concentrate on strategies for disseminating information, or for persuading people to change their behavior. Behavioral responses to communication depend both on the culture of the audience and on the source of the message.

Cultural theory, with its five cultural types of individualist, hierarchist, egalitarian, fatalist, and hermit, offers one of many ways of understanding diversity. Egalitarians tend not to trust government, the media, or big corporations, but seek out information from NGOs and email lists. Hierarchists are more likely to accept official messages but tend to dismiss environmental campaigners. Both egalitarians and hierarchists are likely to seek to conform to the lifestyle norms of their social groups, so achieving collective change in those groups is key. Communication with individualists and fatalists needs to appeal more to self-interest. For individualists, this may be connected more to status and success. For fatalists, it may be more a matter of cost-minimization and ease. Hermits are social non-conformists but may be influenced by some of the communication strategies adopted for other groups.

Besides these culture-specific communication strategies, this chapter also highlights the importance of reflexivity – collective reflection and discourse on the worldviews, goals, values, and practices entailed in sustainable living. As such, it builds on the emerging trend towards reflexivity already noted by some sociologists. This trend can be seen in some of the organizations seeking to promote more sustainable ways of living, including in Britain and the United States the non-governmental Global Action Plan, the Quakers, and the Permaculture movement. GAP probably has the best ability to engage the general public, while the Permaculture approach may be the most promising for generating new solutions. Quakers have well-developed methods for arriving at collective narratives, and these methods are beginning to be applied to the challenge of sustainable living. There is considerable potential for these organizations to learn from each other and to share ideas and practices with a wider public.

Note

1. See http://www.permaculture.org.uk/mm.asp?mmfile=dipcreatingactionlp; accessed January 10, 2006.

References

Argyle, M. (1987). *The Psychology of Happiness*. London and New York: Routledge.

Argyris, C. and Schon, E. (1996). *Organizational Learning II: Theory, Method and Practice*. Reading, MA: Addison Wesley.

Bateson, G. (1972). *Steps to an Ecology of Mind*. Chicago: University of Chicago Press.

Beck, D. and Cowan, C. (1996). *Spiral Dynamics: Mastering Values, Leadership and Change*. Oxford: Blackwell.

Beck, U. (1988). *Risk Society: Towards a New Modernity*, English translation by M. Ritter (1992). London: Sage.

Brower, M. and Leon, W. (1999). *The Consumer's Guide to Effective Environmental Choices*. New York: Three Rivers Press.

Castells, M. (1997). *The Power of Identity: The Information Age: Economy, Society and Culture*, vol. II. Oxford: Blackwell.

Crocker, D. A. and Linden, T. (eds.) (1998). *Ethics of Consumption*. Lanham, MD and Oxford: Rowman and Littlefield.

Dake, K. and Thompson, M. (1999). Making ends meet, in the household and on the planet. *GeoScience*, **47**, 417–24.

Dennet, D. (1993). *Consciousness Explained*. London: Penguin.

Dickinson, R. (1998). Modernity, consumption and anxiety: Television audiences and food choice. In *Approaches to Audiences: A Reader*, eds. Dickinson, R., Harindranath, R., and Linné, O. London: Arnold, pp. 257–71.

Dietz, T. and Stern, P. C. (1993). *Individual Preferences, Contingent Valuation, and the Legitimation of Social Choice*. Washington, DC: National Research Council.

Eisenstein, E. L. (1983). *The Printing Revolution in Early Modern Europe*. Cambridge, UK: Cambridge University Press.

Giddens, A. (1994). Reflexive institutions. In *Reflexive Modernisation*, eds. Beck, U., Giddens, A., and Lash, S. Cambridge, UK: Polity Press, pp. 56–109.

Hines, C. and Ames, A. (2000). *Ethical Consumerism*. Research study conducted by MORI for the Co-operative Bank. Available at: http://www.co-operativebank.co.uk/; accessed 19 September 2006.

Holmgren, D. (2002). *Permaculture: Principles and Pathways beyond Sustainability*. Daylesford, Victoria, Australia: Holmgren Design Services.

Inglehart, R. (1997). *Modernization and Postmodernization: Cultural, Economic and Political Change in 43 Societies*. Princeton, NJ: Princeton University Press.

Jackson, T. and Michaelis, L. (2003). *Policies for Sustainable Consumption*. Report to the UK Sustainable Development Commission. Available at: http://www.sd-commission.org.uk; accessed 19 September 2006.

Jaeger, C. C., Renn, O., Rosa, E., and Weber, T. (1998). Decision analysis and rational action. In *Human Choice and Climate Change, Vol. 3 Tools for Policy Analysis*, eds. Rayner, S. and Malone, E. L. Columbus, OH: Battelle Press, pp. 141–215.

Kasemir, B., Dahinden, U., Gerger Swartling, Å., *et al.* (2000). Citizens' perspectives on climate change and energy use. *Global Environmental Change*, **10**, 169–84.

Kempton, W., Boster, J. S., and Harley, J. A. (1995). *Environmental Values in American Culture*. Cambridge, MA: The MIT Press.

Liebes, T. (1988). Cultural differences in the retelling of television fiction. *Critical Studies in Mass Communication*, **5**, 4, 277–92.

Michaelis, L. (1997). *Policies and Measures to Encourage Innovation in Transport Behaviour and Technology*. Paris: OECD.

Michaelis, L. (2000). *Ethics of Consumption*. Mansfield College, Oxford: Oxford Centre for the Environment, Ethics and Society.

Michaelis, L. (2004). Consumption, reflexivity and sustainable consumption. In *Ecological Economics of Consumption*, eds. Røpke, I. and Reisch, L. Cheltenham, UK: Edward Elgar Publishing.

Michaelis, L. and Lorek, S. (2004). *Consumption and the Environment in Europe: Trends and Futures*. Report to the Danish Environmental Protection Agency. Available at: http://www.mst.dk; accessed January 16, 2006.

Morley, B. (1993). *Beyond Consensus: Salvaging the Sense of the Meeting*. Pendle Hill Pamphlet 307, Wallingford, PA: Pendle Hill Publications.

Nakicenovic, N. and Swart, R. (eds.) (2000). *Emission Scenarios*. A special report of Working Group III of the Intergovernmental Panel on Climate Change. Cambridge, UK: Cambridge University Press.

Price, L., Michaelis, L., Worell, E., *et al.* (1998). Sectoral trends and driving forces of global energy use and greenhouse gas emissions. *Mitigation and Adaptation Strategies for Global Change*, **3**, 263–319.

Quaker Earthcare Witness (2005). Quaker Earthcare Witness (formerly Friends Committee on Unity with Nature). Available at: http://www.fcun.org/; accessed January 16, 2006.

Sathaye, J. and Bouille, D. (2001). Barriers, opportunities, and market potential of technologies and practices. In *Climate Change 2001: Mitigation*. Contribution of Working Group III to the Third Assessment Report of the Intergovernmental Panel on Climate Change, eds. Metz, B., Davidson, O., Swart, R., and Pan, J. Cambridge, UK: Cambridge University Press, pp. 345–98.

Schor, J. (1998). *The Overspent American*. New York: Basic Books.

Simon, H. A. (1955). A behavioral model of rational choice. *Quarterly Journal of Economics*, **69**, 99–118.

Stern, P., Dietz, T., Ruttan, V., Socolow, R., and Sweeney, J. (eds.) (1997). *Environmentally Significant Consumption*. Washington, DC: National Academy Press.

Stoll-Kleemann, S., O'Riordan, T., and Jaeger, C. (2001). The psychology of denial concerning climate mitigation measures: Evidence from Swiss focus groups. *Global Environmental Change*, **11**, 107–17.

Thompson, M. and Rayner, S. (1998). Cultural discourses. In *Human Choice and Climate Change, Vol. 1, The Societal Framework*, eds. Rayner, S. and Malone, E. L. Columbus, OH: Battelle Press, pp. 265–343.

Thompson, M., Ellis, R., and Wildavsky, A. (1990). *Cultural Theory*. Boulder, CO: Westview Press.

17

Educating for "intelligent environmental action" in an age of global warming

Tina Grotzer
Harvard University

Rebecca Lincoln
Harvard University

> If you want to understand nature, you must be conversant
> with the language in which nature speaks to us.
> *Richard Feynman*[1]

What does it mean to be "conversant with the language in which nature speaks to us," and how do we help others develop this capacity? From a pedagogical perspective, if we want to address climate change and help people become a part of the solution rather than the problem, we must answer this question. One of the authors watched as her two-year-old son took the hand of another little boy on the playground and brought him over to the fence to "see the pretty sunset." His sense of wonder and enthusiasm for sharing it inspires hope for the future, and yet there is strong evidence that people of all ages understand little of the language or patterns of nature. Too often, as children grow up, they lose their appreciation for and sense of connection to the natural world. This is, in itself, a deep loss. But even if we retain an appreciation for the beauty of nature, few of us ever develop an understanding of the inherent complexities and dynamics of our environment. To solve environmental problems, an intuitive appreciation for nature is certainly necessary, but it is by no means sufficient. How do we learn the patterns of nature? How do we encourage the development of "environmental intelligence" and, more importantly, "intelligent environmental action"? With urgent concerns such as climate change on the horizon, the answers to these questions form an educational imperative.

This chapter is based in part upon the work of Understandings of Consequence Project, which is supported by the National Science Foundation, Grant No. REC-9725502 and REC-0106988 to Tina Grotzer and David Perkins, Co-Principal Investigators. Any opinions, findings, conclusions or recommendations expressed here are those of the authors and do not necessarily reflect the views of the National Science Foundation.

Research on people's understanding of global warming sends a clear and compelling message. People, even those deemed well educated, do not hold effective mental models of global climate change upon which to base decisions about their actions (e.g., Bostrom *et al.*, 1994; Sterman and Booth Sweeney, 2002; see also Bostrom and Lashof, Chapter 1, this volume). This research continues to find that people tend to confuse global warming with ozone depletion, do not understand the causes of global warming — rarely mentioning energy use and automobile emissions as causes and naming pollution in general as the most commonly cited cause — and do not realize that even if we act now, we will continue to see an increase in global warming. For example, Sterman and Booth Sweeney (2002) found misconceptions amongst MIT Sloan School of Management students, such as the belief that it would be sufficient to stabilize emissions at current rates, and that we can wait to see what happens and then act. They write that while it is not surprising that school children do not understand the processes governing climate change, it is more disturbing that highly educated adults do not. Disturbing, yes; surprising, well, maybe not . . . .

In this chapter, we suggest that given the complexity of the concepts and lack of opportunity to learn them, there's no particular reason that non-scientists would understand or be able to act upon climate change issues, and that the problem needs to be reframed in terms of how we can help people of all ages learn what they need to know. The current adult population grew up in a time when the curriculum did not offer the understandings necessary to enable people to understand the language or patterns of nature in general, or climate change issues in particular. Educational standards are just beginning to reflect the growing awareness of the importance of global warming as a topic of science education, and current curriculum standards in many states have just begun to address climate change directly. This raises a number of questions: "What can we do *now* to educate the present adult population about climate change?" and "How can we help tomorrow's adults better understand environmental issues?" What *is* involved in learning the language or patterns of nature and achieving a deep understanding of global climate change? And even if people hold deep understanding, what else do they need in order to choose actions that support decreases in greenhouse gas emissions and increase the ability to respond to climate change?

Two lessons on educating for intelligent action from education research

There is deep interest in the question of what people need to learn to understand the biosphere, and many ideas from diverse and eloquent perspectives on what educational components need to be included (e.g., Thomashow, 2002). While it is an important first step, as others have argued (e.g., Bostrom *et al.*, 1994), to assess what the public's knowledge looks like in contrast to scientific determinations of what they need to know, the educational problem for achieving understanding extends far beyond that. We need to consider the terrain between lay and expert mental models and figure out what it takes to help learners traverse that terrain. This involves understanding the patterns in how people perceive and think about environmental issues – cognitive and perceptual assumptions or default patterns that are likely to impede understanding and the ability to act. Further, if we set educated action as opposed to inert knowledge as the bar for success, then the educational challenges are extensive. We need to help the public develop the ability to understand climate change but also the sensitivity to perceive opportunities that invite action and, subsequently, the inclination to act.

This chapter introduces two lessons, relevant to global warming and to public education efforts as well as learning in schools, which can be drawn from a broader survey of the literatures in science education, cognitive development, and learning theory. As we elaborate below, this is just a start. The educational challenges surrounding climate change are many, and the particular cognitive issues in understanding global warming invite some of the most difficult problems in education to rear their ugly heads. This suggests a fair amount of humility about the endeavor, but also the importance of mining contributions from diverse disciplines. Finally, we consider that the path to making educational decisions based upon what we want people to perceive, know, and do is anything but straightforward. However, we also suggest areas where we believe our collaborative energies are well spent seeking answers. So what are some lessons from the educational research for teaching the complex issues involved in global warming?

1. People have a tendency towards default perceptual and cognitive patterns that impede understanding and acting on global warming. We need to find ways to address these.

When dealing with complexity, people reveal a set of "reductive biases" (Feltovich, Spiro, and Coulson, 1993) and they tend to make certain simplifying assumptions (Grotzer, 2004; Perkins and Grotzer, 2000) about the structure of the causality involved. Understanding global warming involves grasping a number of concepts that challenge these default patterns of thinking and reasoning.

Educational research shows, for example, that students tend to search for causes that are spatially local and temporally close to their effects (e.g., Driver,

Guesne, and Tiberghien, 1985; Grotzer and Bell, 1999; Koslowski, 1976; Lesser, 1977; Mendelson and Shultz, 1976; Siegler and Liebert, 1974). In general, people are not good at extended searches to assign cause or agency to factors that fall far away from an effect in space or time. The common tendency is towards efficiency. By contrast, the causes and effects of global warming are often separated in space and time (see also Dilling and Moser, Introduction, this volume). An increase in greenhouse gases in one part of the world leads to changes in climate that come decades after, and often thousands of miles away from, these initial emissions.

People also tend to overlook causes that are non-obvious in favor of those that are obvious (e.g., Brinkman and Boschhuizen, 1989; Leach *et al.*, 1992). Greenhouse gases are not visible or physically tangible, making it easier to dismiss them when analyzing the primary causes of climate change. Even in the case of easily observable outcomes, as in the sunset example in the introduction, non-obvious causes are typically overlooked. While not related to greenhouse gases, brilliant sunsets result from another non-obvious cause — particulate matter associated with smog that alters the refractive angle of sunlight. Particulate matter is not easily visible with the naked eye and so it is not usually accounted for as a cause of brilliant sunsets. Greenhouse gases pose an even greater challenge because the causes *and* the outcomes (at least at this stage) are difficult to detect without long-term attention to patterns and careful measurements. The combination of spatial gaps, temporal delays, and non-obvious causes makes it all too easy to ignore the causes of global warming.

The temporal aspects of global warming are difficult for other reasons as well. In order to realize that global warming is occurring, we need to be able to track patterns over time. People are not necessarily good at reasoning about patterns over time, and too often extrapolate from the moment (Dorner, 1989). They typically do understand systems that involve simple accumulation, for instance, increasing trash in landfills, or — more visible to the general population — along highways. However, once complications are introduced such as exponential growth or variations in rate of change, people find it harder to track the patterns. Positive feedback loops, where one event triggers other events that increase the cause of the initial event, introduce further complexity into rates of change. For examples, global warming causes a decrease in the polar ice cover, which reflects solar radiation. When the ice melts, the dark-colored land or water that lies beneath it absorbs more solar radiation, serving to exacerbate the initial problem by increasing local surface heating. Understanding these feedback patterns requires an understanding of patterns that fall outside of the typical accumulation models.

Beyond this, natural systems often have the ability to "absorb" a certain amount of change until enough change has accumulated so that it becomes noticeable, or smaller effects interact so that at a given threshold, the system's equilibrium "tips" and profound changes occur. For instance, while climate change is commonly viewed as a slow and steady process, it is now generally accepted that a sufficient disruption of the balance of temperature and salinity in the ocean as a result of melting polar ice could trigger massive climate shifts over a very short period of time (National Research Council, 2002). Patterns that have no obvious effects early on are easily ignored until it is too late. These have been referred to as "creeping environmental phenomena" (Glantz *et al.*, 1999). In the social realm, systems that change in non-linear ways are sometimes said to cross a "tipping point" – a concept that is beginning to find its way into public discourse (Gladwell, 2000). However, most people still analyze evidence using a simpler accumulation model, so when there are no visible effects, it is all too easy to think nothing is happening.

People tend to give linear or narrative causal explanations that are story-like in the form of "first this happened, then it made that happen," and so on. These have a domino-like quality to them (Grotzer, 1993, 2004). Global warming does not fit well with linear, narrative causal forms. It is the result of multiple intertwined causes, and results in multiple effects at many different levels. Instead of serving only one role in the causal story – as the effect at the end of the chain of causes – effects act as further causes.

When thinking about the origin of an observed phenomenon, most people envision a centralized cause, often with a single agent (like a sergeant in charge of a platoon) or with a coordinated set of agents (like a town decision-making board). People have a harder time envisioning multiple causes that are dispersed. The causes of global warming, however, are exactly that: spread out or decentralized. The effect emerges due to the collective actions of many people – it is an emergent causality. The behavior followed by many individual actors gives rise to what can be dramatic effects (e.g., Resnick, 1994). This runs counter to the default assumptions of most people which tend towards what Resnick (1996) has called the "centralized mindset." It assumes an orchestrated leader or some pre-existing, built-in "inhomogeneity" in the environment is responsible for complex patterns. Working with a computer program called StarLogo, Wilensky and Resnick (1999) found that students are typically shocked to see how individual rules of interaction at one level lead to emergent effects at another level, and that they find it very difficult to predict macro-level properties that emerge in systems as a result of micro-level interactions (e.g., Penner, 2000).

Notice that centralized causes also typically have a centralized and easily identifiable intent. In emergent causality, the effect emerges due to the collective actions of many agents whose actions have no coordinated intent. Agency is distributed and non-intentional (in terms of the specific outcome). Because agency and intent at one level are not easily connected to outcomes at the emergent level, it is difficult for people to see their individual role in the process — resulting in diffusion of responsibility. One person's effort, for instance, to reduce their own greenhouse gas emissions by driving less or turning off the lights at home does not directly result in an observable impact, and it is difficult to encourage behavior that seems futile at the individual level without a sense of the collective, emergent effects (see the discussion on urgency in Moser, Chapter 3, this volume). In emergent causality, there can be intention at the level of the individual actors, yet the broader effect does not necessarily stem from a broadly defined social intention. Understanding the role of intention in systems with trigger effects where the magnitude of the resulting effect may have little to do with any one particular agent, or in self-organizing systems where there is no clear internal or external agent, is also conceptually challenging. The relatively lower salience of many fine-grained actions compared to more dramatic individualized ones is a barrier to recognizing distributed agency. It is very difficult, from a cognitive stance, to juggle the actions of many individual parts and imagine or predict outcomes.

What does it mean to attend to these default patterns when educating about global warming? How might we use the information on people's reductive biases in our attempts to educate students or the current adult population? Becoming aware of people's reductive biases should enable us to find better ways to get our messages across. For instance, we could use computer technology to display simulations of global environmental problems or to explain causal connections between our actions and climate change and so on in public places such as malls or libraries. These could serve to make non-obvious effects more obvious. In the design of programs in the popular media, we can seek ways to reveal how the causal patterns can "sneak up on us" and, through simulated time lapse or dynamic imagery, help the public perceive and attend to these hard-to-perceive patterns. We can also make listeners aware of their default tendencies (e.g., through short, fun games on touch computer screens) and how these tendencies can eventually lead to unanticipated and potentially hurtful consequences.

As we educate for the future, we need to make today's students aware of the structures of complexity so that they are less likely to reduce complex patterns to simplistic ones. This is a problem that education researchers are working to tackle. For instance, researchers have developed a curriculum designed to restructure the ways that students think about causality within given science concepts (e.g., Grotzer, 2002), electronic building blocks that simulate causal and systems concepts to allow students to discover the complex behaviors related to particular structures (Zuckerman, 2003), and computer programs, such as StarLogo (Resnick, 1994) that allow students to experiment with how rule-based interactions of individuals give rise to complex system-level effects. This work has demonstrated significant improvements in students' ability to reason about complex causalities and systems behaviors (e.g., Grotzer, 2003; Grotzer and Sudbury, 2000, Resnick, 2003; Wilensky, 1998; Zuckerman, 2003).

2. We must attend to and develop the public's sensitivity, ability, and inclination to behave intelligently and consider how these three aspects of behavior interact to challenge our efforts.

Environmental education has long sought to engender environmental awareness and appreciation as well as the inclination to care for the environment. In many respects, this recognition surpasses what happens in other areas of education. Environmental educators (e.g., Thomashow, 1995) have designed thoughtful curricula for making people aware of their connection to and dependence on the Earth, and to help students develop an ecological identity and sense of ecological citizenship. These efforts are of the utmost importance.

The educational literature underscores another type of sensitivity and inclination that requires attention. If we want intelligent environmental action, we need to help the public learn how to act on climate change, to perceive opportunities that invite action, and subsequently, to be inclined to act — at a particular moment. Behavioral psychology suggests this can be aided by giving people very specific instructions on the action, frequent prompts, peer support, and recurring positive reinforcement (Clark, Kotchen, and Moore, 2003; DeYoung, 1996; Frahm *et al.*, 1995; Kollmuss and Agyeman, 2002; McKenzie-Mohr, 2000; see also Tribbia, Chapter 16, this volume). Perkins, Jay, and Tishman (1993) define three dimensions of the disposition towards intelligent performance: sensitivity, ability, and inclination. Sensitivity involves the ability to recognize occasions to apply a particular skill, understanding, or piece of knowledge. Ability refers to having that skill, knowledge, or understanding in one's repertoire. Inclination refers to being motivated to apply the particular skill, knowledge, or understanding in the given instance.

In terms of global warming, one can know quite a lot about the issue and can even care about it, but if on a moment-to-moment and day-to-day basis, one cannot recognize opportunities to use that knowledge, then it doesn't do a whole lot of good. While that sounds obvious, education efforts often focus on ability — leaving out sensitivity, and sometimes inclination, too. Perkins *et al.* (2000) found that of the three dimensions, sensitivity appeared to be the largest stumbling block for students. Even when students had the ability to understand particular patterns, they didn't identify instances of them. In part, this may be related to the reductive biases discussed above. One's default assumptions hinder one's ability to recognize alternative patterns. We've probably all had an experience where we realized in hindsight how we might have done something in a better way. The challenge is how to help people become mindful of everyday opportunities to change their behaviors in real time. The experience of 20/20 hindsight is only helpful if it changes what we do next time.

Moving from having the ability to engaging in new action patterns is a notoriously prickly problem in education that is subject to all sorts of situational variables. However, there are some straightforward ways to increase the likelihood that people will recognize action opportunities. What does it mean, then, from a public education stance, to attend to sensitivity, to growing the awareness for "actionable" opportunities?[2] Here are some examples. Articles on global warming could include a list of typical everyday actions and then list an alternative set of "choices for the environment" so that when people are engaged in the actions, the moment becomes a trigger for considering an alternative choice. Similar to campaigns that encourage parents to read to their children or know where their children are, ads could ask, "It's 10 p.m. Have you planned out your day tomorrow to minimize the driving you have to do? Do it for the environment — for yourself and your family." Public service announcements could focus on key decision points; for instance, the choice of a new car and the big environmental consequences that choice has over time.

While sensitivity, ability, and inclination are critical preconditions for action, they can also interact in ways that exacerbate the problem of inaction. For instance, the perceptual and cognitive challenges outlined above relate most directly to the ability to perceive particular kinds of patterns. However, they also affect people's awareness of the problems of global warming and, subsequently, their inclination. So in assessing risk, people attend to and are more easily stirred to action over a risk when its cause is personified or involves intentional agency, is centralized rather than decentralized, is immediate rather than cumulative, and is obvious rather than non-obvious. This can be so despite the magnitude of the effects. Global warming, despite the potential for massive catastrophic effects, sits on the wrong side of each of these tensions. Public education campaigns could make people aware of their tendency to focus on one side of these tensions to the exclusion of others. Imagine a magazine ad that poses the question, "What would you do if a terrorist were working silently to disrupt our weather systems, make our world inhabitable, and destroy life as we know it? Would you act? Would you want your government to act? Well, it is happening, and that terrorist is called Global Warming. It has recruited you as one of its agents – every time you get into your car." The ad could then explain the connection. Making the analogy forces the association in people's minds and makes it more difficult to ignore the non-obvious, cumulative causes of potentially massive catastrophic effects.

One could argue that such approaches play on people's fears. This is certainly true and, ideally, one would want to develop inclination in the most positive ways – that our actions as global citizens matter – as part of developing an environmental appreciation and an ecological identity. This should certainly be a part of the long-term educational process that schools and other communicators engage in, in addition to helping students understand that certain types of causes command their attention more readily than others. However, in terms of public education, there are so many messages competing for people's attention and for limited resources that such comparisons may be needed to help people realize the urgency of a non-obvious, potentially catastrophic problem. (For caveats and concomitant messages and framings, see the chapters by Moser; and Dunwoody, Chapters 3 and 5, this volume.)

In the introduction to this chapter, we argued that if we set educated action as the bar for success, then the educational challenges are extensive. We then elaborated two aspects of the "what should be taught" piece of that challenge: (1) the need to map the terrain between lay and expert mental models in terms of cognitive and perceptual patterns and figure out what it takes to help learners traverse that terrain, and (2) the need to help students and the public develop sensitivity, ability, and the inclination to understand and act on climate change. However, this is really just the tip of the iceberg. For instance, in order to deal well with the challenges of a changing world, people will need to be able to tackle fuzzy, ill-structured problems. They will need to be able to think flexibly and to develop creative solutions to multifaceted, novel issues. They will need to learn the skills of inquiry to find answers to new questions and to apply their understandings to authentic problems at the intersection of science, social and/or public policy, health, communication, and so on (see Bateson, Chapter 18, this volume). When one tries to take on the whole scope of related educational challenges, the magnitude of the problem can be overwhelming — only underscoring the urgency of the call to action.

Achieving "intelligent environmental action": pedagogical challenges

As we set about deciding *what* to teach, we also need to pay close attention to *how* to educate students and the public to understand various causal and perceptual patterns, and how to develop sensitivity, ability, and inclination. Certain pedagogical challenges become central to the endeavor. For example, in thinking about global warming, there is a critical connection between action at the level of the individual and collective action, and it is natural to think about the problem through this lens. This volume takes that approach. At the same time, as discussed above, the ability to make this critical connection is a highly challenging cognitive task. People have great difficulty reasoning at different levels. At different levels, the definition of what constitutes an "object" and the inherent forms of causality acting upon that "object" vary. For instance, in a traffic jam, at one level the cars are the objects and a focus on their actions leads to an analysis of what to do about the jam. At another level, the jam becomes the object of focus, and rather than analyzing the actions of individual cars, one analyzes the patterns relevant to the jam itself (Wilensky and Resnick, 1999). Similar difficulties have been seen in reasoning about ecosystems where students often extend the outcomes of the interactions between individual organisms to the population level — thus missing population-level effects such as balance and flux (e.g., Driver *et al.*, 1994; Grotzer, 2002; Grotzer and Basca, 2003; Wilensky and Resnick, 1999). Our collective efforts would be well spent identifying and addressing central pedagogical challenges such as these.

Another reason that it is important to give careful thought to "how" is that the material we need to teach does not automatically imply a certain way of teaching it. For example, a common approach that scientists take to teaching scientific concepts to the general population is to figure out what mental models they themselves hold as scientists, and then try to teach those models to the public. However, this strategy ignores the fact that scientists hold a wealth of assumptions that provide the context for those models. For instance, while members of the public might argue that you can't prove with certainty that certain outcomes will occur, scientists assume that the enterprise of science rests on the best available evidence. They recognize that the explanatory models that we use today are the best interpretation that we have based on the current evidence. We are not arguing that the population isn't capable of grasping the concepts, or that they are "scientifically illiterate" and that it would take too long to educate them. We are saying that the lay population does not hold the same set of assumptions as scientists do, and that translating messages for the public involves analyzing, from the public's perspective, how those messages will be heard and understood, and then helping them to transition to more scientifically accurate or complete understandings. Then decisions can be made about what contexts should be offered before energy is spent on miscommunication (see Dunwoody, Chapter 5, this volume). Often there are intermediate causal models (White, 1993) that hold sufficient explanatory power to enable intelligent action on behalf of novices and that are more easily grasped than a full-blown expert model would be. For instance, in teaching about density to young children, educators often use particle models or "dots-per-box" models instead of models that explain atomic mass or bonds and how they account for spacing. The simpler model functions as a useful bridge for those students who do pursue deeper understanding and go on to learn the atomic explanation. Neither educators nor scientists can design these models alone. They need to be the result of a collaborative effort between educators and scientists. We strongly encourage the scientific and education communities to collaborate on defining what those might look like.

As new problems arise in education, there is often a call for extensive research. In deciding what we need to know from a pedagogical standpoint, it is important to carefully mine what we already know from related contexts. One of the best ways that we can move forward is to use the information at our disposal. Thomashow (2002: 193) writes that "we know very little about the cognitive origins of ecological learning and biospheric perception." However, spread across the different literatures in cognitive development, learning theory, science education, environmental psychology, and sociology, there is vast information on how children understand concepts relevant to ecology at different ages (Grotzer, 2003). We have to be willing to look across the typical boundaries of our fields. On balance and in support of Thomashow's assertion, we certainly need to expand upon what we know, and this includes greater awareness of the tacit knowledge and epistemological assumptions scientists hold and how they are learned. Of course there are caveats to borrowing research findings across disciplinary boundaries. It is important to keep in mind the contexts in which information was collected. For instance, developmental research is often carried out so that task demands are carefully controlled for; however, the subjects are seldom, if ever, given optimal educational guidance for learning a set of concepts. Therefore, the research contexts tell us what subjects do with carefully controlled task demands, but not what is possible with optimal educational guidance (Metz, 1995, 1997). Our efforts will be most productive if they build upon the existing research base across disciplines with its limits in mind and an eye towards new possibilities.

We believe that we don't yet know what is possible for helping the public understand global warming with optimal educational support or with purposeful collaboration between educators, scientists, and the many others working on the urgent problem of climate change. But this chapter offers some insights into a few key building blocks: understanding people's default cognitive patterns, recognizing their difficulties understanding complex systems, and developing their sensitivities, abilities, and inclinations to act in environmentally intelligent ways. It is imperative that we help the public become environmentally intelligent and learn to act with that intelligence, rather than just admonish people for not doing so. Dedicating ourselves to that effort inspires optimism that future generations will become "conversant with the language in which nature speaks to us," and that we will be able to admire sunsets with our children.

Notes

1. While Richard Feynman was referring specifically to mathematics, the question can be asked as to what other patterns one must grasp in order to understand the language of nature. He made the statement in various forms, both written and in interviews. A published source can be found at Feynman (1967).
2. We realize that public school education intended to induce behavioral changes involves significant, but not unprecedented, policy considerations (e.g., education about recycling also led to new behaviors among students and eventually their families). To adequately treat the arguments made in the contentious debate about control over educational content would require a far more substantial treatment than space here allows.

References

Bostrom, A., Morgan, M. G., Fischhoff, B., *et al.* (1994). What do people know about global climate change? 1. Mental models. *Risk Analysis*, **14**, 6, 959–70.

Brinkman, F. and Boschhuizen, R. (1989). Pre-instructional ideas in biology: A survey in relation with different research methods on concepts of health and energy. In *Research and Developments in Teacher Education in the Netherlands*, eds. Voorbach, M. T. and Prick, L. G. M. London: Taylor and Francis, pp. 75–90.

Clark, C. F., Kotchen, M. J., and Moore, M. R. (2003). Internal and external influences on pro-environmental behavior: Participation in a green electricity program. *Journal of Environmental Psychology*, **23**, 3, 237–46.

DeYoung, R. (1996). Some psychological aspects of reduced consumption behavior: The role of intrinsic satisfaction and competence motivation. *Environment and Behavior*, **28**, 3, 358–409.

Dorner, D. (1989). *The Logic of Failure*. New York: Metropolitan Books.

Driver, R., Guesne, E., and Tiberghien, A. (eds.) (1985). *Children's Ideas in Science*. Philadelphia, PA: Open University Press.

Driver, R., Leach, J., Scott, P., *et al.* (1994). Children's ideas about ecology 3: Ideas found in children aged 5–16 about the interdependency of organisms. *International Journal of Science Education*, **16**, 985–97.

Feltovich, P. J., Spiro, R. J., and Coulson, R. L. (1993). Learning, teaching, and testing for complex conceptual understanding. In *Test Theory for a New Generation of Tests*, eds. Frederiksen, N. and Bejar, I. Hillsdale, NJ: LEA, pp. 181–217.

Feynman, R. (1967). *The Character of Physical Law: Messenger Lectures, 1964*. Cambridge, MA: The MIT Press.

Frahm, A., Glavin, D., Gensler, G., *et al.* (1995). *Changing Behavior: Insights and Applications. Behavior Change Project – Final Report*. Seattle, WA: King County Water Pollution Control Division.

Gladwell, M. (2000). *The Tipping Point: How Little Things Can Make a Big Difference*. Boston: Little, Brown and Co.

Glantz, M. H., *et al.* (1999). *Creeping Environmental Problems and Sustainable Development in the Aral Sea Basin*. Cambridge, UK: Cambridge University Press.

Grotzer, T. A. (1993). Children's Understanding of Complex Causal Relationships in Natural Systems. Unpublished Ed. D. dissertation. Cambridge, MA: Harvard University.

Grotzer, T. A. (2002). *Causal Patterns in Ecosystems*. Cambridge, MA: Harvard Project Zero.

Grotzer, T. A. (2003). Learning to understand the forms of causality implicit in scientific explanations. *Studies in Science Education*, **39**, 1–74.

Grotzer, T. A. (2004). Putting everyday science within reach. *Principal Leadership*, October, 16–21.

Grotzer, T. A. and Basca, B. B. (2003). How does grasping the underlying causal structures of ecosystems impact students' understanding? *Journal of Biological Education*, **38**, 1, 16–29.

Grotzer, T. A. and Bell, B. (1999). Negotiating the funnel: Guiding students toward understanding elusive generative concepts. In *The Project Zero Classroom: Views on Understanding*, eds. Hetland, L. and Veenema, S. Fellows and Trustees of Harvard College, pp. 59–75.

Grotzer, T. A. and Sudbury, M. (2000). Moving beyond underlying linear causal models of electrical circuits. Paper presented at the Annual Conference of the National Association for Research in Science Teaching (NARST), New Orleans, LA, April.

Kollmuss, A. and Agyeman, J. (2002). Mind the gap: Why do people act environmentally and what are the barriers to pro-environmental behavior? *Environmental Education Review*, **8**, 3, 239–60.

Koslowski, B. (1976). Learning about an Instance of Causation. Unpublished manuscript. Ithaca, NY: Cornell University.

Leach, J., Driver, R., Scott, P., *et al.* (1992). *Progression in Conceptual Understanding of Ecological Concepts by Pupils Aged 5–16*. Leeds, UK: University of Leeds, Centre for Studies in Science and Mathematics Education.

Lesser, H. (1977). The growth of perceived causality in children. *The Journal of Genetic Psychology*, **130**, 142–52.

McKenzie-Mohr, D. (2000). Fostering sustainable behavior through community-based social marketing. *American Psychologist*, **55**, 5, 531–7.

Mendelson, R. and Shultz, T. R. (1976). Covariation and temporal contiguity as principles of causal inference in young children. *Journal of Experimental Child Psychology*, **22**, 408–12.

Metz, K. E. (1995). Reassessment of developmental constraints on children's science instruction. *Review of Educational Research*, **65**, 2, 93–127.

Metz, K. E. (1997). On the complex relation between cognitive developmental research and children's science curricula. *Review of Educational Research*, **67**, 1, 151–63.

National Research Council Committee on Abrupt Climate Change Ocean Studies Board, Polar Research Board, Board on Atmospheric Sciences and Climate, Division on Earth and Life Studies (2002). *Abrupt Climate Change: Inevitable Surprises*. Washington, DC: National Academy Press.

Penner, D. (2000). Explaining systems: Investigating middle school students' understanding of emergent phenomena. *Journal of Research in Science Teaching*, **37**, 8, 784–806.

Perkins, D. N. and Grotzer, T. A. (2000). Models and moves: Focusing on dimensions of complex causality to achieve deeper scientific understanding. Paper presented at the Annual Meeting of the American Educational Research Association (AERA), New Orleans, LA, April.

Perkins, D. N., Tishman, S. Donis, K., *et al.* (2000). Intelligence in the wild: A dispositional view of intellectual traits. *Educational Psychology Review*, **12**, 3, 269–93.

Perkins, D., Jay, E., and Tishman, S. (1993). Beyond abilities: A dispositional theory of thinking. *Merrill-Palmer Quarterly*, **39**, 1, 1–21.

Resnick, M. (1994). *Turtles, Termites, and Traffic Jams: Explorations in Massively Parallel Microworlds*. Cambridge, MA: The MIT Press.

Resnick, M. (1996). Beyond the centralized mindset. *Journal of the Learning Sciences*, **5**, 1, 1–22.

Resnick, M. (2003). Thinking like a tree (and other forms of ecological thinking). *International Journal of Computers for Mathematical Learning*, **8**, 43–62.

Siegler, R. and Liebert, R. (1974). Effects of contiguity, regularity, and age on children's causal inferences. *Developmental Psychology*, **10**, 4, 574–9.

Sterman, J. D. and Sweeney, L. B. (2002) Cloudy skies: Assessing public understanding of global warming. *System Dynamics Review*, **18**, 2, 101–7.

Thomashow, M. (1995). *Ecological Identity: Becoming a Reflective Environmentalist*. Cambridge, MA: The MIT Press.

Thomashow, M. (2002). *Bringing the Biosphere Home*. Cambridge, MA: The MIT Press.

White, B. (1993). Intermediate causal models: A missing link for successful science education. *Cognition and Instruction*, **10**, 1, 1–100.

Wilensky, U. (1998). Gaslab – an extensible modeling toolkit for connecting
    micro- and macro-properties of gases. In *Computer Modeling in Science and
    Mathematics Education*, eds. Roberts, N., Feurzeig, W., and Hunter, B. Berlin:
    Springer-Verlag.
Wilensky, U. and Resnick, M. (1999). Thinking in levels: A dynamic systems
    approach to making sense of the world. *Journal of Science Education and
    Technology*, **8**, 1, 3–19.
Zuckerman, O. (2003). A digital infrastructure for hands-on modeling, simulation
    and learning of complex causal processes. Unpublished paper, Cambridge,
    MA: Media Lab, Lifelong Kindergarten Project, MIT.

18

Education for global responsibility

Mary Catherine Bateson
Institute for Intercultural Studies

Introduction

Individual human beings affect the global story: small actions in the micro-cosm have consequences in the macrocosm. The climate change now under way represents a systemic disruption that is the greatest threat on the human horizon, requiring basic changes in habits of thought that link global change to individual and local behavior. Such changes in ways of thinking or acting are costly both economically and psychologically, and will occur only when the threat to the integrity of planetary systems is felt as deeply relevant — as directly connected to the individual. Raising the awareness of climate change in adults and children is not only a matter of conveying the facts of temperature or composition of the atmosphere or the threats that changes in these present. Unless individuals begin to identify themselves as part of the process and identify with it, they will not be willing to change.

Human beings do typically live and act in multiple worlds, each one providing a context for the development of a sense of identity-in-relationship, each one potentially more inclusive. The infant begins with a sense of herself as part of the mother—infant dyad and gradually becomes aware of being distinct within it, developing through households and communities and all the contexts of life. The goal of education for global responsibility must be to give each child a continuing sense of his or her value and responsibility as a part of these larger contexts. Because so many private familial actions have an impact on global systems, it is becoming urgent to maintain a sense of identification with these systems and to think habitually in terms of the causal and ethical relationships between multiple coupled and embedded worlds.

Within the Western tradition, however, we have come to emphasize the individual over the relationships, and we build our ethical systems rather narrowly on the value of individuals, rather than cherishing the multiple emergent identities that consist in being a part of some larger system. This tends to keep us unaware of the broader ethical imperatives of being part of a community or an ecosystem or even a family (see Jamieson, Chapter 30, this volume). We need to develop both an ethic that values relationships as well as individuals and a capacity for identification with all the larger systems of which we are part. Although we are connected to everyone and everything else, we are socialized and educated into viewing ourselves as separate. While we factually are part of multiple worlds, we are socially and culturally entrained to see and value ourselves as separate and to make choices and decisions on the basis of individual and national self-interest. This socialization makes it difficult for us to see that our actions have impacts elsewhere and to respond with a sense of personal pain or loss (see also Moser, Chapter 3, this volume).

For adults, the task is one of relearning. For children, the task is one of building the foundation for future lifelong learning. The tools that will be needed to communicate about the process of climate change have the potential for further broad changes in habits of thought, leading the individual child or adult into a sense of being a part of the biosphere. Such tools include systems metaphors, narratives of connection, cross-overs between disciplines, and cross-overs with ways of knowing such as participant observation. The ultimate goal is an education for global responsibility that unfolds in a pattern of lifelong learning.

Metaphors and systems analogies

Children identify easily across species. A furry mammal? Easy – and children's books are full of them. Children immediately attribute to them feelings comparable to their own and feel concern for their fate. An invertebrate? Consider *Charlotte's Web* and the ladybug books. Because it is possible to identify with a tree, a child can even weave the story of a tree's growth into his own growing and learning. But when identification must go beyond single organisms, it gets more difficult. Dr. Seuss has shown that children can identify with fantasy fauna and flora, using them, in *The Lorax*, to think about pollution and habitat destruction and the human habits that threaten them, but this is not quite the same as identifying with the habitat itself, much less with the oceans, the atmosphere, even smaller ecosystems like a forest or a lake. A children's book long out of print individualized the story of a drop of water moving from cloud to mountainside to reservoir to a child's bathtub, ending when the drop of water reached the ocean. The drop of water recalls the hero's journey, from one adventure to another, which makes identification possible, and the child empathizes with the drop. It is not clear, however, that this kind of identification leads to identification with the great cycles of water and nitrogen and CO_2, yet every living cell shares in these great cycles, echoing them in its metabolism and reproduction.

All living systems are similar in ways that are harder to recognize, and an understanding of planetary systems is potentially linked to an understanding of single organisms – you and me – and small communities, not only because organisms and communities depend upon the larger systems, but because of systemic – organizational – similarities and vulnerabilities. Systemic analogies, then, are an important way of learning to identify with larger and more inclusive systems. Because of the possible analogies between different levels, it is possible to move both up and down the levels of description, back and forth between macrocosm and microcosm: to use global understanding both as a source of local insight and to project local insight onto global issues. When we do make analogies between systemic levels, however, we tend to move upward and project locally learned relationships onto the wider screen, for instance by using the family as a metaphor to extend responsibility and caring (to Mother Earth or the brotherhood of man) or the human body as metaphor for an ecosystem (as in the Gaia hypothesis or the habit of speaking of lakes as sick or dying). Contemporary family therapy depends on understanding the family as a system, drawing on cybernetics and systems theory for a metaphor that can be applied to single organisms, to ecosystems, or to the biosphere. Learning about global climate might also work down the systemic levels by offering tools for thinking about one's own health and one's family or community. A recognition of similarities can be a basis for extending respect, and also lays the basis for a transfer of learning from one context to another.

Narrative

Humans learn and teach through stories, so narrative offers a second approach to teaching awareness and responsibility, one that can also be illustrated through children's books. Narrative introduces elements of process and causation: *The Lorax* is an allegory about consequences.

One challenge of communicating the urgency of responding to global warming is to create narratives that individuals can apply in understanding the consequences of their own behavior.

Storytelling focuses on the local, but does so in a way that can evoke the global. Aristotle, in the *Poetics*, wrote about the necessary unities of dramatic composition, time, action, and plot: a drama, he said, should have one plot, the action of which should unfold on a single day in a single location. These conventions did not inhibit the transfer of ethical ideas to a wider stage, for all of Greek drama was set against a background of shared knowledge of other times and places, and the great tragedies were organized in cycles that followed families (like the House of Atreus) as the crimes of each generation resonated in the next. Characters whose narrow understandings of their choices and purposes are leading them toward tragedy are continually reminded by the chorus of the cyclical continuities of fate and divine will. One might even argue that the tragic flaw of the heroes (and heroines) is that they look at their destinies too narrowly. Because the consequences of local actions are global, we must learn to both think and act globally.

Making the connections across contexts

The Greek tragedies were written in the context of a coherent culture in which each of the classic myths potentially applied to everyone in the audience, but many now read them (if forced to do so) as exotic and irrelevant. In contemporary culture, knowledge is highly compartmentalized and there are vast dissonances between the contexts in which individuals must function, including those of class, race, or ethnicity. The sense of dissonance between contexts means that many graduates feel that school learning is simply irrelevant to the rest of their lives, so that concepts may be seen as only relevant for members of the majority culture or only usable for passing exams. When this happens, a particular metaphor or narrative is not able to inform behavior and remains compartmentalized and encapsulated as a set of test responses. For many children, the discontinuity between the school environment and the worlds of family and neighborhood is so great that it produces an inability to transfer learning from one to the other, or an extreme alienation from one or the other. A few blocks' walk from home to school may represent an uncrossable gulf and a student body simply going through the motions.

The ability to move among multiple coupled and embedded systems is not only critical for making the connection between individual actions and global implications; it is also directly related to the capacity to utilize education by transferring learning, recognizing its relevance in a new setting, and adapting to new contexts (see Grotzer and Lincoln, Chapter 17, this volume). All this separation produces dissonance, like the commonly remarked inconsistency of SUV owners regarding themselves as environmentalists. Actions have consequences at many different levels and we live with an extraordinary range of types of information, bits and pieces of narrative drawn from different eras and cultures, sometimes neatly separated and sometimes messily jumbled together, but rarely integrated. On the evening news there may be reports one after the other of a diplomatic conference in Europe and a Hollywood divorce, a drought in Africa and a football game at a suburban high school, with no framework for making sense of what is going on. The weather we encounter if we walk outdoors is surely affected by global changes, but attempts to draw that connection are more confusing than enlightening. Back indoors, we create for our comfort our own tiny microclimates, using fuel that contributes to ongoing change without making the connections that would promote learning and responsible action. Yet the ability to compare systems of different kinds and make the connections between them, to move between the local and the global, is intellectually and socially essential.

To affect understanding and behavior in relation to climate change, a narrative needs somehow to link multiple contexts. Some time ago I had an experience that forced me to look again at the contrasting contexts of learning as I was whisked in a very short period of time from one kind of setting to another. On Friday morning I spoke to a 6th grade class at a Denver middle school about curiosity and social diversity. I had spoken to an adult audience of parents and teachers in the large school auditorium the night before. In both of these settings I was haunted by what I had been learning on Wednesday and Thursday, sitting in the Advisory Council of the National Center for Atmospheric Research in Boulder, hearing about current work on modeling Earth systems and solar events, simulating thousands of years of climate change on scores of parallel processing computers, linking institutions across the nation and the world in a single research consortium. However different these contexts superficially appeared, it seemed to me important to link them together, to learn from each and to integrate that learning into my thinking about the others, just as in my own kitchen I need to integrate a knowledge of global systems into my handling of trash. Without building these intellectual bridges we cannot adjust our human patterns of adaptation to the world we live in.

For the remainder of Friday and Saturday, I was conversing with faculty and graduate students at the Education Faculty at the University of Colorado in Denver about action research and educational renewal. In retrospect, I felt that these last conversations gave me a way of looking at the extreme disparities of that four-day period. All of these encounters proposed the question of how we, as human beings, can grasp the complexity of the coupled systems, natural and cultural, within which we live, and how we can evoke a sense of responsibility in our actions that affect them. Action research is about combining the roles of local change agent and observer in order to apply what is learned in other contexts. Educational renewal is about modifying schools so that their graduates will be effective participants in the world around them. In both cases, local activity is evaluated in terms of a wider significance. Both evoke the Kantian Categorical Imperative, to act in ways that would be acceptable if generalized throughout society. Recycling a soda can is a minor event on the global scale – but positive if adopted by large numbers of people. Narratives about consequences produce more than empathy – they suggest models for action.

Participant observation

A useful way of combining self-awareness with an awareness of a larger context is participant observation, which underlies action research. Participant observation is traditionally described as a social science methodology, but it is also a way of being in the world that can be taught. The participant observer can often gain access to something she or he wishes to understand only by active engagement. She or he may be researching an industry by taking a job in that industry, both doing the job and observing what goes on, or studying the development of children by joining in their play and taking notes. Self-observation – reflexivity – is a basic element of participant observation and also an essential tool for integrating different kinds of experience. Another term for the orientation of the participant observer is disciplined subjectivity, for without an awareness of one's personal biases and responses, observation will be flawed and faulty assumptions maintained unchanged. Disciplined subjectivity proposes an awareness of several simultaneous layers of experience and learning, reflecting the different layers of attention involved in action research and policy-making. Reflective self-observation is especially important in undertaking actions whose consequences may be unknown – like the behaviors that lead to global warming, which has been described as a massive and unmindful experiment on the very conditions of life.

Changing habits of thought

Metaphor, narrative, and participant observation are tools for learning to connect with a complex world. All of us today experience extraordinary juxtapositions between the microcosm (classroom, neighborhood, seminar) and the macrocosm, the vast planetary setting of our individual actions and interactions, yet at present the necessary connections seem not to be made. School children need to be able to think about the oceans and the atmosphere, not in detail, but as the macrocosm that shapes their lives and for which they must take some responsibility. They must also be able to think about their own roles and the nature of their small community as part of a larger social system. All too often, educational institutions and museums overstress the scale and complexity of the planetary systems of weather and climate that so deeply determine our lives, evoking no more than a "Gee whiz" sense of wonder and experienced as irrelevant, only suggesting the futility of the efforts of individuals and small groups to bring about change where they live.

Teaching about climate change is critically important at the present time as part of a necessary change in *habits of thought* that would strengthen understanding and responsibility in many other contexts: in families and communities; in maintaining a democratic society, evaluating scientific knowledge, seeking social justice, and functioning in a global economy. The study of global systems might actually enrich the capacity to adapt to multiple worlds and to transfer knowledge from one context to another. Most people necessarily go through life acting as if they lived all of their lives in a very small pond, but every pond is set in a wider landscape and microcosms can become models for understanding the larger systems in which they are embedded.

Effective communication about climate change is a vehicle for changes in habits of thought that would include the following:

- *Challenges to think globally*: We see our worlds divided from each other and our planet divided up by multiple boundaries. But studying the atmosphere and the oceans alters the sense of space and place, demonstrating global unity and the sharing of global risk.
- *A time scale extended beyond the present*: An understanding of Earth's climate systems depends on processes extending over thousands of years. As a nation, however, we live longer and think shorter and make all too many short-term decisions. Teaching about climate change is a way of teaching about longer-term patterns and their connection to present actions.
- *Integrating different kinds of knowledge*: We like to keep things separate and neat in our thinking, to put the meteorologists in a different department from the physicists, the anthropologists in a different building from the oceanographers.

By sixth grade, children have different teachers for different subjects and study them in different rooms. Teaching about climate change invites integrative and interdisciplinary thinking.

- *An introduction to systems thinking*: The formal relationships invoked in systemic descriptions, such as positive and negative feedback, thresholds and tipping points, and various kinds of coupling between systems, are then available for application in other contexts. Teaching about climate change is a vivid opportunity to build toward a systemic understanding of interdependence, which can then be applied in other areas of life and ethics (see Grotzer and Lincoln, Chapter 17, this volume).

- *An ability to deal with uncertainty*: We like knowledge to be unambiguous and specific. Although there is now scientific consensus that climate change is occurring, the consequences and timing are still ambiguous and unspecific. Thus, climate change offers a case study of the need for decision-making under uncertainty that arises in all areas of life, and provides a context for discussing probability and the challenge of preparing for events (like earthquakes and terrorist attacks) for which no date can be specified.

- *Knowledge as a path to adaptation as well as control*: It is common to confuse science and technology, to think that the value of studying some phenomenon is the ability to control or manipulate that phenomenon. But in most cases an understanding of weather leads not to manipulation but to adaptation — if the weather prediction is rain, we wear a raincoat or stay home – and similarly for climate. The study of climate change necessarily involves thinking about adaptation, for all too often the vicissitudes of life cannot be "fixed" or solved by some simple intervention.

- *Awareness of the value of diversity*: While climate change will tend to reduce biological diversity in every affected biome, it can highlight the relationships between intraspecific diversity and resilience. Human variation is both biological and cultural, but both forms of variation suggest alternative possibilities. An exposure to diverse ideas and customs is not only a good preparation for a multicultural world – it may prove to be an essential preparation for adaptability and even the invention of whole new ways of being in response to a changing environment. Diversity of intelligences as well as backgrounds in the classroom enriches the conversation about changes in the biosphere.

- *The habit of self-reflection*: Both participant observation and action research, although they have different origins and differ in important ways, serve as models for engaged living in a rapidly changing world by combining the notion of behavior that has *effects* (being an actor, a participant) with the awareness of doing so with incomplete knowledge and for the sake of increasing that knowledge. Each of them, perhaps because of the tension involved, has the potential to evoke reflection and self-examination that can lead to changes in behavior.

Preparing lifelong learners

How do we reach individuals in their local settings? Each setting connects to ever larger and more inclusive systems. Every classroom, home, or workplace is located at one of those addresses children so enjoy putting on envelopes: Main Street, Middletown, Anystate, USA, North America, Earth, Solar System, Milky Way Galaxy, The Universe. Where and when is learning in time? Days or years in the lives of a group of children, a moment in the development of the American understanding of the relationship between universal education and democracy, a flickering instant in the evolution of a species that survives by learning and teaching, and even less in the life of the planet and beyond. What is the plot, the storyline that can somehow project adult participation in an unknown future world and identification with the biosphere?

Human beings are biologically endowed with curiosity and playfulness, the preconditions for learning and adaptation. It may or may not be possible to extinguish these gifts completely, but certainly it is possible to inhibit or suppress them, to make them unavailable as resources in situations where they would be helpful. Even as we encourage the transfer of learning from one discipline to another, we need to promote the capacity for learning outside of formal educational settings, making the habit of learning, and thus the awareness and embrace of change in the self, the thread of a life story. Because learning involves changing, there is a potential conflict between the willingness to learn and a fixed sense of identity — except if learning becomes a basic element in that identity.

Schools and classes within schools, like families, are themselves contexts for learning how the world works, so more effective communication about Earth systems proposes the question of a more interactive, systemic pattern in the process of education as well as the content. Often schools have offered a hierarchical, coercive environment, preparing children for a society in which it is logical for human beings to impose on the natural world a lineal understanding in terms of control and exploitation that is eventually disruptive of natural systems. Our classrooms can be contexts for learning about a different kind of society, pluralistic and democratic. Perhaps the model for classrooms could correspond to ecological patterns that can be observed in the natural world, complex and multifaceted, that just might offer useful analogies for human society and social policy. Perhaps we can reconstruct society to match an emerging sensitivity to the characteristics of systems.

Along with the skills and information set forth in the curriculum, children acquire very abstract understandings about the world that define the nature of their learning and shape the way they apply it — in effect, define its relevance. Deutero learning,[1] sometimes glossed as "learning to learn," should be interpreted as "learning to learn in a given kind of context." Depending on the nature of that context, it will influence the assumptions children bring to new learning for the rest of their lives. If you imagine the classroom context as a model of the universe, children who succeed within the classroom context will transfer their learning outside it as long as the other worlds in which they move and adapt can be seen as having congruent structures. If the classroom is experienced as inimical, so may the world — and survival in that world may replicate a resistance to changing or conforming in school. If the classroom is one in which there is one and only one correct answer to every question, children may go on to be uncomfortable in a world of ambiguity and partial truths and unable to navigate it successfully and mindfully. If the classroom projects a world of direct lineal causation and simple solutions to problems, adults may, for instance, come to regard insecticides as the logical solution to insect pests and prisons as the logical solution to crime. If the classroom is onerous and unpleasant, learning may be the last avenue of approach to unfamiliar situations for its graduates. I say "may" because, if they are lucky, they may encounter other contexts in which other assumptions are learned, but the often unarticulated assumptions acquired in childhood are likely to affect thinking and learning throughout adult life.

It is probable that in our lifetimes we will have to adapt to radical climatic and economic change only moderated by choosing to change sooner rather than later. Responsibility, whether as parents or teachers or citizens, represents commitment not to something fixed and predictable, but to the unknown. This is equally true in the environmental context. We will need to learn along the way, respond to possibilities and events knowing we will make mistakes, improvise, calibrate risk knowing that risk is unavoidable, and put up with a high degree of uncertainty and ambiguity. Making the right choices will depend not on memorized precepts but on transferring knowledge to new contexts and recognizing familiar patterns and processes in a changing world.[2]

Notes

1. For an extended discussion of deutero learning, see G. Bateson (1942).
2. For further discussion of the issues raised in this chapter, see M. G. Bateson (2004). On systemic thinking in ecological education, see especially "Democracy, Education, and

Participation," pp. 285–301; on levels of learning and participant observation, see especially "Learning in Layers," pp. 250–62.

References

Bateson, G. (1942). Social planning and the concept of deutero-learning. In *Steps to an Ecology of Mind*, ed. Bateson, G. Chicago: University of Chicago Press, 2000, pp. 159–76.

Bateson, M. C. (2004). *Willing to Learn: Passages of Personal Discovery*. Hanover, NH: Steerforth Press.

19

Changing the world one household at a time: Portland's 30-day program to lose 5,000 pounds

Sarah Rabkin
University of California–Santa Cruz

David Gershon
Empowerment Institute

> On playing fields and battlegrounds, challenges that would be daunting and impossible if faced alone are suddenly possible when tackled in a close-knit group. The people haven't changed, but the way in which the task appears to them has.
>
> *Malcolm Gladwell (2002: 264)*

In late 2001, denizens of several residential areas in Portland, Oregon, began knocking on doors, inviting their neighbors to take part in a campaign to reduce household carbon dioxide emissions. In doorway conversations, these volunteer "team initiators" emphasized the power of their low-carbon campaign not only to improve environmental quality, but also to promote a sense of community and enhance neighborhood life.[1] Residents who showed interest were invited to attend a block-based information meeting in a neighbor's home.

Altogether, 130 Portland householders opened their doors to peer recruiters, and an additional 22 received invitations to introductory meetings via speakers and literature tables at large public events such as a local conference on sustainability. Seventy-two of those approached were sufficiently intrigued to attend meetings, and all but one of the 72 decided to join carbon-reducing teams. Ultimately, nine block-based teams representing 54 households came together in a pilot program to help each other diminish their impact on global climate.

In short, the CO_2-reducing campaign garnered a recruitment rate of about 43 percent: almost twice that of a similar environmental-action campaign that preceded it in Portland. By community organizing standards, this is remarkable recruitment rate, according to David Gershon, whose Empowerment Institute provided the blueprint and implemented the program.

The outcome was also impressive. With the help of a web-based CO_2 emissions calculator, members of each participating household estimated their baseline CO_2 footprint, then set out to shrink it. Drawing on suggestions provided in a workbook titled *Low Carb Diet: A 30-Day Program to Lose 5,000 Pounds*, they shortened their showers, reset their water heaters, donned extra sweaters, and turned down their thermostats. Some installed energy-efficient appliances or insulated their attics. Others pumped up their tires, tuned up their engines, traded in gas-guzzlers for fuel-efficient cars, or left their vehicles at home.

By adopting these and other carbon-busting practices suggested in the workbook's checklist, the first 31 households to complete the 30-day program succeeded in reducing their household-based CO_2 emissions by an average of about 22 percent, with an average absolute reduction of 6,700 pounds per household.[2] In the process, they strengthened their ties with fellow residents. As participant Amanda Lewis noted, "I like the community-building aspects of this. One of the best things was getting to know my neighbors."

This successful climate control program could potentially inspire city governments everywhere. Its success suggests that groups of citizens can slash residential CO_2 emissions through neighborhood-based initiatives that provide practical carbon-saving tools and that focus on climate protection as a goal. A lack of further funding prevented Empowerment Institute and the City of Portland from expanding the residential low-carb campaign beyond the initial pilot project. It therefore remains to be seen whether CO_2 belts tightened during a 30-day period can stay cinched over the long term. Another open question is whether the program would garner such an enthusiastic response in areas less receptive to climate issues, or whether climate protection would need to be combined with other more tangible benefits in a broader program (for example, see Watrous and Fraley, Chapter 25, this volume).

However, city staffers who helped oversee the project believe that its approach can be exported to a broad variety of communities, including those lacking Portland's reputation for environmental initiative. For municipal governments, grassroots activists, and other interested parties, the "Low Carb Diet" merits a closer look.

Slashing the "ignorance tax"

The "Low Carb Diet" succeeded at mobilizing neighborhoods in part because it focused on individuals and motivation. Portland citizens reduced

their CO_2 emissions as the result of collaboration between the city's Office
for Sustainable Development (OSD) and Empowerment Institute (EI),
a private organization that specializes in behavior change and public participation. Led by founder and CEO David Gershon, EI develops community-
based behavior change programs and builds the capacity of local non-profits
and government agencies to implement them.[3]

The City of Portland's visionary Commissioner of Public Utilities, Mike
Lindberg, initially engaged EI to deliver its Sustainable Lifestyle Campaign.
He was not only its advocate, but participated in the program himself.
He said this about it: "You have the entire household involved in a voluntary
way instead of having a program that is mandated by the government. This
is at the most grassroots level possible, and that makes it more effective."

This initiative enables municipal agencies to help citizens use natural
resources more efficiently. The campaign operates on the premise that
35–85 percent of a community's natural resources are used at the household
level – and that up to 75 percent of those resources are wasted through
inefficiency and lack of awareness.[4]

With its emphasis on grooming savvy citizen resource stewards, the
campaign strives to lower household utility bills and reduce municipal
government service-delivery costs. The Institute provides participants with an
accounting of financial savings resulting from their participation – savings
EI refers to as relief from an "ignorance tax."

The linchpin of the Sustainable Lifestyle Campaign is the Household
EcoTeam Program, which has been adopted by dozens of cities and over
150,000 individuals in the United States and Europe (see also Michaelis,
Chapter 16, this volume). An EcoTeam usually comprises five or six
neighborhood households that meet eight times over a four-month period,
helping each other – with the aid of a step-by-step workbook and a trained
volunteer coach – to reduce waste, pollution, and water and energy use;
to become environmentally conscious consumers, and to bring additional
neighbors into the fold.

"The program is designed," says EI, "to help households systematically
evaluate their environmental impact, learn of actions they can take to lower
it, set up a support group to help them follow through on the choices they
make, and provide feedback to positively reinforce the benefits of the actions
taken so they are sustained over time."[5]

EcoTeams achieve significant average annual resource savings: 35–51
percent reductions in waste-stream garbage; 25–34 percent reductions in
water use; 9–17 percent in energy used; 16–20 percent in transportation
fuel use; and financial savings in the hundreds of dollars per household.[6]

In addition, the process of forming teams and following through on goals can foster social cohesiveness in participating neighborhoods.

EI points out that by refining its management of natural resources and improving local environmental quality, a community may increase its appeal as a place in which to live and work. Strengthening the character and attractiveness of neighborhoods may slow flight out of the community. And increasing the efficiency of citizens' resource use may defer the cost of major infrastructure projects such as water treatment plants and landfills, freeing up funds for other community development projects.

Lang Marsh, Director of Oregon's Department of Environmental Quality, was another strong proponent of bringing EI's environmental behavior change program to Portland. He noted that "We see [this program] as a significant opportunity to achieve citizen behavior change, which has been one of our most difficult challenges in advancing environmental protection."

Success in forming over 200 effective EcoTeams led OSD to request that Empowerment Institute adapt its program methodology specifically for reducing CO_2 emissions – an explicit Portland goal since 1993. EI collaborated with OSD's technical staff to create a tailor-made workbook for this purpose, and thus was launched the "Cool Portland" campaign and the "Low Carb Diet."

Knowing vs. doing

As human beings . . . we can only handle so much information at once.
Malcolm Gladwell (2002: 176)

One essential element of the program was the careful minimizing, selection, and packaging of information conveyed to participants. In observing a variety of municipal citizen-education projects, EI researchers had concluded that while a community may increase environmental awareness via glossy brochures, financial incentive programs, and access to information, these approaches usually fail to engender behavior change. "Citizens are generally willing to cooperate," said Gershon, "but they have a hard time changing ingrained habits."

Overwhelmed with extraneous information, residents may simply bog down in fear or guilt or confusion, and give up on the possibility of making a difference (see Moser, Chapter 3, this volume). Or they may come to see information as an end in itself: "We've observed that information by itself can be an undermining factor in getting people to act," said Gershon. "People think that if they've thought about something, they've *done* something."

Particularly in the case of global climate change — "an issue," said Gershon, "that seems almost unmanageable, out of control" — one element of what citizens need is to have the information broken down into bite-sized actions. The instructions for these actions need to contain concise facts about individuals' role in the problem and step-by-step guidance for how to adopt the new behaviors and practices (see also Tribbia Chapter 15, this volume).

The 31-page Low Carb Diet workbook was designed according to these principles. Attractively illustrated on every page with cartoons and with graceful photographic images of water lilies, it spends just three short paragraphs summarizing, in stark terms, the overarching problem of global climate change ("*the* major environmental threat facing our planet"). By the top of the second page, it is pitching Portland's prospects as a world leader in reducing global climate impact:

> It is now up to the citizens of Portland to take moral leadership by making the lifestyle choices that will lower their CO_2 emissions. If enough citizens step forward it will be noticed and spread to other communities.... If ever there was a time when a community and its citizens could make a difference in the world — this is the time and Portland is the place!

In EI's behavior change workbooks, what Gershon calls the "why-act" for each action is kept to a minimum. "If we can't capture the heart of the matter in an initial sound-bite," he said, "we won't be successful in engaging someone to take action." Recipe-style instructions and checklists for action emphasize doable steps and clear targets. In keeping with its weight-loss analogy, the Low Carb Diet assigned CO_2 savings values to every recommended action.

Like members of a weight-loss club counting calories, participants were able to track and quantify their progress. They could visualize, said Gershon, "where they currently were on the American continuum: from a climate-neutral profile — nobody can really live that way unless they purchase carbon credits — to the high end of the American profligate 100,000-pounds-of-CO_2 lifestyle."

Participants cited the workbook as one of their favorite aspects of the Diet. "I love the workbook," commented Portlander Sergio Diaz. "It's easy to follow with all the information in one place. I thought we would be overwhelmed with information, but we weren't. It's clear and concise. The way it is put together sets up a good challenge."

"It's a pretty cool thing to know your CO_2 footprint," noted participant John Wadsworth. "Bringing my daughter (age nine) to one meeting helped her get on board for a five-minute shower. This inspired me to look into solar hot water heating, and in the normal course of things, I wouldn't have done that."

But a well-designed workbook is only one component of a successful behavior-change program. Participants also need a peer-support network to motivate them to take the actions and celebrate changes in longstanding habits.

Preaching by the choir

Participants were not only supported to make changes in their own lifestyles; they were encouraged to reach out to others to do the same. The rationale is based upon the notion that to truly make a difference, we need to "be the change" and engage others alongside us. Participants were encouraged therefore to pass along their knowledge and enthusiasm to prospective participants.

In helping Portland bring participants aboard the Cool Portland campaign, EI drew on its expertise in the field of social diffusion. Scholars in this discipline attempt to identify what it is that enables new ideas, behaviors, values, technologies, products — all kinds of innovations — to spread through populations. The doyen of the field is Everett Rogers, a Stanford social scientist who has studied 1,500 cases of innovations and their dispersion for the last couple of decades.

Social diffusion's take-home message is straightforward. Innovations do not ripple out evenly across city blocks, apartment buildings, boards of directors, neighborhoods, conference attendees, or any other sort of population. Key to the successful dissemination of an innovation is a category of people that Rogers calls "early adopters": people who are attracted to the innovation and who have a high tolerance for experimentation. If such individuals make up some 10–20 percent of a given population, said Gershon, then "you began to hit a tipping point, a critical mass point where if the innovation is going to take off, it will start diffusing on its own momentum."

"Rogers says that the way the innovation diffuses is from peer to peer, neighbor to neighbor, not outward from an expert," said Gershon. "What I began to learn from that, and to apply, was how to choose my targets for the early adopters. The key is not to try to get everyone anywhere to do something, but to let them come to you. The early adopters self-select. You want to preach to the choir *because the choir will sing loud enough to get everybody into the church*."

Early adopters, or "team initiators," for the Low Carb Diet program came largely from lists of Portland activists compiled at sustainability-related public events. These individuals were trained to use simple talking points to invite neighbors to information meetings led by a paid staff person or a trained volunteer. Those who attended meetings and ultimately joined carbon-reducing teams were not necessarily self-identified environmentalists or sustainability advocates, EI's Gershon pointed out, but rather the neighbors of environmentalists.

The "next cutting edge"

Portland's CO_2 reduction campaign stands apart from other such efforts not only by dint of EI's behavior-change program, but also because of its explicit focus on climate change. For David Gershon, this distinction lies at the project's heart. "The idea that people could be part of something that's directly tackling climate change, the most important environmental issue we face, was very appealing to early adopters," he said.

The emphasis on climate impact represented a natural evolution for EI. "We began our [environmental] work in the early 90s," said Gershon, "when the major UN Earth Summit in Rio hadn't even yet occurred. At that point we decided to use the term 'environmentally sustainable lifestyle' before that phrase was in vogue. We were helping people understand that the environmental choices they made in living their life made a difference."

But the new millennium, in Gershon's view, brought a need for new language. He believes that the power of concepts such as "energy sustainability" and "sustainable lifestyle" to motivate behavior change has reached a kind of plateau — at least in some communities like Portland (but compare to the discussion in Watrous and Fraley, Chapter 25, this volume). "By focusing explicitly on climate change, we were able to appeal to people who were looking for the next cutting edge in environmental responsibility."

Gershon approached the campaign as a research initiative to determine whether citizens could "buy climate change directly" as a motivator, in place of other, seemingly more direct benefits such as cost savings, waste reduction, etc. (see Young, Chapter 24, this volume). "We knew our [EcoTeam] methodology would work," he said, "if we could get people behind the notion of climate change."

What he and his colleagues discovered was that with climate change as the primary motivating issue, neighbor-to-neighbor recruitment yielded 20 percent higher participation rates than Portland's previous EcoTeam campaigns. And the participants not only met the city's original goal of a 10 percent reduction in residential CO_2 emissions; they more than doubled it.

Beyond the "early adopters"

Indeed, Low Carb Diet program participants evinced a degree of enthusiastic, effective participation that surprised and delighted Portland city staff. "It was quite remarkable, the lengths to which people went, and how much each individual household was able to do," said Michael Armstrong, an OSD management analyst who developed the Portland-specific calculations for the Low Carb Diet workbook and served as the city's main liaison for the project.

In a pilot-project evaluation report published in spring 2002, Portland OSD Manager Anthony Roy (a former EI employee) wrote, "It was satisfying to discover so many knowledgeable citizens ready to assist with teams. When people knew the truth about the issue, they wanted to help."

"The people who actually got on the train were really enthusiastic," said Armstrong. "Others said, 'that's great, but not for me.'...I think I agree with Gershon that [the Low Carb Diet] can work....My suspicion is it works great for some people — more than I would initially have guessed — but not at all for others."

While the Low Carb Diet experienced tremendous success in its first 30 days, it is unknown what the limits are to a scaled-up energy conservation campaign based primarily on climate protection goals. The program didn't continue long enough to see how people were "drawn into the church by the choir." As Roy speculates, time and competing priorities may limit how active the "later adopter" category might be. Different approaches may be needed for those initially more reluctant audiences.

"For the program to go to the next level, I think it needs to create a way to make global warming tangible," said Anthony Roy. "For the community to rally around the goal of x [total] tons of CO_2 reduced, there needs to be a directly discernible benefit to the participant for the program to succeed. For instance, each participant gets their name carved on a statue or work of art that is created to symbolize Portland's foresight and its efforts to reduce global warming."

One additional possibility for expanding the participant base, suggested Michael Armstrong, is to consider reaching out not only to neighborhoods but also to organized communities such as religious congregations (see also Bingham, Chapter 9, this volume). He provided as an example the

"1–2–3 Campaign," a CO_2-reduction program spearheaded by Portland-area clergy. "1–2–3" organized churchgoers to reduce their household emissions through three simple actions: reducing thermostat settings by about one degree, reducing driving speed by about two miles per hour, and replacing three incandescent light bulbs with compact fluorescents.

Complementing the basic methodology of neighborhood teams and empowerment for behavior change, such strategies may indeed prove a powerful combination for addressing climate change.

Beyond Portland

Communities that contemplate following Portland's example may be initially daunted by that city's exceptional record for commitment to energy conservation. In 1993, Portland became the first US city to adopt a strategy for reducing CO_2 emissions (see also duVair *et al.*, Chapter 27, this volume). The city's aggressive, multifaceted Local Action Plan on Global Warming has significantly reduced per capita greenhouse gas emissions through a combination of electricity-conservation efforts, waste-biogas fuel-cell power, use of hybrid-electric vehicles in city fleets, replacement of incandescent traffic signals with LED bulbs, and other programs.[7]

Portland's environmental "edge" notwithstanding, however, the city's sustainability experts see the Low Carb Diet as easily transferable to other communities. "I think it's *eminently* transferable," says OSD's Michael Armstrong. "I think it could work anywhere. I think it would work in *Houston* – for some people. I think it's important to acknowledge that it's not going to work for everyone, but it does work for a certain slice [of the population]. The size of the slice may differ from one community to another" (see various community experiences in Young; and Watrous and Fraley, Chapters 24 and 25, this volume).

No matter where they live, activists who launch the Low Carb Diet in their own communities will require some financial resources. In Portland, a US$50,000 foundation grant from the Meyer Memorial Trust secured by EI covered the cost of the pilot program, including staffing, and preparation and production of the workbook, while the city provided office space and staff labor at about 10 percent time over a period of several months. As Michael Armstrong noted, the Low Carb Diet required "a lot of work getting team leaders and coaches up to speed and equipped with the materials they needed."

But needed resources can come from a variety of sources. Much of the support for Portland's pilot project came in the form of in-kind donations from businesses, government agencies, and non-profits. Nine organizations in all, from the federal Environmental Protection Agency to the Portland public transportation network, donated resources, time, and information.

One of the challenges to any new program is transitioning to long-term financial support or institutional practice. City budgets are limited and pioneers must often be creative at finding financial mechanisms to ensure longevity of the program (see Young, Chapter 24, this volume). Funders that make grants available for piloting a new program may not entertain requests for long-term support. In addition, raising financial resources from granting organizations can be time-consuming and uncertain.

"Knowing how to scale [the program] up is a relatively straightforward process," said EI's Gershon. "We tried...but it wasn't a high enough political priority for the City at the time." But a citizen groundswell can influence policy priorities, and Gershon believes that participation in an effective program such as the Low Carb Diet can shift the political seas. "A lot of policy-makers try not to address problems where there's no workable strategy or program in sight," he said.

The Portland initiative provides a clear vision for such a strategy — one that has the power not only to inspire activists, but to "help policy-makers think more imaginatively and boldly" about serious programs for reducing climate impact. "Once you get individuals changing their lifestyle," he said, "they become advocates and start to look for policy-level changes. People become part of advisory councils and start influencing decisions at the municipal level."

A call to early adopters

> The good news is that if people are the problem, people can also be the solution.
>
> *(David Gershon, founder and CEO, Empowerment Institute)*

David Gershon describes the opportunity this way: "Despite its global reach, climate change is very much a local issue. The causes of global warming come in large part from the everyday actions that take place in our communities, with individuals accounting for almost half of a community's emissions and a third of America's total emissions. While progressive local governments and businesses have developed strategies and programs to lower their CO_2 emissions, individual citizens have not."

The Low Carb Diet offers a way to bring individual citizens into the mix — and even if the program does not engage the majority of a city population, it enables those who do come on board to help create a culture of possibility for everyone else. Says Gershon, "I would hope that the success of our Portland pilot program motivates early adopters to say, 'you know what, I'm going to do this — build a groundswell. You have to start somewhere. Why not my block? Why not my community?'"

Notes

1. This chapter is based on various interviews with people involved in the neighborhood campaign described here, as well as additional information from Empowerment Institute and the City of Portland.
2. Figures are from Empowerment Institute's project summary. These self-reported data were gathered from participating households over a period of 30 days and extrapolated to one year.
3. Empowerment Institute has designed programs for dozens of cities in the United States and Europe on topics varying from environmental sustainability to neighborhood revitalization to emergency preparedness. The Institute offers its services to local and national government agencies, non-profit groups, and corporations. See EI's website at: http://www.empowermentinstitute.net/Default.htm; accessed January 10, 2006.
4. EI derives these figures in part from the "factor four" concept developed by Amory Lovins. "Factor four" posits that resource sustainability can be quadrupled using existing conservation technology.
5. See http://www.empowermentinstitute.net/files/SLC.html; accessed January 10, 2006.
6. EI calculates these percentages based on self-reported before-and-after resource-consumption data provided by participants.
7. From http://www.sustainableportland.org/default.asp?sec=stp&pg=glo_home; accessed January 10, 2006.

Reference

Gladwell, M. (2002). *The Tipping Point: How Little Things Make a Big Difference.* New York, NY: Back Bay Books.

20

Changing organizational ethics and practices toward climate and environment

Keith James
Portland State University

April Smith
Colorado State University

Bob Doppelt
University of Oregon

Introduction

> Business organizations play an increasingly important role in the
> world ... Companies that win the public's confidence and trust are
> open, visible, engaging and create business value while delivering
> benefits to society and the environment.
> *William Ford, Chairman and CEO Ford Motor Company*[1]

Rapid technological changes, pressures created by globalization, competition, and increasing resource constraints make the management of organizational change a critical issue for organizations of all types. The scientific study of organizational change is an interdisciplinary endeavor and it has grown in step with the practical importance of change to organizations. Only a limited part of the science of change literature has, however, directly addressed how to change organizations in ways that would benefit the natural environment, in general, or the Earth's climate, in particular. Changing organizations to help address climate change is the focus of this chapter.

The force-field model of change

At any given time, any mature organization is in reality an epiphenomenon of a dynamic balance between its connections to the outside world and its internal components, processes, and functions. For any attempted organizational change, there will be internal and external forces that both support the change, and ones that push against it. For a change to occur, therefore, something must move an organization from its current state of balance to a desirable new and relatively sustainable dynamic balance. We will argue that achieving new and environmentally friendly (hereafter "green") organization states requires broad and deep change in organizations' culture and systems.

Kurt Lewin's (1951; see also Weick and Quinn, 1999) force-field model of innovation and change is an influential approach to studying organizational change. Lewin suggests that organizational change requires that existing dynamic balances of restraining forces and driving forces be disrupted and the organization guided to a new dynamic balance that better supports the outcomes desired. To do so, a force-field analysis needs to be done to intervene effectively in the field of resistors and drivers. A force-field analysis describes the motives and compulsions currently in play in the organization, along with other motives and compulsions that might help sculpt a new dynamic balance. The type and strength of balancing forces will, however, be unique to each particular organization, and they are likely to change during the course of any innovation effort. In fact, since the parts of a complex organization can change at different rates from each other, the organization as a whole is likely to have multiple levels and currents of environmental orientation at any given time. Consequently, a change agent needs to periodically reanalyze forces and potentially modify the change strategy.

Although Lewin did not specifically discuss the problem of transforming organizations for environmental purposes, his research identified some of the common problems that occur in achieving organizational change of any sort. Figure 20.1 lists force types that are most likely to be important in shaping organizational impacts on climate and environment. Our figure modifies Lewin's approach to create a list of the likely sticking points (i.e., resistors) in attempts to change organizations to make them more climate-friendly.

Change resistors and drivers

Change reticence is a natural human response. Moreover, resistance can be healthy. It can force innovators to think carefully through a proposed change and to address legitimate concerns. Thus, evidence of resistance should be taken not as evidence of failure but as an opportunity for change leaders to learn about the dynamics of the organization and improve their concepts and approach, as well as an opportunity to engage and educate

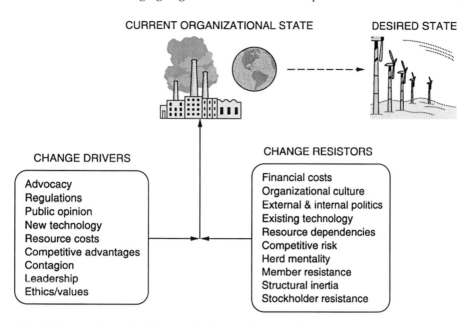

Fig 20.1.   A force-field model of organizational change.
Source: Based on work of the lead author.

resisting organization members. Below we discuss some of the factors in greater detail.

Organizational culture, regulation, public opinion, politics

The factors in this section are grouped together because they are all macro-level influences and most are closely interrelated. Almost all organizational innovation efforts turn into a political process because they affect entrenched interests, individual and group status and identities, and personal and collective values (Frost and Egri, 1991). Organization members who understand and are invested in an existing system will understandably feel anxious about a new system.

One strategy for minimizing political resistance is, therefore, to link the innovation to the values and identities of powerful factions, and to sell it in terms that show it to be a continuation of the company's existing culture and self-image (James, 1993; Bono and Judge, 2003). One public sector organization, for instance, added a "green" (i.e., vegetated) roof to one of its buildings in large part because it prided itself on its aesthetic qualities and was persuaded that the green roof would significantly improve the building's beauty. Similarly, communicating a proposed change in exciting, compelling visionary ways can reduce resistance/increase support (Nadler and Heilpern, 1998). Thus the "pitching" of an innovation, and internal constituency-building and "public relations" help determine success.

External politics also affect internal political resistance. Climate change interests and advocacy groups have typically looked to government as the main mechanism for influencing the behavior of private businesses. Public opinion has, in turn, long been a linchpin for influencing both governments to impose regulations, and businesses to accept them (Sheffrin, 2000). Economic globalization and internal US socio-cultural change have made government both less likely and less able to intervene on environmental issues, however. For instance, globalization allows corporations to shift production outside of countries with restrictive regulatory programs while still accessing the markets in such nations. Corporate interests and allies have become adept at co-opting government leaders and public discourse about the environment. By linking regulation to other public concerns about government officiousness and inefficiency, corporations opposed to environmental regulation have been able to focus the debate around purported negative impacts on the economic well-being of the middle and lower classes (Farey and Lingappa, 1996; Hummel and Stivers, 1998). Moreover, corporate and other anti-regulation interests have also gotten better at framing the debate in terms favorable to their positions (e.g., Rivera and de Leon, 2004; Tilson, 1996; see also McCright, Chapter 12, this volume). Meyer's chapter in this volume (Chapter 28) addresses approaches to using extra-organizational politics and public opinion to support environmentally-friendly organizational change.

Technology

Making a company environmentally friendly can involve changing technology. In some cases, the technology to make green changes exists but is still new enough to be impractical or too costly to implement (James, 1993). Moreover, it takes time, energy, and resources for an organization and its members to adopt new technologies. New technology also displaces experts on the old technology, who will consequently resist technological change for fear of losing power (or their jobs). Some green change efforts may therefore need to minimize technological change or piggy-back on other pressures toward them. For instance, organizations that take advantage of new technology early may gain marketing advantage or influence on industry standards (see Arroyo and Preston, Chapter 21, this volume). Tying environmentally-friendly technological changes to other benefits such as these can be used to support green change. External drivers such as government tax incentives can also help overcome technological resistance within organizations. Providing time, training, and other resources in support of implementing a new technology also helps reduce resistance to it.

Economic resistors, competitive risk, and stockholder resistance

Managers and stockholders are most concerned about the short-term bottom line and that can make them resist green innovation. Innovating in a climate-friendly direction can be initially expensive, especially since green technologies and resources are still relatively expensive. Moreover, potential savings and profits from green change cannot be guaranteed even though they *will* often occur (see the chapters by Atcheson, and Arroyo and Preston, Chapters 22 and 21, this volume). Globalization also militates against organizations' acceptance of changes that impose costs. National or regional costs from environmental regulations yield resistance from organizations if global competitors are not subject to the same costs. Organizations that successfully implement green change are likely to have market security and slack resources that insulate them, at least temporarily, from competition and other forces that can create pressure toward minimizing new costs. However, successful organizations also tend to stick with the strategies that brought them success until there is very compelling evidence that it needs to change. The risks of innovation weigh more heavily in decision-making by successful companies than do its potential benefits (Rajagopalan, Rasheed, and Datta, 1993). Change leaders may, therefore, need to focus attention on looming crises with existing strategies and strengthen perceived benefits of change to overcome the inertia of past successes.

Structural inertia, herd mentality, and contagion

Structural inertia can stem from organizational leadership, organizational designs, governance systems, internal resource allocations and other entrenched elements of an organization's existing system. The whole existing internal structure of an organization creates mass inertia that has to be overcome for any major change to occur (Senior, 2000). Organizations also tend to imitate each other (herd mentality) (Zwiebel, 1995); it is difficult, therefore, to get any organization to adopt a new approach unless many organizations adopt that approach. The combination of herd mentality, structural inertia, and risk aversion means that it will often be relatively new organizations that will be most likely to be major pioneers (Ogbonna and

Harris, 2001). Because they usually cannot compete with existing organizations in established markets or on established technologies, they have to try to find or create a new niche or gain an advantage through new technology. Structural inertia also means that successful green change in an existing organization generally requires systemic organizational change. Finally, the herd mentality could be an advantage if the herd could be encouraged to stampede in a green direction.

The flip side of herd mentality as a resistor to change is contagion as a trigger for it. Contagion is a change driver that can be difficult to generate but can be powerful if activated. Contagion refers to an infectious adoption of an approach or technology that somehow comes to seem the only way to go. It is a momentum for change that makes it seem inevitable and that can sweep resistance aside quickly and thoroughly. Rogers (1983) discusses it at the individual level of technological innovation in describing how the right group of early adopters who are well connected, charismatic, and influential can trigger a chain-reaction of adoption of an innovation that is almost like a fad. The same can occur in the social climate within which businesses operate. In fact, as with the "e-business" bubble of the late 1990s, at times contagion does become a fad and can affect individuals and organizations all at once. Governments, non-profits, and concerned individuals may be able to aid contagion toward green organizational policies and technologies by stimulating, seeking out, and supporting key early adopters, be they new organizations or the unusual existing ones that are able to think and move independently of the herd. Organizations with strong and relatively unique visions for the future may do this best (Baum, Locke, and Kirkpatrick, 1998). As is discussed in more detail below, transformational/charismatic leaders are important to promoting compelling organizational visions. The focus needs to be on those organizations that have the critical connections or the cachet to potentially spark contagion. Sutton's (1997) model of *Catalyst Organizations* identifies the features of organizations with the connections and cachet to exert major influence on other organizations.

Resource constraints

The availability and affordability of key resources is always a key driver of change within an organization. Scarcity can induce a move toward greater resource use efficiency and/or technological change. Fossil fuel and water costs, for example, are projected to increase in coming decades, and shortages of both are possible. Problems with current resources will promote the search for alternatives or conservation measures will help drive proactive innovation.

Pressure from employees and consumers

Consumers sometimes vote with their dollars and, by shifting what they will accept or prefer, push organizations to change. For example, Interface, one of the world's largest producers of commercial flooring, became focused on the environment only after customers began to question the firm's environmental policies (Doppelt, 2003, see also Atcheson, Chapter 22, this volume).

Employees can also create internal pressures on organizations to change. Their opposition to environmentally harmful practices and creativity in developing and supporting green ones (e.g., those that minimize pollution and greenhouse gas emissions; promote environmental justice; use renewable resources; discourage urban sprawl) have the potential to help drive change. When that happens, employees essentially become change leaders. Thus, the discussion of leadership in the section below applies to them as well as to individuals with official organizational leadership roles. At the extreme, employees create public and regulatory pressure through whistle-blowing. Thus, visions of sustainable and beneficial organizational roles toward the environment can be developed by, and need to be addressed to, workers at all levels, not just formal organizational leaders.

Leadership, vision, ethics, and values

While the wrong type of leadership can inhibit innovation and environmental responsibility in organizations, the right type of leadership is crucial in securing the success of organizational change. Research indicates that the presence of an innovation champion (or transformational leader) is the most important factor in success of an innovation (Howell and Higgins, 1990). Leaders can promote innovation by scanning the outside environment for innovative ideas and bringing them into an organization; by stimulating and promoting innovations within an organization; or by providing resources for innovation and protecting them from attacks by entrenched interests.

Transformational leaders promote innovation. They are inspirational, politically astute, and excellent communicators who engender employee support and loyalty by tying individual empowerment with collective organizational goals (Bass, 1996). *Spiritual leadership* can also aid change efforts according to a growing body of organizational-behavior research (for a review see Fry, in press). A spiritual leader in an organization is one who cultivates a climate of ethical responsibility, concern, and humanity toward others, and empowerment and creativity among employees (Fry, in press). They are sometimes spiritual in the religious sense, and sometimes simply emphasize moral ideals without invoking religious concepts. In either case, there is evidence that this type of leadership can effectively motivate many workers to commit strongly to the leaders' vision for the organization. A spiritual style of leadership may be particularly useful for leading all employees to work toward realization of the green (i.e., environmentally friendly and ethical) company as a salient ideal.

Crisis and unintended consequences of change

Crisis is a powerful disruptor of existing organizational dynamic balances, and leaders, especially transformational leaders, sometimes try to create a sense of crisis for that very reason. In spite of organizational leaders', managers', and consultants' best efforts to systematically alter organizations based on rational analysis of changing environments, opportunities, and risks, the reality seems to be that major change to organizations' systems, strategies, or policies rarely happen without the stimulus of a major organizational crisis (King, 1990).

Crisis-driven change must be carefully managed, however, or it can easily go wrong. Change rarely results in a clean movement from an organization's current state to a consciously selected target state. The complexity of change makes peregrinations and unanticipated new states likely. Even if the intended target state is reached, it will almost always yield unintended consequences. The direction and configuration of the organization that will result from crisis-generated change is especially likely to differ from what was planned or expected. Green change leaders must, then, continually assess progress and reassess their strategy. They must use creativity to strengthen and broaden promising trends, while weakening the counter-vailing flows.

The force-field approach is useful for presenting the general literature on organizational change and linking it to the specific issue of changing organizations to be more environmentally friendly. *The key point to note is that the performance of an organization is the emergent product of the interaction of its parts.* The bottom line is that the process of attempting green organizational change is complex, artful as much as scientific, and requires flexibility, creativity, and attention to the multiple constituencies and the often unintended consequences of change-related actions. In the next section, we present an approach to *the applied implementation* of organizational change for sustainability that has shown promise among some early adopting companies.

An approach to organizational change: The wheel of change toward climate and environmental sustainability

Organizations that succeed with going green start in different places and sequence their interventions in different ways, but all tend to follow some version of the approach depicted in Figure 20.2 as the *Wheel of Change Toward Sustainability* (modified from Doppelt, 2003). We then describe each step in some detail and provide examples.

Create a sustainability imperative

Undermining an organization's controlling mental models — what Lewin calls unfreezing the organization — is necessary for development of green

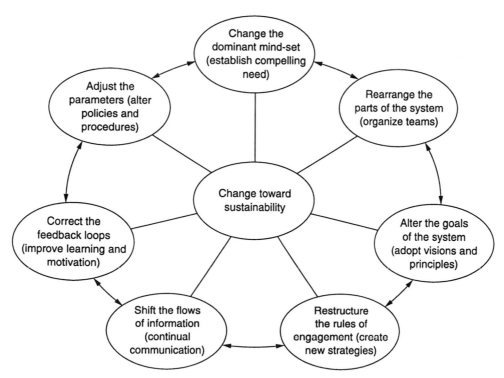

Figure 20.2. Wheel of change toward sustainability.
Source: Doppelt (2003), reprinted with permission from the author.

thinking and practices. That is not easy. A major crisis or strong, long-term charismatic leadership is usually required. For instance, Ray Anderson at Interface challenged employees to think green in an emotional speech and followed it with months of change messages (Doppelt, 2003).

Create a vision for a sustainable state and set goals for achieving that vision

Clear missions and goals for an organization, along with guidelines for how to achieve them, are needed for green change. Those goals and missions must be presented as a compelling vision. Then, a sustained effort over time is required to ensure that they are pursued. For instance, adoption of a sustainability goal such as company-wide use of energy from renewables needs to be followed by supportive decision-making about specific purchases, processes, and practices.

Organize transition teams to chart a course toward sustainability

Rearranging the parts of an organizational system requires broad involvement of organization members and key external stakeholders. A variety of short-term sustainability "transition teams" can be organized. The dynamics of structuring and operating planning/creative teams is a complex issue in itself that has been the subject of a sizable literature. It is beyond the scope of the current work, but the interested reader is referred to Paulus, Larey, and Dzindolet (2001). Teams must change as people analyze, plan, and implement a change strategy toward an organization that supports environmental sustainability. Each team must be clear about what it is striving to achieve, the role of each member, and the rules that will guide its operations.

Alter the processes and rules within the system

Systemic change requires altering the rules about how work gets done in the organization. Successful green-change efforts tend to focus on both operational and governance rules. For instance, in the early 1990s, the Xerox Corporation adopted the vision of becoming "Waste Free." The vision catalyzed changing operations "all the way back to the initial designs." In addition, decision-making was decentralized. By the end of 2001, the equivalent of 1.8 million printers and copiers were reused or recycled, and many other environmental initiatives were under way (Maslennikova and Foley, 2000).

Communicate and reinforce the need, vision, and strategies for achieving sustainability

Even with approaches 1–4, progress will stall without the effective and consistent exchange of information. A constant exchange of information by all employees, up, down, and laterally in the organization's structure about the environmental initiative's need, purpose, strategies, and benefits will promote sustainable change.

Institute mechanisms to stimulate learning and innovation

Even with excellent strategies, obstacles will surface. To overcome them, feedback and learning mechanisms must be put in place or broadened and strengthened so that the identification of sustainability opportunities, the vision, planning skills, any specialized knowledge and expertise needed to implement selected initiatives are developed and shared. Organizations successful at change provide frequent, accurate, timely information on progress and setbacks toward goals and vision. They also reward experimentation and share its lessons.

Make sustainability standard operating procedure

As a change initiative progresses and new practices emerge, the new dynamic must be institutionalized. Efforts must be made to embed environmental sustainability in standard operating procedures and culture. This will be an ongoing process. Change toward sustainability is iterative. The *Wheel of Change* must continually roll forward. The case study of the Herman Miller Company (Textbox 20.1) shows how one company used the steps and strategies just outlined to move toward being a more sustainable company.

We have detailed the internal process of change of organizations toward states that are friendly to the natural environment. The raw-material specifications for suppliers mentioned in the Herman Miller example began to touch on how changes in one organization can begin to influence others. We expand on this idea with a description of an organization that is modeling for and catalyzing changes in many others.

Textbox 20.1 **Steps toward sustainability at Herman Miller Company**

The Herman Miller Company (www.hermanmiller.com) designs, manufactures, and sells furniture systems and products for offices and healthcare facilities. Herman Miller adopted a vision of becoming a sustainable business and a strategy of reducing and eliminating environmental impacts throughout the entire company. The goal was to identify problem areas and incorporate environmentally friendly materials and manufacturing processes into new product designs. Herman Miller adjusted its governance system to further empower employees to take responsibility for the environment. Decision-making was decentralized to the lowest level possible. People were rewarded when they produced better-quality products with reduced environmental impacts. Teams were formed to identify potential environmental impacts of materials and products and to develop green alternatives. The Design for the Environment (DFE) product development team, for example, started the process by "coming up with their own specifications, such as don't glue things together if you can use screws." Institutionalization occurs, for example, by incorporating green approaches into a standard checklist for product development. The checklist is also provided to suppliers so that the company can influence the materials in the feedstocks it purchases. Herman Miller President Paul Murray says: "Because we share the benefits, every employee benefits from increased sales and reduced costs due to less energy use and packaging and other savings our environmental programs produce . . . . Total environmental savings on energy and packaging alone is conservatively estimated to be in the millions of dollars" (cited in Doppelt, 2003:168).

Sutton (1997) presents a framework of types of organizational environmental efforts. His framework starts, at the lowest level of change, with Ethical Opportunist organizations, which is a focus on using supposedly environmentally friendly policies and actions mainly for marketing and public-relations purposes. A Pioneering organization (e.g., Toyota and its Prius) finds a way to make an innovation technically or economically feasible for itself. The most advanced companies in this scheme are Catalysts, which both adopt their own, genuine, environment-friendly policies and practices and also act in ways that encourage other organizations to adopt such policies and practices. Change tends to progress in fits and starts that produce a closer and closer approximation of the target state over time. Thus policies and practices toward the environment, if they change, are likely to progress through some sequence of the stages that Sutton proposes rather than flash directly to the Catalyst state. The case study of

---

Textbox 20.2 **Catalyst of wider change**

Silran[2] Homes (not the company's real name) started in the 1960s as a standard home-building and community development company. In the late 1980s, however, its founder and president became interested, first, in solar energy technologies and then, later, in ideas for creating sustainable communities. Now Silran concentrates on building homes and communities that are largely energy self-sufficient and low in greenhouse gas impacts. Homes in one Silran development, for instance, have four-foot thick rammed earth walls that insulate well enough by themselves to keep average internal temperatures within a 10° Fahrenheit range all year round. All of the homes are also equipped with photovoltaic power cells for electricity generation, as well as passive-solar heat collection designs and heat deflection landscaping for summer. The home designs not only allow buyers to live in energy and environment-friendly ways; they also provide business to solar technology companies and to subcontractors skilled at environmentally friendly building techniques. The community in which those homes sit also has bike and pedestrian pathways that link to retail establishments. That makes if possible for residents to acquire at least some necessities and creature comforts without driving. Because they have a strong base of environmentally conscious customers, many of the businesses in the retail sector tend to feature environmentally friendly products. Thus the Silran approach catalyzes green-friendly changes involving individual customers, corporate suppliers, and retail establishments.

Silran Homes (Textbox 20.2) describes what a catalyst organization can potentially achieve.

Conclusions

Change leaders in organizations, advocacy groups, and government oversight bodies often focus on the driving forces that push change. However, those who have studied organizational change indicate that most failures of change efforts result from unmanaged *resistors* more than weakness of drivers. An obvious conclusion would seem to be that those who want to create green organizations should focus as much or more on weakening or removing the restraints that are keeping an organization's environmental approaches fixed in their current state.

Organizational change is difficult to achieve and change management is a complex processes in any case. This is especially true relative to efforts to get organizations to alter their policies and practices toward climate change

because future potential economic benefits of doing so are pitted against
immediate, bottom-line pressures and costs. As Doppelt's (2003) research
showed, climate advocates and climate scientists need to present visionary,
positive messages; find the best possible organizational leaders and
organizations' allies, rather than waiting for ideal allies; be flexible and use
creative approaches to provoking and guiding organizational change;
and recognize that change is an ongoing process that can result in unintended
consequences. Clearly, there is substantial room to increase the extent to
which private sector corporations and market forces support and contribute
to environmental sustainability. In the reality of the globalized, capitalistic
modern world, market pressures have increased significantly in most
economic sectors, so marshalling other market forces that support sustainability is probably critical to move companies toward green processes and
practices. Climate change appears chaotic on the surface but changes in it are
linked by identifiable underlying patterns. The same seems to be true of
change in organizations toward sustainability. By seeing organizations as
dynamic, chaotic systems and analyzing the patterns of flow and creatively
wielding our paddles, we may be able to guide organizations and the Earth
through the challenge of climate change.

Notes

1. "Inauguration of The Global Reporting Initiative," April 2002, at the United Nations,
   New York City. Quoted in the *Sustainable Investment Research Institute Newsletter*
   (28), August 16.
2. "Silran" is a pastiche that is based on two different home and community development
   companies. It is an original case study composed specifically for this chapter based on
   the first author's personal involvement with the two companies.

References

Bass, B. M. (1996). *A New Paradigm of Leadership: An Inquiry into Transformational
    Leadership*. Alexandria, VA: US Army Research Institute for the Behavioral
    and Social Sciences.
Baum, J. R., Locke, E. A., and Kirkpatrick, S. A. (1998). A longitudinal study of
    the relation of vision and vision communication to venture growth in
    entrepreneurial firms. *Journal of Applied Psychology*, **83**, 43–54.
Bono, J. E. and Judge, T. A. (2003). Core self-evaluations: A review of the trait and
    its role in job satisfaction and job performance. *European Journal of Personality*,
    17 (Suppl. 1), S5–S18.
Doppelt, R. (2003). *Leading Change Toward Sustainability: A Change Management
    Guide for Business, Government, and Civil Society*. Sheffield, UK: Greenleaf
    Publishing.

Farey, K. and Lingappa, V. R. (1996). California's proposition 186: Lessons from a single-payer health care reform ballot initiative campaign. *Journal of Public Health Policy*, **17**, 133–52.

Frost, P. J. and Egri, C. P. (1991). The political process of innovation. In *Research in organizational behavior*, eds. Staw, B. M. and Cummings, L. L. Greenwich, CT: JAI Press: pp. 229–95.

Fry, L. W. (2005). Toward a theory of ethical and spiritual well-being, and corporate social responsibility through spiritual leadership. In *Positive Psychology in Business Ethics and Corporate Responsibility*, eds. Giacalone, R. A. and Jurkiewicz, C. L. Greenwich, CT: Information Age Publishing, pp. 47–83.

Howell, J. and Higgins, C. (1990). Champions of technological innovation. *Administrative Science Quarterly*, **35**, 317–41.

Hummel, R. P. and Stivers, C. M. (1998). Government isn't us. In *Government Is Us: Strategies for an Anti-Government Era*, eds. King, C. and Stivers, C. M. Thousand Oaks, CA: Sage Publications.

James, K. (1993). Perceived self-relevance of technology as an influence on attitudes and information retention. *Journal of Applied Behavioral Science*, **29**, 56–75.

King, N. (1990). Innovation at work: The research literature. In *Innovation and Creativity at Work*, eds. West, M. A. and Farr, J. L. New York: John Wiley & Sons Ltd, pp. 15–59.

Lewin, K. (1951). *Field Theory in Social Science*. New York: Harper & Row.

Maslennikova, I. and Foley, D. (2000). Xerox's approach to sustainability. *Interfaces*, **30**, 226–33.

Nadler, D. A. and Heilpern, J. D. (1998). The CEO in the context of discontinuous change. In *Navigating Change: How CEOs, Top Teams, and Boards Steer Transformation*, eds. Hambrick, D. C., Nadler, D. A., and Tushman, M. L. Boston, MA: Harvard Business School Press.

Ogbonna, E. and Harris, L. C. (2001). The founder's legacy: Hangover or inheritance? *British Journal of Management*, **12**, 13–31.

Paulus, P. B., Larey, T. S., and Dzindolet, M. T. (2001). Creativity in groups and teams. In *Groups at Work*, ed. Turner, M. Mahwah, NJ: Erlbaum, pp. 319–38.

Rajagopalan, N., Rasheed, A. M. A., and Datta, D. K. (1993). Strategic decision processes: Critical review and future directions. *Journal of Management*, **19**, 349–84.

Rivera, J. and de Leon, P. (2004). Is greener whiter? Voluntary environmental performance of western ski areas. *Policy Studies Journal*, **32**, 417–37.

Rogers, E. (1983). *Diffusion of Innovations*. New York: The Free Press.

Senior, B. (2000). Organizational change and development. In *Introduction to Work and Organizational Psychology: A European Perspective*, ed. Chmiel, N. Malden, MA: Blackwell, pp. 347–83.

Sheffrin, S. M. (2000). Regulation, politics, and interest groups: What do we learn from an historical approach? *Critical Review*, **14**, 259–69.

Sutton, P. (1997). The Sustainability-Promoting Firm: An Essential Player in the Politics of Sustainability. Paper presented at the Ecopolitics Conference XI, Melbourne, Australia, October.

Tilson, D. J. (1996). Promoting a 'greener' image of nuclear power in the US and Britain. *Public Relations Review*, **22**, 63–79.

Weick, K. E. and Quinn, R. E. (1999). Organizational change and development. In *Annual Review of Psychology*, eds. Spence, J. T., Darley, J. M., and Fos, D. J. Palo Alto, CA: Annual Reviews, Inc., **50**, pp. 361–86.

Zwiebel, J. (1995). Corporate conservatism and relative compensation. *Journal of Political Economy*, **103**, 1–25.

21

Change in the marketplace: business leadership and communication

Vicki Arroyo
Pew Center on Global Climate Change, Arlington, VA

Benjamin Preston
CSIRO Marine and Atmospheric Research, Aspendale, VIC Australia
(previously at the Pew Center)

In recent years, there has been a shift in the way companies are thinking about climate change and the policies needed to curb it. This change has been heralded by the popular press — including a cover story in *Business Week* in August 2004 entitled "Climate Change — Why Business is Taking it Seriously" (Carey and Shapiro, 2004). Climate change has also been featured in recent issues of *Fortune* (Stipp, 2003) and *The Economist* (2004). In addition, the influential Conference Board[1] recently issued a statement emphasizing the need for firms to evaluate their contribution to the problem, and a number of shareholder resolutions and lawsuits are driving companies to examine and explain their strategies and plans for conducting business in a carbon-constrained world (Bennett, 2004; Leahy *et al.*, 2004; Carlton *et al.*, 2004; see also Atcheson, Chapter 22, this volume).

This shift did not happen overnight. In the years leading up to negotiation of the Kyoto Protocol, a large number of companies participated in a vocal industry group called the Global Climate Coalition (GCC). This coalition opposed any binding restrictions on GHG emissions, questioning the scientific basis for concerns about human-induced climate change and arguing that climate policy would be costly and result in job losses (see McCright, Chapter 12, this volume).

Although the GCC helped to polarize the climate change debate, it was not necessarily representative of the entire business community's perspective on climate change (Climate Change Task Force, 1997). Other companies preferred to engage more constructively in hopes they could help shape policy outcomes. In 1998, the non-profit Pew Center on Global Climate Change was formed to forge a cooperative approach in addressing climate change: one that incorporates credible economic, scientific, and policy analyses while involving the business community in the search for both technological and policy solutions. A central feature of the Pew Center is its

Business Environmental Leadership Council, which started with 13 companies that pledged to reduce their own emissions and to promote pragmatic policy approaches. Specifically these firms agreed to four principles:[2]

1. We accept the views of most scientists that enough is known about the science and environmental impacts of climate change for us to take actions to address its consequences.
2. Businesses can and should take concrete steps now in the United States and abroad to assess opportunities for emission reductions, establish and meet emission reduction objectives, and invest in new, more efficient products, practices, and technologies.
3. The Kyoto agreement represents a first step in the international process, but more must be done both to implement the market-based mechanisms that were adopted in principle in Kyoto and to more fully involve the rest of the world in the solution.
4. We can make significant progress in addressing climate change and sustaining economic growth in the United States by adopting reasonable policies, programs, and transition strategies.

This advisory council of the Pew Center has since grown to 41 companies representing a wide range of sectors making a variety of products including fuel, timber, electricity, cars, and appliances. A number of these firms have taken on targets to reduce their emissions. Other NGOs have also partnered with companies to secure reductions (e.g., Environmental Defense and World Wildlife Fund) or to establish consistent reporting protocols (e.g., World Resources Institute and World Business Council for Sustainable Development, World Economic Forum, and the International Organization for Standardization). In February 2003, the United States-based Business Roundtable announced its Climate RESOLVE (Responsible Environmental Steps, Opportunities to Lead by Voluntary Efforts) initiative,[3] which challenges it members to adopt President Bush's voluntary emissions intensity reduction target. In March of 2004, members of the California-based Silicon Valley Manufacturers' group set a goal of reducing CO_2 emissions 20 percent below 1990 levels by 2010 (Rogers, 2004).[4,5] In the meantime, faced with increasing scrutiny of affiliated companies, the GCC experienced a number of defections during the late 1990s and ultimately shifted its membership from individual companies to trade associations in 2000. The Coalition disbanded entirely in 2002, claiming victory in the policy debate by citing the Bush Administration's decision to abandon the Kyoto Protocol (but more likely the result of waning membership).

This chapter examines the evolution of business positions on climate change and the interaction among businesses, consumers, and public policy.

We first draw from the experience of an earlier global agreement — the Montreal Protocol — to gain insight into how business positions evolved to support mandatory policy and how that policy was forged. In doing so, we discuss motivations of companies that acknowledge and take steps to address climate change — both in their own facilities and products and through their constructive engagement on policy development. Throughout, we also highlight the role of and implications for communication, among businesses, the public, and policy-makers in building momentum for addressing global climate change.

Leaders, innovators, and obstructionists: the role of businesses in change

To gain insight into the role that businesses may play in the issue of global climate change, it is helpful to examine the role of the private sector on analogous issues, such as the Montreal Protocol, which led to the international phase-out of ozone-depleting chlorofluorocarbons (CFCs). While climate change is an even more complex issue, the challenge of addressing CFCs shares a number of common elements: a global environmental hazard requiring international solutions, with the root cause firmly embedded in global industry and commerce. What is often overlooked in making such comparisons, however, is the fundamental role of the business community in the evolution of the CFC phase-out. The 1978 US ban on CFCs affected an $8–9 billion industry (Maxwell and Briscoe, 1997). Predictably, industry first responded negatively and attempted to head off future regulation. For example, in 1980, DuPont initiated the formation of the Alliance for Responsible CFC Policy, an industry group representing CFC manufacturers and consumers, which first lobbied against further regulation of ozone-depleting substances.

By 1985, however, increasing scientific evidence further implicated CFCs in ozone depletion, and the Convention for the Protection of the Ozone Layer had launched international discussions on CFC restrictions. Recognizing growing scientific evidence and policy interest signals as well as the potential market opportunities associated with being a leader in the development of CFC-free products and CFC substitutes, industry shifted its position (Maxwell and Briscoe, 1997). In 1986, DuPont, for example, announced that CFC substitutes could be developed within five years, given sufficient market and regulatory incentives, and the Alliance for Responsible CFC Policy voiced its support for international limits on CFC production. Following the Montreal Protocol, companies rapidly began examining ways for adapting to a phase-out of CFCs.

Industry innovation, particularly by DuPont, ultimately led to the development of substitutes, which helped facilitate the global phase-out (Parson, 2003) (see also Ungar, Chapter 4, this volume).

What lessons can the history of CFC regulation provide for addressing the issue of climate change? Despite early opposition to regulations, businesses proved responsive to emerging scientific information, and, more importantly, to participating constructively in policy dialogue regarding action at the national and international level. In addition, DuPont as well as a host of other companies played an important role in either developing substitutes for CFCs or developing alternatives to their use. Such industrial innovation was more than simply responsible environmental stewardship. It was also good business. The profitability of CFCs had been waning since the US ban in 1976, and thus the development and patenting of CFC alternatives would provide companies like DuPont with a proprietary technology, with potentially greater demand and higher profits (Maxwell and Briscoe, 1997).

Clearly, climate change is a vastly more complex challenge — so many activities and emissions across the globe contribute to the problem; however, there is hope that with growing awareness and action, the world can follow a similarly successful path in addressing climate change. This path could also provide for new technologies and fuels that have benefits beyond fighting global warming. The demise of the GCC and the emergence of the progressive voice of business suggest that the business community is undergoing an important transition. However, the community remains divided. Some businesses are quite vocal about the risks of climate change and the need for action while others continue to obfuscate — at times questioning the science and warning of economic hardship from climate policy. Emerging climate change policies are changing the dynamics of the marketplace and signaling the eventual emergence of carbon constraints, to which some businesses are already responding. Business pioneers are once again proving to be innovators, developing technologies to improve efficiency and reduce emissions at the individual firm level (see also James *et al.*, Chapter 20, this volume).

These steps, while significant for certain companies, are still small relative to those that are required on an economy-wide or global basis. More is needed — in terms of policy certainty — to drive investments in cleaner technologies that will yield emissions reductions at a level that will curb global warming. In the absence of national policy to reduce greenhouse gas emissions, it is instructive to evaluate some of the motivations behind companies taking the lead on climate change even in this policy vacuum.

Motivations behind corporate commitments

In the United States, businesses are currently operating in a rather opaque environment offering little clear policy certainty. On the one hand, the consequences of unmitigated climate change pose economic challenges in and of themselves. The reinsurance company Swiss Re and other insurers, for example, have expressed concern about costs of climate change and partnered with the United Nations Development Program to assess its potential impacts on the insurance sector (Michaels *et al.*, 1997; Berz, 2001). The private sector will also be affected by policies designed to address climate change. At present, businesses are challenged to function within a complex patchwork of emerging mitigation policies such as mandatory regional and international cap-and-trade programs, efficiency and renewables standards, inconsistent technology incentives, and other programs – all subject to future changes (see chapters by Tennis; duVair *et al.*, Chapters 26 and 27, this volume). Making long-term (and even some short-term) investment decisions under such policy uncertainty is difficult. In spite of this uncertainty, a number of firms are beginning to assess and address their contribution to climate change. The rest of this chapter explores some possible rationales for their leadership and their implications for communication.

Leadership has its rewards: Gaining a market edge

A prime catalyst for action is potential cost savings that arise from making a business more efficient. This has been readily demonstrated by a number of businesses in the United States that have reduced annual GHG emissions while simultaneously reducing costs through savings in energy and materials consumption (Table 21.1). For example, by using 9 percent less total energy in 2002 than it did in 1990 (despite an almost 30 percent increase in production), DuPont saved $2 billion in cumulative energy costs between 1990 and 2002. Utilizing energy efficiency and conservation measures, IBM saved $791 million between 1990 and 2002 from reduced energy costs. In 2002, BP announced that it had achieved its 10 percent reduction target (from 1990 levels) utilizing an internal emissions trading system – creating an estimated $650 million of value for the company. And, in 2002 alone, medical products and services company Baxter International reduced its generation of non-hazardous waste and use of packaging, saving the company $2.9 million in the process.

The cost savings achieved by these business leaders illustrate opportunities that exist throughout the economy. Furthermore, in "learning-by-doing" they develop real knowledge regarding methods, costs, and benefits that

Table 21.1. *Corporate emission reductions: Measures, achievements, and cost savings — status 2005*

Company	Measures	Reduction	Cost-savings
3M	Manufacturing process and product improvements Energy efficiency	35% reduction from 1995 levels; targeting a 50% reduction by 2010	US$200 million since 1973
Alcan	Energy efficiency (smelter improvements)	Globally Alcan reduced total emissions by 2.9 million tonnes over 2001 in 2002; Alcan UK achieved 65% absolute reductions	Not reported
Alcoa	Energy efficiency Reduced Waste	25% reduction in greenhouse gas emissions between 1990 and 2003	US$100 million environmental & energy cost savings projected by 2006
Austin Energy (city-owned utility)	Energy efficiency (DSM for customers) Renewable energy (sales)	Saved 500 MW since 1982 — removed the need for an entire coal-fired power plant	Not reported
BMW	Energy efficiency	14% reduction in CO_2 emissions (per unit of production) between 1990 and 2003	Not reported
BP	Methane capture Energy efficiency	18% reduction in greenhouse gas emissions between 1998 and 2001	US$650 million
Canadian forest products industry	Energy efficiency Biomass fuels	62% reduction in CO_2 emissions between 1991 & 2004 because of improved energy efficiency; 38% reductions in CO_2 between 1992 & 2004 because of improvements in transport	Industry reports greater competitiveness as an industry; numbers not reported

Table 21.1 (cont.)

Company	Measures	Reduction	Cost-savings
Deutsche Telekom	Energy efficiency	Reduced CO_2 emissions by 52% between 1995 and 2002	€10 million (US$12.5 million)
DuPont	Reduced N_2O emissions from nylon production (80%) Energy Efficiency (20%) Use of Renewables (<1%); 10% goal for 2010	69% reduction in greenhouse gas emissions from 1990 levels	US$2 billion (efficiency) US$10–15 million (renewables)
Enbridge Gas Distribution	Customer energy efficiency (DSM) (100%)	Avoided 2.5 million tonnes of CO_2 between 1995 and 2003	Can$700 (US$588 million)
IBM	Energy efficiency Reduced PFC emissions from semi-conductor manufacture	65% reduction in CO_2 emissions from 1990 levels; IBM has a 4% energy efficiency goal per year; for last five years firm has achieved annual average efficiencies of 6%	US$791 million
Kinko's	Use of recycled fiber in paper (44%), use of renewables (13%), recycling (43%)	9% reduction in greenhouse gas emissions in 2002	2002 was Kinko's most profitable year to date
Kodak	Energy efficiency	17% reduction in CO_2 emissions between 1997 and 2003	Senior energy manager reports this has been economically beneficial to firm; numbers not reported

Table 21.1 (cont.)

Company	Measures	Reduction	Cost-savings
Lafarge	Energy efficiency (40%), use of cementitious materials (56%), use of biomass fuels (4%)	11% reduction in CO_2 emissions (per unit of production) from 1990 levels	Environment manager reports this has led to savings making Lafarge more competitive in cement sector
Norske Canada	Energy efficiency (25%), fuel switching (from bunker oil to natural gas) (35%), use of biomass fuels	59% reduction in greenhouse gas emissions from 1990 levels	Can$5 million (US$4.2 million) per year
Sony	Reduced product emissions (82%), efficient production (non-CO_2) (17%), use of renewables (1%)	3% reduction in greenhouse gas emissions between 2000 and 2002	Not reported
STMicro electronics	Energy efficiency, switch to fuel cells & co-generation (65% of energy), renewables (5%)	Goal is zero net emissions by 2010	US$900 million during 1994–2010
Swiss Re	Energy efficiency (100%) Plan to become carbon neutral by offsetting 85% of emissions	Goal is zero net emissions by 2010; 10% reduction in greenhouse gas emissions between 1999 and 2001; committed to 15% reduction goal by 2010; 85% to be offset	Not reported

Table 21.1 (cont.)

Company	Measures	Reduction	Cost-savings
Tembec	Energy efficiency Biomass fuels	30% reduction in greenhouse gas emissions from 1990; committed to halving remaining emissions by 2008; and to eliminating all fossil fuel use in its plants by 2010	Senior energy manager reports this has been economically beneficial to firm; numbers not reported
Toyota	Energy efficiency, thermal emissions recovery Purchasing wind power	19% reduction in CO_2 emissions from 1990 levels	Not reported
Unilever	Energy management	8% reduction in CO_2 emissions between 1999 and 2003 (on a kg/tonne of production basis); 2% annual energy efficiency goal for company	Unilever Bestfoods (UK) saved £1.34 million (US$2.5 million) since 2001
United Technologies	Energy efficiency	37% reduction in greenhouse gas emissions from 1997 levels; goal is 40% by 2006	Cost-savings not available
Westpac	Energy efficiency Reduced paper consumption	26% reduction in CO_2 emissions between 1996 and 2002	Cost-savings not available

Source: The Climate Group, 2004, 2005.

can be shared. Toyota, for example, has licensed patented hybrid technology to Ford for production of Ford's hybrid Escape SUV, which likely reflects Ford's strategy to reduce GHG emissions from its automobiles by 2030 (Hakim, 2004). Yet barriers remain to the exploration of similar opportunities. Without increased awareness or incentives, companies may not scrutinize certain aspects of their operations to find and develop these opportunities – a leverage point where better communication could make a difference. Even Fortune 100 firms have been surprised by the cost-saving opportunities available to reduce GHG emissions.

Responding to market opportunities

Businesses can also affect their bottom lines by effectively responding to market signals and opportunities. The private sector responds to consumer demand: given a sufficiently large market for a particular product or service, some businesses will pursue an opportunity by offering goods and services that fulfill that market niche. Opinion polls indicate that the public favors a broad range of actions to increase US energy supply, but renewables and efficiency standards are more widely preferred over expanding the existing infrastructure or building new power plants (Steiner, 2003) (see also discussion in Dilling and Farhar, Chapter 23, this volume) For example, most states have voluntary green power markets that enable consumers to purchase electricity derived from renewable sources. High fuel costs and concerns over US dependence on foreign oil during 2004 contributed to a rapid increase in consumer demand for fuel-efficient vehicles. Sales of Toyota's hybrid Prius were up 225 percent during the first three quarters of 2004 compared to the first three quarters of 2003. Meanwhile, total sales of Toyota's conventional passenger cars increased just 3 percent while total Toyota SUV sales declined 0.5 percent.[6] This response has driven Toyota to expand its productions of hybrid vehicles and to incorporate hybrid technology into other vehicle models.[7]

Aircraft design at Boeing is following a similar pattern, with the new 7E7 passenger aircraft consuming 20 percent less fuel than its predecessors, while increasing cargo and cabin space, and reducing take-off cabin noise by 60 percent.[8] Reduced fuel use represents a cost-saving to airlines while improved cabin comfort offers a competitive advantage in attracting customers. Such progressive product design demonstrates that market competitiveness does not have to be sacrificed in the design of climate-friendly products. Quite the contrary, innovation can enable environmental and economic goals to be achieved simultaneously.

Yet businesses don't have to simply respond to consumer demand. They can also play a key role in shaping that demand. Advertising can be used to showcase green products and attract consumer interest, thereby influencing product demand. In addition, effective marketing can achieve simultaneous goals of attracting consumers, educating the public, and shaping corporate image. Surveys have consistently demonstrated that a sizable segment of socially/environmentally conscious US consumers does exist. While core drivers of consumer behavior still include price, quality, aesthetics, and convenience, marketing can assist in building an environmental ethic and awareness in consumers and guide consumers toward environment-friendly products (see also Michaelis, Chapter 16, this volume).

Government activities

As seen with the Montreal Protocol and subsequent efforts to phase out CFCs, government policies can provide the incentives necessary to secure emissions reductions and increase investment in technology research and development. For example, large industry investments in research and development (R&D) for CFC substitutes corresponded with the regulatory climate – the 1978 US domestic ban on non-essential aerosols (Maxwell and Briscoe, 1997). Those investments declined during the early 1980s when ozone depletion faded from the political agenda, and then surged again in the wake of the Montreal Protocol.

With respect to climate change, US Federal government policies to date have largely comprised of voluntary mitigation efforts, occasional subsidies for deployment (e.g., production tax credits), and government support for technology R&D. Although entry into force of the UNFCCC's Kyoto Protocol results in binding emissions reductions targets for participating countries, the United States did not ratify and thus is not subject to these restrictions. The Bush Administration has opted to continue with voluntary measures driven by a weak, intensity-based reduction target (Executive Office of the President, 2002).[9] This intensity target is comparable to previous "business-as-usual" improvements in GHG intensity and is expected to yield actual emissions increases of 12–14 percent above 2000 levels by 2012.[10] These projected trends indicate that while some voluntary reductions have been significant at the individual firm level, voluntary efforts in the United States have not been sufficient to halt, much less reverse, the growth in US emissions. For example, a number of US states have voluntary green power markets that enable consumers to purchase electricity derived from renewable sources. However, these voluntary markets that are tied to consumer demand have to date not expanded the use of renewable power close to the levels called for by most renewable portfolio standards (RPS) requiring that a specified percentage of the state's electric power generation come from renewable sources.[11] As of September 2006, 22 states and the District of Columbia had such RPSs in place.[12]

For many companies, voluntary emissions reductions actually represent a potential risk. Firms that are not given credit for emissions reductions achieved today may face higher mitigation costs under a future mandatory reduction policy. As such, some companies choose to delay action until mandatory policies are in place, or at least until they are assured that credit can be gained for early voluntary action.

A mandatory reporting program would be a first step to mitigation policy and help establish a baseline while motivating additional voluntary reductions. Providing "baseline protection" can help insure companies against the kinds of risks posed by early action (Loreti *et al.*, 2000). The US Department of Energy has been updating current voluntary reporting guidelines, but proposals to date fail to register reductions achieved before Bush was in office and do not specifically provide for credit towards a future regulatory program (DOE, 2006). The Administration has also not sought legislation to ensure protection of such early action. Nevertheless, business and industrial organizations at the global, national, sectoral, and local level have announced efforts to register and/or reduce GHG emissions of their members (see duVair *et al.*; and Tennis, Chapters 27 and 26, this volume).[13] Although participation in such programs does not ensure government recognition of mitigation efforts, it provides some quantitative baseline enabling firms to track emissions and reductions and to learn by doing. Ford, for example, recently joined the Chicago Climate Exchange (CCX),[14] and has committed to reduce its North American facility emissions by 4 percent between 2003 and 2006 over the average 1998–2001 baseline.[15] The CCX registry enables Ford and other member companies to report emissions reductions to the public, while establishing a carbon market.

Experience with reporting in other cases shows that mandatory, rather than voluntary, reporting influences consumer choice and opinion and thus provides incentives for corporate self-examination in search of inefficiencies and mitigation opportunities. For example, the US Toxic Releases Inventory (TRI) program has led to substantial reductions in emissions of toxic chemicals, simply by requiring companies to report their emissions (Davis and Mazurek, 1995; Gottlieb, 1995). Just prior to TRI's first reporting deadline, the Monsanto Corporation volunteered to reduce its emissions by 90 percent within five years (Graham and Miller, 2001).

Due to their inherent link with industrial activity and energy use, reducing GHG emissions is a far more difficult challenge than addressing emissions covered by TRI. Thus, securing sizable emissions reductions will require linking reporting systems with mandatory GHG reduction policies. Such mandatory mitigation policies achieve a number of goals. First, they secure direct reductions in GHG emissions; second, they promote corporate innovation in improving efficiency and investing in technology research and development

(Goulder, 2004). Finally, they help address corporate concerns regarding regulatory uncertainty.

Despite the introduction of proposed legislation such as the Climate Stewardship Act (S. 139) in 2003 by Senators Lieberman and McCain and subsequent versions of their bill and others,[16] the implementation of mandatory mitigaton policies is currently limited to state and several regional initiatives. These efforts include the Northeast and Mid-Atlantic's Regional Greenhouse Gas Inititaive, the West Coast Governor's Global Warming Initiative, and the executive orders by Governors Schwarzenegger of California and Richardson of New Mexico that set GHG emissions targets for their states (see Tennis, duVair *et al.*, Chapters 26 and 27, this volume).[17] In September 2006 California enacted legislation mandating that the state reduce its emissions to 1990 levels by 2020. Although these state-level policies represent a step forward in addressing greenhouse gas emissions, they are not an efficient substitute for national policies that establish a uniform playing field and certain regulatory environment. Companies with operations throughout the United States face different regulatory frameworks (or none at all) in different regions. Hence, in anticipation of future mandatory mitigation measures, there is an incentive for businesses to work with federal policy-makers to implement national policies that establish a level playing field and consistent regulatory framework.

Government policies can also facilitate emissions reductions over the short and long term by augmenting consumer demand or subsidizing climate-friendly activities or technologies (see chapters by Atcheson, and Dilling and Farhar, Chapters 22 and 23, this volume). Tax incentives targeting efficient or renewable technologies can help make these technologies more competitive in the marketplace. For example, purchasers of hybrid automobiles receive a federal tax break that helps offset the added cost of hybrid technology. Meanwhile, owners of such automobiles are allowed to use high-occupancy vehicle lanes in Virginia and California. Similarly, state-level renewable portfolio standards represent government policies designed to improve net efficiency while giving consumers choice regarding energy purchases. Government technology policies that provide funding for research and development can also accelerate the pace of technological change, augmenting private sector investments and contributing to long-term adoption and dissemination of climate-friendly technologies.

Corporate sustainability

Another driver influencing progressive business action on climate change is an implicit desire to ensure long-term sustainability in the marketplace.

Competitive businesses operate with considerable attention not just to quarterly returns but also to long time scales, and sustaining a business over long periods of time necessitates anticipation of future trends in technology, policy, and the marketplace. Multinational oil companies such as BP and Royal Dutch Shell, for example, have emerged as strong proponents of addressing climate change and developing green fuels and technologies. Although there is considerable uncertainty and debate over the future of oil and other fossil fuel production, oil and coal companies, which represent some of the largest corporations in the world, must consider how their businesses will operate in the future.

Different companies approach this issue of corporate sustainability in different ways (Skodvin and Skjaerseth, 2001). Royal Dutch Shell and BP advocate for addressing climate change, yet quite clearly green energy, such as renewables, represents only a small fraction of their energy portfolios and both companies continue to invest heavily in exploration of traditional fossil fuels. Still, these companies are working to establish themselves as international leaders in the long-term transition of the energy industry to a more sustainable energy supply. Meanwhile, Exxon-Mobil has actively and publicly touted scientific uncertainty as a reason to delay policy action. Even while maintaining this public posture, Exxon-Mobil is hedging its bets with investments in research, such as its investment of up to $100 million in the Global Climate and Energy Project, a research and development endeavor for future energy technologies based at Stanford University.[18] Despite differences between these companies with respect to public communication regarding climate change and support for global warming policy, all face the same long-term challenge of maintaining competitive interest in a leading economic market while satisfying a growing energy demand in an era of increasingly likely constraints (through policy and/or supply) on traditional uses of fossil fuels.

Doing the right thing

Environmental and social, not just corporate or financial, sustainability is another motivation for progressive business practices (see James *et al.*, Chapter 20, this volume). Although driven largely by self-interest, companies are also compelled to change behavior given knowledge that existing practices have harmful social or environmental externalities. There is a moral imperative to act to prevent such externalities once their existence has been demonstrated (Maxwell and Briscoe, 1997). In order to be recognized as good corporate citizens and to maintain employee morale, firms often seek to "do the right thing" (Margolick and Russell, 2001).

Corporate perceptions of responsible behavior are also influenced by public interactions. For example, organizations such as the Coalition for Environmentally Responsible Economies (CERES) bring investors, environmental organizations, and other public interest groups together to "establish an environmental ethic with criteria by which investors and others can assess the environmental performance of companies."[19] Interface, a modular carpeting company, became the first company to comply with CERES' performance criteria in a corporate transformation that founder and CEO Ray Anderson attributed to customer questions about the company's environmental impact (Anderson, 1999). Similarly, the largest coal-burning utility in North America, American Electric Power (AEP), responded to stakeholder questions in August 2004 with a report that acknowledged AEP's contribution to GHG emissions and outlined steps to comply with voluntary targets, invest in new technologies, and place the company in a good position to face future GHG policy (AEP, 2004). Cinergy followed with a report to stakeholders and an annual report touting the importance of addressing climate change and noting their favored policy approaches and their own intentions to do so (Cinergy, 2004). Thus, not only do businesses respond to public demand for goods and services, they also respond to public scrutiny of how those goods and services are produced.

In the climate area, a company's views about the implications of climate change are central to any decision to take on a climate-related target (Margolick and Russell, 1997), and the evolution of scientific knowledge regarding climate change may have contributed to the defection of companies from the GCC during the late 1990s. CEOs from a number of sectors – oil companies (BP and Shell[20]), manufacturing (Alcoa,[21] GE), utilities (Entergy, Cinergy, Duke), and insurance (Swiss Re) have articulated personal concerns for the future of the planet and have incorporated environmental considerations in their internal strategies while calling for swift policy action.

In an August 2004 *Foreign Affairs* article entitled "Beyond Kyoto," BP CEO John Browne highlighted the critical roles that both business and government must play in addressing the climate change challenge, writing:

> Taking small steps never feels entirely satisfactory. Nor does taking action without complete scientific knowledge. But certainty and perfection have never figured prominently in the story of human progress. Business, in particular, is accustomed to making decisions in conditions of considerable uncertainty, applying its experience and skills to areas of activity where much is unknown. That is why it will have a vital role in meeting the challenge of climate change – and why the contribution it is already making is so encouraging.
>
> *Brown (2004:4)*

If businesses lead, will government follow?

The inherently diverse and competitive nature of the business community ensures that different businesses will play varied roles on the issue of climate change: some will opt to be leaders, others followers, and some obstructionists. This heterogeneity may obscure the fact that these companies are often invoking different corporate strategies to respond to common challenges. The exploration of corporate motivations to act on climate change, as presented in this chapter, yields some promising options to engage more businesses on this issue through effective communication.

One clear implication from the experience so far is the importance of champions. They may well be the most credible communicators to businesses not yet engaged. As leaders who have taken the risk of doing something new and different, as leaders of successful firms that have achieved positive results, and simply as knowledgeable colleagues in business, they serve as important catalysts, role models, and resources.

Closely related to the role of such leaders is the kind of information they may share with other companies in the business community. The business environment, of course, is a competitive one where information frequently implies a strategic advantage. The nature of such strategic, and often proprietary, information may prohibit free sharing of information in certain settings. However, our experience at the Pew Center is that firms are often happy to share information regarding their strategies and successes in reducing greenhouse gas emissions. (For a comprehensive look at how corporate decision-makers take account of climate change in formulating effective business strategies, see Hoffman, 2006). As firms begin to see the need to address climate change as not just a moral imperative but also as an opportunity for new technologies, products, and approaches, more will incorporate climate change into their long-term planning and work to build relationships with like-minded firms – including customers and suppliers. Communication regarding climate change science and impacts to and within the broader business community must be sustained, and accompanied by useful, practical, and specific information on what businesses and industries can do in support of solutions. Communication that illustrates with hard facts (emissions, type of activities, and cost-savings) what some companies have done will be persuasive to those that are not yet engaged. Independent organizations, the media, and others can provide for such information-sharing where direct competitors may otherwise not be so inclined.

As the science becomes more certain, both businesses and scientists are more likely to speak out. In June 2005, four leading firms testified before the House Science Committee on their actions to curb their own contributions to climate change,[22] and several others submitted comments for an April 2006 Senate hearing on designing a mandatory climate policy. In June 2005, the US National Academy of Sciences took the unusual step of joining with ten other science academies from major emitting nations to issue a statement of concern about climate change and a call for policy action.[23] This kind of proactive communication by leading firms and respected scientific bodies makes it easier for others to both acknowledge and support climate change and to take steps to address it.

BP, with its 2000 "beyond petroleum" rebranding effort, has effectively created an image of a socially and environmental responsible company. And, Royal Dutch Shell has also contributed to raising the public consciousness about climate change, highlighted most recently when Lord Oxburgh, Chairman of Shell Transport and Trading, stated that the threat of climate change makes him "really very worried for the planet" (Adam, 2004). Meanwhile, Exxon-Mobil's more recalcitrant views toward climate change have led to its own shareholders organizing in protest.[24] Communication by business leaders on the short- and long-term benefits of climate policy, particularly mandatory mitigation policies, is likely to dispell myths about the intractability of the climate problem and costs of addressing it, and motivate political and policy movement.

As social awareness of climate change and its consequences grows and as climate policies continue to emerge at the state, national, and international level, it is imperative that the business community take a leadership role. At the grassroots level, the private sector will play a key role in responding to the growing market for energy and/or climate friendly goods and services and for the innovation necessary to transition the world to a low-carbon future. Further, the commitment to combat climate change that is demonstrated by businesses influences the public's perception of the issue. Although the scientific community is associated with greater credibility, when businesses go public about integrating climate change into their bottom lines, it sends a powerful message about the realities of climate change and the means of addressing it. Contrary to more obstructionist voices, these pioneering businesses illustrate that climate change can be addressed without harming — and, in fact, often boosting — their economic standing.

Business engagement is also important, at the highest levels of government in crafting policy solutions. Business leaders can of course speak from their experience in achieving their business objectives while reducing their emissions. But they can also weigh in regarding preferred policy approaches that provide the right incentives and the obstacles they will encounter in the absence of regulatory certainty or if each state designs its own, individualized program to reduce emissions. Their input into the design of policies to provide for economy-wide reductions in a cost-effective, pragmatic manner is invaluable in yielding sustainable, successful policy outcomes. Indeed, it is difficult to imagine a postive outcome without their expertise, resources, and support.

Notes

1. The Conference Board is a not-for-profit organization that creates and disseminates knowledge about management and the marketplace to help businesses strengthen their performance and better serve society. See http://www.conference-board.org/aboutus/about.cfm; accessed January 9, 2006.
2. See the Pew Center website at: http://www.pewclimate.org/companies_leading_the_way_belc/; accessed January 9, 2006.
3. The Business Roundtable "is an association of chief executive officers of leading U.S. corporations with a combined workforce of more than 10 million employees in the United States" and can be found at: http://www.businessroundtable.org/; accessed January 9, 2006.
4. SVMG is organized to involve principal officers and senior managers of member companies in a cooperative effort with local, regional, state, and federal government officials to address major public policy issues affecting the economic health and quality of life in Silicon Valley. See http://www.svmg.org/; accessed January 9, 2006.
5. The companies include Hewlett-Packard, Oracle, Calpine, Lockheed, ALZA, Life Scan and PG&E – along with the city of San Jose, NASA Ames Research Center, and the Santa Clara Valley Water District.
6. See http://pressroom.toyota.com/photo_library/display_release.html?id=20041001; accessed January 9, 2006.
7. See http://pressroom.toyota.com/photo_library/display_release.html?id=20040803b; accessed January 9, 2006.
8. See http://www.boeing.com/commercial/787family/index.html; accessed January 9, 2006.
9. The Administration's target – an 18 percent reduction in emissions intensity between 2002 and 2012 – is based on improving the ratio of GHG emissions to economic output expressed in gross domestic product, and is comparable to autonomous intensity improvements in recent decades.
10. See http://www.pewclimate.org/policy-center/analyses/response.bushpolicy.cfm; accessed January 9, 2006.
11. For instance, examples of state RPS that require a certain percentage of renewable energy by a certain date include 2.2 percent by 2011 in Wisconsin, 20 percent by 2017 in California, and 10 percent by 2010 in Connecticut (see http://www.pewclimate.org/what_s_being_-done/in_the_states/rps.cfm; accessed January 9, 2006). In contrast, Green-e reports that in 2003 2.9 million MWh of certified renewable electricity was sold through 102 marketers selling 65 Green-e certified products. The figure of 2.9 million MWh represents 0.12 percent of the roughly 2.5 billion MWh sold by utilities in 2003 nationwide (including states with no standards and RPS states). Information on Green-e data is available at: http://www.crs2.net/HTMLemails/2005/VerificationRelease_1.20.05.htm; accessed January 9, 2006. See EIA "Electricity Monthly: Table 1.2. Net Generation by Energy Source: Electric Utilities, 1990 through October 2004." Available at: http://www.eia.doe.gov/cneaf/electricity/epm/table1_2.html#RANGE!A64; accessed January 9, 2006.
12. See http://www.pewclimate.org/what_s_being_done/in_the_states/rps.cfm; accessed January 9, 2006.
13. See http://www.weforum.org/; accessed January 9, 2006.
14. See http://www.chicagoclimatex.com/; accessed January 9, 2006.

15. See http://www.ford.com/en/company/about/corporateCitizenship/report/articlesClimate Targets.htm; accessed January 9, 2006.
16. Modified version of Climate Stewardship Act (S. Am. 826 to H.R. 6) failed in the June 2005 Senate floor debate. However, the Senate passed a nonbinding "sense of the Senate" resolution that Congress should enact a mandatory program to address greenhouse gas emissions. (S. Am. 866 to H.R. 6, sponsored by Senator Bingaman).
17. See http://www.pewclimate.org/what_s_being_done/in_the_states/news.cfm; accessed January 9, 2006.
18. See http://gcep.stanford.edu/; accessed January 9, 2006.
19. See http://www.ceres.org; accessed January 9, 2006.
20. See Royal Dutch Shell: Summary. Pew Center on Global Climate Change. Available at: http://www.pewclimate.org/companies_leading_the_way_belc/company_profiles/shell/; accessed January 9, 2006.
21. See Alcoa: Summary. Pew Center on Global Climate Change. Available at: http://www.pewclimate.org/companies_leading_the_way_belc/company_profiles/alcoa/; accessed January 9, 2006.
22. See http://www.house.gov/science/hearings/full05/june8; accessed January 9, 2006.
23. The "Joint Academies statement of global response to climate change" is available at: http://nationalacademies.org/onpi/06072005.pdf; accessed January 9, 2006.
24. See http://www.campaignexxonmobil.org/; accessed January 9, 2006.

References

Adam, D. (2004). *Oil chief: My fears for planet: Shell boss's 'confession' shocks industry. The Guardian.* June 17, Available at: http://business.guardian.co.uk/story/0,, 1240715,00.html; accessed January 9, 2006.

American Electric Power (AEP) (2004). *An Assessment of AEP's Actions to Mitigate the Economic Impacts of Emissions Policies.* Available at: http://www.aep.com/environmental/performance/emissionsassessment/; accessed January 9, 2006.

Anderson, R. (1999). *Mid-Course Correction: Toward a Sustainable Enterprise-The Interface Model.* Atlanta, GA: The Peregrinzilla Press.

(2004). Climate change and business. *The Economist*, October 7. Available at: http://www.economist.com/opinion/displayStory.cfm?story_id=3262974; accessed January 9, 2006.

Bennett, C. J. (2004). *Climate Change: Clear Trajectory-Haze in the Details. Executive Action Number* 107, The Conference Board, August.

Berz, G. (2001). Insuring against catastrophe. *Our Planet*, **11**, 3. Available at: http://www.ourplanet.com/imgversn/113/berz.html; accessed January 9, 2006.

Brown, J. (2004). Beyond Kyoto. *Foreign Affairs*, **83**, July/August, 4.

Carey, J. and Shapiro, R. (2004). Climate change – why business is taking it seriously. *Business Week*, August 16, 60–9.

Carlton, D. M., DesBarres, J. P., and Fri, R. W. (2004). *An Assessment of AEP's Actions to Mitigate the Economic Impacts of Emissions Policies.* Columbus, OH: American Electric Power.

Cinergy Corp. (2004). *Air Issues Report to Stakeholders: An Analysis of the Potential Impact of Greenhouse Gas and Other Air Emissions Regulations on Cinergy Corp.*, December. Available at: http://www.cinergy.com/pdfs/AIRS_12012004_final.pdf; accessed January 9, 2006.

Climate Change Task Force. (1997). *U.S. Environmental and Business Leaders Agree on Climate Change Action; President's Council on Sustainable Development*

338   *Arroyo and Preston*

Releases "Climate Principles", Press Release of November 26. Available at: http://clinton4.nara.gov/PCSD/tforce/cctf/press.html; accessed January 9, 2006.

Davis, J. C. III and Mazurek, J. (1995). *Industry Incentives for Environmental Improvement: Evaluation of U.S. Federal Initiatives*. Washington, DC: Global Environmental Management Initiative.

Department of Energy (2006). Guidelines for Voluntary Greenhouse Gas reporting. Federal Register 71, no. 77 (April 21, 2006): 20784.

Executive Office of the President (2002). *Global Climate Change Policy Book*. Available at: http://www.whitehouse.gov/news/releases/2002/02/climatechange.html; accessed January 10, 2006.

Graham, M. and Miller, C. (2001). Disclosure of toxic releases in the United States. *Environment*, **43**, 8–20.

Gottlieb, R. (ed.) (1995). *Reducing Toxics: A New Approach to Policy and Industrial Decision-making*. Washington, DC: Island Press.

Goulder, L. (2004). *Induced Technological Change*. Arlington, VA: Pew Center on Global Climate Change.

Hakim, D. (2004). Ford lays out a move to cut auto emissions. *The New York Times*, October 2, p. C1.

Hoffman, A. (2006). Getting Ahead of the Curve : Corporate Strategies that Address Climate Change, Arlington, VA: Pew Center on Global Climate Change.

Leahy, K., McElfresh, R., and Stowell, J. (2004). *Impact of Greenhouse Gas and Other Air Emissions on Cinergy Corp*. Cincinnati, OH: Cinergy Corporation.

Loreti, C. P., Foster, S. A., and Obbagy, J. E. (2000). *Overview of Greenhouse Gas Emissions Inventories Issues*, Arlington, VA: Pew Center on Global Climate Change.

Margolick, M. and Russell, D. (2001). *Corporate Greenhouse Gas Reduction Targets*. Arlington, VA: Pew Center on Global Climate Change.

Maxwell, J. and Briscoe, F. (1997). There's money in the air: The CFC ban and DuPont's regulatory strategy. *Business Strategy and the Environment*, **6**, 276–86.

Michaels, A., Malmquist, D., Knap, A., and Close, A. (1997). Climate science and insurance risk. *Science*, **389**, 225–7.

Parson, E. A. (2003). *Protecting the Ozone Layer: Science and Strategy*. Oxford: Oxford University Press.

Rogers, P. (2004). Valley firms to fight global warming. *Mercury News*, March 29. Available at: http://www.climateark.org/articles/reader.asp?linkid=30525; accessed January 9, 2006.

Skodvin, T. and Skjaerseth, J. B. (2001). Shell Houston, we have a climate problem! *Global Environmental Change*, **11**, 103–6.

Steiner, E. (2003). *Consumer Views on Transportation and Energy*. Office of Planning, Budget Formulation, and Analysis. Washington, DC: Office of Energy Efficiency and Renewable Energy, US Department of Energy.

Stipp, D. (2003). The pentagon's weather nightmare. *Fortune*, January 26. Reproduced at: http://www.independent-media.tv/itemprint.cfm?fmedia_id=5548&fcategory_desc=Environment; accessed January 9, 2006.

The Climate Group (2004). *Carbon Down, Profits Up*. Available at: http://www.theclimategroup.org/assets/TCG_Emissions%20Charts%2004.pdf; accessed January 9, 2006.

The Climate Group (2005). *Carbon Down, Profits Up*. Available at: http://www.theclimategroup.org/assets/TCG_Emissions_Charts_25_01.pdf; accessed January 9, 2006.

22

The market as messenger: sending the right signals

John Atcheson

US Department of Energy

Introduction

At first blush, a chapter on economic and fiscal tools would seem to be out of place in a book about communicating climate change. And yet, markets and prices are essentially little more than structures which exchange information about what we value, and how much we value it. As Nobel prize-winning economist Kenneth Arrow put it, markets "can be viewed as information and decision structures in which communication takes the form of prices.... [P]rices...according to the pure theory, are normally the only communication that needs to be made" (cited in Babe, 1996).

For many economists, prices effectively communicate value *automatically* and most attempts to intercede in this self-organizing system are usually perceived as ill conceived and counterproductive because they interfere with the operation of Adam Smith's "invisible hand." This theory holds that relatively unconstrained markets allocate wealth more efficiently than would any top–down attempt to manage markets. Economists have long recognized, however, that many market exchanges leave out a great deal of information. British political economist A. C. Pigou, writing in the 1920s, noted that smoke from a factory, for example, imposed costs on third parties that were not reflected in exchange prices (Pigou, 1932).[1] These unaccounted costs could be conceived of as information failures; economists call them "externalities."

For the most part, economists have used two devices to improve price signals so that they more accurately include information and value that are normally left out of market exchanges. One method uses Pigouvian taxes and other subventions designed to internalize externalities by applying

The opinions expressed in this chapter are those of the author and may not reflect those of the US Department of Energy.

a "surcharge" to compensate for the costs left out of the transactions (e.g., a carbon tax), or a subsidy to reflect benefits not mediated by the market (e.g., tax credits for wind energy or hybrid vehicles). The second strategy seeks to create transferable property rights in previously non-monetized costs (or benefits) and to exchange these within the market, thus embedding the information that was previously "external" into market transactions (e.g., emissions trading). These approaches share a common perspective grounded in traditional economic theory – they are designed to get prices right, so that the economic framework can function to communicate value more or less automatically. To date, however, it has been politically difficult for the public sector to apply these tools in any coherent or systematic way.

The strategies outlined in this chapter start from a different premise. They focus on actively using information to improve communication about climate change within the market. As such, they will address both public and private sector initiatives that focus on using information and self-interest in real-world financial transactions, rather than on preserving the integrity of abstract theories. The tools, techniques, and strategies outlined flow from the more general theory of market failure identified by Pigou *et al.*

The magnitude of missing information

There are four key pieces of climate-related information left out of market transactions (or only beginning to be included), and each is large enough to potentially dramatically affect consumer and business choices if they were embedded in prices. Their cumulative value could profoundly change investment practices – and therefore behavior – at all levels of our economy. They are:

- the current and future costs attributable to global warming;
- the ancillary costs of relying on fossil fuels, including health and ecological effects, securing access to Middle East oil, and the price penalty for uncertainty and volatility;
- the value of carbon as a commodity in a global trading system; and
- the foregone opportunity cost for not participating in the market for clean technologies in a carbon-constrained world.

Although not a direct cost, the fact that we currently subsidize fossil fuels and fossil fuel infrastructure profoundly distorts price signals and compromises the ability of the market to communicate the true costs of using fossil fuels (e.g., NDCF, 2003).

Each of these areas offers opportunities to use policy tools to improve the ability of the market to communicate the value of mitigating global warming. Below I summarize briefly what we know about the monetized market value of this missing information.

The cost of global warming and climate-related events

One of the critical aspects of climate change that is not well communicated by the market is the increasing cost of climate-related hazardous events. In 2003, weather-related natural disasters caused about $70 billion in damages (~$18.5 billion insured) (Innovest, 2004). In Europe, there were at least 27,000 deaths from a heatwave in 2003. Analyses suggest that it is highly likely that global warming contributed to the severity of the heatwave and that further warming will cause such heatwaves to occur more frequently (Stott, Stone, and Allen, 2004). Finally, in the summer and fall of 2004 and 2005, a succession of powerful hurricanes pounded the Caribbean and Gulf of Mexico region. Debates about attributing these storms to climate change miss the point. While it is impossible to say whether any one of these storms were *caused* by global warming, experts believe them to be *representative* of a world in which climate change has been allowed to progress without serious, immediate intervention, and that these kinds of extreme weather events will escalate absent such intervention (Stott, Stone, and Allen, 2004; IPCC, 2001: 4).

Swiss Re, one of the world's largest reinsurers, estimates that the costs of disasters will reach more than $150 billion per year in less than a decade, and that global warming is a significant contributor to that increase (Blair, 2004). According to the Association of British Insurers, storm and flood damages are projected to triple by 2020 absent decisive action to halt global warming (BBC, 2004; Hall, 2004). Clearly, this is not only the result of a potential increase in the frequency or severity of extreme weather events, but also the result of greater value exposed to the hazard (Pielke and Landsea, 1998). NASA climate scientist James Hansen suggests that several trillions of dollars of infrastructure are at risk from coastal flooding due to global warming and that we are much closer to dangerous anthropogenic interference to the climate than models suggest (Hansen, 2004). Finally, ecological economist Robert Costanza and others attempted to attach monetary value to 17 key ecosystem services. Their study suggested that the annual value of natural capital and ecosystem services was about $33 trillion in 1997 dollars, roughly equivalent to the entire annual global economy. Climate models suggest that these "services" will be compromised by global warming (IPCC, 2001). Thus, some part of these freely given, but extremely valuable and essential natural services will be lost. To date, the market treats them for the most part as if they did not exist or had no value.

Ancillary costs of fossil fuels

An ancillary benefit of mitigating climate change would be avoiding the cost of fossil-fuel-related health and environmental effects. The US Environmental Protection Agency (EPA) estimated that the monetized benefits of implementing the Clean Air Act requirements amounted to $21.7 trillion between 1970 and 1990, not including additional savings derived from avoiding certain ecosystem effects, and certain health effects from toxic pollutants. The benefits of clean air have risen over time as more stringent requirements have been put in place, and as the number of people who are protected increases. In 1999, EPA estimated that the monetized benefits from the Clean Air Act would be $71 trillion by 2000 (US EPA, 1997, 1999).

To the extent that the vast majority of these benefits are achieved by retrofitting treatment and control devices onto power plants, industrial facilities, and vehicles, EPA's current strategy does not address carbon emissions from these sources (in fact, scrubbers and similar end-of-pipe technologies often increase carbon emissions because they reduce overall system efficiencies). Yet action taken to mitigate global warming by using cleaner fuels and more efficient technologies to cut carbon emissions would achieve many of these same benefits, thereby recouping some or all of the trillions in avoided costs as a side benefit of an integrated climate/clean air strategy.[2]

Oil represents a special case in that it is heavily subsidized, and it is traded in a highly structured market that maintains artificially high prices. As a result, it imposes substantial economic burdens aside from the environmental and climate-related costs. For example, a recent study found that the cumulative cost of OPEC market manipulations to the United States had reached about $7 trillion (Greene and Tishchishyna, 2000). In addition, the annual military costs assigned to our dependence upon Persian Gulf oil (exclusive of the recent expenses in the second Iraq war) have been estimated to be $50 billion per year (Copulos, 2003; NDCF, 2003).

There is also an economic penalty associated with the volatility of prices and the uncertainty of supplies. Companies routinely protect themselves against volatility and uncertainty through a variety of pricing strategies including derivatives and other risk-sharing techniques. This "hedging" against uncertainty partly accounts for oil prices in excess of $55 a barrel since the fall of 2004 (Crawford and Fredricks, 2004).

Value of carbon

The value of carbon is no longer entirely "missing." In recent years, a market for carbon has been emerging (e.g., the Chicago Climate Exchange, the northeastern and western consortia of states in the United States trying to establish regional carbon markets, as well as emerging markets in Europe and Japan) (see also chapters by Tennis; and duVair *et al.*, Chapters 26 and 27, this volume).[3] While it is impossible to say with confidence how important carbon trading will become in the future, the market for carbon credits has doubled for two years in a row, and was estimated to reach $480 million in 2004 (Innovest, 2004). With Russia's ratification of the Kyoto Protocol, the size and value of a carbon credits market should expand even faster over the next several years, as more formal trading regimes are established.

The foregone opportunity cost for not participating in the clean technologies market

The market for environmentally clean technologies – primarily clean energy technologies – was expected to reach $607 billion in 2005 (Mohiuddin, 2004). Wind energy is now the fastest-growing form of new energy supplies in the world (DOE, 2005b). With the Kyoto Protocol now in effect, demand for clean energy technologies is expected to increase. To the extent that the United States does not participate in serious multilateral, international climate mitigation efforts, American companies may not be eligible to take part in trading. As the Progressive Policy Institute (Innovest, 2004) points out, companies in both Japan and the European Union have actively sought to develop cleaner technologies in anticipation of the Kyoto Protocol. Meanwhile, the United States has lost its lead in wind and solar technology to Germany and Japan, and not a single American car is listed among the ten most fuel-efficient cars in the world. Moreover, companies headquartered in the United States will face a bewildering array of state, local, national, and international rules governing carbon emissions.

An even greater disadvantage to the American economy can develop if the United States fails to develop better clean technologies, expected to become one of the fastest-growing export markets in the world within the decade (Mohiuddin, 2004). China, for example, recently passed an automobile fuel efficiency standard more stringent than the United States', placing

US automakers at a severe competitive disadvantage to European and Japanese manufacturers in the fastest-growing automobile market in the world. In Canada, automobile manufacturers recently agreed to cut carbon emissions from vehicles by 5.3 million tons by 2010 by increasing fuel efficiency by about 25 percent, and California's Pavely Clean Car Law imposes similar requirements, recently adopted by several other US states (Becker, 2005).

The hidden costs of climate outlined above are not a complete catalogue of the real cost of failing to control GHG emissions. They lead to two conclusions. First, the frequent claim that attempts to control climate change will cost us money and hurt the economy is a profound distortion of what the market – with full information – would communicate. And as Nobel prize-winning economist Joseph Stiglitz points out, absent full information, markets simply don't work as advertised. They will not produce optimal results; indeed, they will – in effect – lie to us (Stiglitz, 2002, 2003).[4]

Second, the real question is not simply "How much will it cost us to avoid (or reduce) climate change?" Rather, it is "How much will it cost us if we don't?" Only by asking and answering both of these questions can we arrive at an economic "truth" that can form the basis of communicating about climate within the context of markets.

Thus, we must demonstrate specific areas in which policies, practices, procedures, and strategies can be used to more effectively communicate the embedded value of climate mitigation strategies in development decisions. The following section will explore a number of tools and techniques for improving our ability to communicate risks, benefits, and rewards for mitigating global warming in different market contexts.

Shareholders communicating to business and the market

Increasingly, Americans participate in market transactions as shareholders, either directly or indirectly through managed funds and institutional funds (Luskin, 2004). The rules of the road for the securities market offer two opportunities to influence the market's ability to more effectively communicate about climate change.

The Securities and Exchange Commission (SEC) requires all publicly traded companies to disclose "financially material conditions" to shareholders or prospective investors. Generally, companies do this through their annual 10-K and 20-F filings – essentially annual reports for domestic and international companies traded in the United States.

Environmental liabilities are considered "financially material," and a growing body of case law and guidance suggests that even *potential* environmental liabilities should be listed as financially material conditions (Sanford and Little, 2004). As yet, there is no specific guidance for when, how, or whether such liabilities should be listed.

For many companies, global warming represents one of the single largest potential liabilities because, either its products or services contribute to global warming and may be affected by tort or regulatory law (see Averill, Chapter 29, this volume), or the company's activities are directly and adversely affected by global warming.

Yet because it is difficult to specifically quantify to what extent a company's long-term value may be affected by climate-related issues, there is no standard way of communicating the risk that climate change poses to a company. As a result, many companies make no mention of climate change in their annual 10-K or 20-F reports (FOE, 2003). Liabilities that are uncertain or difficult to quantify, but which may be significant, are required to be disclosed, however, in a section of the annual report called Management's Discussion and Analysis of Financial Conditions (MD&A). This suggests two approaches for companies to integrate climate information into their accounting and reporting activities.

Communicating from the top down – SEC Guidance to Companies

The SEC could issue a guidance requiring companies to list liabilities related to global warming in their annual reports, either as specific obligations where data are sufficient to characterize their exposure or as disclosures in their MD&A, where data are not sufficient to precisely quantify liabilities, but where climate change has a clear potential to influence a corporation's bottom line in some way.

Communicating from the bottom up – ensuring shareholder value and corporate responsibility

Increasingly, shareholders are active in holding CEOs, CFOs, and boards accountable for considering their long-term fiduciary responsibilities as well as their social and environmental responsibilities. Several not-for-profits have begun using shareholder action specifically as a distinct tactic to move industry and businesses to greater climate awareness and accountability.[5]

Grassroots organizations and shareholder coalitions can make (and have made) progress in encouraging individual companies to take notice of global warming. Even at Exxon-Mobil – to many the Darth Vader of energy companies – 20 percent of the shareholders voted in 2003 in favor of a proactive global warming policy and renewable energy strategy. Notably, the shareholders were *not* the typical crunchy granola crowd; Institutional Shareholder Services Inc. and other groups that advise mutual fund and pension managers recommended that Exxon shareholders support these proposals, in part because they believed it was good for the company's long-term bottom line to take a more strategic look at the opportunities and threats presented by global warming (*Planet Ark*, 2003). In fact, shareholders could argue that failure to disclose potential corporate liabilities associated with global warming would be a violation of Section 302 of the Sarbanes–Oxley Act of 2002.[6]

There is a pure fiduciary case to be made for evaluating and communicating corporate financial risks associated with climate change in annual reports. A growing body of evidence suggests that so-called socially responsible companies and funds outperform the average stock portfolios in value and growth (e.g., Baue, 2004; Innovest, 2004). Institutional investors have taken note of this trend and are increasingly factoring climate considerations into their decisions. Recently, a group of institutional investors with over $800 billion in assets established the Investor Network on Climate Risk (INCR), in part because they believed a responsible corporate climate policy was an indicator of financial success (Baue, 2004; Innovest, 2004).

Both institutional and individual shareholders can use data from studies by analysts such as Innovest to force companies to adopt more responsible – and more strategic – approaches to climate change, by communicating not just the risks it poses, but also the business opportunities it offers (Cogan, 2003; Baue, 2004; Innovest, 2004; see also note 7).

Communicating through government policies and regulations

Aligning clean air and clean energy policy

The federal government could improve market signals by better aligning clean energy and climate policy. If companies (or perhaps citizens acting in aggregate) were allowed to take credit under the Clean Air Act for emission reductions achieved through energy efficiency, distributed energy systems, or renewable energy technologies, the federal and local air regulations could be transformed from a system that impedes clean energy choices to one that acts as an incentive for using them. While there are challenging implementation issues, there are substantial environmental and economic benefits from such an approach.

Clean Air Act regulations typically favor a treatment device designed to handle one contaminant at a time. This misinforms the market about the optimal method for achieving clean air objectives, because it encourages the regulated community to assess costs incrementally based on removing, say − sulfur one year, followed by nitrous oxides (NO$_x$) several years later, followed by particulates several years after that, perhaps followed by mercury, other heavy metals and finally carbon − rather than to consider the ultimate net cost of removing all contaminants (US EPA, 1997, 1999). Independent analyses suggest that until carbon is considered, the regulated community will pursue a piecemeal approach addressing contaminants one at a time with a series of end-of-pipe control technologies rather than a more comprehensive approach relying on more efficient generation and use of energy and use of cleaner fuels (Collamer, 2003). Clean energy strategies can reduce specific air pollutants as well as carbon emissions (and other resources, such as water) − often at a significant savings compared with end-of-pipe command and control technologies (NREL, 2002).

Allowing companies to capture, bank, and trade emission reductions accomplished through energy efficiency and renewable energy would result in significant carbon reductions achieved as an ancillary benefit of clean air regulations. This would also provide incentives for currently unregulated end users to use more energy-efficient technologies and cleaner fuels by making their emission reductions fungible assets that could be exchanged in an emissions trading market.

Aligning economic development policies with climate policies

Cities, counties, and states control many of the policy levers that can mitigate global warming, and it is often in their interest to use them. For example, permitting, zoning, building codes, transportation patterns, and development choices are among the most effective ways to affect reductions in greenhouse gas emissions. Even so, many communities resist making such choices, as they have yet to realize that unabated sprawl imposes enormous demand on municipal services and hence skyrocketing service costs, while leading to more vehicle miles traveled, growing traffic congestion, and growing auto emissions. Whether and how communities and states use these tools to signal the implied cost of development would also go a long way toward communicating the value of climate-mitigating activities (see also the chapters by Young; Pratt and Rabkin; Watrous and Fraley; duVair *et al.*; and Tennis, Chapters 24, 6, 25, 27, and 26, this volume). There is

Textbox 22.1 **Why it's good to grow green**

Both states and communities use tax credits, low interest loans, guaranteed loans, and a variety of other fiscal incentives to attract and retain businesses, create jobs, or otherwise enhance the local economy. According to economist Kenneth Thomas, states, counties, and communities give out $48.8 billion in development subsidies per year (Thomas, 2000). Historically, states and communities have typically doled out these funds without sophisticated financial analysis. If subsidies attracted or retained businesses, they were deemed a good investment.

The advent of box stores such as Wal-Mart has caused governments to take a closer look at what they are buying for their subsidies. A recent study by the minority staff of the US House Committee on Education and the Workforce found that a 200-employee Wal-Mart cost the taxpayer $420,750 per year (Miller, 2004). Another study found that California taxpayers subsidized $20.5 million worth of medical care for Wal-Mart employees. Yet another study found that Wal-Mart has received at least 244 subsidy deals totaling more than $1 billion from state and local governments (cited in Mattera *et al.*, 2004). In short, development strategies designed to fill community coffers are, in many cases, actually emptying them.

This sort of "box store" development is a symptom of a larger problem – unplanned and unmanaged growth, urban sprawl, and the associated notion that economic growth, *per se*, is the *sine qua non* of public policy. From the perspective offered in this chapter, these problems reflect misinformed markets. For example, according to the Texas Transportation Institute (2003), urban congestion cost $69.5 billion in 2003, and urban sprawl partly explains why transportation is the fastest growing source of greenhouse gas emissions in the United States.

Moreover, as one influential urban planner pointed out, one of the most effective economic development strategies for cities is displacing "imports" and retaining capital (Jacobs, 1984). The Rocky Mountain Institute offered a specific example of this general observation that has profound implications for both climate policy and development strategies: the value of energy efficiency and small-scale, climate-friendly alternative energy systems. The Institute found that 70 cents of each dollar spent on energy immediately left the community (Hubbert and Fong, 1995). Retaining some portion of this energy-related money within the community by encouraging efficiency or developing community-based energy systems also had a multiplier effect of about 3 to 1. In short, each dollar saved on energy generated three dollars in income at the local level rather than losing 70 cents. It also cut greenhouse gas emissions by reducing the amount of energy used.

a growing recognition that economic development polices to date range from sub-optimal to downright counterproductive in terms of their ability to create a positive cash flow (see Textbox 22.1) (Texas Transportation Institute, 2003; Mattera *et al.*, 2004; Miller, 2004).

One of the earliest attempts at aligning developmental and environmental objectives – and still one of the most comprehensive ever put in place – was a short-lived pilot plan developed by the state of Louisiana. In 1990, alarmed that the state's industries had one of the highest per capita emissions of toxic chemicals in the country, the governor's office issued an executive order that used an innovative point system which conditioned half the state's economic development resources (~$300 million) to give preference to cleaner industries and cleaner companies.[7] This point system defined criteria such as emissions per unit of product produced, amount of energy used per unit of product produced, and infrastructure burden. It then rated a company seeking business development funds against these criteria in evaluating subsidy requests. The same approach could be used to make development subsidies a means of encouraging high-efficiency buildings, combined heat and power systems, eco-industrial parks, clean vehicle fleets, and smart growth.

Establishing clean development funds

In market transactions, the cost and availability of money and credit is a primary means of communicating perceived risk, intrinsic value, uncertainty, and long-term liability. Therefore, one of the most effective ways to communicate preferences for climate-friendly growth is to develop programs and strategies that provide low or no-cost money for them.

Governments use guaranteed loans and revolving funds to do this, frequently funded through system benefit funds, which are typically established through a small fee (tenths of a penny) applied to energy generation, use, or transmission. Some states, such as Texas, used dedicated funding authorized by the legislature to set up their program (DOE, 2005a). While numerous other states have system benefit funds, all too few dedicate this mechanism for innovative fiscal policy.

Revolving funds are useful for states and communities seeking to finance energy efficiency improvements in publicly owned or operated buildings. A few corporations such as Dow Chemical have used them in similar ways. Revolving funds use revenue from the energy saved by energy efficiency improvements to pay back the loan, so projects can be funded off ledger (i.e., outside normal capital and operating budgets, thus simplifying red-tape and decision mechanisms). The paybacks can be set so that the fund grows over time, or so that it maintains a constant balance. Either way, once established, they can be self-generating (see also Young, Chapter 24, this volume).

The Texas Loan STAR program is one example. Since the program began in 1989, 127 loans to public institutions have generated over $150 million in documented energy cost-savings, while cutting more than 1.5 million tons of carbon (DOE, 2005a). Based on better-than-expected energy savings, the Loan STAR Program projects it will save Texas taxpayers more than $500 million over the 20 years between 2005 and 2025, while cutting greenhouse gas emissions (Texas SECO, 2005).

Another incentive mechanism can be attached to guaranteed loans. Guaranteed loans use government funds to underwrite a lender's risk and buy down interest rates on investments such as energy-efficient technologies or cleaner, more efficient on-site power systems. Since they reduce lender risk, guaranteed loans encourage commercial lenders to devote a higher proportion of their resources to the activities underwritten by the guarantee, and to charge a lower interest rate. Thus, relatively small amounts of public finances can be used to leverage large increases in lending from commercial institutions. Often guaranteed loans can stimulate $5 or more lent for each dollar dedicated to the program (IPMVP, 2005). Guaranteed loans have been used in a number of state and community energy efficiency programs, including New York and Phoenix.

The New York State Energy Research and Development Authority (NYSERDA) established a guaranteed loan program called the $mart^{SM} Loan Fund (NYSERDA, 2005). It is designed to provide an interest rate reduction up to 4.0 percent off a participating lender's normal loan interest rate for a term up to 10 years on loans for certain energy efficiency improvements and/or renewable technologies. NYSERDA has achieved an overall leverage of $4 of private sector funding for each dollar of public money and lowered the cost of borrowing money. Several other states, notably Iowa and Nebraska, have similar programs with similar results (see DOE, 2005a, for additional state summaries).

State programs typically include audits designed to identify cost-effective energy efficiency opportunities and for lenders and funders to assess a portfolio of energy-savings options that represents low-risk capital investments. As with standardized monitoring and measuring protocols, this increases the amount of funds lenders are willing to devote to such loans, as well as reducing their costs.

Greening the screeners

Organizations and institutions responsible for screening investments exist at several levels. Each has the capacity to send accurate market signals about the costs and benefits of mitigating climate change.

Insurers

As noted earlier, insurance companies and insurance underwriters are in the process of revising underwriting rules and insurance rates due to being in the front lines of climate-related financial vulnerability. To protect themselves against increasing exposure, insurance and reinsurance companies have already begun to raise deductibles and premiums. Some have also written exclusionary clauses into purchase agreements. These changes have a cascading effect on industries and economies. Commercial lenders will resist loans which are deemed high-risk by underwriters, effectively limiting the amount of commercial loans for activities at increased climate-related risk such as tourism in coastal areas, agricultural ventures in less predictable climate, and so on.

To the extent insurers will refuse coverage, economic activity may become restricted in those areas. Those left uninsured will have to bear the brunt of climate change impacts or become a bigger tax burden as Congress bails out people and communities affected by extreme events. Due to the central role that the insurance and reinsurance industry plays in the economic life of the national and global economy, insurers, developers, lenders, and governments have a strong interest and role to play in pushing for climate change mitigation.

Bond underwriters

A second set of screeners includes those who underwrite the creditworthiness of states and municipal governments as well as corporations. Organizations such as Moody's, Standard and Poor's, and Fitch ICBA assign values from highly secure to highly risky or speculative bonds. The cost of money increases significantly as a state or municipality goes down in the rating. It can often mean a difference of tens of millions of dollars to the borrower or bond applicant.

As outlined above, ignoring climate-related liabilities can impose enormous economic risks and burdens upon a city or state through increased health-related expenses, lost productivity due to traffic congestion, and other hidden costs. Conversely, policies that are designed to cut emission-generating activities and potential climate impacts can help avoid or minimize these costs while offering economic benefits such as retained capital, increased real-estate values in downtown areas, and reduced vulnerability to energy price volatility and supply uncertainties. Yet bond underwriters so far ignore both the benefits to be gained and the vulnerabilities avoided from climate-conscious air quality, energy, and development policies and practices, and thus do not use such criteria when rating creditworthiness.

To begin fixing this market failure, bond underwriters, insurers, municipal and state financial officers, representatives from the accounting community, as well as climate, air quality, and energy experts, could convene to develop a set of criteria designed to factor in the costs of unsustainable developmental policies into risk ratings for states and communities. This would set enormously powerful incentives for climate-conscious development and business activity.

Corporate screeners

Corporations are leading the way toward financial analyses that factor in the advantages of cleaner, more efficient energy systems. Several researchers have documented literally hundreds of cases where companies have cut carbon and saved money, by internal accounting methods that better capture the returns in labor productivity, lower energy costs, reduced environmental compliance costs, and improvements in workflow, product quality, among other savings (Romm, 2000; Lovins *et al.*, 1999; see also Arroyo and Preston, Chapter 21, this volume).

Companies accomplished this by getting their internal economic signals right. Their corporate financial screeners were able to identify energy costs and capture energy-savings opportunities. Typically, a corporation may use simple payback, internal rate of return, or net present value calculations to evaluate whether a project is fundable. For some companies, energy reliability and energy quality are critical issues. For example, for certain batch processing plants (i.e., plants that produce a variety of goods at different times), financial processing centers, chip manufactures, and users of clean rooms, the cost of power disruptions is measured in millions of dollars per incident. In such cases, on-site energy systems may be cost-effective. According to Joseph Romm, a bank in Omaha that installed two fuel cells in order to assure that they had at least "six nines" of reliability and high-quality energy ("six nines" means the system is 99.9999 percent reliable compared to the typical utility supply, which ranges between three and four nines). The fact that the system produced less greenhouse gas emissions than grid-supplied electric power was a side benefit to the improved reliability.[8]

For most companies labor is the highest cost. Energy-efficient lighting and air systems are documented to provide improvements in productivity and to reduce employee absenteeism (Romm, 2000). For a typical office building, capital construction costs are about 5 percent of the total life-cycle costs, operating and maintenance costs slightly more, and the remainder (nearly 90 percent) is labor (Romm, 2000). Thus energy investments that provide even small improvements in productivity often pay for themselves, even when conventional financial analysis would suggest otherwise.

Performance contracts and ESCOs

Companies and municipalities have used a variety of schemes to take advantage of the fact that energy efficiency investments produce a stream of money from future savings on lower utility bills. With the advent of the International Performance Measurement and Verification Protocol (IPMVP, 2005), commercial lenders have increasing confidence to accept these savings as a revenue stream or as collateral. This has allowed companies and cities to fund efficiency projects off ledger: they can often fund improvements outside the context of hurdle rates, returns on investments, and complex internal funding approval processes with no out-of-pocket expenses.

Energy Service Companies (ESCOs) are one of the more common ways of accomplishing this by providing up-front funding for energy efficiency investments for their clients by sharing money saved through these investments. For example, an ESCO might save a commercial building about 25–50 percent or more on energy bills by increasing the influx of daylight, relighting, better heating and air conditioning, and improved building insulation. The ESCO would take some portion of the savings for a period of years until it had recouped its costs and made a profit. Since about 1995, ESCOs have had a difficult time as electricity restructuring introduced uncertainty into energy markets and dried up utility demand side management programs. Moreover, US federal policy seemed to encourage a business-as-usual approach, substantially discounting the value of energy efficiency in the country's long-term energy strategy.

A few commercial lenders have begun to help fill the funding gap from this slump in ESCO activity by specializing in energy retrofit loans. Banks and lenders with experience in such loans are attracted to them, because they provide low-risk returns, particularly in times of increasing energy prices. As climate change gains in importance and energy prices exhibit increased volatility, ESCOs and other strategies using future savings to fund current efficiency investments are likely to increase.

Conclusions

This chapter argued that markets and prices are essentially means through which we communicate what we value and how much we value it. To date the messages we send through the markets for the most part ignore the costs of a changing climate and overstate the costs of mitigating climate change. This tendency to lie to ourselves through the language of commerce has an enormous potential to cause irrevocable damage to some $33 trillion worth of natural capital. But the reverse is also true: we can set the signals straight through a collective commitment to valuing those things that make life possible – the air, climate, land, water, and species with which we share this planet.

A starting point is to recognize the costs and benefits of preservation and exploitation. How we structure our taxes, the business development investments we make, our lending policies and insurance decisions, the nature of investment portfolios we compose with public and private monies can all signal the value of preservation and conservation – or neglect it.

Some of these decisions are in consumers' hands: each purchase is a vote. Other decisions are in the hands of shareholders and investors. Yet others are in the hands of public officials at various levels of government: those who shape the tax system, make zoning decisions, or develop or elevate standards (see also Dilling and Farhar, Chapter 23, this volume). Realizing the power of markets to drive social, economic, and environmental changes, and identifying the people who can make such changes in the market, this chapter pointed to key audiences and to critical foci for message development. Led by the insurance industry and socially responsible investment funds, the rest of the financial community must now take a common-sense approach to climate change.

One way of reaching out to the economic development community, for example, is through multilateral groups such as the National Coalition for Economic Development (DOC, 2005). Those in the environmental community willing to learn and speak the language of commerce could initiate the dialogue. Scientists and policy analysts can help federal, state, and local environmental regulators recognize that addressing one pollutant at a time compromises long-term air quality and the stability of our climate, while forcing a massive misallocation of resources on a series of Rube Goldberg treatment devices.

The investment community would make great progress if it accepted the growing body of evidence demonstrating that smart climate policies go hand in hand with smart business practices. Similarly, if the SEC recognized the economic vulnerabilities of ignoring the strategic opportunities and long-term threats posed by climate to companies and investors, it could require companies to communicate these risks in their annual reports. Shareholders, both institutional and individual, can continue organizing and exerting pressure on the companies they own to consider impacts of climate on the bottom line in the short-term and more strategic context.

Bond underwriters, commercial lenders, and others with a long-term fiduciary responsibility also have an opportunity to communicate through their rating and lending criteria the risks inherent in ignoring the costs of climate change, the benefits of escaping the risks, and the costs of relying on an increasingly volatile fossil fuel market.

Finally, surprisingly few schools offer programs that require a basic modicum of ecological or scientific literacy of their public policy graduates or MBAs. Academic and professional educators and trainers in the public policy and financial fields are in the position to adopt a curriculum that encourages their graduates to look beyond abstract theories of neoclassical economics and address real-world costs in real time with real financial tools.

If the market is a key "communicator" of essential information affecting our interaction with the planet's bounty, then one principal challenge now is to communicate more effectively to those who make the market "speak." Those who set the signals have a key role to play in assuring that markets communicate the full range of information embedded in our energy production and consumption choices. There is great truth behind the old phrase "money talks." Money tells us what we value, what we cherish. The challenge for us over the next several decades is to "teach" our markets to speak with wisdom on the issue of climate change.

Notes

1. Pigou first explored externalities in his 1912 book *Wealth and Welfare*. He was roundly attacked by most of the mainstream economic community for his advocacy of government interventions in markets, but his work has gained in credibility over time.
2. Note also that regulating carbon dioxide along with other criteria air pollutants (sulfur dioxide, mercury, and nitrogen) can be more cost-effective than regulating only the latter three pollutants, and that fragmented regulatory requirements would have the highest compliance costs (Repetto and Henderson, 2003).

3. More information on some of these can be found at: http://www.chicagoclimatex.com/; http://www.rggi.org/; http://www.europeanclimateexchange.com/index_flash.php, and http://www.pointcarbon.com/; accessed February 10, 2006.
4. Stiglitz won the Nobel prize in economics for his work on how information – or lack of it – influences markets.
5. Two examples are Co-Op America (http://www.coopamerica.org) and CERES, the Coalition for Environmentally Responsible Economies (see http://www.ceres.org/ our_work/principles.htm). CERES recently released a report specifically focused on taking climate change into account in business and financial practices (see Cogan, 2003).
6. The full title of the law is "The Public Company Accounting Reform and Investor Protection Act." PL 107-204, 116 Stat 745. The Act sets up specific standards for accounting and reporting, including legal consequences for the CEO and CFO of a corporation who knowingly falsifies or otherwise misinforms investors.
7. This program was developed by Maurice Knight, Director of Pollution Prevention under Governor Buddy Roemer, and the implementation and demonstration of it was accomplished under a small grant administered from a program I directed in the Pollution Prevention Division at EPA. It ended in 1992 when then-Governor Roemer left office.
8. The net amount of emissions displaced by a fuel-cell system such as the one used at the Omaha bank depends upon the source of the hydrogen used in the fuel cell and the type of grid power being displaced. Since grid power may be dispatched from any of several dozens of power plants over the course of a day, the net emission reduction varies. However, for any grid supplied power relying on either coal or natural gas, fuel cells always derive more power per unit of fuel used, and therefore generate fewer emissions per unit of power provided.

References

Babe, R. E. (1996). Economics and information: Towards a new (and more sustainable) worldview. *Canadian Journal of Communication*, **21**, 2, 161–78. Available at: http://www.cjc-online.ca/viewarticle.php?id=360&layout=html; accessed February 10, 2006.

Baue, W. (2004). Institutional investors can now hedge climate risk with new investment strategy. Available at http://www.socialfunds.com/news; accessed February 10, 2006.

BBC (2004). Climate threat to home insurance. June 8. Available at http://news.bbc.co.uk/2/hi/business/3785619.stm; accessed January 2006.

Becker, D. (2005). Global warming: Sierra Club statement on historic clean cars deal. *Progressive News Wire*, March 23. Available at http://www.commondreams.org/news2005/0323–06.htm; accessed January 2006.

Blair, T. (2004). Remarks of the Prime Minister to His Royal Highness's Business and the Environment Programme, September 15. Available at http://www.number10.gov.uk/output/page6333.asp; accessed June 2005.

Cogan, D. G. (2003). *Corporate Governance and Climate Change: Making the Connection*. Available at http://www.ceres.org/pdf/ceres_cg_rprt.pdf; accessed March 2005.

Collamer, N. (2003). Multi-pollutant outlook for control technology, investment, allowance markets and coal generation. Presentation to The Institute of Clean Air Companies, 2003 Forum, October 14. Available at http://www.icac.com/files/public/Session1_Collamer.pdf; accessed January 11, 2006.

Copulos, M. R. (2003). The real cost of imported oil. *The Washington Times*, July 23, Opinion Section.

Crawford, P. J and Fredricks, E. H. (2004). The crude facts about the price of oil. *Graziadio Business Report*, **7**, 3. Available at http://gbr.pepperdine.edu/043/crudeoilprices.html; accessed January 2006.

Department of Commerce (DOC), Economic Development Administration (2005). National Economic Development Organizations. Available at http://www.eda.gov/Resources/NationalEDOrganizations.xml; accessed March 2005.

DOE, Office of Energy Efficiency and Renewable Energy (2005a). Loan STAR and other state run financing programs. Available at http://www.eere.energy.gov/state_energy_program/feature_detail_info.cfm/fid=45/start=2; accessed April 2005.

Department of Energy (DOE), Office of Energy Efficiency and Renewable Energy (2005b). Wind & Hydropower Technologies Program. Available at http://www.eere.energy.gov/windandhydro/; accessed January 2005.

Friends of the Earth (FOE) (2003). Second Survey of Climate Change Disclosure in SEC Filings of Automobile, Insurance, Oil & Gas, Petrochemical, and Utilities Companies. Available at http://www.foe.org/camps/intl/corpacct/wall street/secsurvey2003.pdf; accessed March 2005.

Greene, D. L. and Tishchishyna, N. I. (2000). *Costs of Oil Dependence: A 2000 Update*, ORNL/TM-2000/152, Oak Ridge, TN: Oak Ridge National Laboratory.

Hall, J. (2004). *A Changing Climate for Insurance*. Association of British Insurers. Available at http://icnewcastle.co.uk/; accessed September 2004.

Hansen, J. (2004). Defusing the global warming time bomb. *Scientific American*, **290**, 3, 70–7.

Hubbert, A. and Fong, C. (1995). *Community Energy Workbook*. Snowmass, CO: Rocky Mountain Institute.

Innovest (2004). *The Carbon Disclosure Project*. Available at http://www.cdproject.net, accessed May 2004.

Intergovernmental Panel on Climate Change (IPCC) (2001). *Climate Change 2001: The Scientific Basis*. Summary for Policymakers. Contribution of Working Group I to the Third Assessment Report. New York: Cambridge University Press.

IPMVP (2005). International Performance Measurement and Verification Protocol. Available at http://www.ipmvp.org; accessed January 2006.

Jacobs, J. (1984). *Cities and the Wealth of Nations: Principles of Economic Life*. New York and Toronto: Random House.

Lovins, A., *et al.* (1999). *Creating the Next Industrial Revolution: Natural Capitalism*. SnowMass, CO: Rocky Mountain Institute.

Luskin, D. (2004). Ownership has its privileges. *SmartMoney.com*. Available at http://www.smartmoney.com/aheadofthecurve/index.cfm?story=20041210; accessed February 10, 2006.

Mattera, P., *et al.* (2004). Shopping for subsidies: How Wal-Mart uses taxpayer money to fund its never-ending growth. Washington, DC: Good Jobs First, Inc.

Miller, G. (US Representative) (2004). Everyday low wages: The hidden price we all pay for Wal-Mart. A report by the Democratic Staff of the Committee on Education and the Workforce, US House of Representatives, February 16.

Mohiuddin, S. (2004). How America lost its clean technology edge. Washington,
    DC: Progressive Policy Institute. Available at http://www.ppionline.org/
    ppi_ci.cfm?knlgAreaID=116&subsecid=150&contentid=253040; accessed
    February 10, 2006.
National Defense Council Foundation (NDCF) (2003). *The Hidden Cost of
    Imported Oil.* Summary available at http://www.iags.org/n1030034.htm;
    accessed February 10, 2006.
National Renewable Energy Laboratory (NREL) (2002). A Comparative Analysis of
    the Costs Achieving Clean Air Using Pollution Prevention and Energy
    Efficiency vs. Treatment and Control Technologies, 1970 to 2000.
    Golden, CO: NREL.
New York State Energy Research and Development Authority (NYSERDA) (2005).
    New York Energy $mart^{SM} Program Evaluation and Status Report. Report to the
    System Benefits Charge Advisory Group. Available at http://www.nyserda.org/
    default.asp; accessed January 11, 2006.
Pielke, R. A. Jr. and Landsea, C. W. (1998). Normalized hurricane damages in the
    United States: 1925–95. *Bulletin of the American Meteorological Society*, **13**,
    September, 621–31.
Pigou, A. C. (1932). *The Economics of Welfare*, 4th edn. New York:
    Macmillan and Co.
Planet Ark (2003). Exxon shareholders vote down environmental proposals.
    Planet Ark daily news summaries, May 30. Available at http://www.planetark.
    com/dailynewsstory.cfm/newsid/20976/story.htm; accessed January 2006.
Repetto, R. and Henderson, J. (2003). Environmental exposures in the U.S. electric
    utility industry. *Utilities Policy*, **11**, 2, 103–11.
Romm, J. (2000). *Cool Companies.* Washington. DC: Island Press.
Sanford, L. and Little, T. (2004). *Fooling Investors and Fooling Themselves: How
    Aggressive Corporate Accounting and Asset Manangement Tactics can Lead to
    Environmental Accounting Fraud.* Oakland, CA: Rope Foundation.
Stiglitz, J. (2002). *Globalization and Its Discontents.* New York: W.W. Norton.
Stiglitz, J. (2003). *The Roaring Nineties.* New York: W.W. Norton.
Stott, P. A. Stone, D. A., and Allen, M. R. (2004). Human contribution to the
    European heatwave of 2003. *Nature*, **432**, 610–14.
Texas State Energy Conservation Office (2005). The Loan STAR Revolving
    Loan Program. Available at http://www.seco.cpa.state.tx.us/ls.htm; accessed
    February 10, 2006.
Texas Transportation Institute (2003). *The Annual Urban Mobility Report.*
    College Station, TX: Texas A&M University.
Thomas, K. (2000). *Competing for Capital: Europe and North America in a Global
    Era.* Washington, DC: Georgetown University Press.
US Environmental Protection Agency (EPA) (1999). *Final Report to Congress on
    Benefits and Costs of the Clean Air Act, 1990–2010*, Report # 410–R–99–001.
    Washington, DC: US EPA.
US EPA (1997) Final Report to Congress on the Benefits and Costs of the
    Clean Air Act, 1970–1990, Report # 410–R–91–002. Washington, DC:
    US EPA.

23

Making it easy: establishing energy efficiency and renewable energy as routine best practice

Lisa Dilling
University of Colorado—Boulder

Barbara Farhar
University of Colorado—Boulder

We all depend on energy to meet our basic needs and support our lifestyle. From cars to cell phones to microwave ovens, most products we use require energy. In the United States most of this energy comes from non-renewable and fossil fuel sources (Table 23.1). Interestingly, for more than 30 years, a majority of Americans has stated that they prefer energy-efficient products and renewable energy over other sources of energy (Farhar, 1994; Coburn and Farhar, 2004; Greene, 1998). And yet, products and services often do not reflect those preferences, while energy use continues to rise in the United States. For the most part, energy profligacy — rather than energy efficiency — has been routine in the United States. In the sections below, we discuss some of the reasons for Americans' high energy consumption and the disconnect between expressed wishes and consumption reality. We also show how seamlessly embedding energy efficiency and renewable energy into products and energy services is necessary if society wishes to reduce energy waste as well as greenhouse gas emissions. "Making it easy" for companies and households to use sustainable energy products and services is a critical element of a comprehensive strategy toward greater sustainability. "Making it easy" means investing the resources to make energy-saving and sustainable-energy choices routine through a variety of mechanisms, which we discuss below. If the use of efficiency and renewable energy is made routine, the time spent on potentially complex decisions can be lessened or avoided altogether. Doing so would also make the communication task easier: instead of trying to persuade individuals one by one to purchase more efficient products and services (likely to be an inefficient and less successful endeavor), campaigns could be directed toward those who build, provide, or legislate efficient and renewables-based products and services.

Table 23.1. *Energy use in the United States*

Sectoral energy use	Total energy consumed (%)	Electricity consumed (%)
Buildings	40	71
Industrial processes	33	28
Transport	28	1

Energy source	Energy consumption by source (%)	Electricity production by source (%)
Coal	22	48
Natural gas	23	20
Nuclear power	8	18
Renewable resources*	6	12
Oil	41	2

*Includes hydroelectric power, biofuels, geothermal, solar and wind.
The use of energy in the United States is approximately evenly divided between buildings, industrial processes and transport. Electricity is predominantly used in buildings, with a significant proportion in industrial processes. Overall, oil is the dominant energy source consumed in the United States, while coal is the country's major source of electricity.
Source: Data from US Department of Energy (2005); US DOE–EIA (2005b).

Energy in the United States

Since the industrial revolution in the mid-1800s, energy use in the United States and globally has skyrocketed (Smil, 1994). Primary energy consumption in the United States has increased over the past 30-odd years from 75.7 to 100.3 quads[1] of energy consumed (US DOE–EIA, 2005b). Tremendous changes in society, in living standards, leisure time, life expectancies, and the fundamental nature of human endeavors have been driven by the use of energy derived mainly from fossil fuels – coal, oil, and natural gas (Table 23.1). Along with these benefits have come unanticipated consequences such as critical dependencies, geopolitical struggles over oil resources, and negative effects such as air pollution and CO_2 emissions.

Although energy dependence is a fundamental characteristic of our economy, we tend to take energy availability for granted in our lives until there is a problem – a blackout, fuel shortage, a colder-than-normal winter. US institutions work hard to ensure that there is sufficient energy, and utilities are regulated by the public sector because of their fundamental support of living conditions and the economy in general. It is in the context of "surprises" or interruptions in our supply of energy, or during "price spikes" that energy efficiency has most gained traction as a societal priority.[2]

The case for policy intervention

Given that consumers have consistently stated that they prefer their products to be efficient and to rely on renewable energy, one would think that the power of the market alone would be sufficient to ensure an energy-efficient society. However, markets do not work perfectly in responding to all aspects of consumer preference. Several types of market failures stand in the way (see Atcheson, Chapter 22, this volume; Brown, 2001).

The first is the "principal–agent problem." This refers to the fact that an individual or a company other than the end-consumer is responsible for making decisions about energy efficiency of a product, and that such decisions may not be in the best interest of the consumer – thus limiting consumer choice. For example, builders tend to build homes to minimize up-front costs, and the long-term energy efficiency of the home can suffer as a result (although see Textbox 23.1 for innovations in this area). It would be in many home buyers' best interests to be able to select equipment in their home that minimizes the full cost of energy and equipment over the lifetime of use, not just the up-front costs (Brown, 2001). Similarly, commercial and residential renters are often responsible for paying utility bills but not able to exercise much choice in the energy efficiency of their unit's appliances or building characteristics.

Another problem explaining market failure is insufficient or incorrect information about energy use of appliances and homes. For example, home occupant behavior profoundly affects energy consumption; yet feedback on how energy is used in the home and in businesses is limited to the monthly utility bill. This aggregated price signal provides insufficient information for occupants to understand the ways in which their behavior is affecting their consumption. It goes beyond turning off lights and adjusting thermostats; the effect of hidden electricity uses such as instant-on TV sets,[3] older windows, older refrigerators, lack of weather-stripping and caulking, and time-of-day uses are also powerful factors affecting energy bills.

A further difficulty is that energy efficiency is not a service in and of itself – it only comes into play as a component of the primary service or product that a customer is buying, such as transportation, a warm house, or access to the Internet. People have limited choices and limited time to make a decision about the choices offered. It takes considerable time to research solar water heating and solar electric systems to retrofit homes. It is unrealistic to expect people to get to work via mass transit when there is no mass transit to use (see Tribbia, Chapter 15, this volume). Finally, the proportionately low cost of energy in comparison with other aspects of a

Textbox 23.1 **Innovations in building and permitting solar homes**

The city of San Diego has, as part of its sustainability initiative, recently started an Expedite program to accelerate the permitting process for building homes with photovoltaic (PV) systems. The city had adopted a broad goal of achieving 50 MW of self-generated renewable electricity within the next five years. As a tangible step to encourage builders to include energy efficiency and PV in their new homes, the city's Expedite program, in essence, puts builders who plan to include solar PV as a standard feature in their homes at the front of the line to receive building permits.

Usually, the permitting process for an undeveloped piece of land takes approximately 24 months. The city cuts that time down to six to nine months if the PV system provides 50 percent of the estimated electricity needs of the home, based on Title 24's estimate of electricity use per square foot. Although it costs the city nothing to implement, the Expedite program results in a small revenue stream because the city charges a US$500 fee for an Expedite application.

The builder benefits from the program because builders pay finance charges on loans for property they own and are waiting for a building permit to develop. These charges vary, of course, but can often range from US$2000 to US$4000 per month. If it takes 24 months to obtain a building permit, the finance charges at the high end of this example could total US$96 000. If the Expedite program results in a building permit in only 6 months, the builder would pay only US$24 000 instead of US$96 000 in finance charges. In addition, the San Diego builder can recoup half of the installed system cost because of the CEC (California Energy Commission) rebates; therefore, the costs that need to be passed along to the home buyer are only 50 percent of the actual installed PV system cost. This price premium ranges from US$8000 to US$10 000 per home. Because housing demand in San Diego is high, buyers readily pay this premium; in fact, one builder's homes are 100 percent pre-sold. The company believes that PV may add value to its marketing campaign.

Of course, with houses in the Expedite program, the PV home buyers don't have the option of selecting a PV system for their homes; it comes as a standard feature. Buyers are always free to purchase a conventional rather than a zero-energy home (ZEH). But, all other things being equal, the buyers benefit from purchasing a home with PV as a standard feature because: (1) they don't have to decide about an unfamiliar but expensive technology, (2) they should enjoy significant savings on their electricity bills for at least 20 years, (3) they will probably realize appreciated resale value for their homes when they sell, and (4) they can roll the cost into their home mortgage, the monthly incremental cost of which should be less than their electricity savings (Farhar *et al.*, 2004).

household budget (except for low-income families) makes it unlikely for energy efficiency to be a priority, or for consumers to take the time to seek out additional information on which to base a decision about efficiency or renewables products and services (Brown, 2001; Greene *et al.*, 2005).

As long as market failures such as these exist, stated consumer preferences for energy-efficient products and services alone cannot supersede the incentive structure of many corporations to manufacture goods and provide energy services in conventional, less efficient ways. Absent high costs for energy or societal pressures and policies, corporations whose products rely on electricity or gasoline do not have a strong incentive to make those products more efficient, and indeed, may have a disincentive if they make more profit from less efficient products or services. In these cases, consumer, corporate, and governmental preferences for different degrees of energy efficiency and sustainability may be in direct conflict.

Because energy is critical to the American economy and way of life, communities, organizations, and individuals often feel powerless, or markedly constrained, with the choices in energy sources and levels of efficiency that others have made for them. To make prices and products reflect true costs more clearly, and to allow consumers to follow their preferences on energy efficiency more easily, public policies can help. Such policies frequently find broad public support.

Policies for energy efficiency

Attempts to promote energy efficiency have grown out of social priorities and policies, rather than from the desire to perfect energy markets (see also Atcheson, Chapter 22, this volume). During the Second World War in the United States, fuel and gasoline were rationed so that sufficient supplies would be available for the war effort. After the war, energy use exploded as more electrical labor-saving devices and appliances were introduced, more people had disposable income, and the population became larger and more mobile. Promoting energy efficiency as a non-wartime priority first gained attention in the United States after a major multi-state electrical blackout in 1965 and environmental concerns over the siting of new power plants on the West Coast (Nadel, 2002). Interest in creating policies that promoted energy efficiency arose at the state level and propagated up to the national level, gaining strong support during the 1970s as a result of the oil embargo and associated price shocks and energy shortages.

To energy analysts energy efficiency typically means *getting more work out of a given unit of energy* (more output for the same or less input), while energy conservation refers to *using less energy* to heat and cool homes, to get from point A to point B, and so on. President Carter wearing a cardigan sweater as he discussed the oil crisis in the early 1970s on television vividly exemplified the idea of conservation. As a result, people began associating conservation with sacrifice and discomfort. Efficiency, on the other hand, does not suffer from this image problem; it implies greater levels of comfort and convenience at the same or a lower price. The implications of the two definitions are quite different as well. Energy efficiency built into appliances does not require lifestyle change, while conservation involves behavior change (see Tribbia, Chapter 15, this volume).

Programs to encourage energy efficiency and renewable energy

Policies that have attempted to encourage or require energy efficiency have had to recognize the wide range of uses of energy, and how structures and incentives are currently being implemented to facilitate energy use (see Table 23.2 for examples). A wide variety of policy instruments has been developed, as "one size fits all" approaches will not work on an energy-use market that includes everything from light bulbs to refrigerators to automobiles, homes, office buildings, energy production and transmission, and large industrial plants. Programs fall into three general categories – incentives, labeling, and mandatory standards. These are meant to address several of the market failures mentioned above and include lowering initial cost barriers and providing additional consumer information, as well as building wise energy use into products and services.

Incentives

In general, incentive programs are designed to overcome the market barriers currently posed by higher initial costs for energy-efficient products, so that lifetime energy savings are realized by more customers than would otherwise choose them. Incentives include tax credits, rebates, and amortization of up-front costs. Rebates can be offered through any level of government, through the local utility or private sector, and have covered, for example, efficient appliances, renewable energy systems, energy audits, and hybrid-electric vehicles.[4]

Amortization of up-front investments in efficiency and renewables is another way in which standard practice encourages the use of sustainable

Table 23.2. *Sample policies and programs related to enhancing and encouraging built-in energy efficiency and renewable energy*

Name of policy or program	Government or business	Mandatory or voluntary	Description/website or other source
Green-pricing programs	Investor-owned utilities, municipal utilities, electric cooperatives	Voluntary	Green pricing is an optional utility service allowing customers to purchase renewable electricity by paying a premium on their electric bills to cover the incremental costs of developing renewable electricity sources. In 1999, 17 states and 33 utilities offered green pricing programs. In 2004 – only five years later – green-pricing programs were offered in 33 states by more than 578 utilities. At the initial green-pricing program offering, 1% to 2% of retail electricity customers opt to participate. The highest participation rate reported so far in a more mature program is 11%. (http://www.eere.energy.gov/greenpower/markets/pricing)
Green-power marketing	State, private companies	Permitted by legislation	Green-power marketing refers to selling certifiable green power in competitive markets, where this is permitted, to retail or wholesale electricity customers. Green-power marketing occurs, for example, in the District of Columbia, California, Illinois, Maryland, New Jersey, New York, Pennsylvania, Texas, Virginia, and several New England states. (http://www.eere.energy.gov/greenpower/markets/marketing)
Renewable energy certificates (RECs)	Private companies	Voluntary	Also called green tags, RECs represent the environmental attributes of the power produced from renewable energy projects. These attributes can be sold separately from the physical electricity. Customers can buy RECs without having to switch electricity suppliers. REC retail products are widely available. (http://www.eere.energy.gov/greenpower/markets/certificates)
Federal appliance standards	Federal	Mandatory	A wide range of innovative products have been developed in the last two decades following efficiency standards. Appliance standards have saved consumers and businesses an estimated US$50 billion on a cumulative basis from 1990 to 2000. (http://www.aceee.org)

Table 23.2 (cont.)

Name of policy or program	Government or business	Mandatory or voluntary	Description/website or other source
Building codes	State/local	Mandatory	The State of California, for example, established energy efficiency standards for residential and non-residential buildings in 1978 in response to a legislative mandate. The California Energy Commission administers "Title 24" standards, which are mandatory for all new buildings, and are updated periodically to incorporate new energy efficiency technology and methods. (http://www.energy.ca.gov/title24)
Renewable portfolio standards (RPSs)	State	Mandatory	RPS are a public policy approach to mandating the use of renewable resources – such as biomass, geothermal, solar, and wind – by requiring sellers of electricity to have a certain percentage of renewable power in their mix. At least 22 states and the District of Columbia have RPS laws, and in 2004, Colorado voters passed the first citizen-initiated RPS in the United States. (http://www.eia.doe.gov/cneaf/solar.renewables; http://www.pewclimate.org/what_s_being_done/in_the_states/rps.cfm; http://www.energyjustice.net/rps/; http://www.green-energy.org/rps.html)
Energy efficiency and renewable energy R&D	Federal	Research and outreach	US Department of Energy's Office of Energy Efficiency and Renewable Energy (DOE/EERE) uses an annual budget approximating $1 billion to conduct research, development, and deployment programs for energy efficiency and renewable energy technologies. (http://www.eere.energy.gov)
State energy program (SEP)	State/federal	Mandatory	SEP provides funding to states to carry out their own energy efficiency and renewable energy programs. SEP funding (from DOE) enables state energy offices to design and implement programs for the needs of their economies, the potential of their natural resources, and the participation of local industries. (http://www.eere.energy.gov)

Program	Level	Type	Description
Million solar roofs (MSR) initiative	Federal/state/local	Voluntary	MSR's goal is the installation of one million solar energy systems by 2010 through education and outreach activities. The program brings together business, government, the energy industry, and community organizations to forge commitments to install solar energy systems. (http://www.millionsolarroofs.org)
States and localities	State/local	Voluntary, mandatory	All 50 states and many localities have programs to promote or require the use of energy efficiency and renewable energy. The Database of State Incentives for Renewable Energy (DSIRE) is a comprehensive source of information on state, local, utility, and selected federal incentives that promote renewable energy. (http://www.dsireusa.org)
Western Governors' Association's Clean and Diversified Energy Initiative (CDEI)	Regional	Policy analysis	CDEI will examine the feasibility of and actions needed to develop 30,000 MW of clean energy in the West by 2015, and to achieve a 20% increase in energy efficiency by 2020. (http://www.westgov.org/wga/initiatives/cdeac/index.htm)
State implementation plans (SIPs)	State	Mandated by the Clean Air Act	States are required to demonstrate compliance with national ambient air quality standards. Three proposals will integrate energy efficiency and renewable resources in the SIP process (Vine, 2002).
Conservation and Renewable Energy Reserve (CRER)	State/federal	Mandatory	Encourages the use of energy efficiency and renewable energy as a compliance strategy to limit the amount of emissions. A total of 300,000 sulfur-dioxide bonus allowances (about 3% of the total emission cap of 8.9 million tons) have been set aside to be allocated to utilities for energy-efficiency measures and renewable energy development (Vine, 2002).

Table 23.2 (cont.)

Name of policy or program	Government or business	Mandatory or voluntary	Description/website or other source
Energy Efficiency/ Renewable Energy Set-asides	State/federal	Mandatory	A pool of allowances coming from with a state's NO_x budget to award energy-efficiency and renewable-energy projects in the states that reduce or displace electricity generation (Vine, 2002).
Net-metering Programs	State	Mandatory	In states with net-metering legislation, such as Colorado and California, utilities must offer building owners the opportunity to tie their solar electric systems to the utility grid. Net metering enables customers to use their own generation to offset their consumption over a billing period by allowing their electric meters to turn backwards when they generate electricity in excess of their demand. This offset means that customers receive retail prices for the excess electricity they generate. (http://www.eere.energy.gov/greenpower/markets/netmetering.shtml)

energy. For example, energy-efficient mortgages (EEMs) take into account the savings on utility bills from efficiency measures (that are documented by a home energy rater) in qualifying buyers for a home mortgage. Similarly, in California, new home buyers benefit by rolling the cost of solar-electric systems into their home mortgages while enjoying electricity cost reductions in the range of 60 percent (Hammond, 2005).

In evaluating the success of incentive programs, it is important to measure whether new customers were won for energy-efficient products, or whether those customers who were already planning to purchase the products are the main beneficiaries (a form of "free-riding"). Studying the effect of free-riding is difficult, and estimates of free-riders benefiting from utility incentive programs range from 5–70 percent (Morgenstern, 1996).

In addition, before consumers can take advantage of these incentives, they must first be aware of them. Communication of the availability of incentives is a key factor in increasing their utility. Cities and utilities offer incentives, but their main communication tool is generally to place inserts in customer bills, which are often disregarded (e.g., Dunwoody, Chapter 5, this volume). If salespeople are aware of incentives, they can mention them to customers, but the diverse nature of retail sales makes it a difficult task to reach all the potential outlets for customer purchases. Sales people are sometimes not sufficiently trained in the energy features of the products they sell, nor do they get incentives for selling more energy-efficient products.

Labels

Another attempt to increase voluntary purchases of energy-efficient products is to require energy-consumption information to be affixed to the product so that customers presumably can take energy-use characteristics into account in their purchasing decisions. These tend to fall into two types of categories – mandatory labels that summarize the energy use of the product using agreed-upon components and strictly monitored tests, and voluntary labeling programs that provide a "seal of approval" that indicates that a product meets a particular minimum level of energy efficiency as determined by the labeling organization.

Examples of mandatory government consumer labeling programs include the fuel economy label (miles per gallon, mpg) required for every passenger car sold in the United States, and the EnergyGuide label, which is required to be placed on certain new home appliances such as refrigerators, freezers, and water heaters.

The intent behind mandatory information labels is to address one of the "imperfections" of the market for efficient appliances – lack of full information on energy costs. However, in practice, these labels have limited success except among very sophisticated consumers predisposed to seeking out such information. Studies of the EnergyGuide appliance label among consumers show mixed or even low utility (reviewed in Banerjee *et al.*, 2003). For example, customers could not necessarily tell which products were more efficient, or confused the labels with a "seal of approval" for higher efficiency. In general, manufacturers and marketers found that the label was not having its intended effect, because energy efficiency was a low priority for most customers and the information was not very understandable (Egan *et al.*, 2000, as cited in Banerjee and Solomon, 2003).

In addition, most consumers cannot ascertain from energy labels the value of energy-efficiency attributes relative to other features. For example, as Greene (1998) points out in the case of automobiles, most buyers find it impossible to calculate the value of long-term energy efficiency relative to initial costs, the costs of efficiency relative to other features of the car, and for the impact of these attributes on his or her pocketbook over the lifetime of owning the car.

The voluntary ENERGY STAR™ labeling program run by the Environmental Protection Agency (EPA) and Department of Energy (DOE) has proven more successful than EnergyGuide labels. The ENERGY STAR™ label certifies that a product meets the energy efficiency criteria for the product category, so it functions more as an "energy-efficiency seal of approval" rather than a neutral informational label. Testing and information from participating manufacturers are used to set the product standards. Surveys show that a higher percentage of customers knew what the label stood for, even if they had not seen it before (Banerjee and Solomon, 2003). ENERGY STAR™ has achieved great success with office equipment, especially computers and peripherals (Nadel, 2002; see also Textbox 23.2). Now, ENERGY STAR™ even extends to buildings, with the ENERGY STAR™ Homes program, although market penetration is still small.

A similar "seal of approval" effort – the Leadership in Energy and Environmental Design (LEED) program – is a voluntary consensus-based national standard developed by the US Green Building Council designed to advance the cutting edge of building techniques for energy efficiency and environmentally friendly design (Hammond, 2005; Keesee, 2005; Hering, 2005).[5] The number of LEED-certified buildings is growing and the program shows good promise for pushing forward innovation and consumer awareness (see also chapters by Watrous and Fraley, and duVair *et al.*,

---

Textbox 23.2 **Office equipment – an ENERGY STAR™ success story**

The ENERGY STAR™ program has set voluntary energy efficiency standards for many types of office equipment – personal computers, computer monitors, computer printers, facsimile machines, and photocopiers. To meet the standard, manufacturers of office equipment must include power management features that put the device into low-energy "sleep" mode when inactive. The ENERGY STAR™ program has been remarkably successful in this particular arena – largely because the costs of compliance with the standards are generally low and the manufacturers can promote their products with the ENERGY STAR™ label (Nadel, 2002). As a result, 80–99 percent of personal computers, monitors, and printers sold in 1999 were ENERGY STAR™ models (Brown *et al.*, as cited in Nadel, 2002). The use of power management features can reduce annual energy use of these appliances by 50 percent.

Chapters 25 and 27, this volume). LEED has particularly influenced commercial builders and owners of commercial buildings.

Unfortunately, even with good recognition, labels are unlikely to influence the purchasing decisions of most consumers (Feldman and Tannenbaum, 2000, as cited in Banerjee and Solomon, 2003). Nonetheless, the "seal of approval" approach tends to be more effective than the neutral informational label (Banerjee and Solomon, 2003).

Standards

Since the 1970s, the United States and many other countries have enacted "minimum efficiency standards" for appliances and vehicles. These standards establish a minimum specific efficiency that products must meet, regardless of how the product is designed, in accordance with any other requirements such as safety standards. For reasons pre-dating public interest in climate change, energy-efficiency standards for vehicles, buildings, and appliances have become more and more common worldwide over the past three decades. As of 2001, the United States had the most appliance standards of any country (18) – and 16 other nations and the European Union also had mandatory equipment efficiency standards in effect (Nadel, 2002).

The complex process of negotiating and determining standards involves manufacturers, government agencies, and other interested advocates. Often, standards and regulatory requirements first emerged at the state level, later moving to the national scene. Manufacturers then called for national standards in some cases (such as refrigerator standards) to avoid having to manufacture products to meet many different state standards. Political pressure, legal processes, and world events can also play a role in whether or not efficiency standards are boosted over time. For example, although the requirement for national efficiency standards was first passed into law in the late 1970s, in the 1980s the Reagan Administration did not propose any new standards until the combination of a lawsuit and action by states and appliance manufacturers resulted in the National Appliance Energy Conservation Act (NAECA) in 1987 (Nadel, 2002; for more on the role of legal processes, see Averill, Chapter 29, this volume).

The most common appliance standard enacted worldwide is for refrigerators, followed closely by room air conditioners (Nadel, 2002). Refrigerators have seen the most dramatic drop in energy intensity (the amount of energy used to provide a given level of service) – 70 percent in the years since 1972. Standards are clearly dominant in driving this trend – as seen in Figure 23.1, energy intensity dramatically decreases in the years in which standards become effective (Nadel, 2002) with much smaller changes in the intervening years. In 2000, appliance standards were estimated to have saved 88 terawatt hours (Twh), or 2.5 percent of US electricity use (Nadel, 2002). Altogether, US federal energy-efficiency standards taking effect in the 1988–2007 period have been estimated to produce an overall net cost-saving benefit of US$80 billion by 2015 and $130 billion by 2030 (Meyers *et al.*, 2003). And the energy savings will increase – estimated annual residential primary energy consumption will be reduced by 8 percent in 2020 (Meyers *et al.*, 2003).

Standards for automobile efficiency, called Corporate Automotive Fuel Economy (CAFE) standards, have been similarly successful at curbing the energy intensity of automobiles in the United States. Estimates of its impact vary, but it is clear that passenger fuel economy has increased and light trucks are more efficient than in 1975, without any net loss in performance (Greene, 1998; National Research Council, 2002). As discussed below, however, attempts to raise CAFE standards beyond the targets set in the 1970s have met with effective resistance.

State and local codes and standards affect the energy uses of buildings, automobiles, and appliances. All 50 states have some sort of residential building standard for energy efficiency, although some are more rigorous than others and some are voluntary rather than mandatory.[6] Building standards can require specific equipment to be installed or for specific performance standards to be met.

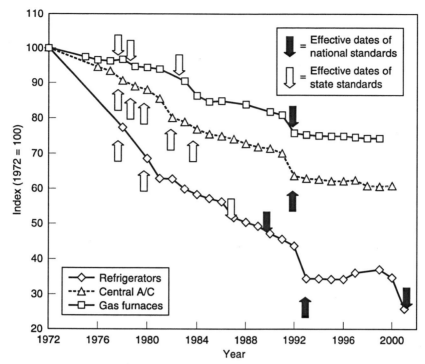

Fig. 23.1.   Energy use in the United States (refrigerators, central air
conditioners, and gas furnaces)
This figure stems from an analysis by the American Council for an Energy-
Efficient Economy (ACEEE) based on data compiled by manufacturer trade
associations. The energy intensity index is based on energy use or energy
efficiency per unit and does not account for modest changes in unit size over
time. The 2001 value for refrigerators is based on the new 2001 federal
efficiency standard.
Source: Nadel (2002), reprinted, with permission, from the Annual Review
of Energy and the Environment, **27**, ©*2002 by Annual Reviews*, http://
www.annualreviews.org.

   Although builders may claim their homes are energy efficient, by and large
they do not yet approach the limits of technical possibilities or the 5-star
ENERGY STAR™ standard. Thus, the potential for effecting reduced
energy use in homes through improved construction practice is large. For
example, the recent additions to the building standards for residential and
commercial buildings adopted by the state of California are estimated to save
500 megawatts (MW) (the equivalent of a normal-sized power plant) of
electricity annually (Anon., 2004). The new additions to the standard include
more efficient fluorescent lighting in all permanent fixtures, improved
glazings in windows, and repair of ductwork when new heating or cooling
equipment is installed.

Building codes may also help stimulate voluntary builder efforts at higher quality and/or affordable homes with efficiency and solar features. Large-production builders in California are now constructing near-zero-energy homes using building-integrated photovoltaic systems with solar or tankless water heating that result in 60 percent lower electricity bills for homeowners (Hammond, 2005; Keesee, 2005; Hering, 2005; Murr, 2005; see also James *et al.*, Chapter 20, this volume).

Renewable energy and standards

Most recent polls show consistent and growing public preference for renewable energy sources (Farhar and Coburn, 2005; Coburn and Farhar, 2004; Farhar, 1993; Farhar *et al.*, 1979). Majorities or pluralities of respondents select solar or solar and wind energy as the energy source "best for the environment," "safest," "most abundant," "least expensive," and so on.[7] In contrast, relatively few respondents (generally fewer than 20 percent) associate oil and gasoline, nuclear power, or coal with positive attributes. Nuclear power, which does not produce any carbon dioxide, is frequently promoted as an energy solution to climate change, but the public still perceives it as unsafe and expensive (Farhar and Coburn, 2005).

Consumer preferences for renewable energy alone have been slow to increase the percentage of power from renewable sources – the percentage of renewable energy consumed in the United States remained relatively constant from the 1970s to 2000 at about 6 percent of total energy consumed (Table 23.1, and US DOE–EIA, 2005a). Participation in voluntary "green-pricing" programs, however, has rapidly grown in recent years (Table 23.2). In addition, Renewable Portfolio Standards (RPS) are relatively new and hold great promise for increasing the percentage of renewable power currently consumed in the United States. RPS are a public policy approach to mandating the use of renewable resources – such as biomass, geothermal, solar, and wind – by requiring sellers of electricity to have a certain percentage of renewable power in their mix. As of February 2006, 22 US states and the District of Columbia have RPS policies as a matter of state law (Table 23.2).

In addition to progress on the policy front to promote renewable energy standards, some utilities have begun to embrace renewables on their economic and risk characteristics alone. As stated by a report of the Lawrence Berkeley National Laboratory (2005),

> Historically, markets for renewable generation have been created and supported primarily by policy efforts. But in recent years, utility resource

planning has emerged as an important driver of new renewable generation — particularly wind power — in some regions. This growing utility acceptance of renewable resources is motivated by their improved economics, and an increasing recognition of the inherent risks (e.g., natural gas price risk, environmental compliance risk) in fossil-based generation portfolios.

Bollinger and Wiser (2005)

Prospects for further gains through energy efficiency and renewable energy

Progress on increasing efficiency standards for vehicles, appliances, and buildings in keeping with what is possible technologically has occurred in fits and starts. Certainly there have been gains in energy efficiency overall. Standards and policies to reduce energy intensity coupled with shifts in the manufacturing sector away from energy-intensive processes have reduced energy use from what it would have been otherwise. One sign of this is that gross domestic product (GDP) has risen at a faster rate over the past few decades than energy use has — which indicates some gains in energy efficiency (Battles and Burns, 1998; Smil, 1994). Overall, however, energy use has risen 30 percent since the 1970s despite these measures. Electricity use grew about 50 percent nationally between 1980 and 1995, reflecting changes in the commercial and residential sectors, such as ownership of increased numbers of electric appliances, changes in household heating appliance choices, and a trend toward generally larger homes, with more area to heat and illuminate, and fewer numbers of persons per home (US DOE−EIA, 1995a, as cited in Morgenstern, 1996; and US DOE−EIA, 1995b: Chapter 3). Finally, while per capita energy consumption has remained fairly constant, total US population has increased by more than 42 percent since 1970 and overall consumption of products has increased, resulting in an overall increase in demand for energy during this time period.[8]

In the transportation sector, although CAFE standards have improved vehicle efficiency from the levels of the 1970s, total energy consumption has increased due to shifts in consumer choices in passenger vehicles (toward bigger, faster vehicles, some of which are not required to meet CAFE standards, such as light trucks and SUVs), increases in passenger miles traveled, and increase in distances that goods are shipped (Plotkin, 2004; US DOE−EIA, 1995b: Chapter 5). CAFE standards for passenger cars and light trucks have not been updated since 1987 (spanning administrations of both political parties) and combined with shifts in the distribution of types of vehicles preferred, the average fuel economy of the US fleet is now only 21 mpg, and has remained relatively stagnant for almost two decades (US Environmental Protection Agency, 2005).

Although it has been state and national policy to develop some mandatory efficiency standards, there is considerable resistance to the notion of imposing any standards at all by stakeholders who perceive their interests could be harmed. Whereas standards have been justified as a response to failures and barriers of the market for energy efficiency, those who oppose standards assert that any interference in the market – i.e., limiting the types of products available to consumers in some way – is detrimental to the optimal performance of the market. Analysts have argued, for example, that standards result in introduction of products into the marketplace that customers have not demanded (Bezdek and Wendling, 2005). Of course, this argument ignores the fact that the market is already out of equilibrium and operating inefficiently, e.g., in not representing externalities. Deciding where and how to adjust the market – whether through higher prices, standards, or public persuasion tactics – is a question of political preference or strategy, and can only be resolved through political processes.[9]

In the case of passenger cars and light trucks, for example, the automobile industry has consistently and successfully opposed raising the CAFE standards on the grounds that consumers have other priorities, namely safety, which allegedly conflict with raising standards (Schmidt, 2002). These claims have been disputed and the debate continues (Greene, 1998).

In summary, the complex picture of US energy production and consumption patterns suggests that efficiency standards for vehicles, appliances, and buildings can certainly go a considerable way toward reducing energy use. Standards, labeling, and incentive programs do play an important role, and more can be done to improve their effectiveness. However, standards and efficiency programs are only part of the picture, which includes consumption behavior, social trends, and market forces. Current trends suggest other approaches will need to be considered if society wishes to reach a much lower level of energy use or carbon dioxide emissions. These include fundamentally new ways of conceiving of building, automobile, and product designs;[10] encouragement of sustainable energy supplies; deliberate community planning to maximize efficient use of energy; and changes in attitudes towards economic and population growth and consumption (e.g., chapters by Michaelis; Jamieson; and Watrous and Fraley, Chapters 16, 30, and 25, this volume).

"Making it easy" to use energy wisely

Experience with past efforts at reducing energy consumption has demonstrated that policies cannot rely solely on consumer education efforts or the market alone. As we have seen time and time again, information destined for the general public is important – but not enough (see chapters by Dunwoody; Rabkin with Gershon; Chess and Johnson; and Tribbia, Chapters 5, 19, 14, and 15, this volume).

With respect to climate change, the goal, ultimately, would be to provide consumers with the widest range of energy products and services that also meet the social values of low carbon-based energy use, energy security, and energy sustainability. We argue that this would require structural changes based on decisions by government and industry to build in efficiency and the use of renewable sources of energy as best practice. Product designers, builders, manufacturers, policy-makers, and consumers all play roles in making this happen. Communication and education is a vital part of the overall strategy for product providers to build in energy-efficient practices and services for consumers without reducing the services that products provide such as comfortable living spaces, convenient transportation, communication, and entertainment. Based on past experience with energy-efficiency and renewable-energy programs, we suggest the following strategies for the future.

Lead by example

Demonstration programs are very important for exhibiting the technical feasibility of a wide variety of technologies. Potential adopters of new innovations need to be shown that the technology, product, or policy is effective at producing results. For example, once green-pricing programs (Table 23.2) were shown to be profitable to utilities, as well as providing them with considerable customer goodwill, hundreds of utilities got on the green-pricing bandwagon and began to offer renewable electricity to their customers.

Nurturing champions and innovators

Innovation champions within each industry play key roles in changing organizational cultures to make them more accepting of energy efficiency and renewable energy. For example, Ryan Green of SheaHomes was instrumental in the inception and initiation of the zero-energy home (ZEH) concept at their 306-home Scripps Highlands development in San Diego. Green had attended a seminar on ZEHs at the National Renewable Energy Laboratory (NREL) in 2000. Enthused about his new knowledge, Green returned to San Diego and became a champion of the ZEH concept, which was adopted by SheaHomes for their San Angelo and Tiempo developments at Scripps Highlands. In each area of expertise, innovation champions make an important difference in whether organizations will move forward with the needed changes to become more sustainable (see chapters by James *et al.*, duVair *et al.*, and Young, Chapters 20, 27, and 24, this volume). Fledgling spin-off businesses need to be incubated. Rewarding and publicizing successful innovations help to spread the word and amplify the effect of role models.

Targeting and tailoring information

Information on building energy efficiency and renewable energy use into daily life *can* be useful if it is tailored and targeted. For example, NREL and the Colorado Energy Science Center regularly hold workshops and briefings that are designed for specific target audiences. For farmers and ranchers in the arid US west, workshops are held at the National Western Stock Show each year. These audiences have unique concerns, such as the need for off-grid electricity or remote-area pumping to water stock. Agricultural organizations often request briefings on the use of wind turbines on their land where grazing and turbines can coexist.

Looking for the "win–win"

Energy characteristics are only part of a suite of overall preferences that product providers and consumers look for. Understanding the economic and market constraints under which businesses operate goes a long way toward designing solutions to build in energy efficiency. Finding options that save costs and improve energy efficiency requires good communication and trusting relationships between groups.

A number of organizations are following such "win–win" strategies. Special programs for builders (e.g., Building America Program) teach them how to achieve much greater efficiency in their new homes and commercial buildings within the costs they would normally spend for a non-efficient new building. The Federal Energy Management Program teaches a multitude of federal agencies with different needs how energy can be conserved on commercial buildings and in fleets, and how to use third-party contractors to pay for the retrofits without large up-front capital investments. Architects and planners are coached on designing for sustainability. Natural Capitalism Solutions, a non-profit, non-governmental organization, advises corporations and governments on cost-effective, and even profitable, approaches to employing sustainable energy. Businesses in energy efficiency and renewable energy products and services receive information on market assessments, cost factors, and economics. In addition to energy prices, more sustainable products and services can be marketed by tapping into other key values, such as personal and family safety, convenience, clean air and water, children's health, a stable climate, and national security.

Making it "cool" to save energy

Product and service choices are influenced by a range of informal as well as less tangible qualities. Take, for example, sales of the Toyota Prius, a hybrid car that is EPA-rated at 55 mpg. Enthusiasts have bought the Prius generally not just because of its environmental benefits and fuel efficiency, but because the car is also "cool." The features and comforts of the car do not suggest sacrifice. Owning a Prius has become for some a status symbol of sorts, an "excessory" to express a certain identity (Heffner *et al.*, 2005; see also Tribbia, Chapter 15, this volume). Friends tell friends. Word of mouth, peer pressure, or peer support can often have more of an impact than advertising campaigns or educational efforts (McKenzie-Mohr, 2000).

Know what works and what doesn't

Programs designed to change energy consumption need to be evaluated carefully and objectively. As shown by experience, even the best-intended programs can be less than effective, and it is important to evaluate, learn, and re-tool so that programs can improve.

Working to "make it easy" to make energy-wise choices in daily life includes efforts from a wide variety of stakeholders in the process, including manufacturers, builders, politicians, consumers, utility companies, advocates, and businesses. Building energy efficiency and renewable energy into our products and power grids depends upon technological innovation, organizational change, consumer and corporate cooperation, and public priority setting (see also James *et al.*, Chapter 20, this volume). It requires new understanding of the constraints and drivers of decisions at many different levels, and a communication strategy whose goal it is to make energy efficiency a built-in part of the menu of choices that consumers face every day. In short, "making it easy" is actually hard work.

Notes

1. 1 quad = 1 quadrillion Btu. Btu stands for British Thermal Unit, or the amount of energy required to raise the temperature of 1 pound of water by 1 degree Fahrenheit.
2. For example, many of the major US policies to address energy efficiency were enacted after the oil shock of the 1970s.
3. Many appliances now use electricity when plugged in even when the appliance is not in use, for example, printers which are turned on even when the user is not printing, appliances with digital clocks, and so on. These passive "plug loads" are increasing, and are sources of efficiency opportunities that are routinely overlooked.
4. See http://www.irs.gov/publications/p535/ch12.html; accessed February 10, 2006.
5. http://www.usgbc.org/DisplayPage.aspx?CategoryID=19; accessed February 8, 2006.
6. See http://www.bcap-energy.org/map_page.php; accessed February 6, 2006.
7. Data compiled by Farhar from the Wirthlin Group (1993) and National Renewable Energy Laboratory (2000).
8. US Census Bureau, 2004–2005, Table No. 2: Population: 1960–2003. Statistical Atlas of the United States, p. 7. See http://www.census.gov/prod/2004pubs/04statab/pop.pdf; accessed February 8, 2006.
9. Thanks to Nick Flores, University of Colorado Department of Economics, for suggesting this argument.
10. See, for example, the Rocky Mountain Institute (http://www.rmi.org/), and Bill McDonough Partners (http://www.mcdonough.com/); accessed February 8, 2006.

References

Anonymous. (2004). New and notable. *Home Energy*, **21**, 2, March/April, 45.

Banerjee, A. and Solomon, B. D. (2003). Eco-labeling for energy efficiency and sustainability: A meta-evaluation of US programs. *Energy Policy*, **31**, 109–23.

Battles, S. J. and Burns, E. M. (1998). United States Energy Usage and Efficiency: Measuring Changes Over Time. Paper presented at 17th Congress of the World Energy Council, Houston, TX, September 14. Available at http://www.eia.doe.gov/emeu/efficiency/wec98.htm; accessed February 6, 2006.

Bezdek, R. H. and Wendling, R. M. (2005). Potential long-term impacts of changes in US vehicle fuel efficiency standards. *Energy Policy*, **33**, 407–19.

Bollinger, M. and Wiser, R. (2005). Balancing cost and risk: The treatment of renewable energy in western utility resource plans. Lawrence Berkeley National Laboratory Report, LBNL 584450. Available at http://eetd.lbl.gov/ea/ems/reports/58450.pdf; accessed February 10, 2006.

Brown, M. A. (2001). Market failures and barriers as a basis for clean energy policies. *Energy Policy*, **29**, 1197–207.

Coburn, T. C. and Farhar, B. C. (2004). Public reaction to renewable energy sources and systems. *Encyclopedia of Energy*, **5**, 207–22. Available at http://www.sciencedirect.com/science/referenceworks/012176480X; accessed February 6, 2006.

Farhar, B. C. (1993). *Trends in Public Perceptions and Preferences on Energy and Environmental Policy.* NREL/TP-461-4857, Golden, CO: National Renewable Energy Laboratory, February. Available at http://www.nrel.gov/docs/legosti/old/4857.pdf; accessed February 6, 2006.

Farhar, B. C. (1994a). Trends: Public opinion about energy. *Public Opinion Quarterly*, **58**, Winter, 603–32.

Farhar, B. C. (1994b). Trends in US public perceptions and preferences on energy and environmental policy. *Annual Review of Energy and the Environment*, **19**, 211–39.

Farhar, B. C. (2001). Sun, wind, and water: A quiet revolution in renewable energy. *Public Perspective*, **12**, November–December, 14–16, 41–5.

Farhar, B. C. and Coburn, T. C. (2005). Energy policy. In *Polling America: An Encyclopedia of Public Opinion*, eds. Best, S. and Radcliff, B. Westport, CT: Greenwood Press, pp. 183–94.

Farhar, B. C. and Coburn, T. C. (2006). *A Comparative Market and Utility Analysis of New High-Performance Homes in San Diego.* Golden, CO: National Renewable Energy Laboratory.

Farhar, B. C., Coburn, T. C., and Murphy, M. (2004). Large production home builder experience with zero energy homes. In *Proceedings of the 2004 ACEEE Summer Study on Energy Efficiency in Buildings, Breaking out of the Box*, eds. Ledbetter, M. and McKane, A. Pacific Grove, CA, August 24, pp. 8-09–8-19.

Farhar, B. C., Weis, P., Unseld, C. T., *et al.* (1979). *Public Opinion About Energy: A Literature Review.* SERI/TR-53-155, Golden, CO: Solar Energy Research Institute.

Greene, D. L. (1998). Why CAFE worked. *Energy Policy*, **26**, 595–613.

Greene D. L., Patterson, P. D., Singh, M., *et al.* (2005). Feebates, rebates and gas-guzzler taxes: A study of incentives for increased fuel economy. *Energy Policy*, **33**, 757–75.

Hammond, R. (2005). The near-zero-energy house. *Solar Today*, **19**, 22–5.

Heffner, R. R., Kurani, K. S., and Turrentine, T. S. (2005). Effects of vehicle image in gasoline-hybrid electric vehicles. White Paper of the Institute of Transportation Studies, University of California, Davis. Available at http://www.its.ucdavis.edu/publications/2005/UCD-ITS-RR-05-08.pdf; accessed February 6, 2006.

Hering, G. (2005). Solarizing the American dream. *PHOTON International*, May 5, 80–85.

Keesee, M. (2005). Setting a new standard: Zero energy homes in the US. *reFocus*, July/August, 26–8.

McKenzie-Mohr, D. (2000). Fostering sustainable behavior through community-based social marketing. *American Psychologist*, **55**, 531–7.

Meyers, S., McMahon, J. E., McNeil, M., *et al.* (2003). Impacts of US federal energy efficiency standards for residential appliances. *Energy*, **28**, 755–67.

Morgenstern, R. (1996). *Does the Provision of Free Technical Information Really Influence Firm Behavior?* Resources for the Future, Discussion Paper 96-10. Available at http://www.rff.org/Documents/RFF-DP-96-16.pdf; accessed February 6, 2006.

Murr, A. (2005). No more electric bills. *Newsweek*, August 15, 43.

Nadel, S. (2002). Appliance and equipment efficiency standards. *Annual Review of Energy and Environment*, **27**, 159–92.

National Research Council (2002). *Effectiveness and Impact of CAFE Standards.* Washington, DC: National Academy Press.

Schmidt, C. (2002). Debate percolates over CAFE standards. *Environmental Health Perspectives*, **110**, A467–8.

Smil, V. (1994). *Energy in World History*. Boulder, CO: Westview Press.

Smil, V. (2003). *Energy at the Crossroads: Global Perspectives and Uncertainties.* Cambridge, MA: The MIT Press.

US Census Bureau (2004–2005). Table No. 2: Population: 1960–2003. Statistical Atlas of the United States, p. 7. Available at http://www.census.gov/prod/ 2004pubs/04statab/pop.pdf; accessed February 6, 2006.

US Department of Energy (2005). *2005 Buildings Energy Databook*. Silver Spring, MD: D&R International, Ltd. (Table 1.1.3). Available at http://buildingsdata book.eren.doe.gov/docs/1.1.3.pdf; accessed February 8, 2006.

US Department of Energy, Energy Information Administration (DOE–EIA) (2005a). *Monthly Energy Review*, January (p. 7, Table 1.3). Washington, DC: DOE.

US DOE–EIA (2005b). *Monthly Energy Review*, August (p. 7, Table 1.3). Washington, DC: DOE.

US DOE–EIA (2005c). *Short Term Energy Outlook*, October 2005 (Fig. 15). Washington, DC: DOE. Available at http://www.eia.doe.gov/emeu/steo/pub/ gifs/Slide16.gif; accessed February 6, 2006.

US DOE–EIA (1995a). *Housing Characteristics 1993*. Pub. DOE/EIA–0314(93). Washington, DC: DOE.

US DOE–EIA (1995b). *Measuring Energy Efficiency in the United States Economy: A Beginning*. Washington, DC: DOE. Available at http://www.eia.doe.gov/ emeu/efficiency/; accessed February 6, 2006

US Environmental Protection Agency (2005). *Light-Duty Automotive Technology and Fuel Economy Trends*: 1975 through 2005. Washington, DC: US EPA. Available at http://www.epa.gov/otaq/fetrends.htm; accessed February 6, 2006.

Vine, E. (2002). Energy Efficiency in Buildings as an Air Quality Compliance Approach: Opportunities for the US Department of Energy. LBNL-49750. Berkeley, CA: Lawrence Berkeley National Laboratory.

24

Forming networks, enabling leaders, financing action: the Cities for Climate Protection™ campaign

Abby Young

ICLEI—Local Governments for Sustainability

> Local governments are well placed to get to the nub of the [climate] problem in a way that global treaties and carbon trading can never achieve. They can get the message across to the residents, shopkeepers and drivers of the world that efficient, cheap-to-run homes, fridges, cars and so on can improve their quality of life while helping a global cause. . . . It's refreshing to see that bottom-up local action still has an important part to play in tackling climate change. Power to the people indeed!
>
> *Jim Gillon (2004)*

Local governments have been the leading champions for climate protection in the United States for over a decade. While not funded or mandated by higher levels of government to do so, local governments have taken on the burden of reducing their communities' contribution to global warming. What motivates this action? In over ten years of working with local governments on climate protection planning, ICLEI—LOCAL GOVERNMENTS FOR SUSTAINABILITY has identified common factors that facilitate local action. These include the presence of a local champion, positive effects on the financial bottom line, and supportive networks of local governments across the country making the same commitments and taking similar actions. The campaign is introduced in the section below, while these key factors are discussed in the remainder of the chapter. Observations from what has worked in these local communities hold valuable practical lessons for communication and social change.

The Cities for Climate Protection™ campaign

ICLEI—LOCAL GOVERNMENTS FOR SUSTAINABILITY operates an international program called Cities for Climate Protection (CCP) that started in 1993.[1]

Through this program, ICLEI engages local governments around the world in developing targets and timelines and implementing local action plans for reducing greenhouse gas emissions from their communities. Over 200 US cities and counties are participating in this program. All of them have committed to completing ICLEI's methodology for climate protection, outlined in five milestones:[2]

1. Conduct a greenhouse gas emissions inventory and projection
2. Establish an emissions reduction target
3. Develop a local action plan to reduce emissions
4. Implement the local action plan
5. Monitor progress and report on results.

By demonstrating the effectiveness of local action at reducing greenhouse gas emissions, ICLEI has brought these local governments into the program. Often, simply drawing connections between the sources of emissions and the policy levers that local governments control (e.g., building codes, land use planning, and zoning) is enough to convince local officials that they do have the power to significantly impact these emissions.

Agreeing that something should be done about global warming and actually finding a way to make that happen within the local government framework are two very different things, however. Getting governments to move from desire to take action to implementing a policy approach is the biggest hurdle in engaging them to act on climate change. In order to achieve this several things need to occur. First, local decision-makers must understand how the actions of their local government can make an impact on this global problem and how those actions can bring many benefits to the community. Second, there needs to be a local champion pushing this issue through the governing body. And third, the local government must feel that it is not taking on this issue alone, either within the community or regionally/ nationally.

Effective communication: making the case to local governments

Why would any local government, beset with layers of unfunded mandates to fulfill, voluntarily take on a challenge as grand as fighting global warming? Somehow, arguments in favor of this course of action have resonated with many local government decision-makers. In trying to communicate the case for local communities to take action on climate change, there are many choices among communication styles and messages to use. What works, what does not, and why?

In order to answer these questions, the "local government condition" must be understood. ICLEI's experience suggests this local government condition has at least three distinct facets: the high degree of officials' accountability to local residents; the considerable financial constraints faced by local governments; and the challenge of a long-term perspective.

State and federal government officials are removed, physically and functionally, from the constituencies they serve. That layer of separation affords state and federal lawmakers a bit more relief from the day-to-day accountability to their electorate with which local officials are faced. As the level of government closest to individuals, families, businesses, and community concerns, local governments do not enjoy the benefits of this separation. Local officials cannot escape the full-frontal accountability in which they are held by their communities on a daily basis. These officials are our neighbors; they eat at the next table in the local restaurant; their children play and go to school with ours; we bump into them at the grocery store. This places tremendous, inescapable demands on the financial, human, and other limited resources of local governments.

Local governments are further challenged by severe constraints in their financial means. Because of a general trend toward government devolution, aggravated by budget constraints at the federal and state levels, local governments are asked, through unfunded mandates or simply because of cuts of support from higher levels of government, to take on greater financial responsibility for issues ranging from disaster management, to hazardous waste disposal, to public education. A common concern among local officials considering whether or not to participate in the CCP campaign thus is, "My local government can't afford to do anything that is not already mandated. How am I supposed to convince my council or city manager to take on global warming on top of everything else?"

Against this backdrop of ever-tightening financial belts, the pervasive argument *for* taking climate change action locally is because it can save money, because it makes economic sense. This has made taking action "because it's the right thing to do" significantly easier, but even without the economic incentives, many local governments simply feel the need to step into the leadership vacuum and tackle climate change (see below for more examples, and the experience of the city of Santa Monica described in Watrous and Fraley, and that of Portland, described in duVair *et al.* and Rabkin with Gershon, Chapters 25, 27, and 19, this volume). However, making a decision to pursue a policy action such as significantly reducing greenhouse gas emissions, switching municipal operations over to renewable power, or protecting the natural environment takes a long-term perspective. And yet, taking that

"100-year view" is often very difficult to do in local governments because of election and budget cycles, economic downturns, and other constraints.

Given these challenges, how can arguments for engaging in climate protection be successfully crafted and delivered? Experience has shown that an effective communication strategy involves answering why local government officials should care, and what they can actually do.

In answering the first question, "why should you, a local official, care?", there is a great deal of scientific information building a convincing case for aggressive climate protection at all levels of government (e.g., IPCC, 2001a,b; National Assessment Synthesis Team, 2001). Information on the potential impacts of global climate change is of particular interest to local decision-makers. For example, officials in coastal cities such as San Diego (see Pratt and Rabkin, Chapter 6, this volume), need to know how sea-level rise will impact stormwater and sewage drainage systems. Policy-makers in Chicago need to be able to prepare for increased severity of heatwaves in order to avoid a repeat of the catastrophic one in 1995 that killed more than 800 people. While studies of observed local changes due to rising temperatures are slowly becoming available, projections of future changes are limited by what science can provide credibly at this time. This leaves a great deal of uncertainty with regard to the nature and significance of climate impacts on a given community. In particular, the science of climate change impacts is not yet precise enough to assign probabilities and dollar figures to that risk. When faced with uncertainty, many decision-makers are hesitant to embark on new policy or reallocate limited resources (see Dilling and Moser, Introduction, this volume). Still, scientific evidence is a powerful tool that can support other arguments in creating an effective message.

The other pieces of an effective message include connecting climate protection with other pressing local issues such as air pollution, traffic congestion, green space, and waste reduction, thus demonstrating important co-benefits to climate protection policies, and providing a peer network of local government officials already taking action to reduce greenhouse gas emissions (see also Lindseth, 2003, 2004).

In ICLEI's experience it is very helpful to communicate that climate protection is not a new policy arena — rather, many strategies to reduce greenhouse gas emissions are commonplace policies that have been around for years: energy conservation programs, bicycle and pedestrian programs, public transportation, and using solar power. In practical terms, this means giving municipalities tools that make that linkage for them. For example,

ICLEI has developed a tool for local governments that harmonizes the quantification of both greenhouse gas and regulated air pollutant emissions. This and other tools demonstrate that many of the actions that reduce greenhouse gas emissions – such as implementing car-pool programs or bike lanes – will also improve local air quality and help the jurisdiction achieve attainment under federal Clean Air Act rules. Moreover, the tools demonstrate how a particular policy action can help the community save energy and thus money, in terms of reduced utility bills.

Finally, peer pressure is another powerful means to motivate action at the local level. Very few local governments want to be the first to try something new. In 1995, only ten local governments in the United States were participating in CCP, and those were largely the "usual suspects" – cities such as Portland, Berkeley, and San Francisco – all known for their progressive politics. After this cluster of groundbreakers began demonstrating cost-effective results with numerous co-benefits (improved local air quality, financial savings, increased community livability), it became easier to make the case to other cities that climate protection was a legitimate policy option for local governments. As of this writing, the national network of cities and counties participating in CCP totals more than 200 jurisdictions, many of which one would not typically expect to pursue progressive environmental policies: Toledo, Ohio; Tucson, Arizona; and Durham, North Carolina. Developing messages that effectively resonate with local officials is only the first step, however. The next is to deliver it to sympathetic ears within a local government.

It takes a champion

Why does one community choose to pursue climate protection and another does not? One of the biggest motivating factors is the presence of a local champion (see also Kousky and Schneider, 2003; Vasi, 2003; Lambright, Changnon, and Harvey, 1996). The highest-performing local governments in the CCP program are those in which there is an elected official of some power who is personally committed to the issue. The champion may be a mayor who is particularly keen on environmental protection, or a council member who has just read a frightening article on the disappearing polar ice caps, or an elected official or senior manager who is very supportive of a related policy path, such as smart growth or energy efficiency. Armed with sufficient information and resources, the champion can make persuasive arguments to his or her colleagues in the governing body, ultimately gaining the endorsement to pursue climate protection as a local policy priority.

ICLEI makes great efforts to identify an influential local champion and work with him or her to develop a unique communications strategy that will be effective for that particular government. Generally, that strategy will draw from all the elements discussed above: scientific evidence of climate change impacts to the region and potential risks to the community; existing policy priorities within the local government that complement or will be supported by the addition of a climate protection program; and examples of co-benefits that the local government and community will likely reap as a result of pursuing climate protection, quantified if possible.

It is important for the local champion to know who (potential) local supporters are (individuals, businesses, community groups, etc.), as well as the existence of regional and national networks of local governments working together on this issue. In states where there is significant policy action at the state level, such as in California, Massachusetts and Connecticut, a growing number of local governments participate in the CCP campaign, perhaps reflecting the more progressive politics in those regions more generally. In New England, most CCP communities joined ICLEI prior to the states embracing climate protection as a regional goal, but, of course, actions at different levels can support and reinforce each other (see duVair *et al.* and Tennis, Chapters 26 and 27, this volume), which has in turn attracted the attention of the funding community (e.g., the US Environmental Protection Agency, and private foundations that fund climate protection activities). In California, this may be more related to the general progressive nature of policy-making in the state (see duVair *et al.*, Chapter 27, this volume).

To participate officially in the CCP campaign, a local government must pass a resolution through its governing body (the city council, or board of supervisors, aldermen, or selectmen) committing to reaching all of the five milestones described above. The local champion works with ICLEI to develop and secure the passage of the resolution. This political act is imperative, as it institutionalizes the commitment of the local government to climate protection. It also provides local government staff with the directive to do the work that is associated with assessing greenhouse gas emissions and developing policy solutions to reducing them.

Once the resolution is passed, the real work begins. Two critical elements to ensuring the success of the local government's climate protection commitment are securing the support of the public, and identifying creative ways to finance the implementation of emission-reducing activities.

Outreach to the public

Most local governments participating in ICLEI's CCP campaign have found, to their pleasant surprise, that there is a large amount of support within the community for the official decision to pursue climate protection (see also Kousky and Schneider, 2003). This support can come from all corners of the community – businesses that want to improve their bottom line by becoming energy efficient (see Arroyo and Preston, Chapter 21, this volume), houses of worship that consider climate protection an issue of global stewardship (see Bingham, Chapter 9, this volume), performing-arts troupes, and, of course, environmental groups. The challenge lies in reaching the general populace that may not be represented by these groups.

ICLEI works with local governments to create outreach programs and strategies for communicating the urgency of global warming to its businesses and residents. In cities such as Austin, Texas, and Fort Collins, Colorado, town hall meetings with speakers from academia, the local government, and community groups have been very effective in building public support for climate protection. In Burlington, Vermont, the Climate Alliance, a group of local governments and community organizations, launched the "10% Challenge"[3] through an interactive website, a commissioned musical on global warming produced by a local theater group, and direct mailings to households, encouraging businesses and individuals to match the city's goal and pledge to reduce emissions by 10 percent. In Portland, Oregon, the city's BEST program (Businesses for an Environmentally Sustainable Tomorrow)[4] provides assistance to area businesses to save energy and greenhouse gas emissions and recognizes their accomplishments.

Creative financing of local action

Once a local government has committed to pursuing an aggressive climate protection policy agenda, how does it turn its good intentions into action? As mentioned above, the biggest stumbling block for any local government is finding the resources to: dedicate staff time to coordinating the climate protection program, develop new policies and programs, implement them in practice, and assess the effectiveness of the local government's approach.

Local governments in the CCP campaign have taken a variety of approaches to financing their climate protection programs (for additional examples and discussion see Kousky and Schneider, 2003). The rest of this chapter provides examples of the financing strategies these local governments have used to put their policy intentions into action. They fall into six broad categories: no-net-cost policies, longer-term investments in energy efficiency, pollution charges, creative new funds, financing partnerships, and bond measures.

No-net-cost policy approaches

Many local governments can achieve financial savings in reduced energy and operations costs by making no-net-cost policy changes. These savings can then be used to finance additional emission-reducing activities or public outreach campaigns.

Energy efficient purchasing

Together, state and local governments in the United States spend an estimated US$12 billion per year on energy bills and another $50–70 billion per year on energy-related products (Harris, no date). The simplest, cheapest way for local governments to reduce these costs is to adopt purchasing policies that favor the most energy-efficient products available. Many local governments across the country have implemented such policies. In Massachusetts, where the state has a comprehensive "Environmentally Preferable Purchasing" policy (see also Tennis, Chapter 26, this volume), many local governments save money by buying through the state purchasing contracts. In Burlington, Vermont, over 80 percent of all equipment purchased is rated ENERGY STAR® or higher in terms of energy efficiency.[5] This equipment is usually of comparable price to more energy-intense products, and yet has a much shorter payback period.

Denver "Green Fleets" policy

The City and County of Denver, Colorado operates a combined fleet of 3,500 vehicles. Faced with rising fuel costs, increased air pollution, and federal mandates to clean the city's air, Denver adopted a "Green Fleets" policy.[6] Under this policy, fleet managers must purchase the most cost-effective and lowest-emission vehicles possible, while meeting operational requirements of the department. In order to accomplish this goal, fuel efficiency standards are included in procurement specifications. The Green Fleets process also includes reducing vehicle size and eliminating old and underused vehicles. As a result of this no-net-cost policy change, the city has reduced the number of vehicles in its various fleets, is purchasing smaller, more fuel-efficient vehicles, and is saving US$40,000 per year in reduced operations and maintenance costs and an additional US$100,000 in reduced capital expenditures.

Energy efficiency savings fund longer-term investments

Many local governments approach their climate protection work in phases. The first phase typically focuses on the "low-hanging fruit," or actions that result in quick payback periods and long-term financial savings. These usually take the form of energy efficiency upgrades.

Tucson sustainable energy building standard

The City of Tucson, Arizona, established a "Sustainable Energy Standard" that requires all construction and major renovation of municipal buildings meet energy efficiency standards 50 percent greater than those of the national Model Energy Code.[7] The Sustainable Energy Standard suggests various conservation measures but allows architects freedom in choosing exactly how to meet the higher efficiency requirements. Designers must detail conservation strategies and perform an energy analysis early in the design process. The city then monitors energy efficiency throughout the contracting, inspection, and testing phases. Builders have found the savings surprisingly easy to achieve. So far, the Standard is saving the city nearly US$200,000 annually and reducing 1,600 tons of CO_2 emissions each year.[8]

Philadelphia LED traffic signals

Red LED (light-emitting diodes) traffic signals were installed in all 2,900 intersections (28,000 light fixtures) in Philadelphia, Pennsylvania. The new signals use 83 percent less energy and require six times less maintenance than incandescent lights, given their much longer lifespan. This one simple project is saving the city over US$800,000 annually in reduced energy and maintenance costs. These savings are being reinvested in additional energy efficiency projects that have heavier up-front costs and longer payback periods.[9]

Charging for pollution

An approach gaining popularity is to generate funds for emission reduction programs by imposing fees on polluters. Several local governments have come up with creative strategies to penalize and generate income from sprawl-inducing activities.

Sprawl generally implies that people need to travel longer distances to jobs, generating emissions. Wise land-use planning and development can make a difference in pollution and energy use as well as transportation problems.

Lancaster distance surcharge

Lancaster, California uses impact fees to reduce urban sprawl by encouraging the location of new urban growth within or adjacent to the urban center. The impact fees reflect the additional cost to the city of expanding infrastructure to service each new development. In this manner, the fee structure provides a rational incentive for developers to build closer to the urban core, thus reducing sprawl. Under its Urban Structure Program (USP), Lancaster imposes a distance surcharge on development fees. The surcharge is calculated based on the distance between the proposed project and the existing urban center. The fees are assessed to new development based on expected fiscal impacts on the city, including infrastructure, utilities, sewer systems and other public services. The surcharge is levied for a period of 20 years, and the farther from town the project is, the higher the charge.[10]

San Francisco Transit Impact Development Fee

Impact fees are commonly applied to developers for parks and schools, but transportation impact fees are rare. The City and County of San Francisco adopted a development fee for all new downtown office construction to provide funding for transit services that such new employment centers would require.[11] The San Francisco's Transit Impact Development Fee (TIDF) moneys have funded increases in transit services to meet peak demand generated by new downtown businesses. The fee of US$5 per square foot is assessed on new office construction and conversions to commercial office space within a designated downtown district. Funds are paid directly to the Municipal Transit District, which operates San Francisco's light rail, cable car, and bus services. Since inception in 1981, the program has generated US$85 million in fees, covering 1.5 percent of the Municipal Transit District's annual operating costs. Clearly, this fee system was not put in place back then for climate change reasons and long before the city's climate change action plan was released in September 2004,[12] but it is a good example for how communities are creatively raising the funds needed to implement climate-relevant programs.

Alameda County, California, waste surcharge

The waste deposited in landfills produces two types of greenhouse gas emissions − carbon dioxide and methane, which has a global warming potential 25 times stronger than carbon dioxide. Anything local governments can do to minimize the amount of waste going to landfills through recycling and reuse programs goes a long way to reducing greenhouse gas emissions.

At the heart of Alameda County's "Source Reduction and Recycling Initiative" is a surcharge imposed on each ton of waste sent to landfill in the county. The over US$7 per-ton surcharge is a dedicated revenue source for stable, long-term funding of waste reduction programs.[13] This surcharge has generated approximately US$100 million in revenues since inception in the early 1990s.[14] The Initiative emphasizes "closing the loop" by building local markets for recycled materials. The Initiative established a US$2 million revolving loan fund and a grant program for non-profits and private businesses. Projects benefiting from this funding include a demonstration project that trains low-income youth to dismantle buildings in ways that preserve lumber, structural steel, cables, and other materials for reuse.

Setting up creative funds

Local governments can use creative financing to jump-start the implementation of energy efficiency projects. In the early 1990s, the City of Saint Paul, Minnesota, for example, negotiated with its utility company for a 0 percent loan to finance its LED project. In addition, city staff coordinated a group purchase with neighboring municipalities, obtaining the lowest LED signal prices in the country.[15]

Los Angeles Rideshare Trust Fund

The City of Los Angeles and its employee bargaining units agreed to a unique arrangement regarding commuter benefits and employee parking. The new arrangement rewards those who rideshare and penalizes solo drivers. Parking fees from solo drivers are used to support rideshare programs. Parking permit fees go to the interest-earning Rideshare Trust Fund. Unlike typical "use-it-or-lose-it" budgets, unspent funds in one fiscal year carry over into the following fiscal year. The Commuter Services Office then applies these monies to its entire program; the initiative is thus relatively insulated from the effects of year-to-year tax revenue shortfalls in the City's General Fund. Trust Fund expenditures are primarily directed toward subsidizing van-pools and employee transit passes. They also support car-pool match lists, purchase and installation of bicycle lockers, and office expenses related to these programs.

Ann Arbor revolving loan fund

In 1998 the City of Ann Arbor, Michigan, set up a Municipal Energy Fund to finance internal energy efficiency upgrades and improvements.[16] The Fund was created with US$100,000 in one-time funding from an unrelated annual fee the city was paying that was about to expire without further action. The funds are available for city departments on a competitive basis. Funding helps pay for upgrading or replacing energy equipment or systems with more efficient technologies. The department or facility receiving the funding repays the loan with 80 percent of the energy savings over a maximum of five years. After that, 100 percent of the energy savings stay with the department or facility. In this manner, Ann Arbor has established a self-sustaining fund to increase efficiency and reduce emissions far into the future.

Building financing partnerships

Large-scale projects that require significant up-front investment often require the pulling together of funding from multiple sources. Local governments have become expert at identifying and attracting potential, even if unconventional, funding sources.

Oregon fuel cell project

The City of Portland installed the world's first city-sponsored, anaerobic digester gas (ADG) fuel cell in its wastewater treatment plant in 1999. The Fuel Cell Power Plant converts methane, a primary constituent of ADG, into electricity, generating power in a virtually pollution-free operation. Portland installed a 200 kW hydrogen fuel cell to help utilize its waste methane and reduce power plant air emissions. The result of this pilot installation is a net reduction of 694 tons of greenhouse gas emissions annually – 14,000 tons over the life of the fuel cell. The fuel cell displaces the need for emergency generators or uninterruptible power supply valued at US$150,000.[17]

The net cost of the project was about US$790,000. The city built a coalition of funding partners for the project, including Portland General Electric, the State of Oregon Business Energy Tax Credit Program, Oregon Office of Energy, Oregon's Biofuels Program, local lender Western Bank, and the US Department of Defense. Portland's long-range plan is to add additional hydrogen fuel cells or other technologies to produce clean, green power.

Bond measures

For large-scale investment projects, local governments can also use bonds as a financing mechanism, payable over the long term, often through energy savings.

Toledo municipal retrofit program

In order to reduce energy use and comply with air quality and chlorofluoro-carbon (CFC) regulations, Toledo, Ohio made comprehensive retrofits of 37 city buildings and facilities. Energy efficiency upgrades were completed for the municipal court, garages, sewer maintenance, health, police, and fire departments.[18] The city sold bonds to finance the program, contracting with a systems controls company to do the improvements. The contract guarantees that energy savings will pay back the bonds — the controls company covers any shortfalls and any net savings accrue to the city. This financing system has allowed building improvements of over US$10 million to be installed with no out-of-pocket expenses. Annual results of the program have already exceeded the contractor's guaranteed energy savings: US$710,000 in reduced energy costs and 5,250 tons of CO_2 reduced in the first year alone.[19]

San Francisco solar bond initiative

Renewable energy is one of the most difficult emission-reducing actions for local governments to implement on a large scale. This is due to the high up-front costs of large-scale distributed generation projects. The City and County of San Francisco found a way around this problem when, in 2001, 73 percent of the city's voters approved a US$100 million bond measure to finance solar, wind power, and energy efficiency projects throughout the city. The bond measure pays for itself through the reduced energy bills resulting from the utilization of alternative energy and decreased energy load through aggressive efficiency upgrades. The first solar project funded by the bond measure has been completed — a 675 kilowatt system at the Moscone Convention Center. The system, when combined with energy efficiency upgrades, is saving the Center US$210,000 per year in reduced energy costs.[20]

Conclusions: global change through local action?

Although local governments are powerful actors in the fight against global warming, there are limits to what local policy-making can accomplish (e.g., Betsill, 2001; Victor, House, and Joy, 2005; Kosloff, Trexler, and Nelson, 2004). As Kousky and Schneider (2003: 370) noted, "Municipalities are taking a powerful first step, but due to their constraints, they must be mirrored in their efforts by top—down incentives from higher scales for larger reductions to occur." The effectiveness of local government climate protec-tion strategies is hampered by inactivity at the federal level to control vehicle fuel efficiency or to set meaningful requirements for renewable energy in the electricity sector. For all that local governments do to promote public transit and provide alternatives to single-passenger car trips, those gains in reducing gasoline consumption are completely eaten up by the falling efficiency of the country's vehicle stock as sport utility vehicles and mini-vans continue to grow in popularity and size (see Dilling and Farhar, Chapter 23, this volume). There is a similar problem with electricity consumption. Local conservation and efficiency programs can be very effective at reducing the demand for electricity. Their effectiveness could increase dramatically if the carbon content of the nation's electricity were significantly reduced. This would require strong mandates at the federal level for electric utilities to reduce the amount of carbon-heavy coal used to produce electricity in favor of cleaner renewable energy sources such as wind (see also Kousky and Schneider, 2003). Local government powers do not reach automakers or electric utilities, save for the small number of municipal utilities in the United States, making these ongoing challenges to local climate protection efforts.

The more than 200 local governments participating in the CCP provide a shining example of local action to solve a global environmental problem. Based on ICLEI's internal compilation of data, these communities are collectively working to reduce an identified 23 million tons of greenhouse gas emissions each year. ICLEI communities estimate that the actions they are taking to achieve these cuts are saving approximately US$535 million annually in reduced energy and fuel costs.[21] The existence of this climate protection network and the results of its actions demonstrate to communities across the country that climate protection is a viable and tested policy route for local governments.

The growing CCP network reflects a trend among communities choosing to take on climate protection as a municipal policy priority, despite, and maybe because of, inaction on this issue at the federal level. Local governmental leaders are stepping into this leadership vacuum and are becoming national champions on climate protection. In July 2005, in Sundance, Utah, the first national mayors' meeting on global climate change took place. Inspired and co-sponsored by Mayor Rocky Anderson of Salt Lake City, ICLEI convened the Sundance Summit: A Mayors' Gathering on Climate Protection[22] to motivate a new cadre of mayors from communities not normally associated with innovative environmental policy-making to take action on climate protection. Also during the summer of 2005, Mayor Greg Nickels of Seattle launched a national sign-on campaign to get mayors across the country to commit to reducing greenhouse gas emissions in their communities. ICLEI is working with mayors who attended the Sundance Summit and those who signed on to the Mayors' Agreement on Global Warming to begin implementing actions to support their commitments.

The growing number of communities participating in the CCP campaign, together with the recent mayoral efforts, reflect a groundswell of opinion in the United States that global climate change puts communities at risk, and that climate protection is an appropriate and beneficial policy direction for local governments. Even if local communities can't solve the global warming problem on their own, they are gathering important experience with policy options (e.g., McKinstry, 2004), while sending an ever-stronger signal to state, federal, and international policy-makers that action is possible, cost-effective, and politically supported.

Notes

1. A pilot project called the "Urban CO_2 Reduction Project," launched in 1992, preceded the CCP.
2. See http://www.iclei.org/index.php?id=800; accessed January 17, 2006.
3. See http://www.10percentchallenge.org; accessed January 17, 2006.
4. See http://www.sustainableportland.org/energy_com_best.html; accessed January 17, 2006. For additional information on Portland, see the chapters by Rabkin with Gershon, and DuVair *et al.*, Chapters 19 and 27, this volume.
5. Personal communication to ICLEI from Burlington city staff, 2001.
6. See http://www.denvergov.org/SDIMO/2049aboutus.asp; accessed January 17, 2006. See also a more extensive discussion of Denver's climate change action plan and transportation strategies in Bulkeley and Betsill (2002).
7. See http://www.tucsonmec.org/ses.htm: accessed January 17, 2006.
8. Reported to ICLEI by City of Tucson Energy Office staff, November 2005.
9. Reported to ICLEI by City of Philadelphia staff in 2000 for its report "U.S. Communities Acting to Protect the Climate."
10. See http://www.newrules.org/environment/lancaster.html; accessed January 17, 2006.
11. See http://depts.washington.edu/trac/concurrency/lit_review/tcrp31_b.pdf; accessed January 17, 2006. See also http://www.spur.org/documents/010801_article_01.shtm; accessed January 17, 2006.
12. See http://www.sfenvironment.com/aboutus/energy/cap.htm; accessed January 17, 2006.
13. See http://www.stopwaste.org/docs/measure-d.pdf; accessed January 17, 2006.
14. Personal communication with Tom Padilla, Alameda County Waste Management Authority, November 2005.
15. Personal communication to ICLEI staff.
16. See http://www.ci.ann-arbor.mi.us/CityAdministration/EnvironmentalCoordination/EnergyFund.html; accessed January 17, 2006.
17. See http://egov.oregon.gov/ENERGY/RENEW/Biomass/FuelCell.shtml; accessed January 17, 2006.
18. See http://www.usmayors.org/uscm/best_practices/bp_volume_2/toledo.htm; accessed January 17, 2006.
19. Reported to ICLEI by City of Toledo staff for its 2000 report "U.S. Communities Acting to Protect the Climate."
20. See http://www.powerlight.com/case-studies/state/san_francisco.shtml; accessed January 17, 2006.
21. See also Betsill (2001) and Kousky and Schneider (2003) for partial, but independent, assessments of CCP cities' cost and emissions savings.
22. See http://www.sundancesummit.com; accessed January 17, 2006. See also Griscom Little, A. (2005). The revolution will be localized. *Grist Magazine*, July 20, 2005. Available at http://www.alternet.org/story/23608/; accessed January 17, 2006.

References

Betsill, M. M. (2001). *Acting Locally, Does It Matter Globally? The Contribution of U.S. Cities to Global Climate Change Mitigation.* Paper presented at the Open Meeting of the Human Dimensions of Global Environmental Change Research Community, Rio de Janeiro, Brazil, October 6–8, 2001.

Bulkeley, H. and Betsill, M. (2002). Denver: climate protection, energy management and the transport sector. In *Cities and Climate Change*, eds. Bulkeley, H., and Betsill, M. London: Routledge, pp. 123–36.

Gillon, J. (2004). People power against climate change (Correspondence). *Nature*, **430**, 15.

Harris, J., *et al.* (no date). Energy-Efficient Purchasing by State and Local Government: Triggering a Landslide down the Slippery Slope to Market Transformation. Berkeley, CA: Lawrence Berkeley National Laboratory. Available at: http://www.dc.lbl.gov/LBNLDC/publications/Energy%20Efficient%20Purchasing%20By%20State%20and%20Local%20Government.pdf; accessed January 17, 2006.

Intergovernmental Panel on Climate Change (IPCC) (2001a). *Climate Change 2001: The Scientific Basis.* Contribution of Working Group I to the Third Assessment Report of the IPCC. New York: Cambridge University Press.

Intergovernmental Panel on Climate Change (IPCC) (2001b). *Climate Change 2001: Impacts, Adaptation and Vulnerability.* Contribution of Working Group II to the Third Assessment Report of the IPCC. New York: Cambridge University Press.

Kosloff, L. H., Trexler, M. C., and Nelson, H. (2004). Outcome-oriented leadership: how state and local climate change strategies can most effectively contribute to global warming mitigation. *Widener Law Journal*, **14**, 173–204.

Kousky, C. and Schneider, S. H. (2003). Global climate policy: Will cities lead the way? *Climate Policy*, **3**, 359–72.

Lambright, H. W., Changnon, S. A., and Harvey, D. (1996). Urban reactions to the global warming issue: Agenda setting in Toronto and Chicago. *Climatic Change*, **34**, 463–78.

Lindseth, G. (2003). The Framing of an International Concerted Action for Local Climate Policy: The Case of the Cities for Climate Protection (CCP). VF-rapport, November. Sogndal, Norway: Western Norway Research Institute.

Lindseth, G. (2004). The Cities for Climate Protection Campaign (CCP) and the framing of local climate policy. *Local Environment*, **9**, 4, 325–36.

McKinstry, R. B. Jr. (2004). Laboratories for local solutions for global problems: state, local and private leadership in developing strategies to mitigate the causes and effects of climate change. *Penn State Environmental Law Review*, **12**, 1, 15–82.

National Assessment Synthesis Team (2001). *Climate Change Impacts on the United States: The Potential Consequences of Climate Variability and Change, Overview.* US Global Change Research Program. New York: Cambridge University Press.

Vasi, I. B. (2003). Organizational Environments and Compatibility: The Diffusion of the Program against Global Climate Change among Local Governments in the U.S. Paper presented at the American Sociological Association, Atlanta, GA, August.

Victor, D. G., House, J. C., and Joy, S. (2005). A Madisonian approach to climate policy. *Science*, **309**, 1820–21.

25

Ending the piecemeal approach: Santa Monica's comprehensive plan for sustainability

Susan Watrous
Independent

Natasha Fraley
Independent

> One thing is clear: the fate of cities will determine more and more
> not only the fate of nations but also of our planet. We can afford to
> ignore the issue of the sustainable management of our cities only at
> our own peril.
> *Elizabeth Dowdeswell (1996), former United Nations*
> *Under-Secretary General and Executive Director,*
> *United Nations Environment Programme*

One California city has made an unblinking appraisal of its fate, and created a comprehensive plan for sustainability. It's a blueprint that links climate change with other environmental issues, economic development, and social equity — all in the larger context of the community's quality of life. Certainly, the effects of climate change already touch the ocean-front municipality of Santa Monica where sea-level rise sends higher-than-historical winter storm waves surging against the base of the Sustainable City Program office, and these effects will continue to change other aspects of life. Extended heatwave days, for example, could dramatically increase the risks to human health (Hayhoe *et al.*, 2004a,b). Yet rather than just focus on energy use and emissions, Santa Monica has chosen a systems approach, addressing climate change with strategies aimed at its cultural and physical causes.[1]

Surrounded on three sides by Los Angeles County, the pint-sized municipality of Santa Monica — just 8.3 square miles on the map, but with global vision — has consistently been at the leading edge of a movement for sustainable cities. In 1999, Santa Monica was the first US city to flip the switch to 100 percent renewable electricity for all municipal facilities, which have also been upgraded for energy efficiency. Of the city's Public Works' fleet, 78 percent currently run on reduced-emissions fuels, up from 10 percent in 1993.[2] Since 2000, the Santa Monica Urban Runoff Recycling Facility (SMURRF) reclaims 95 percent of the water – much of it polluted – that previously flushed directly into the Pacific Ocean at the city's western edge.[3] Over the last decade, Santa Monica has developed a set of rigorous green-building standards that are redefining the notion of healthy, efficient construction, and being used as a model by other communities. Through its efforts since 1994, Santa Monica has shrunk the size of its ecological footprint, a measure of its resource use and waste sinks,[4] by 5.7 percent (Figure 25.1). In a business-as-usual projection, Santa Monica's Sustainable City Program estimated that its greenhouse gas emissions would rise 14 percent above 1990 levels by 2010. Yet between 1990 and 2000, the city reduced its greenhouse gas emissions by 6 percent.[5] And in June 2005 at the

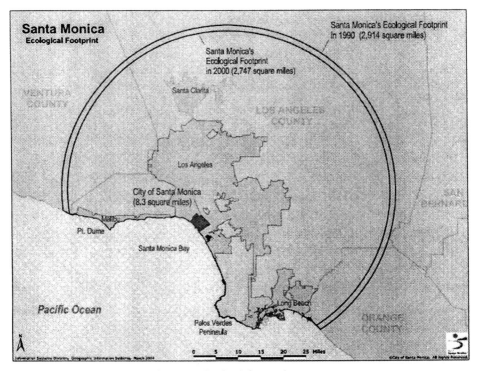

Fig. 25.1.   Santa Monica's ecological footprint
The city's ecological footprint is a measure of its use of resources and its generation of wastes. It considers land use, electricity and fuel use by source, transportation and vehicles, roads, housing, food, products, waste, and recycling. These factors were converted into productive-land-area equivalents. In 1990, the city's footprint was 2,914 square miles or 21.4 acres per resident. By 2000, it was down to 20.9 acres per resident, nearly four acres smaller than the US average.
Source: City of Santa Monica, reprinted with permission.

United Nations' World Environment Day in San Francisco, the city was ranked the fifth most sustainable city in the nation.[6]

So how did Santa Monica get here? Why did the community choose this sweeping approach? What role — if any — did climate change play in this reform? As this chapter will show, it played an integrated, but not a dominant, role. "Santa Monica has been at the forefront of an environmental movement to put climate problems into a broader, long-term vision for their community," says Abby Young, Director of the Cities for Climate Protection Campaign, a program of ICLEI–LOCAL GOVERNMENTS FOR SUSTAINABILITY (see Young, Chapter 24, this volume).[7]

Santa Monica's sustainability story is one of a comprehensive plan, derived from a progressive tradition, launched by visionaries, carried forward every day by a growing number of champions, and one that is dedicated to engaging as many community members as possible. While the city's story has been told before and some of its successes depend on its unique assets, much of its management and goal-setting approach is transferable to other communities.

The evolution of a plan

In 1990, when the Santa Monica City Council created its first Task Force on the Environment, the city tapped a long tradition of progressive politics and an environmentally savvy identity. Yet the task force's initial assessment of the city's environmental performance was mixed.

Dean Kubani, Coordinator of the Sustainable City Program (SCP) and a senior policy analyst with the city's Environmental Programs Division, says that, while the task force applauded some aspects of the city's approach, it criticized the way Santa Monica addressed environmental issues as "piecemeal." The task force recommended that a sustainability plan be developed. At the time, the concept of "sustainability" had just emerged in the public discourse following the release of Local Agenda 21, a framework developed at the 1992 United Nations Conference on Environment and Development (the Rio Earth Summit), which called for "local authorities as the sphere of governance closest to the people...to consult with their communities and develop and implement a local plan for sustainability."[8]

The idea of sustainability fell on fertile soil in this forward-thinking, liberally governed, and well-educated city.[9] It found ready champions in the City Council and some city staff. Moreover, the community had the financial wherewithal to fund the new initiatives and solutions for at least some of its environmental problems. The city's diverse economic base in tourism, and the entertainment and high-tech industries, delivered a healthy municipal income.[10]

In what Kubani describes as "a top–down process," the task force and key city staff drafted the first sustainability plan with some input from community organizations and elected officials. The city also hosted three public meetings to collect feedback on the plan but, according to Kubani, did not "get a flood of public input." Championed by the City Council and staff, the plan steered clear of many political obstacles. "We never had to overcome people who wanted to derail the process. The biggest challenge," says Kubani, "was that people would accept the plan, and then realize that it meant they actually needed to do something differently."

Developed mainly by experts and city staff and adopted by the city in 1994, the original Sustainable City Plan included 16 target indicators in four goal categories for the year 2000.[11] While it included broad goals for the community, much of the implementation focused on changes at the municipal level. "If the city was setting these goals," says Kubani, "it also needed to be leading the charge in helping the community achieve them."

In the target year, 2000, the task force recommended that the city update its objectives – many of which had been fulfilled or surpassed[12] – and the City Council initiated the process. However, over the intervening years, the smooth-sailing approval of the original plan had not been without consequences. "For the first four or five years that we had the original plan in place," says Kubani, "nobody in the community knew about it." The biggest problem with that, of course, is that "there was no real community ownership of the plan." Furthermore, many of the results had come in municipal activities rather than ones more obvious to the residents of the city.

To insure wider community buy-in and associated changes, those involved in the update process vowed to take a different approach. For the revision, the city conducted a broad public process to develop new targets, and – perhaps most importantly – "bring in all the community stakeholders to generate some ownership," says Kubani. The process lasted 15 months, and sought input from neighborhood organizations, schools, Santa Monica College, businesses and business organizations such as the Convention and Visitors Bureau and Chamber of Commerce, and staff from every department in the city. In addition, it included appointed officials from the commissions on housing, planning, social services, and recreation and parks, as well as City Council members. "The words in the goals of the second plan came from members of the community," explains Kubani. Consequently, he notes, "This is now a much more community-wide program."

The updated plan is based on nine principles, philosophical assumptions that underlie the strategic and measurable goals. Guided by the basic concept of sustainability (see Santa Monica, 2003: 5; World Commission on Environment and Development, 1987),[13] the plan reflects the city's long-term vision and dedication to stewardship for its natural environment. Other principles highlight the need for partnerships between government, residents, and businesses; engaging community members in addressing sustainability; and procuring goods and services that emphasize these long-term, socially responsible values. Underlying these principles is the concept of interconnectedness, an understanding that, in the words of the plan, "environmental quality, economic health and social equity are mutually dependent," and that "all decisions have implications to the long-term sustainability" of not only the city, but the regional, national, and global communities of which Santa Monica is an integral part.

The new goal areas include Resource Conservation, Environmental and Public Health, Transportation, Economic Development, Open Space and Land Use, Housing, Community Education and Civic Participation, and Human Dignity, and are measured by a total of 70 main indicators (Santa Monica, 2003). These "comprise the core of the community vision," according to the plan, "and represent what Santa Monica must achieve in order to become a sustainable city" (Santa Monica, 2003: 3).

Mom and apple pie

In response to queries about how the task force enlisted support for the sweeping plan, Dean Kubani has a simple answer: "When you look at our goals, it's mom and apple pie. We want a clean environment, and an economy that provides jobs. We want everybody in the community to be able to meet their basic needs. We want affordable housing. Nobody could argue with this stuff. What's not to like?"

While mom-and-apple-pie makes a good sound-bite, Kubani hastens to point out the basic practicalities at the heart of the strategy: "There are lots of elements of our Sustainable City Plan that would be foolish for cities *not* to implement, because they save money and they're really smart things to do." Kubani cites an energy efficiency upgrade − switching all traffic lights from incandescent bulbs to LEDs[14] − as an example. "They use one-tenth of the electricity, and last 10 to 12 times as long. The cost pays off in six months, and then you're just saving money forever − not only in energy costs, which are substantial, but also in maintenance costs."

While critics might argue that Santa Monica's achievements arise from a particular flavor of political leadership, Kubani insists there are other solid reasons to choose sustainable solutions. "[Cities] that don't have progressive leadership typically aren't exposed to these kinds of solutions, so they have a misconception that this is an environmental thing − it just doesn't work, it's too expensive." On the contrary, he says, "These cities just haven't been presented with it in dollars-and-cents reality − it's simply a smart thing to do."

Strolling along the promenade in Santa Monica, some of the most desirable ocean-front in Southern California, it's hard to imagine who would deny the wisdom of sustaining all this. Palm fronds slap in balmy breezes off the Pacific, and the city's Big Blue Buses − many powered by liquefied natural gas[15] − roll by on their appointed rounds. In front of City Hall, three Rav4 electric vehicles (EV) − part of the municipal fleet − are plugged in at solar-powered EV parking stalls. Across the street, the Santa Monica Public Safety Facility, a green-building project completed in 2003, houses the fire and police departments. Sprinklers irrigate the xeriscaped grounds with reclaimed urban runoff, while the building's outdoor fountain sheets recycled water into the pond below. In the lobby, a sign touts the facility's energy- and water-efficient features, including an annual savings of 630 tons of greenhouse gases. The facility holds a Silver certification from the US Green Building Council's Leadership in Energy and Environmental Design (LEED).[16] For those interested in the city's green efforts, educational signage explains the value of solar panels and green construction features. But while other cities may only highlight the energy and economic savings, Santa Monica not only mentions the concept of climate change, but also sustainability. Why did Santa Monica choose a comprehensive approach − one that's also more difficult to explain to staff and citizens − when simply signing off on these efforts project by project might have been so much easier?

Vision, credibility, and clear responsibilities: ingredients of successful leadership

Drawing together elements as diverse as reducing sewage generation and increasing the number of women, minorities, and people with disabilities in leadership positions, the updated Sustainable City Plan takes into account nearly every aspect of the quality of life. City staff acknowledge it might have been simpler to solve the problems singly, but there's a price in that methodology, too. "If you take a piecemeal approach," explains Kubani, "a lot of things fall through the cracks." While local government might excel at creating a recycling program, for example, the municipality could still be using too many toxic chemicals. "A comprehensive approach says, as a community we need to become more sustainable, and then we get everybody in every department working toward that same goal," says Kubani.

Using an integrative method also steers clear of the pitfalls that employee departure can open. "In a piecemeal approach," says Kubani, "if the person leading the effort leaves, the whole program tends to go away. In a more comprehensive approach, the efforts are integrated and will persist when individuals move on."

One practical aspect of implementation that helps with communication and participation is the Sustainability Advisory Team, composed of about 40 mid- to high-level staff drawn from all the departments of city government. Team members, all of whom were existing employees, were educated about sustainability and how it translates to their areas. These team members then communicate sustainability initiatives to their departments and relay requests back to the Environmental Programs Division.

It took a decade of vision, persistent education, outreach – in both government and the community – and frank trial and error to understand what infrastructure and communication efforts are required to make the plan work. "Even five years ago, if I went to a department head and asked for five staff people to participate in meetings on sustainability, I would have gotten resistance from some of them, saying 'No, they're too busy, they can't do that,'" maintains Kubani. "We've done a lot of work to get where we are now." Although the process now includes systematic communication and outreach, much of the vision, clear goal-setting, and initiating of internal discussions still arise from the home-team champions of the Sustainable City Program (SCP).

One of the people championing the big-picture effort is Brian Johnson, manager of the Environmental Programs Division (EPD), which oversees the SCP. Johnson points out a handful of crucial elements in spearheading such an integrated approach. "One of my jobs has been to frame the expectations of the stakeholders," he says. "There are times when we will achieve milestones, and times when we will achieve inchstones, and other times when we will be fighting hard just to maintain the status quo. That's a part of social movements, and this is a social movement. It has its technological and scientific elements, but it's a social movement at its core" (see also Meyer, Chapter 28, this volume).

The EPD takes the high – and long-term – view of this social movement from the top floor of a historic building on Santa Monica pier.[17]

Offices overlook the wide strand of sand, the Pacific, and the wooden pier with its carnival atmosphere. The combination of history, quintessential Golden State coastline, and re-purposed grandeur provide an apt backdrop for Johnson's reflections on leadership.

He acknowledges the value of idealism to the EPD team, because "it provides so much energy and fuel," and he also maintains a solid commitment to other elements as well. "Much more progress has been built on credibility. ... It's very important that my staff can answer questions, or know how to get the answers, honestly, accurately, and quickly." Johnson also emphasizes respect in the development of the EPD's programs, "respect toward the folks we're going to be dealing with, respect for their thoughts and opinions. All these things are intertwined, of course."

It isn't surprising to hear that the theme of interconnectedness not only informs the city's sustainability goals, but also its management approach. Yet enlisting support for the city's far-reaching vision and the high-profile — as well as the less-visible — projects took a remarkable amount of focused outreach.

Taking it to the streets

Since the 2003 plan update, outreach through active engagement, the media, and public education has been cardinal in the city's approach. During the update process, the 35-member Sustainable City Working Group, which included representatives from nearly every stakeholder group in the community, set the outreach gold standard by generating maximum public engagement. With the new plan in place, the next phase is developing an implementation strategy to meet the latest goals. To do this, the City Council appointed a new 11-member task force with expertise in environmental issues, ecological economics, municipal planning, education, transportation, neighborhood/community organizations, labor, and business. "The task force is working in two directions," Kubani explains. "They're helping develop the implementation plan, but they're also outreaching it. They're doing some networking, and creating more of a buzz in the community about sustainability — so these people were partly chosen for their connections."

Another effort to generate more buzz focuses on city government, which requires what Kubani calls "a relentless communications effort." He explains, "It's difficult to have a comprehensive approach because it takes a lot of communication. At the same time, in systematizing that communication, you help everybody get up to speed." Kubani notes a significant difference in the level of communication between city staff and departments between now and ten years ago, when he started in the program.

This has been achieved through the quarterly staff newsletter, and by posting sustainability coverage to the staff intranet – information that employees see every day when they boot up their computers. In addition, a new part-time publicist helps the SCP keep a news stream flowing to local print and television outlets. Local television carries features and ad campaigns, and has produced a half-hour documentary about the Sustainable City Program.

Despite what Kubani calls this "steady drumbeat of information," public awareness of the city's Sustainable City Plan has grown slowly. In 1998, only 3–4 percent of the public knew of the plan; in a survey completed in December 2003, about 17 percent of the public did.[18] However, Kubani remains hopeful that public awareness will blossom substantially because of the city's publicity efforts in 2004. "I suspect the next survey we do, it's going to be more like 30–40 percent awareness or even higher because the sustainability news has been really hard to avoid. Even though people are subjected to a deluge of global information now, we've really been hitting them at the very local level."[19]

City staff did not always recognize the need for this level of outreach, as well as what elements would be necessary to swing public behavior. "We're a bunch of bureaucrats developing policies," says Kubani. "We're experts at environmental policy, not experts in behavior change. So what are we going to do? Printing a brochure that says global warming is a terrible thing is not going to work." What has worked are strategies from community-based social marketing (McKenzie-Mohr, 2000). "We're finally getting much more savvy about [this kind of] marketing and PR," says Kubani. "Our ultimate goal is to change people's behavior. How are we going to do that? So we're thinking like Coca-Cola people now. We want the public to 'buy our product,' and change their decisions." While the city doesn't boast an advertising budget like the soft-drink giant's, the SCP has figured out how to maximize its efforts, and by their accounts, those efforts are paying off.

Sustainable Works: education and support for all

Educating and supporting residents, students, and businesses in sustainable choices is another way to enlist public engagement. Sustainable Works, a non-profit organization funded by SCP and Santa Monica College, offers programs that promote sustainable practices in the "daily activities of individuals, institutions, and businesses."[20] Their programs fall into two divisions, Community Greening and Business Greening.

Community Greening runs programs for residents and Santa Monica College students in which participants get hands-on experience in greening their lives. Activities are organized around the *Sustainable Worksbook*, and begin with a lifestyle inventory. As Sustainable Works Executive Director Chantel Zimmerman explains, "Participants measure their trash and gas consumption – which they convert to carbon dioxide emissions – look at chemicals in their houses, water and energy use, shopping habits, and a few other topics." This provides a baseline for progress, which is inventoried again at the end of the program. Using what Zimmerman describes as a "buddy system," as well as some behavior-change techniques borrowed from community-based social marketing, "[Participants] make public commitments, then go home and do them. The next week they talk about successes and failures, so they prop each other up."

The combination of education and peer support seems to be successful. For example, results of the student program in fall semester 2003, which involved 172 students, showed an average decrease in water and energy use of 20 percent; a 50 percent reduction in waste, and a 66 percent reduction in toxic chemicals in the students' homes.[21]

According to Ferris Kawar, who runs the residents' programs, about 1,700 residents and students have gone through the Community Greening course since its inception in 2001. Sustainable Works estimates that average families save between US$80 and $200 a year in energy and water expenses as they go through the program.[22] Yet-to-be-conducted long-term surveys of participants will have to show how persistent the behavior changes are (see also Rabkin with Gershon, Chapter 19, this volume).

Business Greening has adopted a consulting model focused on similar environmental impacts. In this program, which is both voluntary and free, Sustainable Works audits a business's current practices with regard to energy, water, and chemical use; waste; purchasing; and transportation. Based on the findings, the organization recommends more sustainable options and implementation strategies to the business. "One restaurant that sticks in my mind," reports Kubani, "did some simple retrofits and saved US$300 a month on its water bill."

Sustainable Works tracks the impact of its recommendations. Of the 33 businesses that Sustainable Works has consulted with, 75 percent upgraded to energy-efficient appliances; 60 percent instituted a ridesharing program; and 80 percent began buying less toxic alternatives to products they used. Some 16 businesses reduced energy use by 11–31 percent.[23]

As of 2004, about 16 businesses complete the program annually. "It's very labor intensive and takes a lot of resources," says Zimmerman. "But on a shoestring, you can do amazing things. We're not trying to be the greenest of the green, but we're trying to bring the non-green on board."

Rewarding persistence, valuing change

This dedication to progress, one business, one person at a time — as well as a passion for the process itself — is clearly evident when one talks with city staff about their efforts. They may think of themselves as "bureaucrats," as Kubani says, but their dedication, creativity, and engagement counters some common stereotypes for the halls of government.

Success is one element that makes the champions' job a little easier these days. "With the update of our Sustainable City Plan, we've been able to get a lot of people on board," explains Kubani, "because we started really small, and we've been really successful. Santa Monica is known for its environmental successes. People in both the government and the community have heard about the city being the first in the country to buy green power, cut greenhouse gas emissions when everybody else's went up, and have one of the greenest fleets in the nation ... So when we ask people to participate, they say yes, because we have credibility."

It's the same strategy the city follows with climate change. As Susan Munves, Program Administrator for Energy and Green Building, says, "The thing with climate change is it's altogether so awful and so frightening that people go into Chicken Little mode. That's where the idea of focusing on the bigger picture of sustainability makes sense — otherwise it's pretty grim" (see Moser on emotional responses to climate change information, Chapter 3, this volume). She adds that it's easier to identify the successes on which to build: "If you take one step and get a positive response, then it makes it easier to take the next one."

The passion — and the uphill nature of the task — requires something of a delicate balancing act for management at times. Brian Johnson says, "We are as frail as any other organization when it comes to human performance and human weaknesses and human successes ... We have a really strong tradition, though, of being innovative progressive thinkers who keep their eyes on the prize. And that's part of my job, making sure that maybe folks who didn't win the battle today [know that], by god, we're going to win this war sooner or later," he pauses and smiles. "And they buy it so far."

Humor aside, Johnson offers a real sense of what it takes to shoulder this task: "We all have to be eternal optimists to do this kind of work, to come here every day, because the more you learn, the more depressing it is." Johnson explains that managing expectations helps, as well as being part of the solution, as opposed to just part of the problem. "One of the most significant motivators for the people who work [in EPD] is knowing that they can make a change, they can make a difference. Today, we can make some progress – and that will be recognized; it will be valued, and it will be productive. If you've got that opportunity everyday, then you've got a spark."

In addition, the Department of Environmental and Public Works Management that oversees EPD is very supportive. In most organizations, Johnson maintains, there is a philosophy that rejects replacing an unsustainable effort or product with a sustainable one, even if the latter costs the same, is safer for the environment, the workers, and the public, and performs equally well. "Why?" Johnson asks. "Because it's different. It's change, and that's the obstacle. So it's a matter of finding a willingness to take a chance on change" (see also Doppelt, 2003).

Johnson credits Craig Perkins, Director of the Department of Environmental and Public Works Management, with creating an environment where change is valued. "Municipalities are typically pretty risk-averse organizations. But here, if it doesn't work, it doesn't mean you get shipped off to Santa Monica's Siberia. It means you had a good idea, you did your homework and thought it would work, and for *X* reasons, it didn't. Let's take the lesson and apply it to the next opportunity" (see also James *et al.*, Chapter 20, this volume).

One place where Santa Monica had a good idea, did its homework, and took a major leap forward was with its trail-blazing green-building standards.

Greening and energizing the human landscape

"In 1994, when the original Sustainable City Plan was adopted," explains Susan Munves, "it was clear we had no clue how to mitigate the impact of new construction, which had the single greatest environmental impact in Santa Monica. So the question became, how do we 'green' construction?" The basic idea was to mandate sustainable building practices without putting developers out of the financial game.

The city hired a team of Canadian consultants to lead the effort to draft Green Building Design and Construction Guidelines.[24] "At the time," says Munves, "there were only a few examples of this kind of guidelines out there, and we wanted to create something that was specific to the United States, and also to Santa Monica." While increased energy performance and decreased energy use were a vital focus, Munves adds, "The process happened without a whisper of discussion about climate change."

The consultants developed a standard at least 20–25 percent more efficient than California's 1991 Title 24 Energy Efficiency Standards for Residential and Nonresidential Buildings, and the City Council approved it in 1994. By law, the new standards were vetted at the state level for compliance with both energy performance codes and cost-effectiveness. "It cost a lot of money and work to go through that process," says Munves, "but Santa Monica did it." In addition, the guidelines were peer-reviewed internationally to ensure "that what we published was correct and of quality," adds Munves.

Providing strategies for every aspect of planning and construction, the new guidelines apply to offices, light industrial and commercial retail buildings, as well as multi-family residences, and hotels and motels. Santa Monica took its own ambitious mandate and applied it to its new Public Safety Facility, beating Title 24 standards for energy efficiency by 36 percent. One of the city's affordable housing buildings, Colorado Court, is 93 percent energy-independent through the use of passive solar, solar photovoltaics, and power co-generation. The building was funded by the city's utilities as well as state and regional organizations. According to Santa Monica's Green Building Program, Colorado Court's energy generation systems "will pay for themselves in less than ten years."[25]

Besides green-building standards, increased energy efficiency in all sectors, and employment of co-generation technologies, "switching to clean energy is the single most effective means the city has to change greenhouse gas emissions," says Munves. And all of this, of course, requires a financial commitment. "You have to be willing to invest in the community to make this happen," says Munves. "Perhaps this is Santa Monica's greatest difference from other communities: We recognize the need to invest in order to get future payback."

Future-protecting the community with common-sense solutions

In reviewing Santa Monica's achievements, it appears the community is already enjoying its payback. The city has kept over 1 million tons of solid waste out of the landfill. It saves more than 328 million gallons of water annually.[26] Schoolchildren can graze at salad bars serving locally grown, organic produce. And more and more people throughout the community are taking up the green standard.

"When I give talks at conferences," says Kubani, "I run through all the stuff that Santa Monica's done, and people's mouths are open. And then I pose the rhetorical question, 'Are we there yet?' " Kubani characterizes Santa Monica's impressive progress as only the first step in a long journey: "We've done the easy stuff, and now we're poised for the next big leap. The hard stuff is [what] nobody's ever done, which is actually creating a sustainable community, creating all the communication channels and the networking to link the environmental, economic, and social together to achieve all of these goals."

But while Santa Monica contemplates its next move, many US communities have not even begun what officials here call "the easy stuff." So how does Santa Monica's vision and process translate to other cities where pressing problems such as an aging infrastructure or a declining revenue base have robbed city coffers of the means to do more than just stay fiscally afloat? The political will among innovators and early adopters often seems to have roots in deep moral convictions. Santa Monica made these choices because it was, as Kubani acknowledges, "the right thing to do."

Moreover, Santa Monica had the fiscal means and discipline to implement the plan. Stephanie Pincetl, a visiting professor at University of California, Los Angeles' Institute of the Environment, told the *Santa Monica Daily Press*, "The city understands there is a cost—benefit analysis that can be done. There are things that cost money, but are the right things to do, and [Santa Monica] has taken a stand that it will do those things" (Bishop, 2004: 7).

EPD Manager Brian Johnson says, "People ask, what is it about Santa Monica that makes it so that these [changes] can happen and you're so far ahead of the curve? ... We don't do frivolous, arbitrary things. We're thoughtful, we're very mindful of the sanctity of the tax dollars that we are spending ... we are just reframing the questions and the performance values and the efficiency values and our ability to think in longer time spans ... So it's less about why Santa Monica is here, but why we're here alone, and why others aren't here with us."

Kubani ventures a guess about the city's jump-start: "Maybe because we have a progressive council and populace, some of these ideas came on our radar screen sooner than they might have for some place else. ... But the things we're doing are fiscally responsible — even cities without a lot of money can make this kind of decisions. ... Politicians look at issues in a very short term, rather than a longer one." He cites election-cycle timeframes as an important obstacle to sustainable progress; "politicians in Santa Monica are saying, 'We're going to do this now and it'll cost a little more, but it's going to have benefits for the next decade.'" While Kubani doesn't have a quick fix for changing the political mindset, he proffers an interesting – and given the city's comprehensive approach, almost ironic – possibility: "Maybe what needs to be done is remove the term 'sustainability' and the environmental connotations of it, and present this in a more pragmatic way, like 'This is just smart. These are common-sense solutions.'"

For now, Santa Monica may be among the vanguard of a national sustainability movement because of its fortunate combination of political will, ethical conviction, committed champions, and economic stability. But its vision, fiscally responsible approach, persistent outreach, commitment to expertise and learning, and incentive systems all speak to a mindset that is not confined to California. Indeed, similar programs are springing up in the Northeast, Midwest, and Pacific Northwest, across party lines and motivated by different incentives, but sharing some of the goals and vision at work in this city by the sea. At a deeper level, however, Santa Monica's leaders distinguish themselves through creative, innovative ways of thinking. They see the connections between global changes and local impacts, between a range of local problems and comprehensive solutions, between individual small-scale actions now and the cumulative consequences sometime in the future.

Contemplating the link between climate change and sustainability, Susan Munves says, "We're future-protecting the community, making the community independent in a variety of ways – energy use, water supply, and so on. Sustainability is a pragmatic approach, and at some point, the focus is not only on prevention, but on adaptation to a world that is experiencing climate change, as well as a host of other problems." Munves' voice conveys the conviction and passion that she shares with her colleagues when she says, "Cities will be a vital part of the solution. We will be helping people deal with a changing world."

Notes

1. This chapter is based on interviews conducted (by phone and in person) between August and November 2004 with Santa Monica City staff, including Dean Kubani, Coordinator of the Sustainable City Program; Susan Munves, Program Administrator for Energy and Green Building; Brian Johnson, Manager of the Environmental Programs Division; Chantel Zimmerman, Executive Director of Sustainable Works; and Ferris Kawar, also with Sustainable Works. Unless otherwise noted, figures and other information were provided by the interviewees.
2. The percentage given for the Public Works fleet does not include police and fire vehicles.
3. The first facility of its type in the nation, SMURRF won the 2001 Grand Award for Excellence in Engineering Design from the American Consulting Engineers Council and the 2002 award for Outstanding Civil Engineering Project from the American Society of Civil Engineers.

4. For more information, see http://www.redefiningprogress.org/footprint/; accessed January 19, 2006.
5. Information on greenhouse gas emissions provided by Dean Kubani, October 5, 2004.
6. For more information, see http://santa-monica.org/news/releases/archive/2005/epwm20050602.htm; accessed January 19, 2006.
7. For more information on ICLEI, see http://www.iclei.org; accessed January 19, 2006.
8. For more information on Agenda 21, see the UN Department of Economic and Social Affairs website: http://www.un.org/esa/sustdev/documents/agenda21/index.htm; accessed January 19, 2006.
9. More than a third of the 85,000 residents hold a bachelor's degree and nearly 25 percent have a graduate degree. [The US census figures for 2000 give a population of 84,084. Estimated population of 85,991 for 2003 is from a Community Profile published by Rand.] For more information, see http://santa-monica.org/business/demographics/population.htm; accessed January 19, 2006.
10. Among residents, the median per capita income is almost US$43,000 per year, while median family income is nearly US$76,000 — compared to median family income of about US$46,450 in the surrounding Los Angeles County. See Los Angeles Almanac online at http://www.losangelesalmanac.com/topics/Employment/em12.htm; accessed January 19, 2006.
11. These are aggregated indicators; in some cases, an indicator includes more than one sub-indicator.
12. See http://santa-monica.org/epd/scp/history.htm; accessed January 19, 2006.
13. "Santa Monica is committed to meeting its existing needs without compromising the ability of future generations to meet their own needs." This echoes the original definition of sustainability proposed by Gro Harlem Brundtland in her path-breaking 1987 report.
14. To read more about electronic light-emitting diodes (LEDs) and the potential cost savings they represent in traffic-light use, see http://people.howstuffworks.com/question178.htm; accessed January 19, 2006.
15. As of June 2005, 60 percent of the Big Blue Bus fleet runs on alternative, reduced-emissions fuels (personal correspondence, Dean Kubani, October 5, 2004).
16. For more information, see the US Green Building Council's website: http://www.usgbc.org/leed/leed_main.asp; accessed January 19, 2006.
17. The building, constructed in the early twentieth century, still houses a working carousel from the 1920s.
18. Information provided by Dean Kubani, October 5, 2004.
19. In 2004, about six months' of publicity built up to the city's metro-sized party — the 10Fest — which marked the tenth anniversary of Santa Monica's Sustainable City Plan. The 10Fest offered films and speakers, a community sustainability awards ceremony, and tours of sustainably enhanced residences, commercial buildings, and SMURRF.
20. For the complete text of the mission statement and more information, see http://www.sustainable-works.org; accessed January 19, 2006.
21. Information provided by Chantel Zimmerman, October 6, 2004.
22. Information provided by Dean Kubani, August 25, 2004.
23. Information provided by Chantel Zimmerman, October 6, 2004. Note that quantitative information on energy and water savings could not be obtained for all businesses. For example, a business may reside in larger office buildings where its energy/water use is not measured separately.
24. For Santa Monica's Green Building Design and Construction Guidelines, see http://greenbuildings.santa-monica.org/introduction/introduction.html; accessed January 19, 2006.
25. From http://greenbuildings.santa-monica.org/whatsnew/colorado-court/coloradocourt.html; accessed January 19, 2006.
26. From http://www.santa-monica.org/epd/news/pdf/10Fest_TenWays_ad_Success_01.pdf; accessed January 19, 2006.

References

Bishop, K. (2004). A city sustained: Santa Monica celebrates 10 years of being green. *Santa Monica Daily Press*, September 30, **3**, 276, 1–7.

Doppelt, B. (2003). *Leading Change Toward Sustainability: A Change Management Guide for Business, Government and Civil Society*. Sheffield, UK: Greenleaf Publishing.

Dowdeswell, E. (1996). Human settlements. *Our Planet*, June. Available at http://www.ourplanet.com/imgversn/81/editorial.html; accessed January 19, 2006.

Hayhoe, K. Cayan, D., Field, C. B., *et al.* (2004a). Emissions pathways, climate change, and impacts on California. *Proceedings of the National Academy of Sciences*, **101**, 34, 12422–7.

Hayhoe, K., Kalkstein, L., Moser, S., *et al.* (2004b). *Rising Heat and Risks to Human Health: Technical Appendix*. Cambridge, MA: Union of Concerned Scientists. Available at: http://www.climatechoices.org; accessed January 19, 2006.

McKenzie-Mohr, D. (2000). Fostering sustainable behavior through community-based social marketing. *American Psychologist*, **55**, 5, 531–7.

Santa Monica (2003). Sustainable City Plan. City of Santa Monica, CA. Available at http://santa-monica.org/epd/scp/pdf/SCP_2003_Adopted_Plan.pdf; accessed January 19, 2006.

World Commission on Environment and Development (1987). *Our Common Future*. Oxford, New York: Oxford University Press.

26

States leading the way on climate change action: the view from the Northeast

Abbey Tennis
Massachusetts Office for Commonwealth Development

Introduction

As federal climate change policy has failed to keep pace with the need for urgent action, policy-makers across the country, indeed, across the world, increasingly are looking to individual US states for leadership in climate protection. And many states are taking action; more than half of US states and territories have released some form of climate plan.[1] This chapter explores the role of state governments in US climate policy, and particularly the leadership and policy innovation in the Northeast.

While individual states may seem to be too small to make a significant difference in the global effort to curb greenhouse gas (GHG) emissions, even the smallest US state's emission levels are equivalent to those of many countries around the world. Massachusetts, for example, is responsible for only 1.35 percent of the total US GHG emissions (MA State Sustainability Program, 2004; NESCAUM, 2004; US EPA, 2002), yet it emits roughly the same amount of GHGs annually as Chile, Austria, or the Philippines and considerably more than Sweden or Switzerland. In fact, if emissions from all of the states in the Northeast were combined, they would release pollution equal to the eighth largest emitter in the world – more than the entire nation of Canada.

Aside from the direct emissions reductions individual states can make, perhaps most significant is states' role as "policy laboratories" for each other and for eventual federal policy development. In areas as diverse as health care, criminal justice, welfare reform, and the environment, federal policies have

This chapter draws heavily on unpublished work by Jason Grumet, Executive Director of the National Commission on Energy Policy, 2004 and help from Susanne Moser, NCAR, and Susan Watrous, Santa Cruz. The author gratefully acknowledges their insights and Mr. Grumet's permission to use some of his writing in this chapter.
Views expressed herein represent the opinions of the author alone.

often arisen out of successful policy experiments that occurred first at the state level. It is entirely possible that state action may lead to future action at the federal level on climate change policy. In the meantime, the extensive network of states sharing best practices, and joining together to amplify the effect of their individual policies, legal actions, and lobbying efforts, has created an incredibly promising new level of multi-state leadership.

The Northeast states have a 40-year history of working together on air policy issues,[2] and have, along with the West Coast states (see duVair *et al.*, Chapter 27, this volume), taken a key leadership role on climate change since the late 1990s. As these states endeavor to do their part as members of the world community, they are among the strongest leaders in US climate policy.

States as leaders, states as laboratories

The notion of states as "laboratories of democracy" is well established in US legislative history (e.g., Walker, 1969; McKinstry, 2004; Kosloff, Trexler, and Nelson, 2004; Peterson, 2004; Rabe, 2004). "It is one of the happy incidents of the federal system," Supreme Court Justice Louis D. Brandeis wrote in 1932, "that a single courageous state may, if its citizens choose, serve as a laboratory; and try novel social and economic experiments without risk to the rest of the country."[3] Congressional leaders often look to local, state, and regional programs for guidance on design of federal policies, particularly on difficult political or technical issues. Some scholars even argue that the US federal system was intentionally set up to favor initiatives that come from the bottom up, in particular from the states, to drive federal policy (Walker, 1969; Victor, House, and Joy, 2005). Most of the major federal environmental laws in existence were, in fact, modeled on successful state programs. Examples include a California state air regulation that became the basis of the first federal Clean Air Act; regulations on water quality by the Delaware River Basin Commission that served as a model for the first federal Clean Water Act; a Pennsylvania surface mining regulation that served as a model for the federal Surface Mining Control and Reclamation Act; and New Jersey's hazardous site remediation program (codified in that state's Spill Compensation Act of 1976) served as blueprint for the federal Comprehensive Environmental Response Compensation and Liability Act.

While much of the substance and structure of state policies is commonly adopted for federal use, the *process itself* of individual state action creating a "patchwork quilt" of different regulations across the country is also a driving force in the creation of federal policy.

> As [state] efforts [on climate change] expand, it is possible that there could indeed be a tipping point whereby state-by-state variation of policies and standards create operational inefficiencies and lead to a call for some form of federal action to establish a uniform policy for the nation. There is abundant precedent for this type of dynamic in environmental and energy policies.
>
> *Rabe (2004: 139)*

State officials have found, time and time again, that opponents of strong state-level environmental policy (often the industries facing regulation) can sometimes turn into allies in the call for a federal policy when navigating the uncertainty and complications inherent to a "patchwork" of regulations become more costly than complying with the policy itself. Leveraging their joint power also motivates states to act in regional alliances.

Greater than the sum of their parts: states working together in regional coalitions

Many advocates, academics, and policy-makers agree that the greatest strength of the Northeast's approach to climate policy has been the decision to work together towards regional policy, rather than focusing only on policy-making within state borders.

The Northeast's acid rain and mercury work, as well as the history of coordinating certain environmental policies and legal actions, built the foundation for the New England Governors' and Eastern Canadian Premiers' (NEG/ECP) *Climate Change Action Plan* signed in 2001.

The New England Governors'/Eastern Canadian Premiers' Climate Change Action Plan

The Eastern Canadian provinces and Northeastern US states have been informally coordinating policy since 1937.[4] This history of policy coordination rests – in part – on the deep interconnections and interdependence of the region through more than 200 years of political, economic, and cultural ties. In the 1990s, the loose coordination of air quality policy-making up until then was taken to a higher level in the region's unprecedented effort to develop a regional mercury and acid rain program. That work served as a useful model for the joint climate plan.

By 2000, leaders in the Eastern Canadian provinces were thinking hard about policies and strategies for addressing climate change. Anticipating that the Canadian federal government would sign the Kyoto Protocol, the

Eastern Canadian provinces knew that collaborating on climate policy with the Northeastern United States would reduce the potential economic disadvantage that could ensue if the United States did not commit to GHG reductions as well. Thus, the Eastern Canadians (specifically New Brunswick) initiated conversations with their southern counterparts. "Our history of working well together, and our successful regional mercury and acid rain reduction plans with the Canadians, paved the way for the Governors and Premiers to consider joining a regional climate plan," said Sonia Hamel, former Massachusetts Air Policy Director, and lead in Massachusetts' work on climate protection (see also Selin and Van Deveer, 2005, 2006).

In the summer of 2001, these regional negotiations resulted in the governors of Massachusetts, Maine, New Hampshire, Vermont, Rhode Island, and Connecticut joining the premiers of the provinces of New Brunswick, Nova Scotia, Prince Edward Island, Quebec and Newfoundland/Labrador in signing the first-ever US/Canadian regional agreement on climate change – the NEG/ECP *Climate Change Action Plan* (NEG/ECP, 2001). State and provincial leaders committed to cutting GHG emissions to 1990 levels by 2010, 10 percent below 1990 levels by 2020, and over the long term, sufficiently to eliminate any dangerous threat to the climate, stating that current science suggested 75–85 percent emission reductions below current levels (NEG/ECP, 2001: 7).

At the time of the NEG/ECP *Climate Change Action Plan*'s signing, no government official had ever committed to long-term deep reductions in GHGs, i.e., the New England Governors and Eastern Canadian Premiers were the first leaders worldwide to promise to reduce emissions at whatever level was needed to protect the climate, and to put a number on the anticipated extent of necessary emissions reductions. While this monumental step in worldwide climate policy was not attached to a specific timeline, it has nevertheless helped inspire a longer-term perspective on climate change policy-making and has been echoed subsequently in the targets set in other countries.

The NEG/ECP plan's regional targets for emissions reductions became the basis for the targets adopted into individual state climate plans in New England (Table 26.1) and beyond. For example, the West Coast Governors also began to develop climate policy in regional collaboration, following the model created by the Northeast (see duVair *et al.*, Chapter 27, this volume). While the NEG/ECP targets were established collectively, the agreement was set so that each state would be responsible for making its share of the reductions in whatever way made the most sense for them.

Table 26.1. *Climate change action plans of Northeastern states*

State	Year Released	Climate change action plans and more information (in chronological order)
Vermont	1998	*Fueling Vermont's Future: Comprehensive Energy Plan and Greenhouse Gas Action Plan*[a] http://publicservice.vermont.gov/pub/state-plans/cepov.pdf
New Jersey	1999 (2002, revised)	*Sustainability Greenhouse Action Plan* http://www.state.nj.us/dep/dsr/gcc/GHG02revisions.pdf
New Hampshire	2001	*The Climate Change Challenge: Actions that NH Can Take to Reduce Greenhouse Gas Emissions*[b] http://www.des.state.nh.us/ard/climatechange/challenge.pdf
New England States/Eastern Canadian Provinces	2001	*NEG/ECP Climate Change Action Plan* 2001 http://www.negc.org/documents/NEG-ECP%20CCAP.PDF
Rhode Island	2002	*Rhode Island Greenhouse Gas Action Plan* http://righg.raabassociates.org/Articles/GHGPlanBody7−19-02FINAL.pdf
New York	2003	*Recommendations to Governor Pataki for Reducing New York State Greenhouse Gas Emissions* http://www.ccap.org/pdf/04-2003_NYGHG_Recommendations.pdf
Massachusetts	2004	*Massachusetts Climate Protection Plan* http://www.mass.gov/Eocd/docs/pdfs/fullcolorclimateplan.pdf
Maine	2004	*State of Maine Climate Change Action Plan* http://mainegov-images.informe.org/spo/pubs/origpdf/pdf/ClimateReport.pdf
Connecticut	2005	*Connecticut Climate Change Action Plan 2005* http://www.ctclimatechange.com/

Source: Compiled from US EPA (2005a), NEG/ECP (2001), and New England Climate Coalition (2005).
[a]Vermont is in the process of developing an updated plan that would implement the regional goals at the state level.
[b]New Hampshire has yet to produce a plan that adopts and implements the regional action plan.

The region would then reassess periodically to ensure that the overall regional target was met.

Now, the Northeast has begun another groundbreaking joint regional climate policy effort, the Regional Greenhouse Gas Initiative, which aims to reduce emissions through regulatory action in the Northeast power market.

The Regional Greenhouse Gas Initiative

As the New England states were completing work on their regional agreement, in June of 2001, the State of New York initiated its own stakeholder-based Greenhouse Gas Task Force. The Task Force was charged with developing a suite of GHG reduction policy options, for all sectors of the economy, which would build upon New York's already successful energy efficiency, renewable energy, and transportation strategies (Center for Clean Air Policy, 2003).[5]

Out of the work of the Task Force a new initiative emerged. In April 2003, New York Governor George Pataki took the bold step of inviting the other governors throughout the Northeast and Mid-Atlantic region (from Maryland to Maine) to help create a regional cap-and-trade system − the Regional Greenhouse Gas Initiative (RGGI, pronounced "Reggie") − to "reduce greenhouse gas emissions from the power sector, while maintaining economic competitiveness and efficiency across the regional power market."[6]

Nine governors initially signed on to work together in the development of this program: Connecticut, Delaware, Maine, Massachusetts, New Hampshire, New Jersey, New York, Rhode Island, and Vermont (Maryland and Pennsylvania chose to observe the process rather than actively participate). This created a massive and unprecedented multi-state program development effort, joining the environmental and energy agency staff members from each state in a consensus-based process to design and implement a program that would be environmentally meaningful, politically viable, and economically beneficial to the participating states. Before RGGI, it was considered an accomplishment even to get environmental and energy governmental staff together within one state: RGGI not only expanded this cooperation across nine states; it brought together the state agency heads as well as the program and policy-level staff, making the collaboration across states work horizontally between staff counterparts in different agencies, and vertically between program implementation, policy development, and agency leadership levels of government. The RGGI staff also committed to a detailed, transparent, and open stakeholder and public-input process, setting the standard for a new level of public engagement in regional environmental policy.

The framework of RGGI follows the model of earlier successful cap-and-trade systems, such as the one used to reduce acid rain-producing sulfur dioxide emissions.[7] Every cap-and-trade program includes the following basic components: (1) a decision about who is being regulated, (2) the reduction levels, (3) the distribution (or "allocation") of permits (called "allowances" in RGGI) that power generators will need, and (4) the structure of the trading mechanism.

Over two years of intense negotiation, the RGGI states determined a detailed set of provisions. They agreed that RGGI would regulate medium and large power plants and that, at most, 75 percent of the allowances in each state would be distributed among the generators, with the remaining 25 percent (or more) being set aside for various public benefit uses. They also agreed that generators who reduce emissions further than required can bank those allowances for the future or sell them to other generators. In addition, generators who have not made enough reductions can buy allowances on the market, or purchase a limited amount of reductions through "offsets" – or certified emissions reductions achieved outside of the power sector, such as through reforestation or landfill methane recapture. RGGI will cap carbon dioxide emissions from power plants at 2005 levels between January 1, 2009 and the end of 2014.[8] From 2015 onward, the cap will decrease so that by 2019, the states collectively will have reduced their emissions by 10 percent below 2005 levels.

This type of program ensures that emissions reductions are achieved in the most economically efficient way, as each facility pays the least cost to achieve its emissions reductions. RGGI will also begin the inclusion of the cost of carbon dioxide emissions in each facility's financial planning in the future.

On December 20, 2005, seven Northeastern governors signed a Memorandum of Understanding (MOU) agreeing to the RGGI program's essential framework (at the last minute, Massachusetts Governor Mitt Romney and Rhode Island Governor Donald Carcieri chose not to sign). The RGGI signatory states will next agree to a joint model rule (expected in 2006) that each state will then bring back to its own regulatory process to enact. RGGI is slated to go into effect in 2009.

The RGGI MOU offers the opportunity to any "observer" state or state not involved to date to join the regional pact, thus opening the door for future expansion. It also explicitly acknowledges Massachusetts' and Rhode Island's contributions to the development of RGGI and holds open the possibility for both states to rejoin the pact at a later date. Nevertheless, to reach a seven-state agreement is a significant achievement. The seven governors together have created the precedent for a large, multi-state effort to regulate carbon dioxide, which, to date, has not been attempted, let alone accomplished, in the United States.

Sharing information and best practices

What might be considered a side benefit of many states working together regionally is the extensive sharing of information and best practices that occurs when state air pollution regulators and planners meet regularly, as in the case of the RGGI staff working group or the New England Governors' Energy and Environmental Commissioners' meetings. Many of the architects of regional air policy programs and agreements have worked together on these issues for ten, or even 20 years, sometimes understanding the policy workings of their colleagues' states as well as their own. While it would be difficult to *intentionally* plan a process where idea sharing is so natural and productive, the meetings of regional air policy staff provide this exact situation, facilitating exchange outside any formal regional agreement.

Clearly, the idea sharing now travels well beyond regional borders. In the area of conventional air pollution control (i.e., criteria pollutants such as ozone, SO_2, mercury, etc.), the Northeast states have long relied on California's leadership to move their own air pollution policies forward. Because California had passed clean air legislation prior to federal legislation, that state's ability to create certain types of air pollution policy on their own was grandfathered into the federal Clean Air Act.[9] The primary example is the Northeast states' use of California motor vehicle standards, first passed in 1966 (see also duVair *et al.*, Chapter 27, this volume). In the case of vehicle emissions control regulations, states are limited to adopting either the federal standards or California's. Recently, the Northeast states embraced the new California standards which, in addition to covering smog-forming pollutants, are now also regulating GHGs. Nationwide, as of this writing, ten states have adopted the California car standards, creating a large and harmonized market for cleaner vehicles.

Increasingly, states share policy ideas across the country regardless of political leanings, size, or region. This explains why, as more and more states develop climate change action plans, the strategies employed for state emissions reductions look increasingly similar. Renewable portfolio standards, energy efficiency and renewable energy funding programs, transportation policies (particularly "smart-growth" policies), and green-building standards are now common to most state action plans (see also Dilling and Farhar, Chapter 23, this volume), and become better developed and honed as they are adopted by each new state.

It takes all kinds: champions behind state climate protection

State climate protection policy is driven by leaders and champions from all levels of government. Dedicated, high-profile gubernatorial and legislative leaders are essential to moving climate policy forward because of the attention they draw to the issue of climate change, their ability to communicate a vision, and for their role as ultimate decision-makers within their states. Equally important, however, is the leadership that comes from the state policy-makers who design, broker, and implement their states' groundbreaking policies. These internal policy champions provide the experience, institutional memory, creativity, technical expertise, and leadership to take landmark climate policies from an idea to an implemented program.

The combined individual talents, skills, and knowledge of each Northeast state's internal champions are the foundation for the successes of the NEG/ECP *Climate Change Action Plan* and RGGI. Each state has, and indeed requires, many internal champions to make these policies happen – often working in separate agencies, with different perspectives and areas of expertise. These people work away from the limelight to initiate and develop individual state climate plans, but their collaboration truly shines in regional collaboration. "The Northeast is blessed with a skillful group of dedicated and experienced air pollution staff who has a long history of effectively working together," explains Sonia Hamel, coordinator of Massachusetts' climate protection programs.

While each of these champions deserves mention, a few individuals have played key roles in recent state and regional efforts. Within the RGGI process, Franz Litz (NY) has brought legal expertise, management skills, and insightful political instincts to the table, while Sonia Hamel (MA) has brought a sense for effective group process and over 16 years of experience in environmental policy and regional cooperation. Chris Nelson (CT) has provided technical expertise on power plant regulation, while Chris Sherry's (NJ) energy policy background has made him the key architect for the offsets program criteria. Although a relative newcomer to regional efforts of the Northeast, Phil Cherry (DE) has contributed years of policy implementation knowledge gained as policy director of that state's environmental agency.

The NEG/ECP climate work – and the acid rain and mercury programs preceding it – could not have happened without the institutional memory and policy guidance provided by remarkably dedicated career civil servants such as Jim Brooks (ME), Dick Valentinetti (VT), Joanne Morin (NH), and Steve Majkut (RI). Each of these officials has supported their own state's participation in regional efforts as well as the development of their individual state climate policies, and now bring that expertise to efforts like RGGI and the development of the Regional Greenhouse Gas Registry (RGGR). These people represent only a fraction of the internal policy champions in the Northeast, yet illustrate the breadth of skills that combine to bring these landmark climate change policies to fruition.

Recently, a new, promising level of inter-state and inter-agency cooperation has developed in the Northeast. The RGGI development has brought together the environment and energy agency heads from the nine Northeastern states, initiating a new kind of cross-state information sharing. For the first time, on a sustained basis, the secretaries and commissioners are working together to craft strategies between the energy and environmental sectors. Because of the regional nature of the power market, this step is essential for truly successful climate change action in the electricity and energy sectors. This new level of cooperation shows great potential for deepening regional collaboration as well as in-state policy-making.

Working nationally, side by side

The political impact of state efforts on climate change is finally beginning to be felt in the halls of federal government, and throughout the business community (see also Arroyo and Preston, Chapter 21, this volume). The influence stems in part from the "patchwork" quilt of state policies discussed earlier; in part from the sheer political momentum created by many states banding together. That also provides political support for individual states — the "safety in numbers" theory. One stark example of states' banding together for momentum occurred in July 2002, when the Chief Legal Officers of Alaska, Connecticut, Maine, Maryland, Massachusetts, New Hampshire, New Jersey, New York, Rhode Island, Vermont, and California wrote a letter to President George W. Bush, asking his administration to address the issue of climate change through comprehensive national policies on GHG emissions.[10] The bipartisan nature of state action and coordination is also a critical influence in building national support for meaningful climate change legislation.

Climate change leadership within state borders

It is important to note that all of the initiatives and policies created at the state level result in real emissions reductions. The greatest reduction results from states reducing emissions from their own operations. To an as yet

Table 26.2. *Cities for Climate Protection in the Northeast*

State	No. of CCP participants
Maine	3
Vermont	4
New Hampshire	2
Massachusetts	21
Rhode Island	1
Connecticut	12
New York	14
New Jersey	3
Current Total	**60**

Source: ICLEI (2005). *CCP Participants in the Greater Northeast Region*, status February 2, 2006; available at: http://www.iclei.org/index.php?id=1121.

limited extent, they can also serve to influence the actions of cities and towns. Over the past decade or more, 60 Northeast communities have signed up for the Cities for Climate Protection campaign (Young, Chapter 24, this volume) (Table 26.2), recently spurring the US headquarters of ICLEI (International Council for Local Environmental Initiatives) to open a Northeast office to coordinate municipal climate action in the region.

While state and municipal governments, of course, interact on many levels, municipal climate protection assistance programs are still in their infancy across the Northeast. For the most part, municipalities began working for climate protection due to their own champions' efforts, whether they were municipal officials or concerned citizens. These municipal groups provide a crucial constituency for encouraging state climate action, while also benefiting from most state climate policies (see Selin and VanDeveer, 2006; Lindseth, 2004; see also chapters by duVair *et al.*; Watrous and Fraley; and Young, Chapters 27, 25, and 24, this volume). While Northeastern states greatly value municipal leadership, they have yet to offer them much policy direction or funding for climate-related programs to date. The provision of local assistance, policy guidance, and technical support to municipalities is a crucial area for future work.

States leading by example

One area in which states can shape their climate impacts most directly is their own operations. "Greening" state government is an important strategy for climate protection because (a) it reduces emissions significantly within the state due to the magnitude of state operations; (b) it allows the state to experiment with, and demonstrate to the public, new environmentally friendly techniques, such as advanced composting, wind power, and green buildings. Use of these technologies, in turn, can be used by other institutions, such as colleges, universities, municipalities, and in businesses; and (c) it is an important communication tool that the state can use to demonstrate its commitment to climate protection and to prove that making changes in building, driving, and purchasing practices can protect the climate and also, in many cases, save money.

"Leading by example" is an explicit goal in the NEG/ECP plan, and hence in each of the individual New England state plans as well. Massachusetts' plan, for example, sets a goal of reducing GHG emissions from state operations by 25 percent by 2012. Initiatives in Massachusetts under the "lead by example" umbrella include improving energy efficiency, utilizing renewable energy technologies, acquiring energy-efficient vehicles, and using lower-carbon fuels (The Commonwealth of Massachusetts, 2004: 16). A council of representatives from over 20 key state agencies and offices, co-chaired by Massachusetts' Environmental Affairs and Administration and Finance offices, guides and coordinates these activities (all part of Massachusetts' State Sustainability Program). Most states that have intentionally adopted environmentally sustainable operations policies have found that the environmental benefit is accompanied by a significant cost savings, allowing vital taxpayer dollars to be put to other urgent needs.

Capacity building for future state roles

Even when the federal government finally acts, there will still be areas where states must play an important role in the control of GHG emissions. One important current state role is to build infrastructure and technical capacity that will be called upon in the shared federal–state implementation of any future federal program. Land-use regulation, "smart-growth" strategies, and efforts that depend upon the presence of public transportation will remain fundamentally within state and local control (e.g., Dilling, 2007). Indeed, currently these efforts hold a prominent place in the state action plans in Connecticut, Massachusetts, New York, Rhode Island, and Maine (see Table 26.1) and will remain critically important components in the future. States that move forward on these issues today will both gain the climate benefits from such policies sooner, and take action without risk of being pre-empted by federal policies in the midst of, or immediately after, designing a complicated state-level program.

Conclusion

Individual states play a vital role in US climate policy, serving as the front line in policy development in the absence of federal regulatory action. Even when, eventually, federal policy is created that pre-empts the variety of state policies in some areas, states will continue to be leaders, serving as groundbreakers and policy laboratories, sharing successes and case studies, and working together to address common problems. In particular, moving policy development forward regionally is a step forward for states and for the country. Regional efforts increase overall emissions reductions and tend to be even more useful for national adoption than individual state policies.

States have managed to bring about many significant policy changes through ongoing communication and education, diligent forging of alliances, and delicate balancing acts among interests from a wide range of industries, businesses, environmental groups, and citizens. These successes "are also proving that these steps can be taken without triggering political warfare" (Rabe, 2004: 172). All kinds of leaders are important in moving this work forward: the high-profile elected officials who provide the vision and political commitment needed for their jurisdictions to protect the climate, but also the leaders behind the leaders who stay through administration turnovers and political ups and downs, providing a consistent voice on issues, a steady hand in policy implementation, and maintaining the institutional memory that allows these efforts to build on each other over many years.

Policy invented in the Northeast and West Coast states become the leading force in US climate policy. "Indeed," as Rabe (2004: 133) concluded, "virtually any future step that the federal government could conceivably take in coming decades is likely to be borrowed from something already being attempted in one or more states. ... [I]t might be foolhardy for the federal government to ignore state experience and at some future point try to impose a new national strategy of its own design."

Notes

1. For a full list of all US states with climate change action plans see: http://yosemite.epa.gov/oar/globalwarming.nsf/content/ActionsStateActionplans.html; accessed February 28, 2006.
2. The Northeast States for Coordinated Air Use Management (NESCAUM) was founded in 1967 to provide assistance to Northeast states on air policy. For more information, see: http://www.nescaum.org/about.html; accessed February 28, 2006.
3. Supreme Court Decision 285 US 262, 311.
4. For more information, see http://www.negc.org, accessed February 28, 2006.
5. New York was already the most energy-efficient state in the continental United States on a per capita basis, accounting for less than 5 percent of the nation's primary energy use,

although it is home to 7 percent of the nation's population. See http://www.ccap.org/pdf/
04-2003_NYGHG_Recommendations.pdf.
6. Press Release from Governor Pataki of New York, April 25, 2003, http://www.state.
ny.us/governor/. For more information on the details of the Regional Greenhouse
Gas Initiative, see http://www.rggi.org.
7. For a more complete explanation of cap-and-trade systems in general, see US EPA (2005b).
8. The initial agreement affects about 600 power plants in the region of 25 MW in size or
bigger.
9. See Section 177 of the Clean Air Act Amendments of 1990.
10. For the full statement of the attorney generals of these states, see http://maagocms.bx.net/
filelibrary/climate.pdf; accessed February 28, 2006 (see also Averill, Chapter 29, this
volume).

References

Center for Clean Air Policy (CCAP) (2003). *Recommendations to Governor Pataki for
Reducing New York State Greenhouse Gas Emissions.* Report prepared in
collaboration with the New York Greenhouse Gas Task Force. Washington,
DC: CCAP.
Dilling, L. (2007). Toward carbon governance: Challenges across scales in the United
States. *Global Environmental Politics*, accepted for publication.
Kosloff, L. H., Trexler, M. C., and Nelson, H. (2004). Outcome-oriented leadership:
How state and local climate change strategies can most effectively contribute to
global warming mitigation. *Widener Law Journal*, **14**, 173–204.
Lindseth, G. (2004). The Cities for Climate Protection Campaign (CCPC) and the
framing of local climate policy. *Local Environment*, **9**, 4, 325–36.
Massachusetts State Sustainability Program (2004). *Fiscal Year 2002 Massachusetts
Greenhouse Gas Inventory for State Agencies.* Boston, MA: Office for
Commonwealth Development.
McKinstry Jr., R. B. (2004). Laboratories for local solutions for global problems:
State, local and private leadership in developing strategies to mitigate the causes
and effects of climate change. *Penn State Environmental Law Review*, **12**, 1,
15–82.
New England Climate Coalition (2005). 2005 *Report Card on Climate Change Action.*
Second Annual Assessment of the Region's Progress Toward Meeting the Goals
of the New England Governors/Eastern Canadian Premiers Climate Change
Action Plan of 2001. Boston, MA. Available at: http://www.newenglandclimate
.org/files/reportcard05.pdf; accessed February 1, 2006.
New England Governors/Eastern Canadian Premiers (NEG/ECP) (2001). *Climate
Change Action Plan 2001.* The Committee on the Environment and the
Northeast International Committee on Energy of the Conference of
New England Governors and Eastern Canadian Premiers. Halifax, NS and
Boston, MA. Available at: http://www.negc.org/documents/
NEG-ECP%20CCAP.PDF; accessed February 1, 2006.
Northeast States for Coordinated Air Use Management (NESCAUM) (2004).
*Greenhouse Gas Emissions in the New England and Eastern Canadian Region,
1990–2000.* Boston, MA: NESCAUM; available at: http://bronze.nescaum.org/
resources/reports/rpt040315ghg.pdf; accessed February 2, 2006.
Peterson, T. D. (2004). The evolution of state climate change policy in the United States:
Lessons learned and new directions. *Widener Law Journal*, **14**, 81–120.

Rabe, B. G. (2004). North American federalism and climate change policy: American state and Canadian provincial policy development. *Widener Law Journal*, **14**, 121–72.

Regional Greenhouse Gas Initiative (RGGI) (2005). *Regional Greenhouse Gas Initiative (RGGI) Memorandum of Understanding*. Available at: http://www.rggi.org/docs/mou_12_20_05.pdf; accessed February 1, 2006.

Selin, H. and VanDeveer, S. (2005). Canadian–U.S. environmental cooperation: Climate change networks and regional action. *The American Review of Canadian Studies*, **35**, Summer, 353–78.

Selin, H. and VanDeveer, S. (2006). Canadian–U.S. Cooperation: Regional Climate Change Action in the Northeast. In *Bilateral Ecopolitics: Continuity and Change in Canadian – American Environmental Relations*, eds. Le Pestre, P., and Stoett, P., Aldershot: Ashgate, pp. 93–113.

The Commonwealth of Massachusetts (2004). *Massachusetts Climate Protection Plan*. Office for Commonwealth Development, Boston, MA. Available at: http://www.mass.gov/Eocd/docs/pdfs/fullcolorclimateplan.pdf; accessed January 30, 2006.

US Environmental Protection Agency (EPA) (2005a). *State Action Plans for Greenhouse Gas Mitigation*. Available at: http://yosemite.epa.gov/oar/globalwarming.nsf/content/ActionsStateActionPlans.html#actionplans; accessed February 1, 2006.

US EPA (2005b). *Allowance Trading Basics*. Available at: http://www.epa.gov/airmarkets/trading/basics/#what; accessed February 1, 2006.

US EPA (2002). *U.S. Emissions Inventory 2002: Inventory of U.S. Greenhouse Gas Emissions and Sinks: 1990–2000*. Washington, DC: EPA; available at: http://yosemite.epa.gov/oar/globalwarming.nsf/content/ResourceCenterPublicationsGHGEmissionsUSEmissionsInventory2002.html; accessed February 2, 2006.

Victor, D. G., House, J. C., and Joy, S. (2005). A Madisonian approach to climate policy. *Science*, **309**, 1820–1.

Walker, J. L. (1969). The diffusion of innovations among the American states. *The American Political Science Review*, **63**, 3, 880–99.

27

West Coast Governors' Global Warming Initiative: using regional partnerships to coordinate climate action

Pierre duVair
California Energy Commission

Sam Sadler
Oregon Department of Energy

Anthony Usibelli
Energy Policy Division, Washington Department of Community, Trade, and Economic Development

Susan Anderson
Office of Sustainable Development, City of Portland, Oregon

Introduction

The governors of California, Oregon, and Washington launched the West Coast Governors' Global Warming Initiative (WCGGWI) in 2003.[1] In doing so, they joined the New England governors and Eastern Canadian premiers in addressing global warming from a state and regional perspective (see also Tennis, Chapter 26, this volume). The governors instructed their staff to look for strategies that mitigate greenhouse gas (GHG) emissions through measures that also promote long-term economic growth, protect public health and the environment, consider social equity issues, and expand public awareness of climate change. This chapter describes the West Coast regional effort and examines the ways in which local and state-level action have contributed to bringing about this regional initiative, and in turn, how the regionally integrated emission reduction initiative supports states and local communities in fulfilling and going beyond their own individual action plans.

Views expressed in this chapter represent the opinions of the authors and do not necessarily reflect the views of the agencies within which they work.

431

Motivations for the West Coast Governors' Global Warming Initiative

The governors of California, Oregon, and Washington initiated the WCGGWI because they recognized that existing state and federal policies would not stabilize, much less begin reducing, the region's level of GHG emissions. The states also hoped to establish precedents that would spur the development and implementation of climate polices in other states and at the federal level.

In addition, the governors were motivated by the fact that global warming will have serious adverse consequences for the economy, public health, and the environment of West Coast states. They recognized that negative impacts will grow significantly in coming years in the absence of action. They also believed that actions to address global warming would yield substantial economic benefits. The West Coast region is rich in renewable energy resources and advanced energy-efficient technologies.

The governors believe that working together will create great economic opportunities and maximize the environmental benefits of cooperative actions. Likewise, they understand that the causes of global warming are not isolated to one state or region. In fact, energy production and trade in the western United States, for example, are closely interlinked, so any climate-related action taken by one state will impact — and should involve — the other western states. While each state is developing state-specific strategies, there are other areas such as regional goals, consistent efficiency standards for buildings and appliances, motor vehicle GHG emission standards, combined purchasing power, interstate transportation measures, electrical transmission, and regional GHG allocation standards, among others, that are much more effective when implemented with a cooperative effort between multiple states.

The WCGGWI began when senior staff from the three states met in Portland, Oregon, in the summer of 2003.[2] The meeting was supported and hosted by the Energy Foundation. If included observers from other jurisdictions and representatives of some non-governmental organizations. In the absence of a formal three-state regional organization, the Energy Foundation's continuing financial and organizational support has been critical to the success of this effort.

Each state came into the process from different points. Oregon had been working on climate change policy since 1988 and had adopted a carbon dioxide (CO_2) standard for new energy facilities in 1997. In June 2003, Oregon Governor Ted Kulongoski issued his Executive Order for a Sustainable Oregon for the 21st Century,[3] including directing the Oregon

Department of Energy to cooperate in a regional approach to addressing climate change and paving the way for participation in the WCGGWI.

Many policy-makers in California became aware of potential harm to the state's unique natural resources when the Union of Concerned Scientists and the Ecological Society of America issued a report on the topic in 1999 (Field *et al.*, 1999; see also Cole with Watrous, Chapter 11, this volume). In 2002, two bills were passed that highlighted the Legislature's concern about risks of global warming, specifically addressing renewable energy and emissions from motor vehicles. In addition, California's decision to participate in the WCGGWI was based on an acknowledgement that climate change is a global problem requiring action and coordination beyond what a single state could achieve. With broad support from a cross-section of Californians (Baldassare, 2005) and efforts being made to integrate energy policy in the state,[4] the path in California clearly headed in the direction of a larger, regional approach to tackling climate change.

Washington's engagement in the WCGGWI arose from three principal factors: recognition of the negative consequences of global warming on the state's economy and environment; public interest group pressure; and the rush toward new power plant construction arising from the West Coast electricity crisis that began in 2000.

Research since the 1990s at the Climate Impacts Group of the Joint Institute for the Study of the Atmosphere and Ocean (JISAO) at the University of Washington has raised awareness of potential climate change impacts in the state and the need to begin finding ways to reduce GHG emissions.[5] At the same time several non-profit organizations, including Climate Solutions,[6] the NW Climate Connection, and Global Warming Action,[7] helped increase the level of interest of state agency and elected officials in global warming and, in particular, focused on the economic opportunities that the State of Washington and the region might capture by developing climate-friendly, renewable energy and energy efficiency technologies.

Finally, for Washington, the rush of applications for new electric power plants that occurred because of the high wholesale electricity prices during the West Coast energy crisis highlighted the impacts of new fossil fuel generation on the state's total GHG emissions footprint. Washington State has one of the highest percentages of zero-GHG electricity sources – hydroelectric generation – so each new natural gas power plant represents a major increase in statewide emissions. Interest in regional approaches to mitigation of GHG emissions from the power sector also motivated Washington's participation in the WCGGWI.

Initial goals and projects

As the states began formal coordination on global warming, they decided to focus on a set of short-term projects that could demonstrate economically beneficial mitigation of emissions within the first year. There was an understanding of the need for and an interest in developing a long-term regional strategy, but the states wanted to begin immediately with a few short-term actions. They chose a set of actions not as a comprehensive strategy, but as individual projects, which became specific recommendations that staff developed in the first year of the WCGGWI given the limited time and resources available. The public did provide input, but limited resources precluded a full-fledged, multi-state integrated stakeholder process. In November 2004, the governors accepted the *West Coast Governors' Global Warming Initiative Staff Recommendations to the Governors*.[8] The recommendations included many tangible short-term actions, such as improving the emissions of the state motor vehicle fleet, purchasing hybrid-electric vehicles, planning infrastructure for electrification technologies at truck stops along I-5, increasing sales of renewable resources by 1 percent or more annually, adopting energy efficiency standards for products not regulated by federal standards, incorporating aggressive measures for updating state building codes to maximize energy efficiency, and organizing a West Coast Governors' Scientific Conference in 2005.[9]

As the next phase of work under this initiative, the governors agreed to explore more comprehensive regional measures to reduce GHG emissions, while working with stakeholders in each of their states. Four areas were highlighted as holding the most promise:

- Adopt comprehensive state and regional goals for GHG emission reductions from all types of sources;
- Adopt standards to reduce GHG emissions from motor vehicles;
- Develop a market-based carbon emissions allowance program; and
- Expand the markets for energy efficiency, renewable resources, and alternative fuels.

The governors directed their state agencies to continue the WCGGWI and to strengthen links with similar efforts in other states, Canadian provinces, and Mexico.

The following sections of this chapter provide some history and insight into individual actions, leadership, and communication strategies of West Coast governments related to climate change. These impressive efforts form the political and practical foundation for the more recent regional initiative, but also hint at the limits of what any single state can achieve by itself in the fight against global warming and thus point to the need for a regional approach. The final section highlights the important role that local governments can and need to play in changing the way energy is used and supplied, land is developed and infrastructure provided, residents and businesses are informed, and where the mitigation of emissions actually occurs.

Leadership on climate change by West Coast states

The State of Oregon

Governor's Advisory Group on Global Warming

In early 2004, Oregon's Governor Kulongoski appointed a 28-member Governor's Advisory Group on Global Warming to develop a global warming strategy that addressed specific state needs and responsibilities, but which was closely coordinated with the work being done at the regional level. The Governor's Sustainability Advisor was a member of both the Advisory Group and the executive committee of the WCGGWI. The Group was informed by scientific knowledge, in particular through a meeting organized by the Institute for Natural Resources at Oregon State University, which resulted in the report *Scientific Consensus Statement on the Likely Impacts of Climate Change on the Pacific Northwest.*[10] The Advisory Group also received input from about 150 representatives of businesses, utilities, state agencies, local governments, universities, and environmental groups in Oregon as well as through a public comment process. The Advisory Group published its recommendations in the *Oregon Strategy for Greenhouse Gas Reductions.*[11]

The Governor has adopted these goals for emissions reductions, which are based on Oregon's share of a global commitment to stabilize GHG emissions. He also initiated specific actions, including a task force to design a cap-and-trade system to limit carbon emissions.[12] At the Governor's behest, the Oregon Environmental Quality Commission adopted the California GHG emission standards for motor vehicles in 2005. He is also establishing a Climate Change Integration Group to continue developing climate change policy recommendations, with an initial focus on adaptation to climate change.

The Legislature acted on the Advisory Group's recommendation to adopt appliance efficiency standards. Other recommendations that are longer term include continuing to improve building efficiency standards, specific renewable resource actions consistent with the Oregon Renewable Energy Action Plan,[13] greening the fleet measures, and creating a plan for education and outreach. Many of these actions are consistent with, and indeed are informed by, recommendations in the WCGGWI regional context. For example, the Advisory Group in Oregon directly adopted many of the recommendations from the WCGGWI as part of its state strategy.

Historical context

The process and recommendations of the Advisory Group built upon years of policy development and public interest at the state and city level. In 1990, a task force of state agencies in Oregon reported that climate change is a serious threat and that Oregon has a responsibility to address it. The 1989 Legislature also called for a comprehensive emission reduction strategy report. In subsequent years, Oregon developed inventories of its GHG emissions and strategies for reducing them. Portland became the first city in the United States to adopt a goal and strategy to reduce GHG emissions (for further details, see below). The Public Utility Commission began requiring consideration of GHG emissions as part of utilities' least-cost plans. In 1997, the Legislature made consideration of climate change part of the process to site new energy facilities. The Oregon Progress Board[14] also adopted an official benchmark in 1992 to hold Oregon's CO_2 emissions to 1990 levels. It maintains the benchmark, but the state has not been successful in meeting it. The ongoing work of staff in several agencies also strongly supports action on global warming through differentiated responsibilities within each agency's purview. The agencies involved include the Department of Energy, the Oregon Public Utility Commission, the Department of Forestry, the Department of Environmental Quality, the Oregon Progress Board, and the Oregon Energy Facility Siting Council.

Tradeoffs and cooperation

Oregon was the first state in the United States to set a mandatory CO_2 standard for power plants with legislation in June 1997. The Oregon standard regulates CO_2 pollution by requiring new power plants and other large energy facilities to avoid or offset a portion of their CO_2 emissions. The net CO_2 emissions rate required of new energy facilities is stricter than can be achieved by on-site technology alone; therefore it requires provision of offset projects. Over the last nine years of implementing this standard, Oregon has shown that a state can take regulatory action to reduce CO_2 emissions with stakeholder concurrence and without affecting the competitive position of the state's energy industry in the region. A key element of the success of this legislation was creation of an independent, non-governmental organization, The Climate Trust, to purchase CO_2 emission offsets using funds provided by facility compliance with the standard.[15]

In developing a CO_2 standard, a legislatively created task force turned to industry, environmental, and agency stakeholders to propose a specific standard and to craft a workable implementation design. The stakeholders spent about 100 hours in negotiations to craft the specific language that was later drafted into proposed legislation.

Industry was willing to accept the CO_2 standard because it wanted to eliminate a standard that required applicants for a power plant site certificate to demonstrate a need for the facility according to a utility's integrated resource plan. The process provided a direct trade of dropping the "need standard" and creation of the CO_2 standard. Environmental groups were willing to accept the elimination of the need standard in order to achieve a reduction in CO_2 emissions. The Department of Energy and the Energy Facility Siting Council felt the package represented an advancement of environmental standards and appropriate regulatory reform to reflect changes in the market for electric power.

The legislation subsequently passed all committees and both chambers of the Oregon Legislature unanimously because the concept of requiring CO_2 offsets was vetted during an earlier Energy Facility Siting Council proceeding and through the siting task force process. It also succeeded because of the participation by key stakeholders and their unified support for the CO_2 standard.

Formal coordination — better prospects for success

The Oregon process, which is continuing and expanding with regional coordination, is an extension of 18 years of involvement by the public, the Legislature, citizen boards, and state and local agencies. Despite those 18 years of effort, the state's GHG emissions continue to rise. Oregon is now looking for new ways to increase the effectiveness of GHG mitigation efforts by formalizing what had been informal coordination with California and Washington over the years.

The State of Washington

Initial efforts by the State of Washington to address global warming and GHG emissions date from the early 1990s when funding from the US Environmental Protection Agency (EPA) allowed Washington to inventory its GHG emissions and identify potential statewide impacts of global warming. The state, like several others, also produced a climate action plan (Washington State Energy Office, 1996). However, although these efforts provided expanded analytical results and descriptive data, they were not translated directly into specific legislation on climate change or new GHG reduction programs.

Carbon regulation in Washington's energy facility siting process

The first substantive actions related to reduction of GHG emissions arose from within Washington's energy facility siting process. As in Oregon, siting of large thermal power plants is a centralized process run by the state's Energy Facility Site Evaluation Council (EFSEC).[16] In 1997, the EFSEC issued a site license for a new natural gas-fired combined-cycle combustion turbine project that included language requiring the project applicant to develop a GHG mitigation report.[17] The initial certification agreement did not include a specific requirement that the facility offset its GHG emissions, but the issuance set in motion a process whereby all new applications to the EFSEC were expected to include some level of mitigation for future GHG emissions at their proposed facility.

In 2001, when the Chehalis Generating Facility revised its application, the amended certificate included an 8 percent GHG emission offset requirement. Following the Oregon model, project applicants are offered two compliance options: (1) directed offsets made by the owner or, (2) payments to an independent third-party organization to provide offset projects.

The existence of a successful Oregon model was critically important to the State of Washington in at least two ways. First, it provided a proven record of compliance that allowed interveners in the siting process (state agencies, public advocacy groups, and citizens) to counter claims made by project developers that GHG mitigation was not technically or economically feasible. Second, it provided the EFSEC with an existing set of administrative procedures that could be built upon and avoid an often lengthy and contentious legislative process.

As requirements to mitigate GHG emissions were applied to subsequent power plant projects, prospective developers became concerned about the application of such requirements without a specific mandate or direction from the Legislature. In 2001, the EFSEC began to adopt a series of standards for environmental compliance by permit applicants. Prior to that time the power plant siting statutes provided little detailed guidance, which left final mitigation decisions to the EFSEC on a case-by-case basis. The most contentious of the proposed new standards were those for mitigation of GHG emissions. During the 2004 legislative session, power plant developers, utilities, public interest groups, and state agencies reached agreement on GHG mitigation requirements for all fossil fuel power plants greater than 25 megawatts. The agreement was signed into law by Governor Locke in 2004.[18] As a result, new power plants are required to mitigate 20 percent of their GHG emissions or pay into a fund for mitigation at the rate of $1.60 per ton.[19]

Planning at the Puget Sound Clean Air Agency

As noted earlier, Governor Locke joined with the governors of Oregon and California in forming the WCGGWI in September 2003. Nearly coincident with the WCGGWI, the Puget Sound Clean Air Agency, the air pollution control agency for Washington's four most urbanized and populous counties, began a stakeholder process to develop a GHG mitigation plan for the agency. Because the planning and funding for this process was well advanced, the State of Washington decided to use the Puget Sound effort as the forum for its climate change stakeholder process. Based on broad public and business input, its final report recommended actions that would reduce GHG emissions by 6 million metric tons (tonnes) by 2010 and 16.6 million tonnes by 2020. The only recommendation of the group that was not unanimously adopted was one that called for the adoption of California's motor vehicle GHG emission standards.[20]

Based on the preliminary results of the Puget Sound climate change planning process, activities of the WCGGWI, and staff recommendations, Governor Locke proposed several major initiatives linked to climate change. In January 2005, the Governor issued an executive order on sustainability in state agency operations.[21] The order calls for increased purchase of hybrid-electric vehicles, restrictions on purchase of SUVs, and increased energy efficiency in state-owned and state-leased buildings. At the same time Governor Locke introduced four pieces of proposed legislation that would: adopt California vehicle emissions standards for criteria air pollutants and the proposed GHG requirements; adopt energy efficiency standards for products not regulated by federal standards; establish statewide goals for emissions reductions similar to the Northeast states (see also Tennis, Chapter 26, this volume); and require utilities to invest in energy efficiency and renewable energy resources.

Of these four pieces of proposed legislation, the first two successfully passed the Legislature — ESHB 1397 (motor vehicle emissions standards), and ESHB 1062 (energy efficiency of certain products). The Legislature also adopted the nation's most stringent requirements for "green" public buildings, in ESSB 5509. The vehicle emissions standards are estimated to reduce vehicle GHG emissions in the Puget Sound area alone by nearly 3 million tonnes in 2020.[22]

The State of California

Individuals making a difference

Climate change as an opportunity for individuals to show leadership can easily be traced back to legislation in 1988 by then Assemblyman Byron Sher (Assembly Bill 4420). AB 4420 directed the California Energy Commission (CEC) to evaluate potential impacts to the state from changes in climate, as well as develop and recommend strategies to mitigate those expected impacts. Throughout his career in the California Legislature, Byron Sher made a difference on a broad range of environmental issues, but climate change and renewable energy benefited especially from his leadership.

Fourteen years later, the California Assembly had a new climate leader: Fran Pavley. In 2002, Assemblywoman Pavley helped push through the country's first piece of legislation to begin mitigating the state's single greatest source of GHG emissions: the automobile. The legislation (AB 1493) − since dubbed "the Pavley Bill" − garnered enormous efforts dedicated to both its passage and its defeat.[23] Once passed out of the Legislature, new opportunities for leadership on climate change opened up: signing this controversial bill into law.

In 2002, then-Governor Gray Davis showed such leadership when he signed the Pavley Bill as well as a bill setting standards for electricity generated from renewable energy (Senate Bill 1078).[24,25] Signing these cutting-edge pieces of legislation into law signaled California's seriousness in addressing climate change and set an example for other (not just West Coast) governments. One individual's leadership, however, would not have achieved much without the support of a first-rate policy team.

Governor Davis brought Mary Nichols and Jim Boyd to the California Resources Agency where they assembled a large multi-agency climate team that met regularly between 2000 and 2004. In 1999, he appointed Dr. Alan Lloyd as Chairman of the California Air Resources Board (CARB), who was instrumental in the implementation of the Pavley Bill. Another key appointment was that of Dr. Arthur Rosenfeld as a Commissioner at the CEC in 2000, who led both energy efficiency and research and development in new directions to address climate change. Finally, in 2002, Governor Davis appointed Michael Peevey as President of the California Public Utilities

Commission to lead bold new efforts to address GHG emissions associated with the generation of electricity consumed in California (including out-of-state generation serving California).

Gubernatorial leadership on energy and climate change continued under Governor Arnold Schwarzenegger who, on June 1, 2005, signed Executive Order S-3-05, which established ambitious GHG emission reduction targets and called for a coordinated effort and routine reporting on progress towards those targets.[26] In signing the Executive Order on World Environment Day in San Francisco (the first time the event was hosted in the United States), the Governor stated:

> I say the debate is over. We know the science. We see the threat. And we know the time for action is now. Global warming and the pollution and burning of fossil fuels that cause it are threats we see here in California and everywhere around the world.[27]

Governor Schwarzenegger's climate leadership benefited significantly from the trusted advice of then-Secretary of the California Environmental Protection Agency (Cal/EPA) and later Cabinet Secretary, Terry Tamminen. Continuity in knowledgeable and committed leadership at the head of Cal/EPA, the Energy Commission, and other key state agencies has been critical in maintaining and advancing the state's ambitious efforts to address climate change (see also Tennis, Chapter 26, this volume). Finally, the governor's efforts to work with leaders of the Legislature such as Fabian Nunez, Don Perata, and Fran Pavley produced bold new laws in 2006 to address climate change: Assembly Bill 32 and Senate Bill 1368. These laws cap GHG emission levels and prevent long-term investment in electricity generation that has high levels of GHG emissions.[28]

Contributions from research and development

While California has the good fortune of climate leaders at the top of state government, it also has the policies and programs in place to dedicate significant resources to research and development (R&D). In fact, it is the only state in the nation with a state-sponsored climate change research program. A public goods charge, paid by electricity ratepayers, funnels more than $60 million towards R&D each year. The Public Interest Energy Research (PIER) Program of the CEC allocates approximately $5–$8 million per year to climate change research.[29] In partnership with the University of California at San Diego (Scripps Institution of Oceanography)[30] and Berkeley and many of the state's other scientists, the CEC created the virtual California Climate Change Center to provide timely access to a broad range of climate-change-relevant research.

The state's resources dedicated to climate change research and outreach are small compared to the investments needed to achieve an 80 percent reduction in GHG emissions from 1990 levels within the next 45 years. However, with the strong support from state officials, the California Public Employees' Retirement System (CalPERS) and the California State Teachers' Retirement System (CalSTRS) have also become key investors in a low-carbon future. Together they set an example for other industries (see Atcheson, Chapter 22, this volume).[31]

The West Coast Regional Carbon Sequestration Partnership (WESTCARB), led by the CEC, is a public–private partnership of organizations in and beyond Washington, Oregon, and California planning to spend millions of dollars on pilot projects involving terrestrial and geological carbon sequestration.[32]

An engaged public

Strong leadership and world-class science on climate change make for a strong recipe for climate action. But having an informed, well-educated, politically and socially active populace is the other essential ingredient for successful long-term solutions to climate change, including political support and behavioral changes (see Tribbia, Chapter 15, this volume). A 2005 survey, conducted by the Public Policy Institute of California (PPIC), found that 86 percent of state residents believe that climate change will adversely affect current or future generations and 57 percent said changes were already happening (Baldassare, 2005). The high level of concern about global warming, and the strong support for state action was astonishing, as was "the willingness of the public to make lifestyle and financial sacrifices for the sake of improving air quality" (Gaura, 2005).

Political leaders in California are keenly aware of polls and the environmental ethos of many of the state's voters. The state plays an important role in informing the public about issues such as climate change, where the risks to economic and ecological resources can be high, as well as the risks to public health. California state government helps inform the public through media campaigns such as "Flex Your Power" and "Flex Your Power at the Pump,"[33] where a broad range of information is made available to many types of energy users.

These state-led outreach efforts are matched, however, by a very effective communication and advocacy effort of environmental groups. They are adept at taking the latest science and translating it into words and concepts that the average citizen can understand (see also Cole with Watrous, Chapter 11, this volume). In recent years, they have taken a non-adversarial, more coalition-based, solution-oriented approach – representing a broad cross-section of interest groups and offering messages that appeal to decision-makers in the private and public sectors.

Climate activities at state agencies

The Governor's GHG mitigation targets have motivated state departments, boards, and commissions to further develop and evaluate potential strategies for achieving emission reductions. The Climate Action Team (CAT) was formed by Cal/EPA to coordinate these activities among state agencies and to prepare biennial updates to the Governor and Legislature on progress made.[34] The CARB leads implementation of the groundbreaking light-duty automobile GHG standards.

Another important way to keep Californians and state government informed on energy and climate matters is the state's *Integrated Energy Policy Report* (*Energy Report*), which the CEC prepares every other year.[35] To develop additional emission reduction options, the CEC convened a diverse Climate Change Advisory Committee.[36] In addition, the state also maintains an inventory of GHGs and provides technical support for the voluntary reporting of GHG emissions at the California Climate Action Registry.[37]

Finally, in December 2004 the California Public Utilities Commission (CPUC) added a price tag on the regulatory risk of GHG emissions, ranging from \$8 to \$25 per ton of CO_2 to be used by the three independently-owned utilities (IOUs) in California when evaluating electricity procurement commitments of five years or more, as well as in the development of their long-term energy procurement plans.[38,39]

Local government as catalyst for climate action

Cities and counties throughout the United States have an important role to play in addressing climate change because of their unique set of responsibilities that affect issues such as land use, transportation, building codes, energy use, electricity generation, as well as waste management and disposal. Several cities in the United States have set specific targets to reduce CO_2 and other GHG emissions, many coordinated by the International Council for Local Environmental Initiatives (ICLEI), called the Cities for Climate Protection Campaign (see also the chapters by Young; Watrous and Fraley; Pratt and Rabkin; and Rabkin with Gershon, Chapters 24, 25, 6, and 19, this volume).[40] On the West Coast, these include among others Portland, Seattle, Los Angeles, Santa Monica, Olympia, and San Diego. The City of Portland, Oregon provides a good example of the programs and policies that local governments can enact that have a specific and measurable impact towards the reduction of GHG emissions. It also played a critical role in fostering action at higher levels of government.

The City of Portland, Oregon

Over a quarter century ago, as an extension of statewide planning requirements, Portland adopted its 1979 local energy policy. This was the first local energy plan in the United States. In 1993 that plan was expanded and a more comprehensive view was taken to integrate energy with housing, land use, transportation, and business concerns. The result was Portland's 1993 Sustainable Energy and Carbon Dioxide Reduction Strategy.[41] This plan was also a first in the United States.

Portland's action plan proposed efforts to reduce the use of fossil fuels and the resulting GHG emissions. However, it was the many ancillary benefits that became the primary selling point and the basis for success of the plan. These co-benefits include reducing local smog, improving water quality, saving businesses and local governments money, reducing traffic congestion, and cutting residential energy bills (see also Young, Chapter 24, this volume).

In 2001, Portland reviewed its successes and set an overall goal of reducing CO_2 emissions 10 percent below 1990 levels by the year 2010. An evaluation of Portland's efforts between 1990 and 2004 shows some pretty impressive results. CO_2 emissions were reduced by 12 percent per capita − possibly the largest reduction anywhere in the United States.[42] These emission reductions were clearly related to changes in energy use:

- Between 1990 and 2004 per capita gasoline consumption fell by 9 percent
- Electricity use by Portland households fell by 10 percent
- Recycling more than tripled − increasing by 210 percent during this period
- Transit ridership increased by 75 percent

So how did Portland achieve these energy savings? What were the key regulatory, technical, and marketing approaches taken by the City of Portland?

Regulatory actions

Portland worked closely with the State of Oregon to develop new residential and commercial energy codes and established its own policy to cut energy use in city government facilities by 10 percent. In addition, beginning in 2001, all new city facilities and any commercial or multi-family construction project that uses City of Portland funding − such as tax increment financing, tax rebates, or loans – must meet a national green-building standard (LEED – Leadership in Energy and Environmental Design).[43]

Portland also decided to purchase 100 percent renewable energy resources for electricity used in government facilities by 2010. This goal is likely to be reached by 2007. Additional requirements to recycle 60 percent of all commercial and residential solid waste as of 2005 and provide free transit downtown also added to Portland's bundle of rules that reduced GHG emissions.

Most interesting, however, is that almost none of these measures were adopted in response to global warming (see also the chapters by Young, Watrous and Fraley, Pratt and Rabkin, Arroyo and Preston, Chapters 24, 25, 6, and 21, this volume). Local leaders wanted to impact local air pollution problems, traffic congestion, economic development, and rising energy bills for residents.

Technical and financial assistance

Dozens of new programs and services were provided by the City of Portland and state agencies to complement the new regulations. These programs provided the right mix of incentives to overcome many of the informational, technical, and financial barriers to reducing energy use and the associated GHG emissions.[44] A new Green Investment Fund was established with $3.3 million invested by city government and an additional $100 million more leveraged from private businesses and residents. These funds were focused on leading-edge technologies to ensure that green buildings became the standard construction practice in Portland. The funds came from a newly formed coalition of solid waste, water, wastewater, and electric utility sources. Politically the only way to focus on green buildings was to pull together all of these parties and show them how each one could benefit.

Marketing, outreach, and education

Traditionally government and utilities have managed conservation programs with an attitude of "here's the program, y'all come ..." Strangely, agency officials were surprised when nobody showed up. To be successful, cities around the United States are learning that their efforts must be client-based. Social marketing has become the buzz-word (see also Rabkin with Gershon, Chapter 19, this volume). Understanding niche markets and providing services in a way that meets specific needs is the only way to change behaviors of the general public (see also Tribbia, Chapter 15, this volume). Perhaps the most successful effort has been an intensive education campaign aimed at residents to promote waste reduction and recycling. This campaign has resulted in a 55 percent rate of recycling, with about 80 percent of all households participating.[45]

Leadership: Walking the talk

To successfully promote a local plan to reduce GHG emissions, cities focus increasingly on the need for direct investments. Nearly ten years ago, Portland invested in technology that uses waste methane from the sewage treatment plant as a fuel source to generate electricity. New capacity added in 2006 will result in more than $500,000 worth of power produced annually. In addition, investment in new building technologies has saved the city more than $15 million. Solar panels can be seen on maintenance vans, parking meter pay stations, and fire stations. Light rail in all directions covers 38 miles with very high ridership. More than 230 miles of bike paths and lanes have been built and all buses are equipped with bike racks. Walking the talk has meant more than just being a model: it has meant millions of dollars in savings to city government and local residents.

Over the past 25 years, Portland has learned an enormous amount about how to motivate behavior change through carefully adopted regulations and targeted public education. Lessons learned in Portland suggest three components are essential to a local campaign to impact global warming:

1. A political base of support and committed political and business champions.
2. Strong policies and appropriate regulations developed jointly by residents, business, and government.
3. A broad spectrum of well-targeted programs available to business, government, neighborhoods, and individual residents.

Finally, while West Coast states have focused on a variety of global warming policies and programs over the past ten years, Portland and several other cities have taken a more hands-on, practical approach. The local and state levels of government continue to support each other. The state adopts strong policies and legislation that promote financing opportunities and broad requirements related to, for example, transportation, building efficiency, and renewable resources. The cities, such as Portland, use these tools locally and develop marketing programs to effectively reach the business and residential sectors. The partnership between the State of Oregon and City of Portland has served to create an atmosphere of cooperation among businesses, residents, and government that has led to significant and very real GHG emission reductions. Moreover, it is that sort of cooperative spirit and partnership with business that enables political leaders to take bold steps toward stronger emission reduction efforts.

Lessons from the West Coast

Many residents, businesses, and governments in Washington, Oregon, and California are "thinking globally and acting locally" when it comes to climate change. The West Coast is especially vulnerable to adverse impacts from climate change on water systems, which can cause significant economic and ecological harm. This vulnerability is recognized by many as a result of committed leadership over a long period of time, relevant research, and concerted outreach efforts, and visible modeling of action by local communities, businesses, and states. For example, more than 40 local governments have joined the Cities for Climate Protection Campaign in these three states.

Clearly, local and state governments have a great deal to learn from each other, and often take advantage of those opportunities. Creating innovative programs and policies at the local level is a relatively low-risk means of testing new approaches, at least from a national and international perspective (see Tennis, Chapter 26, this volume). Local and state governments have been, and no doubt will continue to be, laboratories for energy, economic, and environmental policy.

While experts argue for more policy experimentation at the local level (Morgenstern and Portney, 2004), policy-makers in both West and East Coast states already recognize the importance of taking action on climate change in the face of uncertainty given the total benefits and costs. Some see getting out in front on actions to mitigate GHG emissions and adapt to expected changes as an opportunity to gain competitive advantages, others see it more in terms of an insurance policy, and still others see it more simply as the morally or ethically right thing to do (see also Arroyo and Preston, Chapter 21, this volume).

One common theme in this chapter is the open process used by local and state governments to develop their climate strategies prior to adopting formal policies, programs, regulations, and standards. The topics of climate and energy bring many stakeholders together. A focus on impacts adds interested parties from agriculture and forestry, transportation and cement production, water and waste management, as well as public health. Each stakeholder could point to the other as the one that should act first, but fortunately on the West Coast, we find a spirit of leadership and cooperation across organizations and jurisdictions. While that does not guarantee success or make everything go smoothly, it does help states and communities attempt new policies, set targets, and make and measure progress. Moreover, what happens at one level informs, puts pressure on, encourages, and frequently enables actions at other levels of government. All of this activity is happening against a backdrop of action in the international realm among governments, businesses, and industries worldwide. They provide their own impetus and forms of experimentation with climate policy. Thus, as West Coast governments test and take the lead on emission reduction strategies regionally and nationally, their efforts are also being watched by international players and are certainly helped along by developments in the global arena.

Notes

1. The popular term "global warming" rather than "climate change" was chosen to make the topic more accessible to the general public. For more information about the WCGGWI see: http://oregon.gov/ENERGY/GBLWRM/Regional_Intro.shtml or http://www.climatechange.ca.gov/westcoast/index.html; accessed September 20, 2006.
2. Governors in California and Washington have changed since the start of the initiative, but all three governors in 2006 maintain strong support.
3. See http://www.governor.oregon.gov/Gov/pdf/ExecutiveOrder03-03.pdf; accessed September 20, 2006.
4. See http://www.energy.ca.gov/2003_energypolicy/index.html; accessed September 20, 2006.
5. For more information on the work of the Climate Impacts Group see http://www.cses.washington.edu/cig/about/about.shtml; accessed September 20, 2006.
6. See Climate Solutions at http://www.climatesolutions.org; accessed September 20, 2006.
7. See Global Warming Action at http://www.globalwarmingaction.org; accessed September 20, 2006.
8. See http://oregon.gov/ENERGY/GBLWRM/Regional_Intro.shtml; accessed September 20, 2006.
9. The first conference was held mid-September 2005. See http://www.climatechange.ca.gov/events/2005_conference/index.html; accessed September 20, 2006.
10. See http://www.inr.oregonstate.edu/download/climate_change_consensus_statement_final.pdf; accessed September 20, 2006.
11. See http://oregon.gov/ENERGY/GBLWRM/Strategy.shtml; accessed September 20, 2006.
12. See http://oregon.gov/ENERGY/GBLWRM/CATF-members.shtml; accessed September 20, 2006.
13. See http://egov.oregon.gov/ENERGY/RENEW/RenewPlan.shtml; accessed September 20, 2006.
14. The Oregon Progress Board is an independent state planning and oversight agency. The 12-member panel, chaired by the Governor, is made up of citizen leaders and reflects the state's social, ethnic and political diversity. See http://www.oregon.gov/DAS/OPB/index.shtml; accessed September 20, 2006.
15. See http://www.climatetrust.org; accessed September 20, 2006.
16. Washington's siting process differs from Oregon's in that Washington EFSEC makes recommendations to the Governor. The Governor approves, demands changes, or rejects the license.

17. Washington State Energy Facility Site Evaluation Council, *Chehalis Generation Facility, Site Certification Agreement*, March 4, 1997, amended March 1, 2001, p. 19. See http://www.efsec.wa.gov/Chehalis/chehalisfinalsca2001.pdf, accessed September 20, 2006.
18. Washington State House Bill 3141, 2004.
19. Revised Code of Washington Chapter 80.70 – Carbon dioxide mitigation.
20. Summary of the report *Roadmap for Climate Protection: Reducing Greenhouse Gas Emissions in Puget Sound* is available at http://pscleanair.org/specprog/globclim/cpsp/pdf/rptexecsum.pdf; accessed September 20, 2006.
21. Executive Order 05-01, Establishing Sustainability and Efficiency Goals for State Operations.
22. Summary of the *Roadmap* report cited in note 20 above.
23. California statutes of 2002, chapter 200; see http://www.arb.ca.gov/cc/ab1493.pdf; accessed September 20, 2006.
24. For the vehicle GHG regulations see http://www.arb.ca.gov/regact/grnhsgas/grnhsgas.htm; accessed September 20, 2006.
25. California statutes of 2002, chapter 516; see http://energy.ca.gov/portfolio/documents/SB1078.PDF.
26. See http://gov.ca.gov/index.php?/executive-order/1861; accessed September 20, 2006.
27. For the full text of the Governor's comments on World Environment Day, see http://gov.ca.gov/index.php?/speech/1885; accessed September 20, 2006.
28. Information about the landmark bill AB22 is available at http://www.arb.ca.gov/cc/docs/ab32text.pdf; accessed September 30, 2006.
29. For more information on research see: http://www.climatechange.ca.gov/research/index.html; accessed September 20, 2006.
30. Learn more about climate science at Scripps by visiting: http://meteora.ucsd.edu/cap; accessed September 20, 2006.
31. In March 2005, CalPERS and CalSTRS partnered with Ceres and the Investor Network on Climate Risk to address investment and risks from climate change. See http://www.calpers.ca.gov/eip-docs/investments/assets/equities/aim/registration-final.pdf; accessed September 20, 2006.
32. See http://www.westcarb.org; accessed September 20, 2006.
33. See http://www.fypower.org and http://www.fypower.org/save_gasoline; accessed September 20, 2006.
34. For information about CAT activities see http://www.climatechange.ca.gov/climate_action_team; accessed September 20, 2006.
35. See http://www.energy.ca.gov/2005_energypolicy/index.html; accessed September 20, 2006.
36. For documents related to the Climate Change Advisory Committee see http://www.energy.ca.gov/global_climate_change/04-CCAC-1_advisory_committee; accessed September 20, 2006.
37. See http://www.climateregistry.org; accessed September 20, 2006.
38. CPUC Decision D.04-12-048 of December 16, 2004; see http://www.cpuc.ca.gov/PUBLISHED/FINAL_DECISION/43224-07.htm#P897_229166; accessed September 20, 2006.
39. See http://cpuc.ca.gov/static/energy/climate±change/060419_ghg_emissionscap.htm; accessed September 20, 2006.
40. See http://www.iclei.org/index.php?id=1118; accessed September 20, 2006.
41. See http://www.portlandonline.com/shared/cfm/image.cfm?id=112110; accessed September 20, 2006.
42. For more information, see the June 2005 Global Warming Progress Report at http://www.portlandonline.com/shared/cfm/image.cfm?id=112118; accessed September 20, 2006.
43. See the US Green Building Council's website at http://www.usgbc.org/DisplayPage.aspx?CategoryID=19; accessed September 20, 2006.

44. See http://www.portlandonline.com/osd/index.cfm?c=42134; accessed September 20, 2006.
45. See recycling report at http://www.portlandonline.com/osd/index.cfm?c=41464; accessed September 20, 2006.

References

Baldassare, M. (2005). *PPIC Statewide Survey: Special Survey on the Environment, July 2005*. San Francisco, CA: Public Policy Institute of California. Available at: http://www.ppic.org/main/publication.asp?i=623; see also: http://www.ppic.org/content/pubs/S_602MBS.pdf; accessed January 23, 2006.

Field, C. B., Daily, G. C., Davis, F. W., *et al.* (1999). *Confronting Climate Change in California: Ecological Impacts in the Golden State*. Cambridge, MA and Washington, DC: Union of Concerned Scientists and Ecological Society of America.

Gaura, M. A. (2005). Poll says legislators should act on climate Californians believe impact of warming has already begun. *The San Francisco Chronicle*, July 21, Available at: http://www.sfgate.com/cgi-bin/article.cgi?f=/c/a/2005/07/21/BAGTBDR0H71.DTL; accessed January 23, 2006.

Morgenstern, R. and Portney, P. (eds.) (2004). *New Approaches on Energy and the Environment − Policy Advice for the President*. Washington, DC: Resources for the Future.

Washington State Energy Office (1996). *Greenhouse Gas Mitigation Options for Washington State*, WSEO 96−28, April. Available at: http://yosemite.epa.gov/oar%5Cglobalwarming.nsf/UniqueKeyLookup/RAMR62FL2W/$File/WA_Action_Plan.pdf; accessed January 23, 2006.

28

Building social movements

David S. Meyer

University of California–Irvine

Maybe a president or Congressional leader will, upon reading a report by the American Academy of Arts and Sciences or hearing a plea from the leader of another country or even glancing at this book, become convinced of the need to take substantial action to address the dangers of global warming. Once convinced, this leader would build broad political coalitions with other political figures based on reasoned appreciation of a real environmental problem, and devise a comprehensive set of policy reforms to reduce America's key role in promoting global warming. Within a few years, the United States would develop and implement a range of policies to reduce the concentration of CO_2 in the atmosphere, including dedicated taxes, regulatory reforms, and large-scale initiatives in renewable energy and energy conservation.

This *could* happen, but that's not the way major changes in American history have taken place before. Reasoned discourse needs a push. Elected officials, even presidents, are subject to an ever-expanding range of pressures and constraints, making substantial policy reform in any area exceedingly difficult. Further, the American Constitution established a governmental structure built to produce stability rather than innovation, solidity rather than responsiveness — in effect, a Humvee rather than a hybrid of government. But substantial, if sometimes unwelcome, innovations in policy have taken place in America, including (among others) the institution of a federal social safety net for the elderly in the 1930s, the establishment of a permanent and globally engaged military establishment in the 1940s, federal intervention in the cause of civil rights and women's rights in the 1950s and 1960s, and a broad retrenchment in government support for the

I first presented the ideas in this paper at the "Climate Change Communication and Social Change" workshop, at the National Center for Atmospheric Research (NCAR), Boulder, Colorado, June 8–11, 2004. I am grateful for the comments of the participants of the workshop on those ideas.

less fortunate, commenced in the 1970s. These fundamental changes were never the result of moral suasion and rational argument alone; rather, they were propelled by organized interests dedicated to promoting social change over a long period of time. Far more often than not, one component of an effective change promotion strategy has involved communicating issues to a very broad public, and mobilizing large numbers of people in a social movement, both to enhance the directed education of the public and policy-makers, and to enhance the pressure on recalcitrant politicians while promising rewards to activists in government.

Even if substantive change doesn't always come exclusively from the bottom up, citizens not in government are virtually always involved in setting a political climate that supports promoters of change, but they do not do so on their own, independently of governments. At the most fundamental level, adopting more world-friendly policies requires at least those of us in affluent countries, as individuals, to change the way we live our lives, reducing how much carbon we emit, and probably how much fossil fuel — directly and indirectly — we consume (see, e.g., Michaelis, Chapter 16, this volume). But individuals don't exist in a vacuum. Rather, while we take pride in the choices we make, like a patron of a restaurant, we don't generally set the menu from which we order (see also Dilling and Farhar, Chapter 23, this volume). To adopt a model that suggests that changes in individual consciousness, from the bottom up, will ultimately manage global warming by spreading the word person to person is not only naïve, but ultimately counterproductive. Convincing Americans of means, for example, to forego driving SUVs to drive less powerful and smaller cars, or to take mass transit, one by one, is quite obviously to multiply infinitely the amount of work that educators and activists would have to do. In fact, government is an essential player in changing people's minds, and even more importantly, changing their conduct. (It really doesn't matter if all — or even most — people understand the science of climate change or alter their aesthetic preferences about transportation so much as that they change their conduct.) Allowing the notion that individual consumer choices can substitute for substantial policy change would not only be a misdirection of resources, but also a way to produce frustration and failure.

Social movements are the form of social and political organization that typically promotes the broadest range of public communication and politi-cal mobilization. I begin with the assumption that public engagement is a necessary, but not sufficient, condition to promote change. Social move-ments are a vehicle for public education, social engagement, and political pressure. In order to deal with the large problem of global warming, activists will need to work on numerous targets and levels of action. An effective social movement on global warming will mean combining the mass mobilization and national profile of movements such as those against nuclear weapons, and the grassroots change in social norms such as those associated with campaigns against cigarette smoking and drunk driving. Although it won't be easy, building such a social movement is the best prospect for effecting large-scale social change.[1]

Interesting, educating, and activating a broad public on a complicated, and apparently distant, issue[2] like global warming seems very difficult, but by assessing the obstacles, and finding models from other, seemingly distant, issues, we can develop effective strategies for communicating not only risks, but possibilities as well. Here, it's important to recognize that public education, that is, communication about specific issues that takes place outside the boundaries of schools, must have an activist component in order to stick. Education about environmental issues generally, and global warming specifically, can't add up to anything unless it is directed to some kind of program, be it individual reform or political action (see also Grotzer and Lincoln, Chapter 17, this volume). In either case, public education has to be directed to mobilization, that is, inducing someone to undertake some kind of directed action.[3] This is important not only for political influence, but also for meaningful education; information about global warming must be useable in order for most people to pay attention.

For most people, participation in a social movement is dependent upon coming to a belief that a problem is (a) urgent, (b) has potential solutions, and (c) that his or her efforts might matter. Global warming is a difficult sell on all three dimensions. When mainstream political figures talk about climate change, they dispute the facts, and they also set a very long time horizon before people will see the immediate effects of global warming (see Leiserowitz, Chapter 2, this volume). It's hard for most people to worry actively about the consequences of a change of a few degrees in temperature seen over the period of decades, particularly when well-established interests work to make sure that they don't see that threat (Moser and Dilling, 2004; McCright, Chapter 12, this volume). Thus it's hard to establish the urgency element of the story. The most commonly proposed remedies seem large, distant, and involve constraints on individual choice, like proscribing driving large cars or paying a carbon tax. Further, meaningful action appears to be contingent on large-scale global cooperation. When this seems dubious, difficult, or hard to effect by any one individual, it's easy for people to treat climate change like, say, sunspots, that is, something that can't be addressed. The necessity of large cooperation, and the absence of specific policy proposals to press for, or proposals that seem ineffective and expensive, make it hard to cultivate a sense of efficacy as well. Thus the issue of climate change is one that seems less urgent, less amenable to action, and less likely to respond to individual action than virtually anything else we can think of. Of course, there are more accessible individual consumer choices, for example, buying a hybrid rather than a conventional car, but each individual choice doesn't clearly add up to a coherent strategy to combat global warming. In addition, some choices are only possible for those most advantaged financially. So, what is to be done? Understanding the development and impact of the social movements of the past can guide us in seeking to create contemporary movements. In what follows, I mean to take lessons from past movements to figure out what activists can do to maximize their impact on the political process. The history of social movements in America is one that features story after story of unimaginable change that suddenly becomes taken for granted. Surely, Rosa Parks and her allies in the civil rights movement 50 years ago were in a far worse position to affect government policy than advocates of adopting more Earth-friendly policies are today, and the civil rights movement fundamentally altered America – even if there is still work to be done. And this is true for "distant issue" movements as well. Activists have, in the past, effectively mobilized citizen concern and action about nuclear weapons, helping to alter fundamental policies more deeply than most would have recognized as possible.

Mobilization strategies

Successful mobilization is a function of both activist choices and of change in circumstances well beyond the control of activists. Thus the civil rights movement was aided by America's participation in the Cold War, because government leaders recognized that race relations in the United States were a liability (Dudziak, 2000), and horrific events like the murder of the teenage African American Emmett Till provided outside sympathy and space in mass media for making claims (Sitkoff, 1981). Till's death provided a nationally recognized instance of outrage that galvanized both activism and support for the civil rights movement. The horrific incident served as an easily understandable window into a much larger world of injustice. Activism against nuclear power grew in the United States in the wake of the accident at Three Mile Island (Walsh, 1981), and people interested in cutting the size of the American nuclear arsenal were aided by reformers in the former Soviet Union (Meyer, 1990). But external changes or threats don't automatically create substantial mobilization or change; activists need to be in position to take advantage of opportunities they cannot always create. This means being prepared to spend time in the political wilderness, relentlessly organizing and pushing a message even when progress is slow.

Lessons from the anti-nuclear weapons movement

In the case of movements against nuclear weapons, the American public remains generally unengaged, uninterested, and willfully uninformed. Nuclear weapons, at least undetonated, touch American lives far less directly than climate change, and it's certainly not clear that individuals can do anything, beyond demanding that government act. To be sure, there have always been small groups of people concerned with weapons issues, constantly trying to educate and mobilize the public, and to influence government policy. Most of the time, however, they are marginal to American politics, and distant from any influence on policy, ignored by most people and by policy-makers.

In four periods since the advent of the nuclear age, however, activists succeeded in making their case to a broad public, and gaining the attention and some response from policy-makers: 1945–47 (scientists' campaign for international control); 1954–63, episodically (ban the bomb, test ban); 1968–72 (anti-ballistic missile campaign), 1980–84 (nuclear freeze). Anti-nuclear weapons mobilization increased when the presidential administration was visibly failing on nuclear arms control, when the cost of military spending was increasing, and when the apparent danger of nuclear weapons was increasing. In short, the broad mobilization took shape when the issue became more obviously urgent, and when the capacity of government to manage the issue was in question. Activists were able to take advantage of unexpected and salient events, including incidents of radioactive fallout, because they had built political organizations, communication networks, and shared understandings of the problem.

Recognizing and creating urgency

Importantly, in all of these cases, it wasn't the grassroots activists who initially recognized the urgency; indeed, the long-time marginal groups saw consistent urgency. Instead, it was when experts in strategy, science, foreign policy, and arms control, people who normally enjoyed some access to government, lost faith in or access to the current administration, and saw the need to go public to press their own agendas, that the broader public became more aware and more engaged. This makes sense: the details of ongoing arms control negotiations or the patterns of military procurement are even more difficult to interpret than the levels of carbon in the air. The condition of a security threat becomes palpable as a problem only when "experts" with broad credibility in a range of fields say it is in fact a problem (see also the chapters by Cole with Watrous, and Warner, Chapters 11 and 10, this volume).

Lack of leadership

On nuclear weapons issues, the expert defection from government policy was always the key precursor to broad mobilization. Scientists, beginning with those physicists actually involved in the creation of the first bombs, strategists, physicians, retired members of the military, and clergy, generally not engaged in public discussion of nuclear weapons issues, turned to the public when convinced that their concerns weren't served by the administration then in power, going public and calling for mobilization almost as a last resort when normal politics had failed. They understood, more or less explicitly, that broad public mobilization would have the effect, minimally, of bringing attention to their concerns, and of forcing policy-makers to justify their policies more frequently and more explicitly, and perhaps to develop policies that were easier to sell to this engaged public.

Connecting organizers, communicators, and action

In effect, these elite or expert defections from either quiescence or institutional politics directed public attention to their issues, and simultaneously legitimated and underscored the ideas of long-time activists, building a bridge between the political margins and the political mainstream. Here, activist strategy matters a great deal. Activists were most successful in mobilizing support when (1) they could claim the active support of dissident experts and elites, who worked to educate and mobilize citizens, as outlined above; (2) they offered a clear and apparently simple policy proposal for action (e.g., "test ban," "nuclear freeze") to organize around; (3) they offered people something concrete to do that might reasonably be seen as part of a larger process that could ultimately change the world; and (4) when they built a broad coalition unifying groups with widely varying analyses, constituencies, and demands. Importantly, the organizations these dissident scientists established, including the Federation of Atomic Scientists and the Union of Concerned Scientists, outlived the mobilization of the moment and provided a base of support for subsequent movements – including the emerging campaign for a response to global warming.

Taking these issues of strategy in turn, we begin with links between long-time activist groups and experts. Obviously, negotiating cooperation takes a commitment from both the experts and the activists. On the activist side, groups need to maintain an interest in employing expert testimony, in creating fora in which scientists and strategists can make their cases, and a flexibility to work with people who may not share all of their interests. Historically, activist groups have been more willing to negotiate these alliances than the experts themselves, who often see public engagement as risking their professional autonomy and, even more significantly, their credibility. In order to build a broad movement to address global warning, however, it is essential for reasonably large numbers of scientists to dedicate some portion of their time to public education, which would start by working with activist groups (for example, Cole with Watrous, Chapter 11, this volume). In this regard it is essential to recognize that the information itself isn't mobilizing or determinative, and that neither technical academic publications nor broad statements by groups of scholars, expressed as cause ads in newspapers or consensus statements by professional organizations, is enough. Scientists need to hit the hustings, to present their views to students on college campuses, to community groups, and under the auspices of religious organizations (see also Bingham, Chapter 9, this volume). Here, activist professional groups have played a critical role in the past. Organizations such as the Federation of Atomic Scientists, the Union of Concerned Scientists, and the National Resources Defense Council have functioned as almost activist halfway houses (Morris, 1984) in anti-nuclear movements, providing expert information in more digestible formats, and linking speakers with groups that need them. Such organizations may be at the center of a new activist campaign to combat global warming.

Mobilizing around policy goals

In terms of policy goals, social movements are better at saying no to something than pressing for something new. Indeed, even movements that make proactive claims (for women's rights, for example, or "slow food," also frame them as against something else: discrimination or fast food). It's easier to mobilize to stop pollution than to promote clean air; to end discrimination, rather than to devise programs that ensure equality and social progress; to stop and reverse the arms race, rather than to maintain an arms control regime; or to end a war, rather than to negotiate international alliances to try to work to reduce violence.[4] Although the latter outcomes are often government responses to activist movements expressing simpler, more fundamentalist, goals, making practical policy reforms is not something movements are well structured to do. For the anti-nuclear weapons movements, crystallized demands, such as a ban on nuclear weapons, a prohibition on anti-ballistic missiles, or a freeze on the arms race, ostensibly policy proposals in themselves, have been rallying cries for public engagement, and have represented a shorthand plea to policy-makers to do something. It is imperative for climate change activists to find a comparable shorthand goal. Of course, activists are constantly floating new initiatives in this regard, and it is impossible to tell in advance what crystallized version of demands will take off with a broader public. Nonetheless, the key is to find a simple and readily accessible term and campaign that signifies both specific policy reforms and a broader worldview (e.g., "women's liberation," "no nukes").

For all social movements, however, both experts and government will ultimately moderate a movement's expressed goals, translating protest into policy, narrowing goals and finding mechanisms to address them. The Department of Labor, the Environmental Protection Agency, and the Equal Employment Opportunity Commission were all responses to social movements, which developed as permanent government habitats for considering a defined set of issues.

A clear action strategy

Third, information without an action strategy promotes a feeling of futility among those who hear it. Effective organizers give recruits something to do: a petition to sign, a badge to wear, someone to call, a demonstration to attend, and even a language to use (take, for example, the creation of the title "Ms."). Climate change activists must employ such practices, while guarding against the danger that defined consumer choices could *substitute* for collective political action. Organizers must try to use an individual's decision to buy a hybrid automobile, for example, as one step in a larger process of political engagement. Further, if meaningful action is defined by expensive consumer choices, the breadth of activist participation will be severely circumscribed. Activists need to find concrete *political* actions to promote. Anti-nuclear activists used ballot referenda, petitions, electoral campaign work, and repeated demonstrations to give newly aware citizens something apparently meaningful to do. Organizers need to find comparable outlets for political action. Such actions have not only their political effect, but define supporters as a distinct group, and build social networks to promote subsequent action.

Importantly, despite all of the activism on nuclear weapons — or on civil rights or workplace safety — the movement never got all it explicitly demanded; movements never do. This does not suggest that they don't make a difference; rather, social movements give institutional actors the cover and pressure they need to effect reforms more modest than what activists demand. It is always a long and difficult battle.

To take another example of movements that linked political and personal change, think about the decades-long campaign against cigarette smoking. All of the components of mobilization were in place, commencing with expert advice and proselytizing to individuals, while simultaneously advancing public health policies. While some individuals were certainly motivated by health arguments about quitting smoking, these arguments grew increasingly powerful when the price of cigarettes increased, primarily through taxes at least partly designed to punish smokers, and when the public spaces in which one could smoke without facing arrest or social sanction continued to decline. In short, public policies drove individual consumer decisions. For global warming, it's imperative for organizers to link individual choices to concrete political demands that make polluting behavior financially or socially unacceptable and offer a politically sanctioned viable alternative.

Finally, effective social movements are defined by broad political coalitions, uniting groups with disparate interests and constituencies. The instrumental tasks, managing a referendum campaign or organizing a demonstration, serve as vehicles for building such coalitions. In the past, anti-nuclear weapons movements have surged when they engaged not only community groups, but also national organizations not generally explicitly concerned with security issues, including the National Association for the Advancement of Colored People, the National Council of Churches, and the Young Women's Christian Association. Rape law reform in the 1970s was effected rather rapidly because of an odd alliance of legal reformers, feminist activists, and conservative crusaders against crime. The contemporary movement against the war in Iraq includes foreign policy isolationists, supporters of the United Nations, environmental activists, vegans, and organized labor. Importantly, activists don't need to agree on all the details of the claims that all of the cooperating organizations made, but had to agree to work on a common agenda. In order to be effective, global warming activists have to reach out to groups not normally engaged in the environmental movement, for example as is happening in the Apollo Project, which links together labor, environmental, and justice concerns.[5] To a large degree,

this means making general choices about alliances, and agreeing to cooperate on issues not directly related to climate change in order to cement
group cooperation (see also Agyeman *et al.*, Chapter 9, this volume).
Coalition politics can also peel support away from opponents' coalitions.
These are two sides of the same coin.

For the most part, successful social movements make political choices that
cluster alliances largely on one side of the political spectrum or the other.
Environmentalists have mostly tried to avoid such choices, and ambitious
Republican politicians have sometimes embarked upon progressive environmental policies in order to cultivate affluent and well-educated suburban
voters (see the case of California discussed in duVair *et al.*, Chapter 27, this
volume). In the world of political movements, however, this is a thin slice of
potential activists. More commonly, activists carve out alliances that link,
say, an anti-abortion position with politically conservative positions on
other issues that are not obviously related – or link moderation on weapons
programs to social spending for education and poor people – again, not
necessarily related. Organizers against global warming need to broaden their
potential base of support by extending a political platform and picking a set
of potential allies. This isn't easy, but it is probably necessary.

The state of social science allows predictions far more contested and far
less precise than contemporary models of climate change. It's impossible to
tell how much organizing is enough to effect change, or when an organizing
effort will reach a sufficiently broad population to effect policy. Further,
it's impossible to know, in advance, what series of events leads to the emergence of global warming on a broad public agenda. Unforeseeable contin-
gencies will matter in unpredictable ways (see Ungar, Chapter 4, this
volume). The important work is to position the climate change movement to
take advantage of them in advance.

The road ahead is surely bumpy, with clear paths to success difficult to
find. That said, it is unrealistic to expect the necessary level of change without
popular engagement and political pressure. More than that, activists working
on causes that seemed far *less* promising have forged successful movements
and begun to change the world. We can't afford to expect anything less in
this case.

Notes

1. Groups of activists are trying to build such a movement. For one example, see
   http://www.whatworks-climate.org; accessed January 17, 2006.
2. This term, "distant issue," is from Dieter Rucht (2002).
3. I realize that this view is not universally held, but I would argue that movements
   that direct their efforts to non-politicized generic consciousness-raising waste their

moment of attention, and ultimately produce more in the way of self-satisfaction *and* frustration than do mobilization directed campaigns.

4. Partly, this is a function of coalition-building: it's easier to get diverse groups of people to agree on defining a social bad, rather than coming to agreement on some ultimate goal and how to get there. It's also partly a function of American political institutions, which are designed to make it easier to stop something than to create a new initiative (Meyer, 2007).

5. See http://www.apolloalliance.org/; accessed January 17, 2006.

References

Dudziak, M. (2000). *Cold War Civil Rights: Race and the Image of American Democracy.* Princeton, NJ: Princeton University Press.

Meyer, D. S. (1990). *A Winter of Discontent: The Nuclear Freeze and American Politics.* New York: Praeger.

Meyer, D. S. (2007). *The Politics of Protest: Social Movements in America.* New York: Oxford University Press.

Morris, A. (1984). *The Origins of the Civil Rights Movement.* New York: Free Press.

Moser, S. C. and Dilling, L. (2004). Making climate hot: Communicating the urgency and challenge of global climate change. *Environment*, **46**, 10, 32–46.

Rucht, D. (2002). Distant issue movements in Germany: Empirical description and theoretical reflections. In *Globalizations and Social Movements*, eds. Guidry, J. A., Kennedy, M. D., and Zald, M. N. Ann Arbor, MI: University of Michigan Press, pp. 76–105.

Sitkoff, H. (1981). *The Struggle for Black Equality, 1954–1980.* New York: Hill & Wang.

Walsh, E. (1981). Resource mobilization and citizen protest in communities around Three Mile Island. *Social Problems*, **29**, 1–21.

29

Climate litigation: shaping public policy and stimulating debate

Marilyn Averill
University of Colorado–Boulder

Introduction

Advocates interested in climate policy are turning to the courts to try to force action on issues relating to human-induced climate change. Climate litigation involves elements of law, science, policy, and ethics in addressing claims relating to global warming. Applying a variety of legal theories, litigants are using climate litigation to clarify existing law, challenge corporate behavior, assign responsibility, and seek damages for climate-related injuries. These cases rely heavily on expert scientific testimony and are likely to affect perceptions about the credibility, salience, and legitimaczy of climate science. They also are likely to encourage public debate and to stimulate political advocacy.

Many of these lawsuits have been filed in the United States, where litigation is widely used to resolve environmental disputes, but these cases have both national and international implications. Climate litigation can help shape US policy relating to climate change, both directly and indirectly; explain specific issues relating to climate change; provide guidelines for establishing causation and assigning responsibility for climate-related injuries; and affect perceptions about climate science that may influence policies and legal decisions beyond the borders of the United States.

Many of these cases are likely to be unsuccessful from the point of view of those hoping for greater action on climate change. Even an unsuccessful case, however, can make climate change more visible to a range of audiences by raising awareness, educating the public, and stimulating public debate on issues relating to climate change.

This chapter will discuss some of the issues raised by these cases and their possible significance for climate change communication and related social change. Societal impacts may not depend on the current status of a case, so the chapter describes claims at issue but does not attempt to provide current status. The intent is to consider the role that such cases can play and not to discuss the specific arguments or decisions made in each case. The stakes are high and appeals should be expected, regardless of rulings in lower courts. Some of these cases may even reach the highest levels of state and federal courts; one already has been accepted for review by the US Supreme Court.[1] Most of the cases discussed here are still active as this book goes to press. New cases are likely to be filed asserting new claims or supporting old claims with new information. Even cases that have been dismissed may have social impacts as published results stimulate public discussion.

The chapter begins with an overview of environmental litigation in the United States. It then discusses possible impacts of current climate litigation, including effects on the interpretation and application of existing law, corporate behavior, determinations of responsibility, public education and debate, public perceptions about climate science, and political advocacy. Possible unintended consequences also are discussed.

Environmental litigation in the United States

The United States Congress has enacted citizen-suit provisions in many federal environmental statutes (Arbuckle *et al.*, 1991).[2] These provisions allow private citizens to go to court to seek compliance with environmental laws and regulations. Environmental advocates have sought to ensure that government agencies comply with their environmental responsibilities, clarify the meaning and scope of environmental laws, prevent environmentally unsound actions, seek redress for harm caused by environmental violations, and prevent public or private actions likely to threaten human safety or the environment.

Other plaintiffs have used these provisions to block environmental protection. Legal actions have challenged environmental rules, blocked federal actions, and protected private property rights affected by environmental regulation.

Sax (1998: 301) describes environmental litigation as "a means of access for the ordinary citizen to the process of governmental decision-making." Others argue that litigation interferes with democratic values by removing environmental decisions from elected officials.[3] Regardless of its implications for democracy, environmental litigation has played an important role in shaping law and policy in the United States. The potential roles legal action can play in the case of climate change are discussed in the section below, along with examples of recent cases.

Potential climate litigation impacts

Clarify exiting law

Several of the current climate-related cases pending in US courts address governmental authority or obligations to take actions relating to climate change. These cases address questions such as whether federal or state agencies have the authority to regulate greenhouse gas emissions under existing law, that is, whether laws are in place to allow or to prevent agency action relating to climate change. If agencies do have the *authority* to act, the next question is whether current laws *require* an agency to take action relating to climate change.

Courts decide cases within the framework of existing law. Betsill and Pielke (1998: 161) stress "the importance of a framework for domestic debate about [an] issue and the establishment of a vehicle for linking science and policy in a meaningful way, including the promulgation of criteria for action." Congress created such a legal framework for the problem of stratospheric ozone depletion and litigants used the courts to compel compliance with that framework.

Congress has not yet provided such a policy framework for dealing with climate change. While some existing statutes such as the National Environmental Policy Act (NEPA) and the Clean Air Act (CAA) may apply to climate change, they do not provide a clear legal mandate for federal agencies to deal with the issue. Nevertheless, several existing laws have components that may allow or even compel action relating to climate change. Lawsuits have been filed to clarify the extent to which existing law applies to climate issues. These lawsuits should clarify the application of current US laws to climate change and highlight areas where additional political action may be needed if the government is to take action to address climate-related issues. Two of the climate lawsuits involve claims against federal agencies, alleging that they have not complied with relevant law, and a third challenges a state law regulating greenhouse gas emissions.

In *Friends of the Earth v. Watson*, two non-governmental organizations (NGOs) joined with three cities to file a complaint against the Export–Import Bank of the United States and the Overseas Private Investment Corporation.[4] The lawsuit alleges that these federal agencies failed to meet the requirements of NEPA because they neglected to take climate effects into account in conducting environmental impact reviews of their programs. If the plaintiffs are successful, these agencies would be required to consider the cumulative effects of their programs on climate in future environmental impact reviews. Other federal agencies may then include climate effects in environmental reviews to avoid becoming the target of a similar action.

Massachusetts v. United States Environmental Protection Agency involves the Clean Air Act.[5] The question addressed is whether the CAA gives the US Environmental Protection Agency (EPA) the authority to regulate greenhouse gas emissions from motor vehicles and, if so, whether EPA has an obligation to regulate. Plaintiffs challenged an EPA decision to deny a petition to regulate greenhouse gas emissions from new motor vehicles.[6] The EPA under the G. W. Bush Administration asserts that it has no authority to regulate greenhouse gas emissions under the CAA and, even if it has the authority, EPA maintains that it has the discretion to decide whether to regulate and would choose not to do so. The ruling in these consolidated cases will determine whether EPA has authority to regulate greenhouse gas emissions under the CAA and could require EPA to reconsider its decision on the petition to regulate.

Litigation also can challenge governmental strategies to mitigate climate change. In, *Central Valley Chrysler-Jeep, Inc. v. Witherspoon*, several vehicle dealers and manufacturers[7] brought suit against the State of California to block a state law requiring reductions in greenhouse gas emissions from new cars by 2016 (Hakim, 2004; Environmental Defense, 2005; see also duVair *et al.*, Chapter 27, this volume).[8] The United States has environmental laws and regulations that range from the local to the national level. Coordination among these laws is often difficult. States and the federal government often have different standards for environmental protection. In some cases the more stringent standard prevails, but in others there are questions as to which standard should apply.[9] The automobile manufacturers claim that the federal government has pre-empted the regulation of vehicle fuel economy by setting national standards, making it inappropriate for states to regulate in this area. They argue that consistency requires a federal standard rather than a patchwork of state standards.[10] The same manufacturers have fought against stricter fuel economy standards at the national level. If the automotive plaintiffs prevail, then the California emissions law will not take effect.

Change corporate behavior

Some of the current cases involve claims against corporations. Litigation can affect a corporation both directly and indirectly and can provide a powerful tool to force change. A court order may direct immediate and fairly specific action, but litigants often take voluntary action prior to a court's decision. Corporations, as well as government agencies, incur immediate legal costs and negative publicity and face the possibility of large additional costs or changes in operations if they lose their case. Defendants may seek a settlement to mitigate such damages. Corporate defendants also may undertake a publicity campaign to manage public perceptions or lobby politicians to change laws. In any event, corporations take litigation risks seriously. A successful case against one company may lead others to re-evaluate corporate practices in light of increased litigation risks or to lobby for changes in law.

Some plaintiffs use tort law to bring claims against corporations. Tort law allows plaintiffs to claim compensation for injuries caused by actions that can be attributed to the defendant or to ask that defendants modify injurious behavior. Tort law includes several different causes of actions, including, but not limited to, product liability, negligence, and public nuisance.[11]

In *Connecticut v. American Electrical Power, Inc.*, plaintiffs used tort law principles to make claims against the six biggest electrical power companies in the United States.[12] The case alleges that emissions of carbon dioxide constitute "ongoing contributions to a public nuisance"[13] and asks that the companies be ordered to reduce their emissions by a specified percentage every year for at least a decade. If the plaintiffs are successful, this case will force the power companies in the lawsuit to reduce emissions, which may indirectly lead to additional reductions by other companies who fear that similar litigation will be filed against them.

Assign responsibility

Much of the recent climate change debate has focused on questions about who has caused the problem and who should be held responsible. Courts traditionally have made decisions about assigning responsibility and allocating liability and will be asked to do so again in the area of climate change.

Proving causation and allocating responsibility set high hurdles for litigants seeking damages for climate-related injuries (see, e.g., Allen, 2003; Allen and Lord, 2004). These cases put the debate squarely where the science is still uncertain, on the question of specific causes and the relative contribution from different causes. In the case of a global issue such as climate change, plaintiffs will find it difficult to prove that a particular defendant was the cause of a specific harm and should be held legally responsible. Even if the science could accurately determine the relative contributions of natural and human influences, it will be extremely difficult to allocate liability among the legion of human influences to decide how much a given defendant should pay (see, e.g., Grossman, 2003). While courts have developed theories such as market share liability to allocate responsibility in areas such as pharmaceuticals,[14] these theories involve liability for the impacts of a particular product. Climate change is infinitely more complex, with causes ranging from electricity generation to transportation to deforestation to farming practices. Deciding how much any one defendant should be held responsible will be exceedingly complicated.

Legal theories, particularly tort theories, must be stretched to accommodate climate change, and complaints must be carefully crafted to avoid dismissal of claims. But legal theories do change to accommodate new factual situations,[15] and the same could happen in climate litigation.

The Inuit of northern Alaska and Canada already face injuries from melting permafrost and sea ice, receding shorelines, and changing ecosystems (e.g., Correll, 2004; see also McNeeley and Huntington, Chapter 8, this volume). In December 2005, Sheila Watt Cloutier, Chair of the Inuit Circumpolar Conference (ICC), with the support of the ICC and working with attorneys from the Center for International Environmental Law (CIEL) and Earthjustice, filed a petition with the Inter-American Commission on Human Rights (IACHR) within the Organization of American States claiming climate-related human rights violations by the United States (Black, 2005; CIEL, 2005; Inuit Circumpolar Conference, 2005; Revkin, 2005; Climate Justice, 2006). While the IACHR lacks the authority to order reductions in GHG emissions, the petition could encourage settlement talks, publicize injuries suffered by the plaintiffs, raise awareness about the effects of climate change, and set the stage for future litigation.

Inform, encourage, and change public debate

Each of the climate lawsuits tells a story based on the particular facts and legal theories presented in the case. Climate litigants will allege specific impacts to individuals and communities and possible economic consequences and will spin stories about both winners and losers. These stories, constructed for understanding by non-experts, may capture the public imagination and stimulate discussion not only about the scientific basis underlying climate change but also about who should be held responsible for the impacts and what actions society should take to cope with climate change.

Information about climate litigation will reach the public through a variety of sources. Print and broadcast media will report cases deemed to be of interest to the public. Media coverage probably will focus on cases involving large damage claims, compelling public interest stories, or major changes in policy. Lawyers and academics will publish analyses of climate cases in a wide variety of professional journals and other magazines. Teachers will discuss climate cases in classrooms. Environmental reporting services will provide reports on case filings and court decisions. Finally, advocacy groups across the spectrum will use information about court cases to promote their points of view and encourage social action through newsletters, websites, and other outlets.

Cases in litigation can illuminate opaque arguments about the science, politics, ethics, and policy of global climate change, and make issues more accessible to the public. Surveys indicate that most Americans already are aware of our changing climate and view it as a societal problem, but recognizing the problem is only a first step to understanding the issue (see also the chapters by and Bostrom and Lashof, and Leiserowitz, Chapters 1 and 2, this volume). Better understanding of the possible causes and effects of climate change will allow more informed debate about appropriate societal responses (Moser and Dilling, 2004). Citizens with a basic understanding of the relationship among human behavior, greenhouse gas emissions, and the greenhouse effect will have a better understanding of proposals to limit industrial emissions or change consumer behavior. Additional information about the possible impacts of climate change on specific areas or populations, about the possible climate winners and losers, can trigger debate about what is fair and whether and how society should protect the most vulnerable, as well as who should be held responsible (see also Agyeman *et al.*, Chapter 7, this volume). Climate litigation also may trigger public debate over how to balance risk and uncertainty as well as costs and benefits.

Climate litigation encompasses many of the big disputes in the climate change debate. How is climate affecting individuals and communities? What is the division between human and natural causes of climate change? Has harm already started to occur? Who is responsible, the corporation that emits CO_2 by burning fossil fuels to produce energy, or the consumer who uses the energy? What costs are associated with reducing greenhouse gas emissions — or with failing to reduce them? What governmental policies are best for the public? Who deserves to be compensated for injuries, both domestically and internationally? Who should pay for injuries to people or property? What allocation of responsibility among governments, corporations, and individuals is fair? Do we know enough about climate change and its effects to be able to answer such questions? How should decisions be made in the face of scientific uncertainty?

Lawsuits can serve as an exercise in civic education, informing "litigants and other citizens, the legal community, and various governmental and non-governmental institutions about the epistemological, social, and moral dilemmas" associated with scientific and technological issues (Jasanoff, 1995: 20–1). "At their most effective, legal proceedings have the capacity not only to bring to light the divergent technical understandings of experts but also to disclose their underlying normative and social commitments in ways that permit intelligent evaluation by lay persons" (*ibid.*, 215). One commenter stated that "It may well be that today much of what most Americans learn about science comes from the coverage of science in our courtrooms" (Leone, 1995: x).

Climate litigation can affect public understanding of and perceptions about climate science. These cases will rely heavily on the testimony of scientific experts. Such testimony, as reported or used by writers or advocates, is likely to affect public perceptions of the credibility, salience, and legitimacy of climate science (Cash *et al.*, 2002). Courts provide a connection between climate science and society by providing a forum in which science must be presented in terms that a layperson can understand. Lawyers and their expert witnesses must make information understandable to non-expert judges and jurors who often have little background in science (Jasanoff, 1995). Cross-examination of expert witnesses allows deep inquiry into the scientific basis for claims and the personal affiliations and interests of the experts themselves. For example, cross-examination may uncover problems such as faulty reasoning, unwarranted assumptions, or conflicts of interests.

Climate litigation will allow experts on both sides of an issue to present their best case, to highlight the science that supports a particular point of view. Experts also need to demonstrate the importance or salience of the science within the context of the facts in a particular case. Both scientific methods and conclusions will be subjected to scrutiny by opposing counsel. Judicial rulings on evidence and final decisions in the case are thus likely to affect public perceptions of the credibility of the science presented.

Courts can shape perceptions of the legitimacy of science by making determinations about which experts are more reliable. They do this by examining expert credentials and rejecting those that do not meet minimum standards. Opposing counsel will inquire into possible biases of the experts and reveal connections to special interests.

Stimulate political advocacy and action

The rising interest in climate litigation may reflect a shift in tactics resulting from the withdrawal of the United States from the Kyoto Protocol.

"Rather than treaties and regulations, litigation may soon be the weapon of choice for those concerned about human-induced global warming" (Mukerjee, 2003: 1). Some see climate litigation as a way "of inducing or even compelling the United States to meet its global responsibilities" (Strauss, 2003: 10185). Others suggest that governments rather than courts should address climate issues (*Connecticut v. American Electric Power Co., Inc.*, 2005). Litigation itself may pressure politicians to take action at the local, state, or national level, although the action may not be of the type desired by plaintiffs.

Possible unintended consequences

Plaintiffs asking courts to force action on climate change may produce unintended consequences that could lead to negative social or environmental impacts. Differently put, climate plaintiffs may risk more than just losing a lawsuit. For example, a court may rule that climate science is still too uncertain to justify the requested action, indirectly legitimizing governmental inaction based on uncertainty. Ineffective expert testimony in one case could undermine perceptions of the legitimacy of climate science in general. In corporate cases, successful litigation against one company could lead industries such as power companies or automobile manufacturers to lobby harder for legislation protecting against such lawsuits.

Plaintiffs seeking to block action on climate change run similar risks. For example, industry complaints about a patchwork regulatory framework could encourage Congress to enact stricter national standards for greenhouse gas emissions (see also Tennis, Chapter 26, this volume).

Indirect effects of climate lawsuits cannot be predicted and there is a risk that they could actually discourage action to prevent climate change. Climate litigation seems likely to stimulate public debate and possibly to increase pressure for political action, but in what direction? Will the public decide that climate change represents real risks to humans and ask the government to take steps to facilitate mitigation and adaptation? Or will lawsuits lead the public to decide that the risks of climate change are too remote and the costs of action too high, and push for governmental inaction or action that protects the status quo? Will consumers begin to question their own contributions to climate change and change their behavior patterns (see Tribbia, Chapter 15, this volume), or will they find ways to ensure their current behavior can continue?

Conclusion

No matter how different actors use the courts to affect action on climate change, litigation may be slow to produce direct social change. Successful litigation is not required, however, for these cases to make climate issues more visible to the public. News stories, professional publications, and outreach by advocacy groups discussing climate cases can contribute to greater public understanding and debate.

Climate litigation may help shape policy and changes in behavior at all levels of society by framing issues, presenting opposing arguments, legitimizing scientific evidence, establishing responsibility, and stimulating public debate. Politicians may have primary responsibility to construct a policy framework for climate change, but once it is provided, courts will be used to enforce proper implementation. Litigation is not a substitute for other approaches to addressing climate change, but climate-related cases provide material that can be used by others to communicate the urgency of climate change and to promote social action.

Notes

1. *Massachusetts v. United States Environmental Protection Agency*, 126 S. Ct. 2960 (2006).
2. The Administrative Procedures Act (APA) provides an additional basis for litigation if federal agencies are acting in an arbitrary and capricious manner. For example, the APA can be used to challenge actions under the National Environmental Policy Act (NEPA), which provides an explicit process for public input in large projects that could potentially impact the environment but does not include a citizen-suit provision.
3. See, for example, *Connecticut v. American Electric Power Co., Inc.* (U.S.D.C. S.D.N.Y., Nos. 04 Civ. 5669 and 04 Civ. 5670), Opinion and Order issued Sept. 15, 2005. Judge Preska dismissed the case saying it involved political questions that should be addressed by the political branches rather than the judiciary. The case is now on appeal (Climate Justice, 2005).
4. The plaintiffs, Friends of the Earth, Inc.; Greenpeace, Inc.; City of Boulder, CO; City of Arcata, CA; and City of Oakland, CA filed suit in Federal District Court in the Northern District of California. *Friends of the Earth v. Watson*, No. C 02-4106 JSW (N.D. Cal.).
5. Eight related cases were consolidated under the name *Commonwealth of Massachusetts v. United States Environmental Protection Agency*, consolidating cases Nos. 03-1361 through 03-1368 (D.C. Cir.), and were heard on appeal by the Federal Circuit Court of Appeals for the District of Columbia. The collective plaintiffs include the states of Massachusetts, California, Connecticut, Illinois, Maine, New Jersey, New Mexico, New York, Oregon, Rhode Island, Vermont, and Washington; the cities of New York, Baltimore, and the District of Columbia; the Commonwealth of the Northern Mariana Islands; America Samoa Government; and NGOs, including The International Center for Technology Development, Bluewater Network, Center for Biological Diversity, Center for Food Safety, Conservation Law Foundation, Environmental Advocates, Environmental Defense, Friends of the Earth, Greenpeace, National Environmental Trust, Natural Resources Defense Council, Sierra Club, Union of Concerned Scientists, and US Public Interest Research Group. The case is now before the US Supreme Court (see note 1).

6. See 68 Fed. Reg. 52,922 (Sept. 8, 2003).
7. BMW, DaimlerChrysler, Ford, General Motors, Mazda, Mitsubishi, Porsche, Toyota, and Volkswagen belong to the Alliance of Automobile Manufacturers, which filed the suit. Honda and Nissan, which are not members of the Alliance, also oppose the regulation but have not joined the lawsuit.
8. Car companies have until 2016 to achieve a 30 percent reduction in emissions. They have to start producing vehicles that comply by 2009, but the regulations will not be fully phased in until 2016.
9. California generally is allowed to set its own emissions standards for new vehicles. The state has special status under the Clean Air Act (CAA) because California had strict air quality standards before the CAA was passed in 1970. Other states may choose between the CAA or California standards.
10. Environmental standards often differ across states, creating an inconsistent regulatory patchwork that complicates compliance for businesses. Industries sometimes have requested federal standards to ensure consistency (see also the chapters by Arroyo and Preston; and Tennis, Chapters 21 and 26, this volume). In this legal case, plaintiff manufacturers seek to prevent a patchwork from forming by opposing any regulation of greenhouse gas emissions at the state or federal level.
11. For an analysis of the applicability of tort claims in the climate context, see Grossman (2003).
12. *Connecticut v. American Electric Power Company, Inc.*, filed in 2004 in the United States District Court for the Southern District of New York. Plaintiffs include the states of Connecticut, New York, California, Iowa, New Jersey, Rhode Island, Vermont, and Wisconsin; and the City of New York. Defendant power companies include American Electric Power Co., Inc.; American Electric Power Service Corp.; The Southern Co.; Tennessee Valley Authority, Inc.; Xcel Energy Inc.; and Cinergy Corp. (see note 2).
13. Grossman (2003) contends that public nuisance is the tort theory most likely to succeed in climate litigation.
14. For an example of how courts have divided liability among several pharmaceutical companies marketing the same drug, see *Sindell v. Abbott Laboratories*, 607 P.2d 924 (Cal. 1980), in which a California court allocated liability for diethylstilbestrol injuries according to each manufacturer's market share.
15. See, e.g., the description of the evolution of tort law in Jasanoff (1995).

References

Allen, M. (2003). Liability for climate change: Will it ever be possible to sue anyone for damaging the climate? *Nature*, **421**, 891–2.

Allen, M. R. and Lord, R. (2004). The blame game: Who will pay for the damaging consequences of climate change? *Nature*, **432**, 551–2.

Arbuckle, J. G., *et al.* (1991). *Environmental Law Handbook*, 11th edn. Rockville, MD: Government Institutes, Inc.

Betsill, M. M. and Pielke Jr., R. A. (1998). Blurring the boundaries: Domestic and international ozone politics and lessons for climate change. *International Environmental Affairs*, **10**, 148–72.

Black, R. (2005). Inuit sue over climate policy. BBC News. Available at: http://news.bbc.co.uk/1/hi/sci/tech/4511556.stm; accessed January 3, 2006.

Cash, D. W., Clark, W. C., Alcock, F., *et al.* (2002). *Salience, Credibility, Legitimacy and Boundaries: Linking Research, Assessment and Decision Making.* Working Paper RWP02-046, Cambridge, MA: John F. Kennedy School of Government Faculty Research, Harvard University.

Center for International Environmental Law (CIEL) (2005). Inuit File Petition with Inter-American Commission on Human Rights, Claiming Global Warming Caused by United States Is Destroying Their Culture and Livelihoods. Available at: http://www.ciel.org/Climate/ICC_Petition_7Dec05.html; accessed January 3, 2006.

Climate Justice (2006). Inuit File Petition with the Inter-American Commission on Human Rights. Available at: http://www.climatelaw.org/media/inuit.iachr; accessed January 3, 2006.

Climate Justice (2005). U.S. Nuisance Appeal. Available at: http://www.climatelaw. org/media/U.S.%20nuisance%20appeal; accessed January 3, 2006.

Correll, R. (2004). *Impacts of a Warming Arctic: Arctic Climate Impact Assessment.* Arctic Council, International Arctic Science Committee. Available at: http:// www.acia.uaf.edu; accessed January 3, 2006.

Environmental Defense (2005). Carmakers Sue California over Groundbreaking Clean Cars Law. Available at: http://www.undoit.org/whatsnew_spotlight. cfm?story=carmakers; accessed January 3, 2006.

Grossman, D. A. (2003). Warming up to a not-so-radical idea: Tort-based climate change litigation. *Columbia Journal of Environmental Law*, **28**, 1–61.

Hakim, D. (2004a). Automakers sue to block emissions law in California. *The New York Times*, December 8, accessed, p. C1.

Inuit Circumpolar Conference (2005). *Petition to the Inter American Commission on Human Rights Seeking Relief from Violations Resulting from Global Warming Caused by Acts and Omissions of the United States.* Available at: http:// www.inuitcircumpolar.com/files/uploads/ICC-files/FINALPetitionICC.pdf; accessed January 3, 2006.

Jasanoff, S. (1995). *Science at the Bar: Law, Science, and Technology in America.* Cambridge, MA: Harvard University Press.

Leone, R. C. (1995). Foreword. In *Science at the Bar: Law, Science, and Technology in America*, ed. Jasanoff, S. Cambridge, MA: Harvard University Press, pp. ix–xi.

Moser, S. C. and Dilling, L. (2004). Making climate hot: Communicating the urgency and challenge of global climate change. *Environment*, **46**, 10, 32–46.

Mukerjee, M. (2003). Greenhouse suits: Litigation becomes a tool against global warming. *Scientific American*, **288**, 2, 14–15.

Revkin, A. (2005). U.S. rapped for stance on climate. *The New York Times*, December 8. Available at: http://www.iht.com/articles/2005/12/08/news/ climate.php; accessed January 3, 2006.

Sax, J. L. (1998). Defending the environment: A strategy for citizen action. In *Law and the Environment: A Multidisciplinary Reader*, eds. Percival, R. V. and Alevizatos, D. C. Philadelphia, PA: Temple University Press, pp. 300–5.

Strauss, A. L. (2003). The legal option: Suing the United States in international forums for global warming emissions. *Environmental Law Reporter*, **33**, 10185–91.

Cases

Central Valley Chrysler-Jeep, Inc. v. Witherspoon, No. CIV-04-6663 REC LJO (E.D. Cal. 2004)

Connecticut v. American Electric Power Co., Inc., 04 Civ. 5669 (LAP), 04 Civ. 5670 (LAP), (S.D.N.Y. 2005); WL 2249748.

Friends of the Earth v. Watson, No. C 02-4106 JSW 2002, N.D. Cal.
Massachusetts v. United States Environmental Protection Agency. 126 S. Ct. 2960
    (2006).
Massachusetts v. United States Environmental Protection Agency. No. 03-1361 (and
    consolidated cases) (D.C. Cir. 2003).
Sindell v. Abbott Laboratories, 607. P.2d 924 (Cal. 1980).

Statutes

Administrative Procedures Act, 5 U.S.C. §§ 551 *et seq.*
Clean Air Act, 42 U.S.C. §§ 7401 *et seq.*
National Environmental Policy Act, 42 U.S.C. §§ 4321 *et seq.*

30

The moral and political challenges of climate change

Dale Jamieson
New York University

Climate change presents us with a complex moral problem that our current political system is not well suited to address. Thus, it should not be surprising that we are failing to address it. In fact, climate change presents us with several distinct challenges. The first and most obvious involves coping with the changing climate itself. For societies that are not well adapted to normal climate variability in the first place, the more frequent and extreme events produced by climate change will be devastating. These effects will ramify through their economic, social, and political systems, spreading out into the international order. In addition, much of what we value about non-human nature will be lost since the clock of evolutionary adaptation runs much more slowly than that of human-caused environmental change. These are the kinds of problems that we can expect to face on the relatively optimistic scenario that the shifts in the earth system caused by climate change will be relatively moderate. Should major ocean or atmospheric circulations fail or sea levels rise catastrophically, the whole idea of adaptation will seem "quaint" at best.

Climate change as a moral problem

While the challenge of coping with a changing climate is daunting, it is one that is widely recognized and discussed. The moral and political challenges of climate change are relatively neglected. Climate change is a dramatic challenge to our moral consciousness, but it is not often perceived this way because it lacks some of the characteristics of a paradigm moral problem.

What are these characteristics? A paradigm moral problem is one in which an individual acting intentionally harms another individual; both the individuals and the harm are identifiable; and the individuals and the harm are closely related in time and space.

Consider Example 1, the case of Jack intentionally stealing Jill's bicycle. The individual acting intentionally has harmed another individual, the individuals and the harm are clearly identifiable, and they are closely related in time and space. If we vary the case on any of these dimensions, we may still see the case as posing a moral problem, but its claim to be a paradigm moral problem will be weaker. Consider some further examples.[1]

> Example 2: Jack is part of an unacquainted group of strangers, each of which, acting independently, takes one part of Jill's bike, resulting in the bike's disappearance.
> Example 3: Jack takes one part from each of a large number of bikes, one of which belongs to Jill.
> Example 4: Jack and Jill live on different continents, and the loss of Jill's bike is the consequence of a causal chain that begins with Jack ordering a used bike at a shop.
> Example 5: Jack lives many centuries before Jill, and consumes materials that are essential to bike manufacturing; as a result, it will not be possible for Jill to have a bicycle.

While it may still seem that moral considerations are at stake in each of these cases, this will be less clear than in Example 1, the paradigm case with which we began. The view that morality is involved will be weaker still, perhaps disappearing altogether, if we vary the case on all these dimensions simultaneously. Consider the final example.

> Example 6: Acting independently, Jack and a large number of unacquainted people set in motion a chain of events that causes a large number of future people who will live in another part of the world from ever having bikes.

For some people the perception persists that this case poses a moral problem. This is because the core of what constitutes a moral problem remains. Some people have acted in a way that harms other people. However, most of what typically accompanies this core has disappeared. In this case it is difficult to identify the agents, victims, or causal nexus that obtains between them; thus, it is difficult to assign responsibility, blame, and so forth.

These "thought experiments" help to explain why many people do not see climate change as an urgent moral problem. Structurally, the moral problem of climate change is largely the same as Example 6. A diffuse group of people is now setting in motion forces that will harm a diffuse group of future people. Indeed, if anything, the harms caused by climate change will be much greater than the loss of the opportunity to have a bicycle. Still, we tend not to conceptualize this as a moral problem because it is not accompanied by the characteristics of a paradigm moral problem. Climate change is not a matter of a clearly identifiable individual acting intentionally so as to inflict an identifiable harm on another identifiable individual, closely related in time and space. Because we tend not to see climate change as a moral problem, it does not motivate us to act with the urgency characteristic of our responses to moral challenges.

Climate change as a challenge to our political system

Climate change challenges our political system in addition to the problems that it poses to our moral consciousness. One way to see this is by distinguishing political action based on values from political action based on interests and preferences. These terms are ambiguous and often used in crosscutting ways, so a certain regimentation is required in order to make some important distinctions.

Values, as I will use the term, are close to the core of a person's identity and are relatively stable: they reflect how someone wants the world to be, not merely what the person may want for himself. Preferences, on the other hand, reflect what people want at a particular moment. Preferences and values can come into conflict in our behavior. Someone may both value an egalitarian distribution of wealth, and prefer to be very rich. This may express itself in her voting for egalitarian political candidates while seeking to make the sharpest possible financial investments. Unless irony is at work, a similar conflict can be seen in people who put Sierra Club bumper stickers on their hummers. The term "interest" is often ambiguous between what a person may currently want and what is good for her. We can speak of someone's interest in health while at the same time noting her interest in smoking. Bringing these thoughts together, we can say that values express people's view of how the world ought to be, interests concern what is good for them either in the short or long term, and preferences express what it is that they currently want.

That the American political system is based on interest-group politics is a commonplace among many political scientists. Indeed, politics is sometimes defined as "who gets what, when, where, and how." To the extent that this is true, it will be difficult to respond politically to climate change. For many of those who will be most harmed by climate change do not participate in the American political system (see Agyeman *et al.*, Chapter 7, this volume). These include non-human nature, future generations, citizens of other countries, and even disenfranchised and alienated American citizens. In reply, it is sometimes said that these interests gain political representation through the active participation of others who care about them and assert their interests. To some extent this is true, but it is obvious that at best these marginalized interests are represented only as shadows rather than in their full vivacity. This can be seen by comparing the case in which my interests are represented by someone with many interests of their own who also cares about me, and the case in which I assert my own interests.

However, it is not entirely true that America is an interest-group democracy. It is often remarked in electoral analyses that voters do not always express their interests in the ballot box. For example, poor people often vote for rich people who will give themselves tax cuts at the expense of their poor supporters; soldiers often vote for leaders who will put their lives at risk; even criminals sometimes vote for candidates who want to crack down on crime. There are many ways of trying to explain this behavior, but one way is to say that people often act politically on the basis of their preferences rather than their interests. This is not surprising since there are many cases outside of political life in which preferences and interests diverge and we find our preferences compelling. For example, I want to eat tiramisu, even though it is not in my interest to do so. Even more strongly, I may want to smoke although it is counter to my interests. And I may want to drive my SUV despite my valuing of nature and future generations.

One reason people act politically on the basis of preferences rather than interests is the power of "branding" (see Popkin, 1994). By and large candidates do not seek to convince the public of the wisdom or justice of their policies; instead, they attempt to make themselves a "brand" with which people want to associate (see Postman, 1985; and Mayhew, 1974).[2] In doing this they exploit deep facts about the psychology of social animals like us who evolved in small societies, largely dependent on emotion rather than reason in guiding their behavior (see Frank, 2004).[3] Since asserting positions and making arguments are at best not part of the branding process and at worst antithetical to it, political campaigns have become the last place to find serious discussion of important public issues. It is tempting to blame politicians and their handlers for this, but we citizens are also to blame. We tend to punish politicians (of whatever political stripe) who take strong, understandable positions on important public issues.

When branding rather than reasoning is the main point of public discourse, it is not surprising that a political system based on preferences and anchored in branding would fail to come to terms with an issue as complex as global warming. How dated is former president Lyndon Johnson's frequent appeal to his father's favorite Bible passage, "Come now and let us reason together" (Isaiah 1:18). Indeed, rather than appealing to reason, some of those who oppose taking action on climate change have consciously adopted disinformation as a political strategy (see McCright, Chapter 12, this volume). Many parties to the debate have treated value statements as lines in the sand rather than as invitations for dialogue (see Regan, Chapter 13, this volume). It is hard not to believe that this way of practicing politics will lead to disaster, whether on this issue or some other. In the end, we have collectively produced outcomes from which many of us individually feel alienated. This is true both in our politics and in our collective production of climate change.

There is another way of thinking about how a democratic political culture should function, one centered on deliberative engagement with values rather than on branding (see Elster, 1998).[4] The deliberative ideal is reminiscent of the Enlightenment views that dominated European and American political thought in the eighteenth century. It is based on the idea that the best society is one that is a democratic expression of the reflective views of its citizens, based on their most fundamental values. These views require constant examination, which is why free speech is important, and also a foundation in our best understandings of the world, which is why education matters (see the chapters by Bateson; and Grotzer and Lincoln, Chapters 18 and 17, this volume).

This sentiment would have been familiar to the founders who recognized that American democracy was tenuous and made stringent demands on its citizens. It is reflected in the following anecdote told about Benjamin Frankin. As he was leaving the hall in Philadelphia on that sunny day in 1787 when the Constitutional convention had finished its work, a woman approached him and asked, "Mr. Franklin, what kind of government have you given us?" He is said to have replied: "A Republic, madam, if you can keep it."

There is much that is important about Franklin's reply. I want to highlight only his sensitivity to the precariousness of the American system of government. To Franklin, and many of the other founders, a political system is not an abstraction delivered by gods. It is a set of institutions designed by people to serve their deepest purposes. Our political system must be one that we can successfully manage. It is no good demanding of ourselves what we are incapable of delivering, and there is no question that our psychologies and nature constrain and condition the kinds of institutional arrangements that are manageable by us. In general, what we need both to keep our republic and to address slow-onset long-term problems like climate change is a sense of ownership and identification with the outcomes that our actions produce. It is this sense of ownership and identification that allows us to overcome the alienation from the collective consequences of our actions.

Climate change and character

How can we gain this sense of ownership and identity? This requires an ideal of character for what is required to live in a highly interconnected, globalized world.[5] Here I can give only a brief sketch of some fragments of this ideal, what might be called "the green virtues." Before sketching these virtues, however, it is important to acknowledge the complex relationships that exist between our character as individuals and the societies into which we are born. Institutional structures deeply affect what kind of people we will be, but what kind of people we are also has profound effects on the nature of our society. We cannot opt for changing ourselves rather than changing the world or the world instead of ourselves: in an important sense of the expression, we are the world.

Humility is a widely shared moral ideal that is not often connected to a love of nature or the importance of living lightly on the earth. Yet indifference to nature is likely to reflect the self-importance or lack of self-acceptance that is characteristic of a lack of humility. A person who has proper humility would be horrified at the prospect of changing Earth's fundamental systems, and would act in such a way as to minimize the impact of their behavior.

Temperance is an ancient virtue that is typically associated with weakness of will. However, conceived more broadly, temperance relates to self-restraint and moderation. A temperate person does not overconsume; he "lives simply, so that others may simply live" (see Jamieson, 2002).[6]

Finally, we can imagine a virtue that we might call mindfulness. Behavior that is rote and unthinking, as is the case with much of our environmentally destructive behavior, is the enemy of mindfulness. A mindful person would appreciate the consequences of her actions that are remote in time and space. She would see herself as taking on the moral weight of production and disposal when she purchases an article of clothing (for example). She would make herself responsible for the cultivation of the cotton, the impacts of the dyeing process, the energy costs of the transport, and so on. Mindful people would not thoughtlessly emit climate-changing gases.

As I have noted, it is easy to see that institutions play important roles in enabling virtue. Many of these roles (e.g., inculcation, encouragement) have been widely discussed from Aristotle to the present. It is also important to recognize that how societies and economies are organized can disable as well as enable the development of various virtues (see chapters by Atcheson; and Dilling and Farhar, Chapters 22 and 23, this volume). For example, in a globalized economy without informational transparency, it is extremely difficult for agents to determine the remote effects of their actions, much less take responsibility for them. Thus, in such a society, it is difficult to develop the virtue of mindfulness.

Concluding remarks

Climate change presents us with many challenges, and many people are working hard to overcome them. In this chapter I have focused on the moral and political challenges of climate change. They are important because seeing an issue as a moral problem can provide the motivation for individual and political action. The moral and political challenges are related because the ideal of a deliberative and reflective politics requires citizens who express particular moral virtues in their behavior.

The language of morality is the language of care, empathy, responsibility, and duty. This language has largely been absent from discussions of climate change. Instead the language of science, economics, and technological development has been dominant. Of course there are important roles for such discourses, but people do not change their lives on the basis of a cost—benefit analysis. Successfully addressing climate change requires long-term, sustainable changes in the way we live. This will only come about when we take responsibility for our actions, and express our concern for future generations and the health of the earth through our everyday actions. The transformation that is required is not only personal, but profoundly collective and political as well. The hope for such a change rests on a new kind of open-hearted dialogue about what we are doing to ourselves and our children in the mindless pursuit of more and more stuff. As the nineteenth-century philosopher John Stuart Mill told us long ago, it is not economic growth for its own sake we should strive for, but rather improvements in the "Art of Living." This, he thought, could only be obtained in a world that to a great extent remained free of human domination (see Mill, 1994).

Climate change is not only a challenge to our ethics and politics, but also has the potential for improving them. Successfully responding to climate change can make us better people and help us to reclaim our democracy. This connection between the state of our souls and the fate of the Earth was clearly seen by Walt Whitman, the sage poetic observer of American democracy, when he wrote: "I swear the Earth shall surely be complete to him or her who shall be complete."

This should give us heart. We must begin from where we are — changing ourselves, changing our leaders, and changing our institutions — but from here we can change the world. Biking instead of driving or choosing the veggieburger rather than the hamburger may seem like small choices, and it may seem that such small choices by such little people barely matter. But ironically, they may be the only thing that matters. For large changes are caused and constituted by small choices.[7] And in the end, however things turn out, it is how we live that gives meaning and significance to our lives.[8]

Notes

1. Some of these examples are inspired by those given by Glover (1975).
2. A wonderfully insightful exposition of this thesis is Postman (1985). For a more scholarly treatment, see Mayhew (1974).
3. The idea that we are primarily emotional rather than rational animals (*contra* Aristotle) is an ancient idea that achieved its fullest philosophical expression in the work of the eighteenth-century philosopher David Hume. It has been explored in great detail by such contemporary psychologists as Daniel Kahneman and Daniel Gilbert, and such moral philosophers as Simon Blackburn and Allan Gibbard. The political consequences of this has been explored in such books as Frank (2004) (see also Moser, Chapter 3, this volume).
4. A vast literature on deliberative democracy has developed in recent years.
5. I have discussed this at greater length in "Ethics, Public Policy, and Global Warming," reprinted as Essay 18 in Jamieson (2002).
6. This expression is attributed to Ghandi. See http://www.dropsoul.com/mystic-quotes.php; accessed January 17, 2006.
7. Beef production is extremely energy and water intensive, and cows are a major source of methane emissions. A molecule of methane has more than 20 times the global warming potential than a carbon dioxide molecule.
8. For a good bibliography on ethics and climate change, see http://rockethics.psu.edu/climate/index.htm; accessed January 17, 2006.

References

Elster, J. (ed.) (1998). *Deliberative Democracy*. Cambridge, UK: Cambridge University Press.

Frank, T. (2004). *What's the Matter With Kansas?* New York: Henry Holt, and Company.

Glover, J. (1975). It makes no difference whether or not I do it. In *Proceedings of the Aristotelian Society*, Supplementary Volume **49**, pp. 171–90.

Jamieson, D. (2002). *Morality's Progress: Essays on Humans, Other Animals, and the Rest of Nature*. Oxford: Oxford University Press.

Mayhew, D. (1974). *Congress: The Electoral Connection*. New Haven, CT: Yale University Press.

Mill, J. S. (1994). Principles of political economy. In *Reflecting on Nature: Readings in Environmental Philosophy*, eds. Gruen, L. and Jamieson, D. New York: Oxford University Press, pp. 29–30.

Postman, N. (1985). *Amusing Ourselves to Death: Public Discourse in the Age of Show Business*. New York: Viking.

Popkin, S. L. (1994). *The Reasoning Voter: Communication and Persuasion in Presidential Campaigns*, 2nd edn. Chicago: University of Chicago Press.

PART THREE

Creating a climate for change

31

An ongoing dialogue on climate change: The Boulder Manifesto

Robert Harriss
Houston Advanced Research Center

> As to methods there may be a million and then some, but principles
> are few ... The man who tries methods, ignoring principles, is sure
> to have troubles.
>
> *Ralph Waldo Emerson*

The rich and diverse perspectives reported in this volume reflect the emergence of a new and unique social learning network on climate change communication and social change. The threads for a new fabric of knowledge have been gathered. The next steps are to weave together these fragments of thought, to move from knowledge sharing to actual dialogue, and to create a new learning ecology that effectively encourages the more environmentally sustainable behaviors necessary to avoid the dangerous impacts of climate change. This chapter presents the Boulder Manifesto that articulates the book's contributors' collective way of seeing the climate change communication challenge and of understanding our part in creating it — as it is and might be. The utility of a manifesto lies in providing a conceptual aperture, an opening through which future social realities may unfold. The nature of the Boulder Manifesto rests on the expertise, spirit of barn-raising and commitment to a common cause that is threaded through the preceding chapters. A continuing and successful dialogue among the diverse stakeholders represented in this volume will be an enormous challenge — dialogue is a state out of which we are continuously falling (Isaacs, 1999).

In the case of sustained local and regional action on threats posed by climate change there are potential pitfalls ahead. The scientific modeling of climate change futures is fraught with uncertainties related to when, where, and how much climate will change at the local scales that strongly influence human perceptions of environmental change. To use a medical analogy, it will be difficult to diagnose the disease, forecast the development, prescribe the right drugs, and determine the correct dosage for treatment — and do so in time to take preventive or even just remedial action. The climate change contrarians will continue to make frequent use of such gaps in scientific understanding and capabilities to delay action. Thus, moving from dialogue to strategy to action cannot rest too heavily on the expectation of a mature climate science. Rather, the Boulder Manifesto implies an action agenda based on the knowledge already available, but more so on the powerful emotional role that hope plays in empowering humans to seek a higher level of creative problem-solving — conceiving of desired futures as well as the real means to achieve them.

Gaining priority for the climate change issue on the public policy agenda hinges on creating a positive shared vision for solutions that mitigate the common perception that the future simply happens to us. The Boulder Manifesto maintains that through dialogue we can begin envisioning a future worth fighting for, a future that we can bring about through conscious choice. When the future can no longer be expected to follow on neatly from the past, then imaginative means must be employed to shape a powerful, sustained, and consistent narrative that can attract and mobilize broad political support for change. The narrative must integrate threats and opportunities posed by climate change with the many other factors, both near term and far out into the future, that compete for attention in business, community life, and households across America.

The spirit and content of the Manifesto was first articulated and discussed at the summer 2004 workshop that initiated this book project. It is inspired by the work of Lyle (1994), McHarg (1995), Ogilvy (2002), and Bradbury *et al.* (2004). The theme of transformative change is common to their work in landscape ecology, business strategy, and individual learning. The Manifesto intends to provide a compass pointing the way for the next steps of the climate change communication and social change research and action enterprise. Its simple statements aim to create a commonly shared frame that will nurture an efficient, adaptive, resilient, and sustainable dialogue on climate change communication in an evolving social action network. The five propositions below are a first step in furthering the process for a future dialogue on this new knowledge that is seeking to emerge.

The Boulder Manifesto

Climate change communication will stimulate a dialogue on hopeful visions of future everyday life through democratic means. The fundamental driving forces of climate change reflect the daily decisions of every citizen of Planet Earth. Human needs and wants result in changes in the ecology of places both near and far from sites where each person lives. A democratic dialogue on hopeful visions for a sustainable future must evolve from defending opinions to listening without resistance, and towards a generative dialogue that invents new possibilities and stimulates collective action.

Climate change communication will create a future vision that addresses immediate societal issues and needs while linking them to larger, systemic climate change issues. Developing this vision involves reflection with a view to clarifying strategic action. Exploring and preparing a course of action to mitigate potential threats posed by climate change implies the following five questions: (1) What is possible? (scenario building), (2) What can we do? (strategic options); (3) What will we do? (strategic decisions); (4) How will we do it? (actions and operational plans); and (5) What is everyone's specific role in these solutions? An effective future prospective and climate change communication narrative will acknowledge the diversity of issues that compete for the public policy agenda (United States Government Accountability Office, 2005) and social complexities such as carbon lock-in (Unruh, 2002), but they are likely to be more effective if they meet current needs while beginning to address or at least not foreclose future options. For example, climate change information is very relevant to the analysis of scenarios and strategies for containing long-term health-care costs. There is growing evidence indicating that climate change will influence future patterns of infectious diseases, the frequency and intensity of urban heatwaves, and the ecology of disease vectors (McMichael *et al.*, 2003; McMichael, 2001). Communicating the relevance of climate change to health and health-care costs in the context of a comprehensive and respected assessment provides an opportunity for integrating high-profile issues on the American policy agenda. Similar opportunities exist for integrating climate change communication into dialogues on long-term energy and national security futures.

Climate change communication will acknowledge and incorporate the diversity of local cultures and practices that contribute to a sense of place. If the metric for success in climate change communication is encouraging more environmentally sustainable behaviors from individuals, families, communities, and businesses across the nation, a global message is unlikely to be effective. A dramatic environmental event or early warning message widely distributed through public media can raise awareness, but seldom promotes long-term transformational change in human behaviors.

The dramatic rise of mass consumption in twentieth-century America was significantly influenced by pictures of the future that connected to personal needs and desires (Cohen, 2003). The marketing of consumer products in the twenty-first century is based on increasingly sophisticated methods of market segmentation that speak to individual and community lifestyles (Twitchell, 2002). Climate change communication must paint pictures of the future that inform individuals of their roles as citizens, consumers, and stewards of the future. The question then becomes: if as citizens, we do not just create our own small personal future but co-create our world future, what future world do we want to create together?

Climate change communication will be concerned with the design of institutions that promote adaptive and resilient, life-enhancing practices and lifestyles. The challenge of climate change is not just to develop new technologies. It requires adjustments to our assumptions, attitudes, and understanding to move from our current lifestyles − that are increasingly separated from nature − to a design for living that is efficient, adaptive, resilient, and that continuously learns from the regenerative strategies characteristic of natural systems (Lyle, 1994; McHarg, 1995; Gunderson and Holling, 2002). The global climate change issue has emerged from the environmental movement of the 1970s that successfully influenced policies and actions for improving and protecting the quality and character of air, water, and landscapes in cities and regions across the nation. Environmental solutions largely addressed pollution at the end of the pipe; climate change solutions and their communication will necessarily confront broader and deeper dysfunctional practices in our social, governance, and economic systems (National Research Council, 1999; Hallsmith, 2003).

Climate change communicators and change agents will embody the change, demonstrating that personal climate-friendly lifestyles can eventually lead to change in the system as a whole. This is the often-ignored Socratic lesson − "know thyself." Individual and organizational role models − in a nation as big as the United States with its nearly 300 million individuals and more than 18,000 communities − that have adopted the numerous strategies and actions congruent with a more sustainable future are still so few that we see them listed in a manner similar to endangered species. Individuals and organizations engaged in climate change communication and social change must "walk the talk" to be credible as living proof that true prosperity, for both economy and environment, can and does arise from innovation rather than compromise and exploitation.

Where shall wisdom be found?

> Some problems are so complex that you have to be highly intelligent and
> well informed just to be undecided about them.
>
> *Laurence J. Peter (1982)*

This final thought takes its title from literary critic Harold Bloom's quest for
sagacity that might solace and clarify the many dimensions of personal
anxiety so common in contemporary America (Bloom, 2004). Bloom's
profound knowledge of writers ancient and modern leads him to conclude that we can know wisdom, whether or not it can be identified with
"the Truth" that might make us free. Climate change belongs to a class of
"wicked problems" where the Truth will always be ephemeral. Horst Rittel,
Professor of Planning at the University of California, Berkeley, first identified a wicked problem as one for which each attempt to create a solution
changes the understanding of the problem. Wicked problems cannot be
solved in a traditional linear fashion, because the problem definition evolves
as new possible solutions are considered and/or implemented (Rittel and
Webber, 1973).

Solving wicked problems associated with climate change communication
and social change must always be treated as an experiment. There is too
much to be known, too many interactions that may be hidden at the outset,
and too many conflicting demands to be satisfied. Our best path forward is
to engage the people who will most likely be affected, build trust with and
among them, and, together, move the entire process toward progressively
better understanding of how to stimulate environmentally sustainable
behaviors in every city, town, and state in America.

While many low-emission strategies and technologies are already available, the climate change problem will not be solved with ready-made
solutions. Nor will those of communication or social change. We should be —
and remain — suspicious of any proposed "silver bullet" solution to such
complex problems, even as they become ever more pressing. Every step
towards transformative change will reveal new problems, new information,
and new opportunities. Success in addressing the communication and social
change challenges may depend most of all on a change in attitudes about the
people who will carry the message and bring about societal transformation.
Inviting such stakeholders to actively participate in forging the agreements
of what constitutes success, in identifying the problems, and in developing

and refining potential solutions is a strategy that is seldom used in academic research. By involving stakeholders as collaborative partners in the generation of climate change communication and social change strategies they will understand what the choices are, what the tradeoffs are among them, and what everyone's role is in implementing them. That way, we all become active partners in achieving success. Perhaps the Boulder Manifesto will serve as a useful tool for bringing new voices and ideas to the dialogue on climate change, on how to communicate it, and how to set the social transformation in motion required to deal with the challenge.

References

Bloom, H. (2004). *Where Shall Wisdom Be Found?* New York, NY: Riverhead Books.

Bradbury, H., Waage, S., Rominger, R., *et al.* (2004). Effecting change in complex systems: A dialogue on overarching principles to inform action. *Reflections*, 5, 4–5.

Cohen, L. (2003). *A Consumers' Republic: The Politics of Mass Consumption in Postwar America.* New York: Alfred A. Knopf Publishers.

Gunderson, L. H. and Holling, C. S. (2002). *Panarchy: Understanding Transformations in Human and Natural Systems.* Washington, DC: Island Press.

Hallsmith, G. (2003). *The Key to Sustainable Cities.* Gabriola Island, BC, Canada: New Society Publishers.

Issacs, W. (1999). *Dialogue and the Art of Thinking Together: A Pioneering Approach to Communicating in Business and in Life.* New York, NY: Currency–Doubleday.

Lyle, J. T. (1994). *Regenerative Design for Sustainable Development.* New York: Wiley Publishers.

McHarg, I. L. (1995). *Design with Nature.* New York: Wiley Publishers.

McMichael, A. J. (2001). *Human Frontiers, Environments and Disease: Past Patterns, Uncertain Futures.* Cambridge, UK: Cambridge University Press.

McMichael, A. J., Campbell-Lendrum, D., Corvalan, C., Ebi, K., Githeko, A., Scheraga, J., and Woodward, A. (2003). *Climate Change and Human Health: Risks and Responses.* Geneva, Switzerland: World Health Organization.

National Research Council (1999). *Our Common Journey: A Transition Toward Sustainability.* Washington, DC: National Academy Press.

Ogilvy, J. A. (2002). *Creating Better Futures: Scenario Planning as a Tool for a Better Tomorrow.* New York: Oxford University Press.

Peter, L. J. (1982) *Peter's Almanac.* New York, NY: William Morrow & Co.

Rittel, H. and Webber, M. (1973). Dilemmas in a general theory of planning. *Policy Sciences*, 4, 155–9.

Twitchell, J. B. (2002). *Living It Up: America's Love Affair with Luxury.* New York, NY: Simon & Schuster Publishers.

United States Government Accountability Office (2005). *21st Century Challenges: Reexamining the Base of the Federal Government.* GAO-05-325SP, Washington, DC: GAO.

Unruh, G. C. (2002). Escaping carbon lock-in. *Energy Policy*, 30, 317–25.

32

Toward the social tipping point: creating a climate for change

Susanne C. Moser
National Center for Atmospheric Research

Lisa Dilling
University of Colorado–Boulder

Introduction

Over the course of the project that culminated in this book, the landscape of climate change science, communication, and related societal responses has changed remarkably — both in the United States and elsewhere. Considering the entirety of what political scientists call the global warming "issue domain" (Jenkins-Smith and Sabatier 1999; Clark *et al.*, 2006), we have witnessed considerable movement, including the growing number of actors involved, and shifts in their "goals, interests, beliefs, strategies, and resources; the institutions that enable and constrain interactions among those actors; the framings, discourse, and agenda related to the issue; and the existing policies and behaviors of relevant actors" (Clark *et al.*, 2006).

In this concluding chapter, we have five goals. First, we reflect on where we are in the evolution of the climate change issue domain, and then develop a simple conceptual framework to integrate the many perspectives offered in preceding chapters on the role communication can play — in principle — in facilitating social change. Next we dispel a number of myths still prevalent among communicators and social change agents that we believe hinder change. Fourth, we extract larger lessons from the chapters that could improve climate change communication and advance the evolution of this issue domain. Finally, we suggest questions for future research and action steps. The collective experience represented in this volume suggests that these research directions and action steps can further support effective communication and responses to climate change in ways that help move all levels of society toward an environmentally, economically, and socially more sustainable future.

How close to the tipping point?

Only hindsight allows us to know for sure how far along a society is on the trajectory of social change. Clearly, the judgment of "where we are" also depends on the type and scale of change on which one chooses to focus. We might observe social change at the individual, organizational, and societal levels, reflected in a myriad of social characteristics, such as economic or media trends, the growth of social movements or counter-movements, and so on.

In recent years, scientists have begun integrating various theories of the dynamics of social change in the context of the daunting task of navigating a transition toward a sustainable world (Gunderson and Holling, 2001; Gunderson, Holling, and Light, 1995; Kates *et al.*, 2001; NRC, 1999; Raskin *et al.*, 2002; Schellnhuber *et al.*, 2006; Speth, 1992). Reducing greenhouse gas emissions and adapting to the unavoidable impacts of climate change are important components of this sustainability transition (Smit *et al.*, 2001; Toth *et al.*, 2001; Yohe, 2001). As we argued in the introduction to this volume, awareness of the need for this fundamental transition – and for climate change policies in particular – has grown, but many social systems are also highly resistant to change. Efforts to mitigate climate change will eventually reach deeply into the workings of society and may well require difficult policy choices.

Certainly, some types of social transformation are faster and more easily achieved while others take longer (e.g., Bamberg, 2003; Michaelis, 2003). Moreover, various characteristics of society transform simultaneously, albeit at different rates and in ways that may or may not be positively reinforcing. The diverse literature on these social change processes has no easy answers on how to move smoothly through this transition, yet experts share a common conceptualization of how such change typically unfolds (Figure 32.1) (e.g., Kemp and Lorbach, 2003; Kemp and Rotmans, 2004; Moyer *et al.*, 2001; Raskin *et al.*, 2002; Rogers, 2003; Rotmans, Kemp, and van Asselt, 2001).

Where are we, then, on this prototypical S-curve of social change? We argue that the US climate protection movement with all its different components is quietly building and beginning to emerge, but – as of early 2006 – is still in a phase prior to the take-off or tipping point (e.g., Moser, 2007). Tipping points are commonly defined as moments in time where a normally stable or only gradually changing phenomenon suddenly takes a radical turn (Gladwell, 2000), e.g., when something unique suddenly becomes common. Once a movement reaches that point,

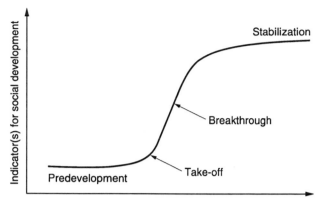

Fig. 32.1.   The typical stages of social transformation
This conceptual diagram will be interpreted in different ways depending on
the social change in question. For example, the invention, spread, and
eventual saturation of the market with a new energy-related technology may
follow such a development and might be measured by sales (vertical axis)
over time (horizontal axis).
Source: Kemp and Rotmans (2004), reprinted, with permission, by René
Kemp.

public and political engagement, policy change, and the adoption of climate-
friendlier technologies (e.g., renewable energy) will spread rapidly from the
pioneers and early adopters to the vast majority of the population, industry,
and other civic and political actors. We cannot predict what event or
convergence of processes will trigger this take-off, but it could emerge from
either within the issue domain or from a societal development that has
nothing to do with climate change. By changing the ways in which climate
change is communicated and reframed, people can help the movement take
off if and when such a trigger event occurs (Moyer *et al.*, 2001).

We are already seeing heightened attention to climate change. Increasingly
urgent scientific findings, together with stories on local to international
policy developments, the entry of important players into the climate change
issue domain, and government interference with how scientists communicate climate change, have kept global warming on the public radar screen.
With the growing number of increasingly diverse players now involved,
climate change communication is also changing – at least in some corners –
in tone and focus. The examples in this book illustrate that communication
can facilitate social change and move us along the S-curve of societal
transformation. Below, we offer a simple conceptual framework that helps
us understand this relationship and reflects the dynamics described in
preceding chapters.

Conceptualizing the link between communication and social change

In the introduction to this book we broadly defined communication as a *"continuous and dynamic process unfolding among people that facilitates an exchange of ideas, feelings, and information as well as the forming of mutual understanding and common visions of a desirable future."* Here, we link this dynamic conception more fully with social change.

The oft-repeated call for more and better communication to support behavior and other social changes assumes that *motivation* (typically in the form of more information) is the missing ingredient, that all we need to achieve more sustainable policies and practices is stronger and more appropriate motivation. Motivators, as discussed throughout this book, can range from psychological and moral to economic and legal ones. For different audiences and in different contexts, these motivators play an indispensable role, but typically, motivation is only half the story.

The other half is *resistance* or *barriers*. Again, our contributors have written at length about barriers in a wide variety of contexts, from a lack of infrastructure to self-limiting mental models. These very real barriers are rarely consciously considered and addressed at the outset of a communication and social change campaign. In short, if there is one "simple" formula that can capture the link between communication and social change, it is this:

> For communication to be effective, i.e., to facilitate a desired social change, it must accomplish two things: sufficiently *elevate and maintain the motivation* to change a practice or policy and at the same time *contribute to lowering the barriers* to doing so.

Too many communicators and change agents still do not pay attention to both sides of this equation. The simplicity of this conceptual formula – while fairly obvious – hides the fact that there is nothing "simple" about implementing it. Simply trying to motivate people to change behavior without acknowledging the real barriers in the way of change will have little success. While just talking about the lack of infrastructure will not put it in place, it may shift the focus of the campaign. Alternatively, communication can achieve its desired outcome if it addresses existing obstacles explicitly and provides useful solutions. Looking for synergies among actors and institutions across scales is essential for addressing the many situations where barriers are outside the sphere of influence of a single actor.

Our mental models or habits of thought are among the most critical barriers to change. Those we hold about communication and social change are no exception. The first challenge before us, then, is to help change the mental models that communicators and social change agents hold about the ways in which they can make a difference. The challenge is one of dispelling common myths about communication and social change.

Dispelling common myths about communication and social change

We call the many ineffective ways people use to promote public and political engagement with climate change "myths," since they still pervade many of the communication strategies practiced today. What allows these myths to persist is that they often contain a kernel of truth. If taken to the extreme, however, they are counterproductive and actually become hurdles to effective communication and social transformation.

"If only people understood the problem, they would change their behavior"

This is maybe the most common myth among many communicators, especially scientists. It is known in the social sciences as the knowledge or information "deficit model" and claims that people simply don't have enough information; therefore we must fill in this deficit. Once people understand, they will be motivated to act on the problem (Sturgis and Allum, 2004; Blake, 1999; Bak, 2001; Schultz, 2002). Researchers have repeatedly shown the deficit model to be an inadequate explanation of human behavior. However, this insight has not yet fully permeated the natural and physical sciences community nor, indeed, much of the non-governmental or outreach communities. Of course, when necessary, correcting incorrect mental models of the causes of a problem is critical to predisposing people to actions that protect the climate (Bostrom and Lashof). However, "more information" typically does not generate action on issues of societal import, and, as Dunwoody, and Rabkin with Gershon report, it is possible for information and understanding to become substitutes for action: individuals absorbing information sometimes feel that they have actually "done something" simply by having learned about a problem.

"Communicating climate change means convincing people of the reality of the problem"

This closely related myth assumes that people are just not yet convinced of the reality of climate change. However, polls show clearly that the majority of Americans already believe that climate change is real (see Leiserowitz). Even a strong conviction that there is a problem may not result in different behavior or action, however, as Tribbia, Moser, Chess and Johnson, and others explain. A corollary is that public communication on climate change has emphasized global temperature trends, emissions, and other changes in the climate system. So long as the focus stays here, we remain trapped in conversations about physical science and unable to move to the next phase of communication: the social changes needed to respond to climate change. Moving the conversation beyond science then enables discussion of solutions, values, and visions of the future, a more appropriate and effective step toward social action on climate.

"Maybe if we just scare people more, they'll get how urgent climate change is"

Clearly, climate change offers plenty of reasons to believe it is a critically urgent issue (see especially Moser). Yet there are two reasons to be wary of appeals to fear to communicate urgency. First, climate change by itself can provoke a strong fear response, possibly resulting in maladaptive behaviors. Second, attempts to communicate global warming by emphasizing its scary aspects often serve only to reinforce these counterproductive responses. The temptation, therefore, to "make climate change more frightening" is not generally going to be effective in motivating real long-term behavior change, absent practical, doable alternative courses of action.

"What we really need is a big disaster"

Several leading politicians and global change experts express the sentiment that a disaster or two would move us to action. Indeed, a disaster offers a "window of opportunity" that grabs people's attention, serves as a "teachable moment," could be seen as a harbinger of future increases in disasters caused by climate change, and typically affects – directly or indirectly – large numbers of people (i.e., voters). Sometimes disasters can indeed bring long-neglected actions to the front burner. But, as Bostrom and Lashof warn, disasters may also activate inappropriate mental models (e.g., disasters as "acts of God"), or aggravate already latent fear responses and – depending on circumstances – provoke fear rather than danger control reactions. Moreover, in some cases, disasters actually close windows of opportunity, as government agencies and disaster victims focus on immediate needs – and rightly so, using available staff and financial resources for recovery. As example after example shows, post-disaster may not be the best time to implement climate mitigation or adaptation policies (e.g., Moser, 2005), and actually distract people from climate change.

"Let's wait for a 'knight on a big white horse' to lead us forward"

Numerous examples in this book illustrate the critical importance of leaders to initiate or sustain change (e.g., James *et al.*, Watrous and Fraley, DuVair *et al.*, Tennis, Young, and Meyer). These examples confirm a deeply held belief — especially among Americans — in the necessity of individuals taking charge of a situation. While history has produced great leaders doing just that, this mental model can also induce complacency: until a leader appears, there is nothing to do. Simultaneously, we can blame heads of industry or government for stalling action. Hoping that someone will eventually take care of this problem in the future ignores the obvious — the many actors already preparing the ground for social change. The questions are, what works for the present situation? Where can change happen? And who are the necessary change agents?

"(Scientific) uncertainty is the main obstacle to action"

This common myth assumes that behavioral change or political action has not yet occurred because climate change is wrought with such big scientific uncertainties that it precludes action. It is undoubtedly true that the climate system is enormously complex and uncertain. What this myth ignores, however, is the fact that we almost never have certainty about the future in other areas of personal or public concern: real estate or stock markets, population changes, national security risks, international conflicts, or our personal health. Despite the uncertainty we act in these arenas all the time, seeking dialogue, policy, and other strategies to manage risks. It is not scientific uncertainty, but human choice in the face of uncertainty that makes our response to climate change different.

"Climate change is a unique social challenge — we've never had to deal with anything like it"

The lure of this myth is strong, since climate change involves such long time periods and lags, complexities, and uncertainties in its physical and social aspects (Field *et al.*, 2004). However, if one looks at social change throughout history, other issues such as slavery or fundamental changes in people's worldviews (e.g., from an Earth-centric to a solar-centric view of our place in the universe) have had similar path dependencies, decadal commitments, and expensive capital implications. The problem with accepting the "uniqueness" myth is that it tends to paralyze individuals and institutions rather than empower them. Pointing out how we have made similarly long-term, radical social transitions in the past can help mobilize people and put the challenge into proper perspective.

"Appealing to people's rational side is the most effective way to communicate"

Western rationalist mythology would want us to make decisions based on rational thought alone. Yet people make decisions based on a myriad of influences, including "irrational" beliefs and emotions (e.g., Kahneman, 2003). Emotions play a powerful and necessary role in decision-making (Leiserowitz, Moser), as do preferences such as convenience, prestige, and so on (Chess and Johnson, Tribbia, Agyeman *et al.*, Michaelis, Rabkin with Gershon, and Jamieson). Communicators need to appeal to these emotional, belief-, value-, and identity-driven aspects of individuals, especially the "empowering" emotions, rather than the ones that tend to promote apathy, denial, and disengagement, as well as to their rational side.

" 'Good' values will produce 'good' outcomes for the climate"

It is easy to fall into the myth that sees a direct and causal link between our most revered values and a positive environmental outcome. Taking this view can miss critical contradictions, however. The value of "justice for all" is a good example. We staunchly believe in equity and fairness for all people as a sacred right, but potentially run into a values conflict when we apply this principle to climate change. As Agyeman *et al.* and McNeeley and Huntington argue, in the past, industrialized countries have been the major contributors to anthropogenic climate change. In the very near future, rapidly developing countries will come into their fair share of economic development, energy use, and related greenhouse gas emissions. Without "leapfrogging" technology that allows these countries to develop economically while avoiding negative environmental and climate impacts, our highly cherished value of fairness will be in direct conflict with mitigating climate change.

Having identified some of the most critical myths impeding effective communication for social change, we can now ask: what might work better?

Communication in true support of social change: Insights from the chapters

As we stated above, communication can more effectively facilitate social change the more it elevates and maintains the motivation to change a practice or policy, while lowering the barriers to that change. Below we synthesize important insights on each component of this communication—social change link.

Elements of effective communication of climate change

Our contributors understand the communication process not merely as one-way information transmission but as an intricate and many-faceted form of mutual engagement.

Audience choice is the first and most important strategic step in that process. With an eye to a particular desired outcome, communicators must carefully select the right audience, i.e., those people who need to know about the issue, can make a desired change, pull the necessary political levers, or are in a position to influence those who can. For example, certain audiences such as manufactures, designers, builders, planners, and policy-makers make decisions that affect the actions of many people, not just those of an individual (Dilling and Farhar, Cole with Watrous).

Next, it is critical to learn as much as possible about the audience's (or interpretive community's) pre-existing mental models and levels of understanding (Bostrom and Lashof); its interests, values, and concerns; and the channels through which the audience is likely to receive the communication — a briefing, a public talk, a mailing in the monthly electricity bill, or the evening news (Dunwoody, Leiserowitz). The channels themselves must be evaluated carefully to ensure they can achieve the desired outcome (Dunwoody). For example, interpersonal communication is typically required to persuade someone of the importance of an issue or of a needed behavior change.

Further, effective communication has to achieve a match between message content, framing, and the concerns and values with which audiences resonate (Bostrom and Lashof). Pratt and Rabkin describe how neglecting to do so prevented previous communication efforts to reach residents of San Diego. Similarly, Agyeman *et al.* argue convincingly that the justice frame (which emphasizes a fair distribution of environmental burdens across class and ethnic differences) may serve well to reach minority and working-class audiences, until recently largely ignored and unmoved by "mainstream" climate communication. Bingham, McNeeley and Huntington, Michaelis, Leiserowitz, and Arroyo and Preston make parallel arguments for different audiences.

In the majority of chapters, our contributors find that in order to make global climate change relevant to their audiences, global warming has to be made "local," whether directly by focusing on impacts that matter to them, or indirectly by focusing on the co-benefits of climate-friendly action. One useful option is to tap into lay people's own observations of change to jump-start a conversation about what climate change might mean locally, and help people relate to global patterns of change already well documented. This "conversation starter" helps to counter perceptions that climate change will happen "in a hundred years from now" in far-away places to other people (Leiserowitz). Leiserowitz argues that communicators should highlight those climate change impacts that can be linked to people's persistent concerns, such as human health or the future of their children. Cole and Watrous emphasize the importance of describing potential climate change impacts on places that rank highly in the public imagination. Ungar carries this argument further by suggesting that climate change has to be linked to culturally resonant symbols to gain wider traction. In some instances "making global climate change local" means not talking about climate change at all, but about related issues (e.g., air pollution, traffic congestion, energy, and, hence, cost savings) or embed it in larger concerns such as sustainability (Watrous and Fraley), which may find broader appeal.

While we have argued that the climate conversation has to move beyond science, science will not disappear, nor will scientists stop being important messengers. The challenge remains how to make global warming locally relevant and personally salient in scientifically credible ways. Indeed, this tension carries throughout this book and we expect it to persist for the foreseeable future. Here, Cole and Watrous, Warner, and McCright provide useful insights: contrary to most scientists' impulses, they suggest to lead with certainty, i.e., with those issues where the science is strongest (e.g., already documented changes, human contributions to climate change). Clearly, as Leiserowitz argues, uncertainty should never be hidden and scientific understanding never overstated. Such a strategy will only backfire.

Moreover, some frames through which we look at climate change make the issue of scientific uncertainty less important. For example, Bingham's and Agyeman *et al.*'s frames of climate change as a moral or justice issue don't require perfect understanding to see why caring for creation or defending social equity is important.

Effective communication also has to achieve a match between the audience and what it believes to be a credible messenger. For example, Cole and Watrous describe how California state policy-makers find local scientists more credible, accessible, and persuasive than "just any" scientist. Chess and Johnson, Leiserowitz, Young, Watrous and Fraley, and Tennis equally emphasize that whatever the message, it has to come from a trusted source.

Importantly, many of our contributors argue for broadening the range of "messengers." As climate change moves beyond being a purely "scientific" or "environmental" issue, industry leaders, political champions, religious leaders, and "PLUs" (people like us)[1] in other communities are joining the public discourse as equally legitimate communicators. Because messengers are part of the framing, the right speaker also suggests to the audience how to interpret climate change (Bostrom and Lashof).

Message reception is a frequently neglected element of the communication process. While the relative importance of the emotional aspects of communication may differ among audiences, to many, climate change is scary, overwhelming, and can quickly evoke dark images of the future, resulting in maladaptive responses to information (Moser). The remedies she and others (e.g., Tribbia, Chess and Johnson, Rabkin with Gershon, and Bingham) suggest include tapping into culturally resonant, positive, empowering values and personal aspirations such as innate goodness, responsibility toward others and the Earth, leadership, innovation, respect, caring, and stewardship (see also Schultz and Zelezny, 2003).

Our authors also argue that effective communicators need to be more solution- (rather than just problem-)oriented: giving people specific ideas of what to do, how to do it, creating a sense that these actions can effectively contribute to the solution, and, in fact, that individuals are part of larger efforts. They have to evoke hope, empower people, and help them envision an obtainable, desirable future.

A thread of thinking runs through this volume that is not at all revolutionary or new by itself but contrary to much past practice: the call for a shift from one-way "message delivery" to more engaging, dialogic forms of communicating (Regan, Harriss, Agyeman *et al.*, Rabkin with Gershon). Dialogue offers a forum in which value differences, conflicts, misunderstandings, and the communal visioning, and search for solutions can be addressed directly. Young, DuVair *et al.*, and Tennis also discuss how one-on-one and small-group communication in social networks is essential for knowledge exchange, resolution of differences, social learning, diffusion of innovation, persuasion, and mutual support.

Elevating and maintaining motivation for change

Social change-oriented communication, then, is a form of social capital building, and itself benefits from pre-existing social capital, which supports societal transformation at all levels.[2] Because this transformation is difficult and − certainly in the case of climate change − will require long-term engagement, the question of how to elevate and maintain motivation to implement the necessary changes becomes centrally important.

Almost all contributors to this book discuss a range of motivations that communicators and social change agents can use. While maybe not the most critical prompt, knowledge can be motivating when it helps actors understand the effective and appropriate level of change necessary, as Bostrom and Lashof, Leiserowitz, Grotzer and Lincoln, and Chess and Johnson argue.

At a more fundamental level, however, climate change communication must reach into deeper and more persistent beliefs, concerns, social norms, aspirations, and underlying values to generate motivation. Grotzer and Lincoln, and Bateson describe ways to instill these deeper values and beliefs in children through an education that prepares young minds for the changing and globally interconnected world they are inheriting. Leiserowitz and Moser suggest tapping into affect while others advise that deep and persistent motivations lie in people's social identities and relationships (e.g., Tribbia, Michaelis, Arroyo and Preston, Dilling and Farhar, Tennis, Meyer, Jamieson). Showing how an unsustainable behavior is contradictory to people's self-conception or how they want to be seen by others can motivate them to adjust their behaviors to create greater consistency.

The role of social influence and power in human behavior has long been recognized by social psychologists and others (e.g., French and Raven, 1959; Raven, 1993; Bagozzi and Lee, 2002) and provides another important reason to expand the range of communicators. Social influence is exerted in a variety of ways, including through expert knowledge, the ability to reward or coerce, by moral standing, or by reputation that confers legitimate power. Communicators will have different, or perhaps multiple, sources of influence and may thus be able to reach different audiences in a powerful way.

Of course, our actions and aspirations are never entirely consistent. Numerous social influences act on us at all times, resulting occasionally in outright contradictory actions or underlying values. For example, emitting massive amounts of greenhouse gases by jetting around the world to partake in important climate meetings is hard to reconcile with our long-term intention to leave our children a world safe from dangerous climate change.

To some audiences, the language of the bottom line and the risk of financial loss can be strongly motivating. Hence Atcheson's urging to build these economic signals into "doing business." Several of our contributors describe how this is already occurring – at least among the pioneers: Arroyo and Preston, James *et al.*, Dilling and Farhar, Young, and Watrous and Fraley discuss how taking early action can be economically beneficial for businesses and municipalities, giving them a market advantage, protecting their shareholders from climate-related risks, or simply saving money. Similarly, leading US states have found ways to save money by reducing emissions, and thus illustrate that "it can be done" without hurting businesses (duVair *et al.* and Tennis).

Economic incentives can easily translate into political and legal incentives for politicians and lawmakers. Many view "being a leader" as politically advantageous vis-à-vis their various constituencies and their political ambitions (e.g., California Governor Schwarzenegger as described by duVair *et al.*). The political calculus, however, is quite vulnerable to changing circumstance, and may thus be a less reliable motivation (Tennis). Averill points to threats from litigation that can exert a strong motivation to engage constructively on climate change issues, while Dilling and Farhar point to other legal mechanisms (standards, mandates) to act in "climate-friendly" ways.

For mid-latitudinal audiences, all these motivations serve to connect climate change to the individual, making it more salient, more personally relevant and in some cases more visible and urgent. In Alaska – as McNeeley and Huntington show – climate change itself has become the unequivocal motivation as impacts already directly affect people's lives. Their direct experience, in fact, is motivating Alaska Natives to alter the conversation and focus of social action.

None of the motivations discussed above unleashes its full power until it becomes integrated into a common problem understanding and a common vision of a desirable future. Such visions play a critical role in maintaining people's engagement, especially when "the going gets tough," when people encounter set-backs, or simply cannot see immediate results from their efforts. There is good reason to believe that even as we make great strides and good progress in reducing our emissions, global atmospheric GHG concentrations and temperatures will increase before they stabilize, much less decline. Volatile commitment to climate action is a reasonable scenario of societal response. It is thus all the more important that in small and large communities we develop visions of positive, desirable futures, and identify clear metrics of progress toward these goals, so that even if the full vision is not yet realized, people see a path for getting there and recognize themselves — as hope theory (Moser) suggests — as being on the way.

Overcoming barriers

It would be easy to suggest that barriers are just the opposite of all the things that could motivate us. But they are not always mirror images, and in fact, neglecting the real-world obstacles in a communication campaign may cause many change efforts to die a silent death. Communication and social change campaigns need to explicitly consider and address these barriers to increase their chances of accomplishing their intended goals.

In an information-overloaded world, our filters have become very selective. People may simply block out yet another climate change story that follows an all-too-familiar pattern, uses the same old "talking heads," triggers the same old associations, or worse, says one more time (in pictures or words) that "the sky is falling." Thus, the task of climate change communication is to break through the sound barrier, have news value, create a "need to know" (as Dunwoody argues), and maybe even be odd or surprising (e.g., humorous, or using the creative arts). In short, the communication must give people a reason to pay attention and then sustain the listener's engagement.

People's mental models (Bostrom and Lashof) or "habits of thought" (Bateson) are highly resistant to change; they are the well-trodden paths in our brains that we use to try to understand even unfamiliar issues. Just as it is critical to acknowledge and rectify these mental models, it is equally difficult to break behavioral habits and procedures. James *et al.* describe this situation for businesses and other organizations, and the same is true even in such pioneering governments as those in Santa Monica (Watrous and Fraley) or on the West and East Coasts (duVair *et al.* and Tennis). It is not just *beginning* a new behavior but also *stopping* an old action that involves social or financial costs (see James *et al.*, Arroyo and Preston, Young, Watrous and Fraley, Atcheson, and Tennis). The remedy these authors suggest is to understand more fully the hidden benefits of acting a certain way as well as those of changing to another, and then to address them explicitly: whether through training, financial incentive, verbal encouragement, or a different expectation from a supervisor. But ultimately, habits are changed with patience, creativity, down-to-earth practical thinking, flexibility, and stern commitment.

Commonly, as people begin to understand and emotionally relate to the risks of climate change, they want to do something about them. This reflects their desire to not just be part of the problem, but part of the solution (see, e.g., Moser, Dunwoody). People want to know what they can do, that they are able to do it, and that others are doing their share as well. The insights from this volume suggest that having solution information at the ready is just as important as, if not more important than, problem information to sustain an audience's active engagement, an often difficult task for scientists. The successful collaboration and division of labor between scientists and an advocacy group described by Cole with Watrous suggests one possible model for resolving scientists' need for maintaining their credibility while being responsive to audience needs.

Merely identifying possible actions, however, is inadequate for all but the most self-motivated, committed, and sophisticated audiences. For most individuals, what is required to actually implement these actions is empowerment through a sense of self-efficacy (Moser, Tribbia), social support or peer pressure (Chess and Johnson, Rabkin with Gershon, and Watrous and Fraley), or modeling by others (James *et al.*, Arroyo and Preston, and Young). In groups people help each other learn, offer assistance, but also produce accountability – all of which can overcome resistance and barriers. Finally, people can fail to implement possible actions if the lack of infrastructure or other logistics make it impossible (Tribbia, Dilling and Farhar).

It is equally important to recognize that some social norms and values may be antithetical to the innovative thinking necessary for climate-beneficial changes or to change in general. For example, we deeply value *stability* (social, political, climatological, even ecological), but we also value *growth* (particularly in economic terms). We tend *not* to teach our children to expect or how to effectively deal with constant change (Bateson).

Moreover, social and political institutions help stabilize a society, and, thus, by their nature are resistant to change. This makes political institutions in some ways less responsive to a constituency's changing political "mood" and may – in part – explain the slow response of US federal lawmakers to the increasing level of pressure from the states, the business community, and municipalities. Jamieson contends that the US electoral system (including such supporting systems as campaign financing and the media) makes it increasingly difficult to bring about social change through this pathway. Meyer, on the other hand, describes how political movements have succeeded in aligning bottom–up pressures to bring about political and policy change, and Averill shows how advocates use the legal system to do the same. Tennis and duVair *et al.* add to this debate that lower levels of government can serve as "laboratories" for political and institutional innovation as well as often forcing the hand of federal policy-making.

In summary, at the individual and group levels, many facts of social systems, including social norms and narrow interests, can hinder change, as can organizational culture and herd mentality (see James *et al.*). Communicators would do well to help people find higher common ground and identify ways to meet their diverse goals so as to help reduce or at least not increase their emissions impact.

Fostering social change

Writing about social change as it happens never provides more than a snapshot in time while the ground beneath us shifts. The last few years have seen an encouraging increase of diverse activities in many sectors of society, which defy the image of the United States (in particular) as a merely obstructionist, laggard country.

Much as Atcheson, James *et al.*, and Arroyo and Preston describe in their chapters, we now observe a growing number of players in the business community who take seriously the climate risks to their investment portfolios and business strategies. Many cities beyond those described by Young, Watrous and Fraley, Pratt and Rabkin, Rabkin with Gershon, and duVair *et al.* have committed themselves to meeting the Kyoto Protocol emission reduction goals.[3] A growing number of US states are also committing to GHG reductions and mandatory renewable portfolio standards – duVair *et al.*, Tennis, and Dilling and Farhar describe only the most visible ones. Several high-profile climate-change-related lawsuits, as Averill describes, are also exerting some political pressure, while conscious movement-building efforts (see Meyer), concomitant with the break-up of previously monolithic opinion blocks (e.g., the evangelical community) are under way (Moser, 2007). The fledgling political shifts and policy changes at the US federal level must be seen as the result of these political pressures from within and from international players. In fact, recent years have seen a remarkable increase in political engagement on climate change. The question, then, is whether there are any areas where it is most important to focus future communication and social change efforts. Our suggestions may frustrate some and encourage others.

Based on our observations of this burgeoning activity and recognizing the important impacts these activities have on other areas of society, one answer to the question of what matters most is simply to say: it all matters for the climb up the S-curve of societal transformation. There is *no one* key leverage point or obvious scale to focus all our attention ("if we only did X, the problem would be solved").

Instead, what the chapters help us understand is a critical cross- and across-scale story underlying social change. Activities at the small scale — the individual, local, and state levels — might be limited in actual emission reductions, but spread a symbolic message that engages others. These small-scale actions are critical, as Arroyo and Preston describe in industry or Young in municipalities, for the emergence and spread of innovation and social learning, and illustrating to the more risk-averse in society that there are "low-hanging fruit." These early and relatively easy "solutions" are ways to get a first commitment, on which bigger commitments can be built. They also build bottom—up political and economic pressures that begin to take off when heretofore separate social groups or interests form coalitions and thereby leverage their respective strengths for greater impact (duVair *et al.*, Tennis, Meyer). The small-scale actions and successes slowly change the political climate, which in turn enables larger policy and political changes. As we learned from the state chapters by duVair *et al.* and Tennis, policy changes at lower levels of government can also create direct or indirect pressure on higher levels of government to level the regulatory playing field. Other chapters tell a complementary top—down story to help "make it easy" for consumers to choose climate-friendly products and services (e.g., Dilling and Farhar, Tribbia, Atcheson, and Young).

Our contributors also help us understand that social change happens on a variety of temporal scales. The deeper the sought social change, the longer it will take to bring about. There is an unresolved tension running through our chapters regarding how deep societal changes in response to climate change need be, and we don't try to resolve it here. Resolving these tensions involves normative choices that point again to the need for broad public dialogue.

Those who feel the greatest sense of urgency may quickly dismiss many of the small beginnings described in this book. In fact, the changes described here are all of the incremental variety, and only Meyer and Jamieson even hint at the necessity for more radical changes. Yet an integral view of the chapters suggests that the small as much as the larger changes are needed. Quick, superficial changes may be necessary to ready a society for larger, deeper changes when the window of opportunity suddenly opens. None of us knows on which pathway societal transformation will eventually converge. Novel ideas, innovations, and momentum-building, rather than premature selection and channeling, are required at this stage. We also know that decisions that create long-term path dependencies will serve as hindrances to societal transformation.[4] Thus, any decisions or actions that lock us in, rather than open opportunities and maintain flexibility (ranging from the building of power plants to how we educate our children) should be assessed very carefully, conscious of their long-term impact on climate and society.

Outlook

The growing body of scientific (and popular scientific) literature on how societies throughout human history have fared in the face of emerging large-scale crises (e.g., Tainter, 1988; Diamond, 2005; Lovelock, 2006; Schellnhuber *et al.*, 2006) places today's encouraging societal trends as well as the discouragingly slow changes in an appropriate light. Whether or not any one civilization sustains itself over time appears to be linked to its ability to remain resilient in the face of environmental stresses such as climatic change, maintain good neighborly relations, and respond effectively and in time to signals of deterioration or threat from the environment. These, in turn, all deeply depend on social capital (e.g., Ostrom, 1990; Adger, 2003; Lehtonen, 2004; Pelling and High, 2005). This book gives ample reason to believe that we can further build momentum and plow the ground for even bigger shifts in how we communicate and act in the face of momentous social and environmental change. Thus, despite serious signs emerging from climate science, for us, this is *not* the time for hopelessness or despair, but a time for deep concern, creative engagement, and informed, committed and forward-looking action.

Conclusion: promising research directions and action steps

In preparation for writing this chapter, we asked our contributors to suggest critical research questions and action steps emanating from their chapters. Their diverse, helpful, and hopeful input suggested to us categories of research directions and practical steps that seem fruitful pursuits in the future.

Promising research directions

The challenge, need, and opportunity for greater multi- and interdisciplinary research on communication and social change

Multi- and interdisciplinary endeavors such as ours share one common challenge: some of the research questions we ask simply reflect that we lack certain insights that another discipline may already hold, and that would greatly advance our own if integrated. The benefits of mutual education and eventual integration are beyond measure. While in-depth disciplinary research will always be needed, more interdisciplinary research along the communication–social change continuum would greatly speed up our understanding.

The need for practice-oriented social science research, communicated effectively

If we have learned one thing during this project, it is that social science is not any better than physical science in sharing its insights with the practitioner community. Not surprisingly, then, several of the questions our practitioner contributors asked reflect what they need from social science and economics. In short, there is a great need for "use-inspired research" (Stokes, 1997) in the social sciences, and for social scientists to effectively communicate those insights to practitioners.

Maybe at the top of their list is economic information, in particular the cost of inaction. When making the case for why to spend money "up front," having specifics on the costs of not acting would help them be more persuasive in the language that communities, businesses, and even individuals speak. Other practitioners request related information that would quantify the co-benefits of taking action. Incentives for action on climate often overlap with other benefits to communities, such as reduced traffic, greater cost savings and efficiency, or better quality of life. These co-benefits are often of primary importance to motivating action on climate change (e.g., Young, Watrous and Fraley, duVair *et al.*). Communities do not generally have the resources or staff to analyze and compare the benefits and tradeoffs of various options for action. Thus, research of this sort would be useful for communities and actors on a wide variety of scales.

Comparative and cross-cultural studies on communication and social change

While this book is almost exclusively focused on the United States, several contributors rightly point to the need for cross-national, cross-cultural comparisons of communication and social change efforts, and of societal responses. The overarching intent, of course, is to determine which of the insights presented here hold across contexts, which are more context-specific.

Advancing our understanding of climate change communication

Some of the insights from our contributors converge on fairly consistent messages about communication for social change. Others have not yet been widely applied, and still others are brought to the climate change table from very different contexts; thus we cannot be confident that they apply and work. The next stage of research should test these claims rigorously. For example, what is the right balance of fear appeals versus positive messages, of emotional and rational/cognitive messaging? How much knowledge of climate change is necessary, how much detail is too much, and when does it become a hindrance to action?

We also need rigorous testing in pilot campaigns of the proposed mental models and different frames (including the messengers who might deliver them). Which work best for which audiences? Are there any models and frames that work for all audiences? Do the insights from health psychology hold for climate contexts? Can we say more about the relative impact of different communication channels, including the dialogue format, on understanding, motivation, and behavior change?

Parallel to the cross-national, cross-cultural comparison mentioned above, several contributors suggest a cross-social group comparison on all of these issues. Is belonging to a particular social or cultural group or "interpretive community" more important than standard demographic variables such as gender, age, ethnicity, or economic status?

Finally, our contributors identified the need for a range of appropriate and meaningful metrics by which to judge the efficacy of our communication efforts. With an increasing emphasis on accountability in both government and business, evaluating programs' effectiveness becomes critically important. Moreover, metrics of the success of communication must be linked to the social changes they are meant to motivate.

Advancing our understanding of motivations

Advances in understanding people's motivations will come from rigorous testing of the diverse and differentiated insights generated by our contributors. For example, what combination of factors motivates and sustains behavior change in different social groups, cultural cohorts, interpretive communities, or even specific neighborhoods and localities? The psychological landscape of individuals' responses to the climate change problem has barely been explored and promises fundamental new insight. Our contributors raised another intriguing question, namely "what does it take to make new behaviors 'stick'?" given the long-term commitment necessary for climate protection? Furthermore, are there any patterns as to who the "influentials" are (for different groups in society)? This would help us identify them and what might be most "persuasive" to different audiences.

Advancing our understanding of resistance and barriers to climate action

Practice-oriented research identifying barriers in specific situations is much needed. For example, cities are interested in knowing which mitigation strategies are most important at their scale, given limited resources and limitations imposed from higher levels of government or regional development trends. They need help in tackling these obstacles. They also want to understand how much of the populace must be educated and engaged to move forward with climate change.

Another set of questions raised focuses on patterns of barriers in social change: are there specific barriers that are especially important at different stages of social transformation, and what support is needed to overcome them? How do we build capacity to think about climate change, to communicate it, to empower people, to facilitate change? If fear appeals or traditional outreach campaigns are limited in their effectiveness, which of the suggested alternatives are better in "getting through to people"? Finally, while numerous scientific efforts are under way to develop metrics of progress toward sustainability, most of these have not reached the practitioner community. Metrics are needed that are meaningful to those who are making changes to provide periodic feedback and encouragement along the way and to serve as an evaluative yard stick to allow for mid-course corrections. As one of our contributors argued, "What gets measured gets done."

Useful action steps

What becomes implicitly apparent in our chapters is that the hallmark of pioneers is their political courage, creativity, staying power, and resourcefulness. Moving forward strategically and effectively will continue to require just that. In support of social change agents and communicators, our authors offered the following suggestions.

Clearinghouses and active networks for sharing insights, strategies, challenges, and successes

Our contributors agree that communication in formal and informal networks helps learning, spreading of innovations, and providing mutual support.

The most frequent suggestion for future actions is thus a call for more sharing of insights and resources. "Tell me what worked in your situation!" This was also the greatest benefit to those who participated in the workshop that started this project. Many existing networks already serve this function, or could be connected, but more diverse and unusual encounters will have to be created. Perhaps one such idea would be to develop climate sister city programs (parallel to other sister city programs already widely implemented). They would help raise awareness of climate change impacts among the sister city residents, stimulate discussion, foster amicable bonds across the planet, and serve other educational purposes.

Training communicators, expanding the messenger pool

Small efforts already under way in the United States to train individuals to communicate climate change effectively[5] need to expand significantly, beginning in elementary school and becoming embedded in professional training (e.g., for scientists, journalists, NGO staff, and outreach experts).

Many contributors, however, add that such training efforts should go beyond the typical communicators and messengers of the past. One, for example, suggested training "regular folks" to become "street communicators" in neighborhoods and communities which the traditional communicators just don't reach. Another idea is training "climate (or sustainability) ambassadors" who could carry ideas and solutions (including hard financial figures on what the actions cost and what they saved) from company to company.

Vastly expanding and testing diverse communication efforts

Governmental or non-governmental organizations in other countries are currently developing or launching major communication campaigns, for example in the UK, the Netherlands, and Australia. No such concerted effort is currently under way in the United States but would be desirable. Our contributors suggested not a single-focused traditional mass media campaign, but sophisticated efforts that would consist of various elements, including campaigns focused on health, creating a "brand," or a solution-oriented "action step" campaign. Any campaign would need to carefully test, evaluate, and adjust these different approaches and associated messengers, and then use them extensively over time.

If our ultimate goal is to reach the widest group of people, especially the younger generation, we will need diverse, innovative, and more accessible communication approaches (e.g., various forms of popular, street, and higher art, and the new media). This would also include creating numerous small forums for dialogue and begin the visioning that will be needed to keep going over the long term.

Working across the boundaries of social divides

The most interesting developments in terms of communication and social engagement at present are those that cross common societal divides. Indeed, these "boundary-crossing" efforts spread the word to groups heretofore untouched by finding common ground and common language; they enlarge the engaged population, and allow change agents to tap into a richer set of assets and resources. Through broadening the constituency, they also increase political pressure. Thus, strategically focusing efforts on crossing these divides, consciously reaching out to the "odd bedfellows," will help moving us along the S-curve.

Attending to the need for deeper social changes

Finally, there is a temptation to focus on the smaller easier changes in part because they are useful ways to get a foot in the door with the more skeptical population. Several of our contributors suggest, however, that it is important not to neglect the larger, deeper, more difficult changes, such as our political, electoral, and education systems. Sometimes the most ambitious goals are excellent motivators (e.g., "Let's be the first to put a man on the moon"). For example, under the pressures of attaining educational standards, maintaining funding, and the day-to-day challenges of schooling, many educators and policy-makers have yet to realize how critically important changes in this area might be to adequately prepare young people for the challenges of a world under the pressures of climate change.

At our workshop, Bob Kates reminded us of the old adage that "perfection is the enemy of the good." As we contemplate how to move forward with the wealth of insights and the many more questions that this book holds, we couldn't agree more. The issue of climate change is too important to wait for perfection in how we communicate global warming and facilitate effective societal change to the problem. But there is ample room for improvement and we hope that this book offers some useful steps in that direction.

Notes

1. See Agyeman *et al.* for more discussion on this issue. In some cases, communities trust people who are similar to themselves more than people who are different.
2. Our underlying understanding of social capital is captured well by Daniel, Schwier, and McCalla (2003), who define social capital as a "common social resource that facilitates information exchange, knowledge sharing, and knowledge construction

through continuous interaction, built on trust and maintained through shared understanding."

3. For more information, see http://www.ci.seattle.wa.us/mayor/climate/; accessed February 10, 2006.
4. In fact, many would argue that this dual approach is consistent with the precautionary principle, even if to some of its staunchest defenders precaution in the face of potentially dangerous climate change might require far bigger effort (e.g., Harremoës *et al.*, 2002).
5. See, for example, http://www.greenhousenet.org; accessed March 5, 2006.

References

Adger, W. N. (2003). Social capital, collective action and adaptation to climate change. *Economic Geography*, **79**, 4, 387–404.

Bagozzi, R. P. and Lee, K.-H. (2002). Multiple routes for social influence: The role of compliance, internalization, and social identity. *Social Psychology Quarterly*, **65**, 3, 226–47.

Bak, H.-J. (2001). Education and public attitudes toward science: Implications for the "deficit model" of education and support for science and technology. *Social Science Quarterly*, **82**, 4, 779–95.

Bamberg, S. (2003). How does environmental concern influence specific environmentally related behaviors? A new answer to an old question. *Journal of Environmental Psychology*, **23**, 21–32.

Blake, J. (1999). Overcoming the "value–action gap" in environmental policy: Tensions between national policy and local experience. *Local Environment*, **4**, 3, 257–78.

Clark, W. C., Mitchell, R. B., and Cash, D. W. (2006). Evaluating the influence of global environmental assessments. In *Global Environmental Assessments: Information and Influence*, eds. Mitchell, R. B., Clark, W. C., Cash, D. W., and Dickson, N. Cambridge, MA: The MIT Press, 1–28.

Daniel, B., Schwier, R. A., and McCalla, G. (2003). Social capital in virtual learning communities and distributed communities of practice. *Canadian Journal of Learning and Technology*, **29**, 3. Available at: http://www.cjlt.ca/content/vol29.3/cjlt29-3_art7.html; accessed February 15, 2006.

Diamond, J. (2005). *Collapse: How Societies Choose to Fail or Succeed*. New York: Penguin.

Field, C. B., Raupach, M. R., and Victoria, R. (2004). The global carbon cycle: Integrating humans, climate and the natural world. In *The Global Carbon Cycle: Integrating Humans, Climate and the Natural World*, eds. Field, C. B. and Raupach, M. R., SCOPE Report #62. Washington, DC: Island Press, pp. 1–13.

French Jr., J. R. P. and Raven, B. H. (1959). The bases of social power. In *Studies in Social Power*, ed. Cartwright, D. Ann Arbor, MI: University of Michigan Press, pp. 150–67.

Gladwell, M. (2000). *The Tipping Point: How Little Things Can Make a Big Difference*. Boston and New York: Little, Brown and Co.

Gunderson, L. H. and Holling, C. S. (eds.) (2001). *Panarchy: Understanding Transformations in Systems of Humans and Nature*. Washington, DC: Island Press.

Gunderson, L. H., Holling, C. S., and Light, S. S. (eds.) (1995). *Barriers and Bridges to the Renewal of Ecosystems and Institutions*. New York: Columbia University Press.

Harremoës, P., Gee, D., MacGarvin, M., Stirling, A., Keys, J., Wynne, B., and Guedes Vaz, S. (eds.) (2002). *The Precautionary Principle in the 20th Century: Late Lessons from Early Warnings*. London: Earthscan.

Jenkins-Smith, H. C. and Sabatier, P. A. (1999). The advocacy coalition framework: An assessment. In *Theories of the Policy Process*, ed. Sabatier, P. A. Boulder, CO: Westview, pp. 117–66.

Kahneman, D. (2003). Maps of bounded rationality: Psychology for behavioral economics. *The American Economic Review*, **93**, 5, 1449–75.

Kates, R. W., Clark, W. C., Corell, R., *et al.* (2001). Sustainability science. *Science*, **292**, 641–2.

Kemp, R. and Loorbach, D. (2003). Governance for sustainability through transition management. Paper presented at the Open Meeting of the Human Dimensions of Global Environmental Change Research Community, Montreal, October 16–18.

Kemp, R. and Rotmans, J. (2004). *Transitions Toward Sustainability*. London, UK: Edward Elgar.

Lehtonen, M. (2004). The environmental–social interface of sustainable development: Capabilities, social capital, institutions. *Ecological Economics*, **49**, 2, 199–214.

Lovelock, J. (2006). *The Revenge of Gaia: Why the Earth Is Fighting Back – and How We Can Still Save Humanity*. London: Allen Lane.

Michaelis, L. (2003). Sustainable consumption and greenhouse gas mitigation. *Climate Policy*, **3**, S135–46.

Moser, S. C. (2005). Impacts assessments and policy responses to sea-level rise in three U.S. states: An exploration of human dimension uncertainties. *Global Environmental Change*, **15**, 353–69.

Moser, S. C. (2007). In the long shadows of inaction: The quiet building of a climate protection movement in the United States. *Global Environmental Politics*, accepted for publication.

Moyer, B. with McAllister, J., Finley, M. L., and Soifer, S. (2001). *Doing Democracy: The MAP Model for Organizing Social Movements*. Gabriola Island, BC: New Society Publishers.

National Research Council (NRC). (1999). *Our Common Journey: A Transition Toward Sustainability*. Washington, DC: National Academy Press.

Ostrom, E. (1990). *Governing the Commons: The Evolution of Institutions for Collective Action*. Cambridge, UK: Cambridge University Press.

Pelling, M. and High, C. (2005). Understanding adaptation: What can social capital offer assessments of adaptive capacity? *Global Environmental Change*, **15**, 4, 308–19.

Raskin, P., Banuri, T., Gallopin, G. C., *et al.* (2002). *Great Transition: The Promise and Lure of Times Ahead*. A Report of the Global Scenario Group. Stockholm Environment Institute, Stockholm, Sweden.

Raven, B. H. (1993). The bases of power: Origins and recent developments. *Journal of Social Issues*, **49**, 4, 227–51.

Rogers, E. M. (2003). *Diffusion of Innovations*. New York: Free Press.

Rotmans, J., Kemp, R., and van Asselt, M. (2001). More evolution than revolution: Transition management in public policy. *Foresight*, **3**, 15–31.

Schellnhuber, H. J., Cramer, W., Nakicenovic, N., Wigley, T., and Yohe, G. (2006). *Avoiding Dangerous Climate Change*. Cambridge, UK: Cambridge University Press. Available at: http://www.defra.gov.uk/environment/climatechange/internat/pdf/avoid-dangercc.pdf.

Schultz, P. W. (2002). Knowledge, information, and household recycling: Examining the knowledge-deficit model of behavior change. In *New Tools for Environmental Protection: Education, Information, and Voluntary Measures*, eds. Dietz, T. and Stern, P. C. Washington, DC: National Academy Press, pp. 67–82.

Schultz, P. W. and Zelezny, L. (2003). Reframing environmental messages to be congruent with American values. *Research in Human Ecology*, **10**, 2, 126–36.

Smit, B., *et al.* (2001). Adaptation to climate change in the context of sustainable development and equity. In *Climate Change 2001: Impacts, Adaptation and Vulnerability*, ed. IPCC Working Group II. Cambridge, UK: Cambridge University Press, pp. 877–912.

Speth, J. G. (1992). The transition to a sustainable society. *Proceeding of the National Academy of Sciences*, **89**, 870–2.

Stokes, D. E. (1997). *Pasteur's Quadrant: Basic Science and Technological Innovation.* Washington, DC: Brookings Institution Press.

Sturgis, P. and Allum, N. (2004). Science in society: Re-evaluating the deficit model of public attitudes. *Public Understanding of Science*, **13**, 55–74.

Tainter, J. A. (1988). *The Collapse of Complex Societies.* New Studies in Archeology, Cambridge, UK: Cambridge University Press.

Toth, F. L., *et al.* (2001). Decision-making frameworks. In *Climate Change 2001: Mitigation*, ed. IPCC Working Group III, Cambridge, UK: Cambridge University Press, pp. 601–88.

Yohe, G. (2001). Mitigative capacity – the mirror image of adaptive capacity on the emissions side. *Climatic Change*, **49**, 247–62.

About the authors

Julian Agyeman is Associate Professor of Environmental Policy and Planning at Tufts University, Boston—Medford. His research interests are in the nexus between environmental justice and sustainability. He is co-founder and co-editor of the international journal *Local Environment: The International Journal of Justice and Sustainability*. His books include *Local Environmental Policies and Strategies* (Longman, 1994), *Just Sustainabilities: Development in an Unequal World* (The MIT Press, 2003), *Sustainable Communities and the Challenge of Environmental Justice* (NYU Press, 2005) and *The New Countryside? Ethnicity, Nation and Exclusion in Contemporary Rural Britain* (The Policy Press, 2006). He is a Fellow of the UK Royal Society of the Arts, a member of the US National Academies' Board on the Transportation of Nuclear Waste and is on the editorial boards of *Sustainability: Science, Practice and Policy*, *The Journal of Environmental Education*, and the *Australian Journal of Environmental Education*.

Susan Anderson is Director of the City of Portland Office of Sustainable Development (OSD) – a municipal agency working to ensure the environmental and economic health and prosperity of Portland's neighborhoods and businesses. OSD is responsible for city-wide solid waste collection and recycling, energy conservation, renewable energy resources, sustainable construction practices, electric and natural gas utility regulatory issues and a variety of other environmental programs. OSD is the lead agency for implementing Portland's Local Action Plan on Global Warming – A local plan to reduce greenhouse gas emissions 10 percent by 2010. Over the years, Susan has worked with more than 30 communities to promote resource efficiency, the use of renewable resources and sustainable practices in commercial facilities, housing, transportation, land-use planning and economic development. Prior to her work with the City of Portland, Susan was

Director of an environmental consulting firm. She also held positions with the Oregon Department of Energy, was an environmental land-use planner and a public relations professional. She holds undergraduate and advanced degrees in economics, environmental science, and urban and regional planning.

Vicki Arroyo is Director of Policy Analysis for the Pew Center on Global Climate Change, a position she has held since July 1998. She oversees analysis of science and environmental impacts, economics, and domestic policy issues. She also served as Managing Editor of the Center's book – *Climate Change: Science, Strategies, and Solutions*. Previously, she practiced environmental law with Kilpatrick Stockton and served in US EPA's Office of Air and Radiation and Office of Research and Development. From 1988 to 1991, she directed the Louisiana Department of Environmental Quality's policy office, and briefly served as Governor Buddy Roemer's environmental advisor. She has also worked at the local level on environmental issues in Massachusetts, West Virginia, and Louisiana. She holds a BS in biology (double major in philosophy) *magna cum laude* from Emory University; a Masters of Public Administration from Harvard University (top honors in program), and a J.D. *magna cum laude*, from Georgetown University Law Center, where she was Editor-in-Chief of *The Georgetown International Environmental Law Review*.

John Atcheson has over 30 years of experience in environmental policy with private industry, government, and non-governmental organizations. He helped establish the US Environmental Protection Agency's Office of Pollution Prevention, and directed the Office of Policy and Planning at the Department of Energy's Office of Energy Efficiency and Renewable Energy. He now serves as an Energy Technology Program Specialist in the Weatherization and Intergovernmental Program of the DOE's Energy Efficiency and Renewable Energy Office. He was also the first visiting Fellow at the RAND Corporation's Pardee Center for Longer Range Global Policy and the Future Human Condition. In addition, he served as an adjunct professor at Vermont Law School's Environmental Law Center. He has been responsible for major innovations in national environmental and energy policy, and he has written extensively on these topics. John received his degree in geology from the University of Maryland.

Marilyn Averill is a doctoral student in environmental studies at the University of Colorado at Boulder. Before returning to graduate school, Marilyn was an attorney with the Office of the Solicitor, United States

Department of the Interior, where she provided legal advice to the US Fish and Wildlife Service and the National Park Service. She holds Master's degrees in Public Administration from the Kennedy School of Government at Harvard University, in Educational Research and Evaluation Methodology from the University of Colorado, and a law degree from the University of Colorado. Marilyn's current research interests focus on international environmental governance, the politics of science, and science and technology policy, particularly in the context of global climate change. Her recent work involves the use of science and the treatment of uncertainty in climate litigation, and the effects these cases may have on law, science, and policy.

Mary Catherine Bateson is a cultural anthropologist with a Ph.D. from Harvard University. She has held teaching, administrative, or research positions at seven US universities and in the Philippines and Iran, and is now Professor Emeritus at George Mason University in Virginia, and President of the Institute for Intercultural Studies in New York. Her recent research has dealt with individual and societal adaptation to change and informal learning in adulthood. She has authored or co-authored ten books of which the most relevant to issues of climate change are: *Our Own Metaphor: A Personal Account of a Conference on the Effects of Conscious Purpose on Human Adaptation* (1972; Hampton Press, 2004); *Peripheral Visions: Learning Along the Way* (New York: HarperCollins, 1994); *Willing to Learn: Passages of Personal Discovery* (Hanover, NH: Steerforth Press, 2004); and, with Gregory Bateson, *Angels Fear: Towards an Epistemology of the Sacred* (1987; Hampton Press, 2004).

Sally Grover Bingham is the founder and executive director of The Regeneration Project, a national ministry dedicated to deepening the connection between ecology and faith. The Regeneration Project's (TRP) primary focus is the Interfaith Power and Light Campaign, "a religious response to global warming." This campaign helps congregations serve as models for cutting greenhouse gas emissions. Sally is an Episcopal priest working as the Environmental Minister at Grace Cathedral in San Francisco. She serves on the national board of Environmental Defense and the advisory board of Union of Concerned Scientist. In July 2001 Sally received the Green Leadership Award from the Center for Resource Solutions, the US EPA and the US Department of Energy. World Wildlife Fund recognized TRP as a Sacred Gift to the Planet. The Global Energy Award was presented to Sally by Mikhail Gorbachev in 2002. In July 2003, Sally was honored by the Bay Area Air Quality Board as a "Clean Air Champion." She stood with

Governor Schwarzenegger when he signed the GHG Reduction Bill in 2005 and was included in the Clinton Global Initiative.

Ann Bostrom is an Associate Professor of Public Policy and Ivan Allen College Associate Dean for Research at the Georgia Institute of Technology in Atlanta. She holds a Ph.D. in public policy analysis from Carnegie Mellon University. She completed postdoctoral studies in engineering and public policy at Carnegie Mellon and in cognitive aspects of survey methodology at the Bureau of Labor Statistics before joining Georgia Institute of Technology, where she has been a faculty member since 1992. In 1999–2001 she served as program director for the Decision Risk and Management Science Program at the National Science Foundation. Her research focuses on mental models of hazardous processes (how people understand and make decisions about risks), with application to risk communications. She co-authored *Risk Communication: A Mental Models Approach* (Cambridge University Press, 2001), with M. Granger Morgan, Baruch Fischhoff, and Cynthia J. Atman. Ann is also on the editorial boards of *Risk Analysis* and the *Journal of Risk Research*.

Caron Chess is Associate Professor in Rutgers University's Department of Human Ecology and founding director of Rutgers' Center for Environmental Communication. Caron's research explores evaluation of public participation and the study of the interaction between organizational factors and risk communication. Among her current research projects is a study of how interactions among organizations affected communication about anthrax during 2001. She currently serves on the National Academy of Science's Committee on Public Participation and Environmental Decision-making and has served on the NAS Committee on Risk Characterization, as well as others. She has also been president of the Society for Risk Analysis. In addition to peer-reviewed publications in academic journals, she has written guidance materials that are used widely by government and industry practitioners. Prior to joining academia in 1987, Caron held positions in government and non-profit environmental advocacy organizations.

Nancy Cole is the Deputy Director of the Union of Concerned Scientists' Global Environment Program. She directs UCS's Sound Science Initiative — a special project designed to help scientists present accurate, credible information about global warming and other issues to policy-makers and the media. A grassroots-organizing veteran, Nancy now works with scientists across the country to bring the voice of the scientific community to bear on critical global environmental issues. She joined UCS in 1992 as campaign organizer for a national drive to educate citizens about renewable energy options. Nancy subsequently became director of public outreach with overall responsibility for planning and developing UCS's education, organizing, and outreach efforts. In 1995, she co-authored the book *Renewables Are Ready: People Creating Renewable Energy Solutions* (Chelsea Green Publishing Co.). Before joining UCS, she worked for ten years for the national organization INFACT, a consumer-oriented, corporate accountability group, serving as its executive director from 1984 to 1992. She lives in Cambridge, MA with her spouse and daughter.

Lisa Dilling is a Visiting Fellow of the Cooperative Institute for Research in Environmental Sciences (CIRES) at the Center for Science and Technology Policy Research at the University of Colorado, Boulder. Lisa holds a Ph.D. in biological sciences from the University of California, Santa Barbara. Her career has spanned both research and practice arenas of the science–policy interface, including six years in Washington, DC as a program manager for the US National Oceanic and Atmospheric Administration managing research in the US Global Change Research Program, and three years researching the connection of carbon cycle science to policy, communication for climate change, and scales of decision-making. While in Washington she co-chaired an inter-agency group of six federal agencies working to better integrate carbon cycle science in the United States. Prior to her current position at the University of Colorado, she was at the Environmental and Societal Impacts Group (now the Institute for the Study of Society and Environment) at the National Center for Atmospheric Research. Her current research focuses on the use of information in decision-making and science policies related to climate and, in particular, the carbon cycle.

Bob Doppelt is Director of Resource Innovations, a sustainability and climate change research and technical assistance program in the Institute for a Sustainable Environment, and a Courtesy Associate Professor in the Department of Planning, Public Policy and Management at the University of Oregon. He is also a principal with Factor Ten Inc., a sustainability change-management consulting firm. Bob served on the Western Division of The President's Council on Sustainable Development and on the US EPA National Advisory Council for Environmental Policy and Technology. He is a graduate of the International Program on the Management of Sustainability, Ziest, The Netherlands. He is lead author of *Entering the Watershed* (Island Press, 1993) and author of *Leading Change toward Sustainability: A Change Management Guide for Business, Government, and Civil Society* (Greenleaf Publishing, UK, 2003) deemed just six months after its release as "one of the nine most important publications in sustainability" by a Globescan survey of international experts.

Sharon Dunwoody is Evjue-Bascom Professor of Journalism and Mass Communication and Associate Dean for Social Studies in the Graduate School at the University of Wisconsin—Madison. A former mass media science writer, she earned a Ph.D. in mass communication and now teaches science writing and conducts research on public understanding of science questions. Her current research includes a five-country survey of scientists' perceptions of the mass media and studies of the effects of media risk messages on general audiences. She has served as a member of the Communications Advisory Committee for The National Academies and is currently serving on the Committee on the Public Understanding of Science and Technology of the American Association for the Advancement of Science.

Pierre duVair has led the California Energy Commission's climate change policy program since February 2001. He evaluates greenhouse gas emissions accounting and strategies to address climate change. Pierre leads the Commission's work to provide guidance to the California Climate Action Registry. He also monitors climate policy developments at the local, state, national, and international levels. Prior to his employment at the Energy Commission, Pierre worked for eight years with the California Department of Fish and Game. At DFG, he led interdisciplinary assessments of natural resource damages that result from pollution incidents, managed a research program on damage assessment techniques, and evaluated federal and state legislation on natural resource impacts. He also teaches economics as a visiting lecturer at the University of California—Davis and California State University—Sacramento. Pierre received a bachelor's degree in biology and economics, a master's degree in economics, and a Ph.D. in environmental policy.

Barbara C. Farhar is a Senior Policy Analyst and Senior Social Scientist with the National Renewable Energy Laboratory, in Golden, Colorado and Adjunct Faculty at the University of Colorado—Boulder and the University of Colorado—Denver. She has been directing research on technology—society interactions and diffusion of innovations for more than 25 years. Barbara currently focuses on studies of zero-energy homes and pursues issues in gender and sustainable energy. Before that, she also focused on geothermal energy policy issues, market assessments for renewable energy and energy efficiency technologies, and manages research on perceptions and preferences on energy and the environment, markets for on-site generation, and evaluation research for home energy rating systems and energy efficiency financing. With more than 225 publications, Barbara holds a Ph.D. in sociology from the University of Colorado, where she also received her M.A. and B.A. *magna cum laude*. She is Phi Beta Kappa.

Natasha Fraley is a freelance content developer, science writer, and exhibition planner. She has an M.S. in biology from the University of Oregon. Natasha has written for websites and magazines mainly about marine science. She has been involved in the research, content planning, and writing of exhibitions for aquariums, visitor centers, museums, and science and technology centers since 1994. While working on a project for the Monterey Bay Aquarium, she was involved in many workshops and seminars focused on research about how to get people to change their attitude and behaviors around conservation issues. Natasha has been privileged to work on several projects with those goals. Awakening people to their connection to the planet is Natasha's challenge and what brings her joy.

David Gershon, founder and CEO of Empowerment Institute, is considered one of the world's leading authorities on behavior change and large-system transformation. David applies this expertise to cutting-edge community, organizational, and societal transformation issues. His clients include large cities and organizations wishing to create behavior change among their constituencies. The versatility of the empowerment tools enables them to address issues ranging from environmental behavior change to emergency preparedness; from low-income neighborhood revitalization to organizational culture change. Longitudinal research studies indicate that adopted behavior changes are sustained over time. David has authored eight books including the bestselling *Empowerment*, which has become a classic on the subject. Considered a master personal development trainer, whose work has received many honors, he co-leads the Empowerment Institute Certification Program − a school for transformative change leadership. He has lectured at Harvard and MIT and served as an advisor to the Clinton White House and United Nations.

Tina Grotzer is a principal investigator at Project Zero and an assistant professor at the Harvard Graduate School of Education. Her current research identifies ways in which understandings about the nature of causality impact students' ability to learn complex science concepts. She directs the Understandings of Consequence Project, funded by the National Science Foundation, which identifies default assumptions about the nature of causality that students bring to their learning. Her publications include an extensive review of the research on the development of complex causal understanding, recently published in *Studies in Science Education*. She has published extensively in both academic journals and teacher-oriented publications and is deeply committed to helping teachers acquire and use the knowledge gained through her research. Her career has bridged research and practice and she taught in public and private schools for 14 years. She holds a doctorate from Harvard.

Robert C. Harriss is President and CEO of the Houston Advanced Research Center located in The Woodlands, TX. He also holds adjunct positions as Professor in the Department of Planning and Design, College of Architecture and Planning, University of Colorado—Boulder and in the Department of Marine Sciences, Texas A&M University at Galveston. Bob is a principal scientist in The Institute for Oceans and Coasts. Previously Bob served as Director and Senior Scientist of the Institute for the Study of Society and Environment at NCAR, as Director of the Sustainable Enterprise Institute of the Texas Engineering Experiment Station. He held the Wiley Chair in Civil Engineering. Bob also was Director of the Science Division, Mission to Planet Earth, at NASA headquarters. His current research concerns integrative approaches to planning for disaster-resistant and resilient coastal communities and the design of serious games for learning about disaster dynamics.

Halida Hatic received her M.A. in Urban and Environmental Policy and Planning from Tufts University in May 2005. Her primary research interests are climate justice and sustainability. Her thesis analyzed state-level climate action plans for their consideration and inclusion of principles of equity and justice. She currently works in the New England office of the US Environmental Protection Agency as a transportation planner in the Air Quality Planning Unit, where she focuses on local and regional initiatives to reduce emissions from transportation and construction sources.

Orville H. Huntington is a Native Alaskan who grew up living and still lives a traditional Alaska Native subsistence way of life in the village of Huslia, Alaska. With a B.S. in wildlife biology (University of Alaska Fairbanks) and lifelong learning in traditional Native Alaskan Koyukon Athabascan Knowledge and Wisdom, he currently holds the position of a refuge information technician at Koyukuk/Nowitna National Wildlife Refuge Complex at Huslia, Alaska. Previously he worked with the Bureau of Land Management as an emergency fire fighter and as a self-employed commercial fisherman. In his native Huslia, Orville serves as a tribal and city council member and sits on several other villages' boards of directors, on the Brooks Range Contracting Board of Directors, and the Interior Regional Housing Board of Commissioners. He also chairs the Interior Athabascan Tribal College Board of Trustees and is the Vice-Chair of the Alaska Native Science Commission Board of Commissioners.

Keith James is Professor of Psychology at Portland State University. He received his Ph.D. in social psychology and organizational behavior from the University of Arizona. He has held previous positions at Columbia University, Colorado State University, the University of Alaska, York University, and the University of Calgary. His work focuses on Native American/Native Canadian community sustainability, creativity and innovation in organizations, and social and organizational influences on health. Keith has served on selection committees for Ford Foundation fellowships (administered by the National Academies of Sciences), as well as on grant review and program advisory panels for the Ford Foundation, the National Science Foundation, the National Institute of Occupational Safety and Health, the National Heart, Lung and Blood Institute, and the Social Science & Humanities Research Council (Canada). He has been a Fulbright Fellow and a Fulbright Distinguished Scholar and is a Sequoyah Fellow of the American Indian Science and Engineering Society.

Dale Jamieson is Professor of Environmental Studies and Philosophy, and Affiliated Professor of Law at New York University. He is also an adjunct scientist at the National Center for Atmospheric Research. Since receiving his Ph.D. in philosophy from the University of North Carolina in 1976, he has gone on to teach philosophy and environmental studies in universities around the world. He is the author of *Morality's Progress* (Oxford, 2002), and the editor of *A Companion to Environmental Philosophy* (Blackwell, 2001). He is also the author of many papers on climate change. Dale is currently writing an introduction to environmental ethics and co-authoring an environmental science textbook, in addition to researching various questions in ethics and the environment.

Branden Johnson is a research scientist in the Division of Science, Research and Technology, New Jersey Department of Environmental Protection. He holds a B.A. in environmental values and behavior (University of Hawaii), an M.A. in environmental affairs, and a Ph.D. in geography (Clark University). He was Associate Professor of Science, Technology, and Society (Michigan Technological University), and currently holds adjunct positions in human ecology, public health, and geography (Rutgers University). His research and practice has included risk perception, risk communication, public participation, and environmental policy, including executive direction of the New Jersey Comparative Risk Project, chairing the Risk Communication Specialty Group of the Society for Risk Analysis, and serving as a communication expert on US National Research Council committees. Recent research topics include the bases for trust in risk managers, the effect of information processing on risk decisions, organizational perspectives on risk communication, and informational support for environment-protective behavior.

Robert W. Kates is a geographer and independent scholar in Trenton, Maine, and University Professor (Emeritus) at Brown University. Having failed retirement, he is co-convenor of the international Initiative for Science and Technology for Sustainability, an executive editor of *Environment* magazine, an affiliate of Clark University, College of the Atlantic, and Harvard University, a board member of the Acadia Disposal District and Maine Global Climate Change, Inc., and a review editor for the third and fourth IPCC climate change assessment. He is a member of the US National Academy of Sciences, the American Academy of Arts and Sciences, and the Academia Europaea. His most recent books include *Great Transition: The Promise and Lure of the Times Ahead* (Stockholm Environment Institute, Boston, 2002), and with the AAG Global Change and Local Places Research Group, *Global Change in Local Places: Estimating, Understanding, and Reducing Greenhouse Gases* (Cambridge University Press, 2003).

Daniel A. Lashof is the science director and deputy director of NRDC's climate center. He is active in the areas of national energy policy, climate science, and solutions to global warming. Dan is involved in developing federal legislation to place enforceable limits on carbon dioxide and other heat-trapping pollutants and to reduce America's dangerous dependence on oil. He has followed international climate negotiations since their inception and is a lead author of the Special Report of the Intergovernmental Panel on Climate Change on the role of land-use change and forestry in exacerbating or mitigating global warming. Dan has testified many times before Congress about energy policy and global warming. He holds a bachelor's degree in physics and mathematics from Harvard University and a doctorate from the University of California–Berkeley. He has taught environmental science as an adjunct professor at the University of Maryland and is the author of numerous articles on climate change science and policy.

Anthony Leiserowitz is a research scientist at Decision Research and a principal investigator in the Center for Research on Environmental Decisions at Columbia University. He earned a Ph.D. in environmental science, studies, and policy (focal discipline: geography) from the University of Oregon. His research focuses on environmental risk perception, decision-making and behavior, the human dimensions of global environmental change, and sustainability science. He has conducted studies of public climate change risk perceptions, policy preferences and behavior at multiple scales including individual states (Alaska and Florida), the United States (five national surveys), and internationally (the UK, Germany, Japan, Mexico, Brazil, and Argentina). His work also examines the role of underlying cultural attitudes, values, and worldviews in human decision-making and behavior.

Rebecca Lincoln is a research assistant at Project Zero at the Harvard Graduate School of Education, and is pursuing an Sc.D. in environmental health at the Harvard School of Public Health. She has worked as an environmental educator and an environmental scientist, and has trained teachers in methods of applying the results of research from the Understandings of Consequence Project (http://pzweb.harvard.edu/ucp/) to classroom practice. She plans to work at the intersection of environmental science, public health, and education, to research connections between environmental damage and human health, and to find ways to effectively communicate the results to affected communities. She holds a B.A. from Oberlin College.

Kathy Lynn is Associate Director for Resource Innovations, an organization affiliated with the Institute for a Sustainable Environment at the University of Oregon. Since 2003, Kathy has focused on working with rural communities and Native American tribes in the Pacific Northwest to understand their risk to wildfire, ability to access fire-related programs, and build capacity to reduce wildfire risk. Kathy has a Master's degree in community and regional planning from the University of Oregon. She lived in Haiti from 1996 to 1999, working as a peace corps volunteer and on a community-based disaster preparedness program. Kathy has also worked with the Oregon Natural Hazards Workgroup at the University of Oregon's Community Service Center and with the Federal Emergency Management Agency as a liaison to Native American tribes.

Aaron M. McCright is an assistant professor in the Lyman Briggs School of Science, Department of Sociology, and Environmental Science and Policy

Program at Michigan State University. He holds a Ph.D. in sociology from Washington State University, where he began his specialty in environmental sociology and political sociology. His recent publications include: "To Die For: The Semiotic Seductive Power of the Tanned Body" (with Phillip Vannini) in *Symbolic Interaction* (2004); "Defeating Kyoto: The Conservative Movement's Impact on US Climate Change Policy" (with Riley E. Dunlap) in *Social Problems* (2003); "Politics and Environment in America: Partisan and Ideological Cleavages in Public Support for Environmentalism" (with Riley E. Dunlap and Chenyang Xiao) in *Environmental Politics* (2001); "Challenging Global Warming as a Social Problem: An Analysis of the Conservative Movement's Counter-Claims" (with Riley E. Dunlap) in *Social Problems* (2000). His current research focuses on the relationships among dimensions of power, scientific reflexivity, and environmental problems.

Shannon McNeeley is a graduate student fellow supported by the National Science Foundation IGERT (Integrative Graduate Education Research and Training) Resilience and Adaptation Program in the Anthropology Department at the University of Alaska, Fairbanks (UAF). She holds a Master's degree in international environmental policy from the Monterey Institute of International Studies. Prior to coming to UAF she worked at the National Center for Atmospheric Research in Boulder, Colorado as an associate scientist in the Institute for the Study of Society and Environment (ISSE), where she still holds a visiting scientist appointment. Her work focuses on research, capacity building, and education and outreach in the area of climate change impacts, vulnerability, and adaptation, particularly of minority populations and related issues of equity and social and environmental justice. Her dissertation is a community-based, collaborative project with Athabascan tribes in Interior Alaska who are coping with the impacts of warming and rapid environmental change.

David S. Meyer is Professor of Sociology, Political Science, and Planning, Policy, and Design, at the University of California—Irvine, where he is a Fellow of the Center for the Study of Democracy. His work is most directly concerned with the relationships between social movements and the political contexts in which they emerge. He's interested in why movements emerge when they do, and what influence they have on politics and public policy. He is author of *The Politics of Protest: Social Movements in America* (Oxford, 2006) *A Winter of Discontent: The Nuclear Freeze and American Politics* (Praeger, 1990) and co-editor of four books on social movements. He has also published dozens of articles and book chapters, mostly on social movements in advanced industrialized societies. He holds a Ph.D. in political science from Boston University, and a B.A. from Hampshire College, where he studied literature. Before coming to UCI, David taught at the City College of New York and the City University Graduate Center, the University of Michigan, and Tufts University.

Laurie Michaelis completed his Ph.D. in energy studies at the University of Cambridge, UK in 1986. He has worked since in education, research, and policy on energy, climate change, and other environmental issues. From 1992 to 1999 he was a policy analyst at the International Energy Agency and then at the Organisation for Economic Co-operation and Development in Paris. He was Convening Lead Author on transportation sector mitigation in the Second Assessment Report of the Intergovernmental Panel on Climate Change (IPCC), and was a lead author in other IPCC reports including the Special Report on Emission Scenarios published in 2000. From 1999 to 2002 he was Director of Research for an international research commission on sustainable consumption, based in Oxford University. He now works freelance, developing resources, and facilitating events for Quaker meetings and other groups in Britain, supporting them in developing collective approaches to sustainability.

Susanne C. Moser is a research scientist at the National Center for Atmospheric Research's (NCAR) Institute for the Study of Society and Environment in Boulder, Colorado. A geographer by training (Ph.D. from Clark University, 1997), her work over the past 15 years has focused on interdisciplinary challenges such as the impacts of climate change and sea-level rise on coastal areas, community and state responses to such global change hazards, the interaction between science and policy/practice, and the communication of climate change risks in support of societal responses to climate change. Susi was a post-doctoral fellow at Harvard's John F. Kennedy School of Government, and prior to coming to NCAR worked for the Heinz Center in Washington, DC and as staff scientist for climate change for the Union of Concerned Scientists in Cambridge, Massachusetts. In 2005 she was selected as a fellow of the Aldo Leopold Leadership and the UCAR Leadership programs.

Linda Giannelli Pratt is Chief of the Office of Environmental Protection and Sustainability with the City of San Diego Environmental Services Department. For more than 25 years, she has successfully built a professional career focused on community-based environmental protection. Her experience is broad, and includes positions as a laboratory analyst, regulatory compliance specialist, consultant, adjunct professor at the University of California–San Diego, and director of regional environmental programs. Her reports have been published in three books and many professional journals. Linda received a B.S. in microbiology and chemistry and an M.S. in environmental studies.

Benjamin Preston currently works as a research scientist in the Climate Change Impacts and Risk stream of the CSIRO Division of Marine and Atmospheric Research, where his research focus is the development of risk-based methods for climate change impact assessment, communication, and management. Prior to joining CSIRO, Ben was a senior research fellow with the Pew Center on Global Climate Change in Washington, DC, where he provided scientific advice to policy-makers, the media, and the general public on climate change science and impacts. His background also includes research experience in environmental toxicology, public health, and environmental justice. Ben received a B.S. from the College of William and Mary and a Ph.D. from the Georgia Institute of Technology.

Sarah Rabkin is an award-winning teacher of environmental writing and literature courses in the Department of Environmental Studies at the University of California–Santa Cruz. She also leads outdoor workshops on writing and keeping illustrated field journals for a variety of educational institutions, and freelances as a writer, editor, and writing coach. Her articles, essays, columns, poems, and reviews have appeared in dozens of regional and national publications, and her illustrated journal pages have been exhibited on gallery walls. She has an undergraduate degree in biology from Harvard and a graduate certificate in science communication from UC–Santa Cruz. Sarah loves language and landscapes and the potent places where the two come together; she takes pleasure in helping others find the power in their own voices.

Kathleen Regan is the Laboratory Coordinator for the Department of Civil and Environmental Engineering at Tufts University in Medford, MA. After 15 years of basic ecological research at the Ecosystems Center of the Marine Biological Laboratory in Woods Hole, MA, she returned to school for a bachelor's degree in philosophy. She then worked as Program Coordinator for the Public Conversations Project (PCP) in Watertown, MA, a non-profit that facilitates conversations among groups who are deeply divided over contentious public issues. There, she developed PCP's focus on the communication challenges to environment and climate change conversations.

Kathy's current interests are the relationship of environmental challenges to issues of social justice, the ways non-scientist citizens make sense of highly technical information in order to make political and social policy decisions, and the ethical challenges facing individuals and communities as they grapple with these questions.

Samuel Sadler is a senior analyst in the Renewable Energy Division of the Oregon Department of Energy. Since 1988 he has been the lead staff for developing climate change strategy for the Oregon Department of Energy. He is the Oregon coordinator for the West Coast Governors' Global Warming Initiative. He served as staff coordinator for the Governor's Advisory Group on Global Warming in 2004 and serves as lead staff for the Governor's Climate Change Integration Group, formed in early 2006. He is staff to the Governor's Carbon Allocation Task Force, which is designing a state-level carbon cap-and-trade system. Sam represented the Department of Energy in negotiating a carbon dioxide standard for new energy facilities, which the Oregon Legislature adopted in June 1997. He has been project officer for implementing the CO_2 standard for new power plants and for expanding the coverage of the CO_2 standard to other types of energy facilities. Sam also oversees the implementation of CO_2 offset projects by an energy facility site certificate holder.

April Smith is a doctoral student in the Industrial/Organizational Psychology program at Colorado State University, where she teaches a course in the psychology of creativity. She was a Gipson Scholar in psychology and creative writing at Albertson College, Caldwell, Idaho, and received the Rocky Mountain Psychological Association's Regional Research Award, as well as the Caxton book award for creative writing. April recently defended her master's thesis, entitled "Exploring the Structure of Organizational Creativity." She is currently working on a five-year grant for the Center to Protect Worker's Rights and the National Institute for Occupational Safety and Health, developing training programs to improve the psychological adoption of safety techniques among construction workers. Her research interests include creativity, innovation, organizational culture, safety, and training. April currently lives in Fort Collins, Colorado, where she enjoys outdoor sports and the sunny Front Range climate.

Abbey Tennis currently serves as Climate Change Program Administrator in the Massachusetts Office for Commonwealth Development (OCD). OCD was created to forge coordinated policies and programs between the state's energy, environment, housing, and transportation agencies. Headed by Secretary Doug Foy, the office has led sustainable development in the state, including focusing state investments around projects that promote transit-oriented development, affordable housing, greenfield protection, and brownfield redevelopment. OCD also oversees Massachusetts' climate protection policy, continuing Massachusetts' commitment to the goals of the 2001 New England Governors' and Eastern Canadian Premiers' Climate Change Action Plan, and providing direction for cleaner energy, transportation, government, and buildings in the state. Abbey co-authored Massachusetts' first Climate Protection Plan in 2004, and coordinates the Plan's implementation effort, as well as administers a $9 million partnership between the Massachusetts Technology Collaborative's Renewable Energy Trust and OCD. Abbey has a degree in environmental studies from Oberlin College.

John Tribbia is a research assistant working with Dr. Susanne Moser at the Institute for the Study of Society and Environment (ISSE) in the National Center for Atmospheric Research (NCAR), Boulder, Colorado. He earned a B.S. in Mathematics from the University of Puget Sound in Tacoma, Washington and is presently a Ph.D. student in the Department of Sociology at the University of Colorado—Boulder. His academic interests within environmental sociology include the study of human—environment interactions. John also works at a locally owned bike shop named University Bicycles in his hometown — Boulder. He likes to consider himself an "Alternative Transportation Retailer," as he sells bikes and other cycling related retail. John is an avid commuter by bike, bus, and foot. The smell of gasoline gives him a headache.

Sheldon Ungar is a Professor of Sociology at the University of Toronto at Scarborough. He received his B.S. in sociology from the University of McGill before attending York University, where he was awarded his Master's degree and his Doctorate. He taught briefly at York while he was completing his Ph.D. and joined the faculty of the University of Toronto shortly afterwards. Shelly has researched real-world events that have produced social anxiety, including the nuclear arms race, emerging diseases, and global climate change. Beside his continued interest in the media and climate change, his current work also focuses on an examination of knowledge and ignorance and seeks to identify how popular culture and new media affect "cultural literacy." As part of his study of the cultural production of ignorance, he has examined instances of the "silencing of science."

Anthony Usibelli is the Director of the Energy Policy Division of the Washington State Department of Community, Trade, and Economic Development (CTED). Prior to joining CTED, he worked on energy codes with the Washington State University Energy Program, managed the energy efficiency section at the Washington State Energy Office, and was a staff scientist at the Lawrence Berkeley National Laboratory. Tony has also been an adjunct faculty member at the Evergreen State College where he has taught or co-taught courses in energy, natural resources, and global climate change. Tony holds an M.S. in energy and resources from the University of California−Berkeley and B.A. in geography and classical archaeology.

Lucy Warner is Director of Communications for the University Corporation for Atmospheric Research, which includes the National Center for Atmospheric Research. In this capacity she oversees media relations as well as a variety of publications and websites for the public and the broad community of researchers and policy-makers. She has worked in the field of climate communications for over 20 years. Her program has received numerous awards for its publications, videos, and media products designed to explain weather and climate issues to a broad audience. Prior to joining the field of climate communications, Lucy worked in New York and London for a variety of book and magazine publishers as both a writer and editor.

Susan Watrous is a writer, editor, and writing coach whose work has appeared in national and regional magazines and trade publications. After a stint as a technical writer during Silicon Valley's boom years, she returned to work that is closer to home and to her heart. Currently, Susan teaches writing and journalism at her *alma mater*, the University of California−Santa Cruz, and works as the Print Advisor for student publications there. Formerly editor of a scuba diving/travel magazine, she served as project editor for a book on diving from *Discovery Communications* and *Insight Guides*, and edited *Cities Under the Sea: Coral Reefs*, an educational CD-ROM for children (8−14) produced by Jean-Michel Cousteau. She was the associate editor for the first edition of *The Ultimate Adventure Sourcebook*, and has provided independent research assistance to Professor Anna Tsing at UCSC on environmental politics in Indonesia and non-timber forest products in the United States.

Abby Young is the Director of Strategic Planning for the International Council for Local Environmental Initiatives (ICLEI) − Local Governments for Sustainability. She guides the short- and long-term visioning and development planning for ICLEI's US organization. Prior to her position as Director of Strategic Planning, Abby served for over nine years as Director of ICLEI's US Cities for Climate Protection™ (CCP) Campaign, a program that engages now over 200 US cities and counties in measuring and reducing greenhouse gas emissions from their communities. As the US CCP Director, she oversaw the various programs and projects of the Campaign, including development of training programs and materials, workshops and research to assist local governments in achieving their climate protection goals. Abby has a B.A. in political science from the University of California–Santa Barbara and a Master's degree in International Energy and Environmental Policy from the Johns Hopkins School for Advanced International Studies.

Index

congregation, 153, 160
Congressional Black Caucus Foundation, 121
Connecticut, state of, 388, 419, 421, 466
connectors, 247, 248, 406 (*see also* tipping point)
connotation, 44, 49
consensus, 208
conspiracy theory, 52, 59
consumer
    consumer choice, 328, 329, 331, 354, 407, 452,
        454, 458, 459
consumerism, 256, 258
consumption, 7, 115, 220, 252–4, 309, 375,
    376, 480
    consumer behavior, 468
    consumption behavior, 251–63, 376
    consumption culture, 252, 254–6
    consumption style, 252, 254
    culture of consumption, 252, 263
    ethics of consumption, 252, 259
    patterns of consumption, 249, 251–5, 452, 488
contagion, 307–8
control, sense of, 38, 40, 66, 73
controversy, 97, 100, 210, 219
conversation style, x
conversation, ix, xi, xii, 144, 145, 147, 150, 213,
    215, 286, 288, 500
coping capacity, 3, 7, 20, 121, 128, 136
    (*see also* adaptation)
coping with climate change (impacts), 140, 142–5,
    149 (*see also* adaptation)
coping with climate change, 475
corporate action on climate change, 14, 18
Corporate Automotive Fuel Economy (CAFE),
    372, 375, 376 (*see also* stakeholders)
corporate responsibility, 332, 333, 345, 346, 352–3,
    462, 463
Correll, Robert, 55
cost, 70, 97, 99, 129, 132, 323, 341–2, 344, 347, 470,
    487, 509 (*see also* finances)
cost savings, 106, 107, 307, 314, 323–8, 342, 349,
    350, 352, 353, 362, 372, 374, 378, 385, 387,
    390, 403, 408, 411, 427, 444, 446, 503
    (*see also* finances)
cost-benefit analysis, 354, 481
cost-benefit, 412, 413
cost-effectiveness, 336, 349, 387, 390
Council of Athabascan Tribal Governments, 144,
    146, 149, 150
counter movement, 8, 184, 200, 201, 203, 204,
    206–8, 492 (*see also* social movements)
creation, stewardship of, 91, 155, 161, 500
credentials, 202, 206
credibility, 17, 56, 59, 134, 153, 157, 202, 207, 224,
    225, 227, 228, 230, 248, 335, 386, 406, 409,
    456, 462, 469, 488, 500, 501, 505
    (*see also* trust)
    scientific credibility, 85, 144, 168, 185, 190, 191,
        193, 196, 456, 457
credit (for climate actions), 329
creeping environmental problems, 173, 270

crisis, xiii, 307, 310–12 508
criticize/defend, 215 (*see also* resistance,
    persuasion)
cultural innovator, 260
cultural norms, 66, 256, 259, 263
Cultural Theory (CT), 254–5, 257, 258–60, 263
    egalitarian, 255, 257–9, 260, 262, 263
    fatalist, 255, 258, 259, 263
    hermit, 255, 258, 259, 263
    hierarchist, 255–7, 259, 261–3, 263
    individualist, 255, 256, 258, 259, 262, 263
cultural whirlwind, 10, 84, 85
cultural/culture change, x, 101, 256, 258–9, 306,
    377
culture, 10, 18, 130, 136, 150, 175, 214–16 225, 252,
    258, 259, 284, 285, 399, 500, 506, 510
    American culture, 207, 208, 489
    cultural norm, 228
    culture of consumption, 255, 256
    dominant culture, 10, 142, 146, 258, 284
    indigenous culture, 142, 149
    issue culture, 17
    organizational culture, 304–6, 313, 377
    political culture, 479
    scientific culture, 173–5

damages (from climate change), 341, 354, 447, 462,
    464, 466
danger control, 69 (*see also* fear control)
danger, 66
Davis, Gray, Governor, 440
debate, x, 7, 9, 16, 21, 85, 87, 146, 168, 173, 178,
    203, 207, 213, 214, 253, 256, 306, 376, 462,
    463, 466–9, 471, 479
decision-making, 94, 96, 101, 132, 150, 186, 282,
    301, 312, 314, 354, 359, 361, 369, 371, 487
defector, 456 (*see also* message honing/tailoring,
    credibility)
defense mechanism, 67, 74, 487
deficit model
    information/knowledge, 64, 94, 132, 223, 241,
        295, 452, 495
    knowledge/information, 3, 4, 11, 15
deficit, information/knowledge, 399–413
Delaware, state of, 421
delaying action, 9, 167
deliberation, 479, 481, 482
democracy, 287, 289, 478, 479, 481, 482
democratic processes, 417, 464, 486, 487, 505
demonstration programs, 438
denial, 11, 52, 67–8, 161, 164, 258
Denver, Colorado, 390
depolarization, 140
despair, 12, 66, 508
deutero-learning, 256, 289, 290 (*see also* double-
    loop learning)
dialogue, x, xi, 15, 16, 18, 134, 140, 146, 147, 159,
    164, 208, 213–21, 252, 254, 261, 479, 481,
    485–90, 497, 501, 507, 513
    dialogue, essential elements of, 217–8

energy (cont.)
  energy intensity, 372, 375
  energy market, 328, 329, 362, 363
  energy policy, 129, 433, 444
  energy production, 376
  energy reliability, 115, 352
  energy resources, 105
  energy savings, xiii, 363, 379, 387, 394, 395
  energy security, 110, 377, 487
  energy services, 328, 359, 361, 362, 377, 379
  energy sources, 149
  energy supplies, 376
  energy supply, 85, 182, 353, 360, 432, 436, 466
  energy use, 251
  energy use/consumption, xv, 178, 224, 229, 230
  erergy conservation, 110
energy-efficient mortgages (EEMs), 369
engaging people, 86, 87, 106–8, 177, 254, 255, 263,
  286, 289, 304, 312, 384, 479
environmental defense, 74, 157, 320
environmental group, 142, 205
environmental justice, 119–21, 134
environmental justice and climate change initiative,
  121, 125–7
environmental stewardship, 91, 333
environmentalism, 7, 75, 227, 238, 240, 243, 251,
  285, 298, 442
environmentalists, 460
environmentally preferable purchasing, 107, 113,
  114, 390
environmentally responsible behavior (ERB), 71,
  223, 237–49, 258, 408, 487
equality, xiv
equity, 120, 127–9, 136, 147, 219, 399, 431,
  498, 500
ESCOs (Energy Service Companies), 353–4
ethics, 7, 127, 163, 203, 252, 254, 258, 281, 282, 284,
  288, 309–10, 468, 481, 482
  ethical imperative, 282
evaluation, 510–2
Evangelical Christians, 45, 45
evidence, 1, 3, 59, 156, 201, 202, 206, 208, 209, 388,
  444, 469, 471
  conflicting evidence, 173
expectations, 109, 175, 205, 255, 405, 410
experience
  personal, 94–5
  place-based/personal, 5, 16, 31, 140, 163, 213,
  219, 503
expert, 96, 97, 100–1, 132, 173, 178, 193,
  202, 203, 206, 214, 306, 378, 394, 455,
  459, 469
expert opinion, 173
  academic experts, ix, xii
  academic, 99
  expert opinion, 175
  local expert, 143
  native expert, 140
exposure to risk, 7, 128, 132, 341, 351
externalities, 7, 332, 339, 340, 352

fairness, xiv, 146, 201
faith-based community, 17
  faith based community actions on climate
  change, 65, 153–65
fatalism, 60
fear, xv, 12, 13, 65, 70, 81, 86, 108, 134–5, 161, 169,
  177, 274, 295, 409
  feal appeal, 11, 38, 39, 64, 69–71, 73, 496, 510
  (see also danger control)
  fear control, 67
federal action on climate change (US), 2, 20, 464
Federal Energy Management Program, 378
Federation of Atomic Scientists, 457, 456
feeling overwhelmed, 12, 65, 69, 84, 295, 453
feeling powerless, 10, 12, 86, 242, 246, 296, 362, 454
Feynman, Richard, 266
finances, 349
  bond measures, 394–5
  bonds, 351–2
  bottom-line argument, 107, 112, 115, 307, 346,
  383, 389, 404, 411–13 446, 503

  creative financing, 300, 301, 349, 350, 353,
  383–97, 423, 441, 444, 445
  credit, 349, 394
  financial cost, 85, 172, 263, 300, 306–8, 332,
  361–3, 371, 379, 390, 396, 426, 466
  financial partnerships, 432, 445
  financial screening, 351–4
  financing partnerships, 394
  fiscal tools, 339–55, 412, 446
  loans, 349, 350, 354, 362, 393
  no-net cost policy, 390
  revolving fund, 349, 393–4
fingerprint, 84, 87, 145 (see also attribution)
Fleener, Craig, 144, 146
fleet, greening the, 436
focusing event, xiv, xv
footprint, environmental/ecological, 113
force-field model of change, 303–4, 310–1
Ford, William, 303
Fort Collins, Colorado, 389
forums, 107, 150, 204
  community forums, 107, 108, 110, 111, 114
  youth forums, 114
fossil fuel dependency, 59, 60, 86, 98, 158, 159, 209,
  360
fossil fuel use/consumption, 45, 58, 59, 106, 145, 146,
  147, 150, 171, 207–209, 306
fossil-fuel dependency, 145, 150
frame, 55, 157, 486
  framing, 10, 16, 17, 31–40, 59, 84, 129, 130, 131,
  133–4, 140, 143, 144, 145, 146, 147, 150, 171,
  207–9, 216, 306, 471, 491, 499, 501, 510
  reframing, 493
"free riding", 369
fuel efficiency/fuel economy, 86, 87, 106, 112,
  146, 328, 343, 369, 375, 379, 390, 395,
  427, 452

Printed in the United States
111195LV00004B/41-64/A